Lecture Notes in Energy 7

For further volumes:
http://www.springer.com/series/8874

Juan Rosellón • Tarjei Kristiansen
Editors

# Financial Transmission Rights

## Analysis, Experiences and Prospects

Springer

*Editors*
Juan Rosellón
Centro de Investigación y Docencia Económicas (CIDE) and German Institute for Economic Research División de Economía
Mexico, Mexico

German Institute for Economic Research (DIW Berlin)
Berlin
Germany

Tarjei Kristiansen
Åsegårdsvegen 65
6017 Ålesund
Norway

ISSN 2195-1284 ISSN 2195-1292 (electronic)
ISBN 978-1-4471-4786-2 ISBN 978-1-4471-4787-9 (eBook)
DOI 10.1007/978-1-4471-4787-9
Springer London Heidelberg New York Dordrecht

Library of Congress Control Number: 2013931659



Printed on acid-free paper

Springer is part of Springer Science+Business Media (www.springer.com)

# Preface

Efficient operation and optimal expansion of the transmission network system are some of the most complex and challenging issues that we are faced with in operational market and regulatory policy design for liberalised electricity markets. Financial transmission rights (FTRs) – the theme of this book – represent an interesting and welcome addition to the "box of tools" of market-based, regulatory instruments and mechanisms for effective network operation and regulation.

The actual, real-world situation in most liberalised electricity markets is characterised with insufficient investment in the transmission network, resulting in binding capacity constraints, network congestion, and welfare losses to society, compared to an optimal expansion path. Inadequate operational rules for the handling of such constraints have added to the failure of the regulatory system to effectively cope with network issues of a long-term as well as a short-term nature.

The ambition of the editors of the book – Juan Rosellón and Tarjei Kristiansen – has been to produce a book that can be "an accessible source to researchers and professionals working with financial transmission rights (FTRs) and electricity market regulation". I think that they have admirably succeeded in their objective and task. The collection of chapters presents an up-to-date survey and stocktaking of FTRs as a regulatory policy instrument, with a well-balanced blend of theoretical and practical contributions. The editors should also be credited for not overstating the case for FTRs in electricity market and network regulation, pointing to some of their weaknesses and limitations, e.g. in relation to optimal transmission investment, and emphasising the need for coordinating the use of FTRs with other regulatory instruments to achieve stated policy objectives.

A fundamental question in electricity market design is to what extent market transactions and price formation in electricity markets should be separated from considerations of transmission constraints and network congestion issues. I think there are strong arguments for adopting a two-step procedure: first, establishing prices in efficiently functioning electricity markets and, then, solving the network problems that this market allocation may create, rather than in one simultaneous operation. In particular, if transmission network and transmission system operation issues are allowed unduly to set the agenda for market operations, we may end up with an imperfectly functioning market system as a totality.

In such a perspective, FTRs may be considered as a bridging or "intermediary" instrument between the market and network parts of the system. The primary role of FTRs, as I see it, at least at this stage of development of market design, is to function as a transmission congestion risk hedging instrument, comparable to the role that financial derivatives play as hedging instruments on the market side. To further develop and introduce FTRs in this capacity should be a regulatory policy priority. Further research and refinement seem to be needed before FTRs can be introduced effectively as an instrument for transmission network investment, for mitigating market power, and for dealing with other regulatory issues and challenges arising in a liberalised electricity market system. The book also gives well-founded guidance and direction for such research and refinement.

Einar Hope is Professor Emeritus at the Norwegian School of Economics (NHH) and Past President of the International Association for Energy Economics (IAEE).

**"The essays in this book cover the latest thinking on alternative mechanisms to manage efficiently scarce transmission capacity, to set prices for using transmission networks to recover their capital and operating costs, and to provide good incentives for transmission network operation and investment. Anyone interested in frontier thinking and practice on these issues will benefit from reading the essays in this book."**

*Paul L. Joskow, President, Alfred P. Sloan Foundation (New York, NY) and Elizabeth and James Killian Professor of Economics, Emeritus, MIT (Cambridge, MA).*

**"This book from leading experts in the field finally provides a comprehensive overview of theory and practice of financial transmission rights (FTRs). FTRs are a central element of successful power market design. They allow generation and load to address the financial risk of congestion in the transmission network and thus manage the impacts of increasing deployment of wind and solar power. Read this book to learn how the FTRs can be part of long-term energy contracting, and thus facilitate investment in generation and transmission capacity."**

*Karsten Neuhoff, Head of Department Climate Policy, DIW Berlin*

**"Juan Rosellón, a leading researcher with experience as regulator, and Tarjei Kristiansen, a business person with strong academic records, are the ideal combination of academic background, hands-on experience, and visionary thinking to assemble – jointly with well-known co-authors – this volume that allows the reader to assess issues about financial transmission rights, and to weigh potential future applications but also the limits thereof."**

*Christian von Hirschhausen, Chair of Infrastructure Policy, Berlin University of Technology*

# Acknowledgment

Juan Rosellón acknowledges that his work during the elaboration of this book was carried out using support of the European Union, through a Marie Curie Incoming International Fellowship, and of Conacyt (p. 131175).

# Contents

# Bios

## Chapter 1

### *William Hogan*

**William W. Hogan** is Raymond Plank Professor of Global Energy Policy at Harvard University's Kennedy School of Government and Research Director of the Harvard Electricity Policy Group.

## Chapter 2

### *Ignacio J. Pérez-Arriaga*

**Professor Pérez-Arriaga** teaches at Comillas University, MIT and the Florence School of Regulation.

### *Luis Olmos*

**Dr. Olmos** is a senior researcher at the Institute for Research in Technology of Universidad Pontificia Comillas.

### *Michel Rivier*

**Professor Rivier** teaches at the Engineering School and is a senior researcher at the Institute for Research in Technology, both at Universidad Pontificia Comillas, Madrid.

## Chapter 3

### *Shmuel Oren*

**Shmuel S. Oren** is the Earl J. Isaac Professor in the Science and Analysis of Decision Making in the Department of Industrial Engineering and Operations Research at the University of California at Berkeley. He is also Co-Chair of the Management of Technology Program of the College of Engineering and Haas School of Business at Berkeley.

## Chapter 4

### *Udi Helman*

**Dr. Helman** has a Ph.D. in energy economics from the Johns Hopkins University and worked at the U.S. Federal Energy Regulatory Commission and the California Independent System Operator before his current position directing economic analysis for BrightSource Energy, a leading developer of solar thermal power projects.

### *Ben Hobbs*

**Dr. Hobbs** has a Ph.D. in environmental systems engineering from Cornell University, directs the Environment, Energy, Sustainability & Health Institute at Johns Hopkins University, and chairs the CAISO Market Surveillance Committee.

## *Richard O'Neill*

**Richard O'Neill** is the Chief Economic Advisor at the Federal Energy Regulatory Commission and has a doctorate in Operation Research from the University of Maryland.

## *Michael H. Rothkopf*

Passed away

**Michael H. Rothkopf** was an applied mathematical modeler who studied and wrote about auctions and bidding since 1965. Most recently he was Smeal Chaired Professor of Supply Chain and Information Systems at Penn State's Smeal College of Business.

## *William R. Stewart, Jr*

**William Stewart** is the David L. Peebles Professor of Business at the Mason School of Business.

# Chapter 5

## *Manho Joung*

**Mahno Joung** holds a Ph.D. in Electrical Engineering from the University of Texas at Austin (2009).

## *Ross Baldick*

**Ross Baldick** is Professor at the Department of Electrical and Computer Engineering at The University of Texas at Austin.

## Chapter 8

### *Bastian Henze*

**Bastian Henze** works as an advisor on regulatory affairs for a German Transmission System Operator and holds a Master Degree in Economics from Tilburg University (2008).

### *C.N. Noussair*

**Professor Noussair** teaches experimental economics at Tilburg University and holds a Ph.D. from the California Institute of Technology (1993).

### *Bert Willems*

**Bert Willems** teaches industrial organization and regulation at Tilburg University and holds a Master degree in Engineering (1998) and a Ph.D. in economics (2004) from KU Leuven.

## Chapter 9

### *Richard Benjamin*

**Dr. Benjamin** is a member of the Round Table Group of Experts and has a Ph.D. in Economics from the University of Illinois.

## Chapter 10

### *Guillermo Bautista*

**Dr. Guillermo Bautista Alderete** is the Manager of Market Validation and Quality Analysis in the California ISO and serves as an Editor for IEEE.

## Chapter 11

### *José Arce*

**Jose Arce** runs Maple Analytics, a proprietary trading operation, and has a Ph.D. in Electrical Engineering from the National University of San Juan, Argentina (2001).

## Chapter 12

### *Seabron Adamson*

**Seabron Adamson** is an energy analyst and economist and a research affiliate and adjunct lecturer at the Tulane Energy Institute. He holds a M.S. degree from M.I.T and Georgia Tech, and a M.A. in economics from Boston University.

### *Geoffrey Parker*

**Dr. Parker** is Professor of Management Science at Tulane University and serves as Director of the Tulane Energy Institute. Parker received a B.S. degree from Princeton University and a M.S. and Ph.D. degrees from MIT.

## Chapter 13

### *E Grant Read*

**E. Grant Read** is a consultant on electricity market issues in the Asia-Pacific region, and Adjunct Professor with the University of Canterbury, in Christchurch, New Zealand. He has a B.Sc. (Hons) in Mathematics (1975), and a Ph.D. in Operations Research (1979), both from the University of Canterbury.

### *Peter R. Jackson*

**Peter Jackson** has a B.Sc. (Hons) in Economics (1992) and a M.Com in Management Science (1994). He is currently studying towards a Ph.D. in Management Science at the University of Canterbury.

## Chapter 14

### *Gauthier de Maere d'Aertrycke*

**Gauthier de Maere d'Aertrycke** has been working at the Center of Operations Research and Econometrics under the supervision of Prof. Yves Smeers. His work is focused on quantitative energy economics.

### *Yves Smeers*

**Professor Smeers** works on computational models of electricity and gas markets and on investment models in these sectors. He holds Master Degrees in engineering and economics from the universities of Liège and Louvain (Belgium) and M.S. and Ph.D. degrees in Industrial Administration from Carnegie Mellon.

## Chapter 15

### *Zdenka Myslíková*

**Zdenka Myslíková** holds a Master's degree in Economics from Center for Research and Teaching in Economics in Mexico City (CIDE) (2009), and worked for the Mexican Energy Regulatory Commission.

### *Eric Zenón*

**Eric Zenón** works at the Centro de Investigación en Energía (CIE) in Mexico, and has a master degree in Engineering and Ph.D. in Engineering (Energy) from the National Autonomous University of Mexico (2011).

# Editors

## *Juan Rosellón*

**Juan Rosellón** is tenured Professor at the Department of Economics of the Centro de Investigación y Docencia Económicas (CIDE) and German Institute for Economic Research (DIW Berlin). He has been Visiting Professor at the Technische Universität Berlin (2010–2011), the Technische Universität Dresden (2007–2010), and the Kennedy School of Government of Harvard University (2002–2005). He was awarded in 2012 with a Marie-Curie Fellowship from the European Union, in 2009 with the 4th Reimut Jochimsen Prize by the Deutsche Bundesbank, in 2008 with a Georg Forster Fellowship by the Humboldt Foundation, in 2003 with a Repsol-YPF-Harvard-Kennedy-School Fellowship, and in 2002 with a Senior-Fellow Fulbright Fellowship. Professor Rosellón has published extensively on incentive mechanisms to regulate restructured electricity and natural gas markets, including price-cap methodologies to promote infrastructure investment and addressing strategic behavior by participants with market power. He is member of the Editorial Board of Economics of Energy & Environmental Policy, and is also the Editor of CIDE's SSCI Journal Economía Mexicana, Nueva Época. He was Chief Economist of the Mexican Energy Regulatory Commission, CRE (1995–1997), and faculty member of the Program on Privatization, Regulatory Reform and Corporate Governance at Harvard University (1997–2000), and at Princeton University (2001). In 1994, he received the National Award in Economics from Mexican president Ernesto Zedillo. Professor Rosellón holds Ph.D. and M. A. degrees (Economics) from Rice University, as well as M.Sc (Mathematics), B.Sc (Mathematics) and B.A (Economics) degrees from the National Autonomous University of Mexico (UNAM), where he received Gabino Barreda Medal, the most important student honor granted by UNAM. Contact details: Mexico: División de Economía, Carretera México-Toluca 3655, México D.F., 01210, Mexico, juan.rosellon@cide.edu; Germany: German Institute for Economic Research (DIW Berlin), Mohrenstrasse 58, 10117 Berlin, Germany, jrosellon@diw.de.

## *Tarjei Kristiansen*

**Dr. Tarjei Kristiansen** was a Vice President with JP Morgan in London and responsible for German and Nordic power market analysis. Dr. Kristiansen has also worked for the trading company RBS Sempra Commodities, the utility Statkraft and KEMA Consulting in its markets and regulation team. Dr. Kristiansen holds an M.Sc in Physics from the University of Oslo and a Ph.D. in Electrical Power Engineering from the Norwegian University of Science and Technology. He completed the Fellows Program at the Mossavar-Rahmani Center for Business and Government at the John F. Kennedy School of

Government, Harvard University. His Ph.D. focused on risk management associated with transmission congestion including financial transmission rights. He frequently publishes papers on European electricity markets, modeling and financial transmission rights in the US. When he is not analysing and modelling electricity markets, he spends time jogging or reading finance. He can be reached at Tarjei_Kristiansen_2003@alumni.ksg.harvard.edu.

# Abbreviations

| | |
|---|---|
| AC | Alternating Current |
| ACER | Agency for the Cooperation of Energy Regulators |
| AEMC | Australian Electricity Market Commission |
| AEMO | Australian Electricity Market Operator |
| AMP | Automated Mitigation Procedure |
| ANP | Adjusted Nodal Price |
| AP | Average Participations |
| ARR | Auction Revenue Rights |
| ATC | Available Transfer/Transmission Capacity |
| BC | Binding Constraints |
| CAISO | California Independent System Operator |
| CFD | Contracts for Differences |
| CPF | Constraint Participation Factor |
| CPNode | Node Available for Trading |
| CRA | Cambridge Research Associates |
| CRR | Congestion Revenue Right |
| CRR | Constraint Rental Right |
| CSC | Constraint Support Contracts |
| CSP | Constraint Shadow Price |
| CVaR | Conditional Value at Risk |
| CWE | Central West Europe |
| DA | Day Ahead |
| DC | Direct Current |
| DEC | Contract Settles as the Difference in LMPs Between RT and DA |
| DOE | Department of Energy |
| E | East |
| ECU | Experimental Currency Units |
| ENTSO-E | European Network of Transmission System Operators |
| ERCOT | Electric Reliability Council of Texas |
| EU | European Union |

| | |
|---|---|
| FB | Flow Based |
| FERC | Federal Energy Regulatory Commission |
| FGR | Flowgate Right |
| FNP | Full Nodal Pricing |
| FTR | Financial Transmission Right |
| GB | Gigabyte |
| GDSK | Generation and Demand Shift Key |
| GSK | Generation Shift Key |
| GWAP | Generation Weighted Average Price |
| HEPG | Harvard Electricity Policy Group |
| HVDC | High Voltage Direct Current |
| IDMA | Implicit Dispatch Matching Allocation |
| INC | Contracts that Settles as the Difference in LMPs Between DA and RT |
| IOU | Investor Owned Utility |
| IRSR | Inter- Regional Settlements Residue |
| ISO | Independent System Operator |
| ISONE | Independent System Operator North England |
| ISS | Incremental Surplus Subsidy |
| JETRA | Joint Energy and Transmission Rights Auction |
| LF | Loss Factor |
| LHS | Left Hand Side |
| LI | Long Island |
| LMP | Locational Marginal Price/Pricing |
| LRA | Locational Rental Allocation |
| LRAC | Long Run Average Total Cost of Generation |
| LSE | Load Serving Entities |
| LT-FTR/LTFTR | Long-Term Financial Transmission Right |
| LWAP | Load Weighted Average Price |
| MC | Market Coupling |
| MCE | Ministerial Council on Energy |
| MISO | Midwest Independent System Operator |
| MNSP | Market Network Service Providers |
| MS | Market Splitting |
| MW | Megawatt |
| MWh | Megawatt per Hour |
| NEM | Australian National Electricity Market |
| NEMDE | Australian National Electricity Market Clearing Engine |
| NEMMCO | National Electricity Market Management Company |
| NGC | National Grid Company |
| NR | Net Revenue |
| NSW | New South Wales |
| NW | North West |
| NY | New York City |
| NYISO | New York Independent System Operator |

| | |
|---|---|
| NZEA | New Zealand Electricity Authority |
| OASIS | Open Access Same-Time Information System |
| OPF | Optimal Power Flow |
| OTC | Operating Transfer Capability |
| OTC | Over the Counter |
| P | Price |
| PBR | Performance Based Regulation |
| PJM | Pennsylvania-New Jersey-Maryland |
| PL | Profit and Loss |
| PoI | Point of Injection |
| PoW | Point of Withdrawal |
| PRHSC | Protected RHS Capacity |
| PRR | Participant Rental Right |
| PTDF | Power Transfer Distribution Factor |
| PTP | Point-to-Point |
| PTR | Physical Transmission Right |
| PX | Power Exchange |
| Q | Quantity |
| QLD | Queensland |
| RHS | Right Hand Side |
| RPI | Rate of Inflation |
| RT | Real Time |
| RTO | Regional Transmission Operator |
| SA | South Australia |
| SFT | Simultaneous Feasibility Test |
| SP | Shadow Price |
| SRA | Settlement Residue Auction |
| SW | South West |
| TAS | Tasmania |
| TC | Transfer Capacity/Total Cost |
| TCC | Transmission Congestion Contract |
| TLC | Trilateral Market Coupling |
| TO | Transmission Owner |
| TRANSCO | Transmission Company |
| TSO | Transmission System Operator |
| TSR/TSA | Thunderstorm Alerts |
| UIOSI | Use It Or Sell It |
| UoS | Use of the System |
| USA | United States of America |
| VaR | Value at Risk |
| VB | Virtual Bidding |
| VIC | Victoria |
| VIU | Vertically Integrated Utility |
| VTR | Variable Transmission Revenue |
| ZPTDF | Zonal Power Transfer Distribution Factor |

# Introduction

The aim of this book is to be an accessible source to researchers and professionals working with financial transmission rights (FTRs) and electricity market regulation. It contains contributions from leading experts within the field (both practitioners and academics) that provide overviews, both on theoretical and practical aspects, with detailed discussions. Many of the contributions summarize the current state of knowledge in a specific area related to FTRs and provide insightful comments that seek to be accessible to the target audience (typically final-year undergraduates and postgraduate students studying the economics of energy as well as practitioners in industry and government). FTRs are a relatively new concept and have mainly been implemented during the last decade. To our knowledge, there have not been many academic books published on this subject. We therefore seek to improve the understanding of such financial transmission contracts. We hope that when the readers have completed this book, they will have a more comprehensive understanding of FTRs and possibly integrate this knowledge into their daily work or study.

The allocation of scarce electricity transmission capacity presents a major market design challenge. The electric power system is subject to generation and transmission technology constraints that make it difficult to define tradable property rights for physical transmission. This difficulty has led economists to instead create markets for transmission property rights, which are settled against the congestion price component of locational marginal prices (LMPs) (Hogan 1992). These markets have been increasingly adopted in the United States and several other countries. FTRs were first introduced in northeastern US power markets in the late 1990s. The motivation for the introduction was primarily the management of transmission price risk. However, other motives included the provision of revenue sufficiency for contracts for differences, the redistribution of the congestion revenue that the system operator collects, as well as the provision of price signals for transmission and generation investors.

More precisely, an FTR is a financial contract to hedge source-to-sink (point-to-point) congestion and entitles its holder the right – or obligation – to collect a payment when congestion arises in the energy market. The basic definition of an

FTR consists of: (1) a source and a sink node that identify the point-to-point direction of the contract, (2) a megawatt (MW) award that is constant for the duration of the contract, (3) a settlement period, and (4) a life term which identifies the period of time over which the contract is valid. Nowadays, most FTR markets offer an obligation type, for which the holder has either the right to collect a payment when congestion occurs in the energy market or the obligation to pay when the congestion in the energy market is in the opposite direction of the FTR definition. The payment or charge is computed as the price differential between the sink and source nodes times its MW award. An FTR option, in contrast, provides only the upside benefit to its holder since there is no charge to the holder when congestion is in the opposite direction of the FTR (Lyons et al. 2000).

Since FTRs are only financial contracts, the payments or charges are independent of the actual use of the transmission system by their holders. This separation provides efficiency by not interfering with the optimal operation of the system. The allocation mechanisms are usually auction processes run by an ISO (Independent System Operator). Regardless of the means to issue FTRs, an ISO needs to limit the overall amount of FTRs that can be feasibly issued. A simultaneous feasibility test is the underlying process to determine the appropriate amount of FTR awards. When FTRs are modeled in an allocation process, such as auctions, the source and the sink used to define every FTR represent bilateral trades for which injections and withdrawals of power determine the power flow contributions in the transmission system. Thus, any set of FTRs that can be issued has to be a feasible power flow, in which no transmission constraints are violated. The transmission system used in the issuing processes represents as close as possible the transmission system and configuration that will be used later in the energy market. Since the allocation process is usually driven by an optimization engine, the optimal solution (or set of feasible FTRs) is determined by considering simultaneously all FTRs. Therefore, the optimal set of FTRs is necessarily *simultaneously feasible* as FTRs will provide counterflows to each other. By using a simultaneous feasibility test to determine the optimal set of FTRs to be awarded, revenue adequacy can be ensured. Revenue adequacy is then the condition in which sufficient money from the forward energy market is collected to cover all FTR payments over a given period of time.

These and other more detailed concepts on FTRs will be addressed in this book. The areas covered comprise a wide range of topics related to FTRs. The first part of the book deals with the formal presentation of key theoretical aspects on a variety of issues such as FTRs and their different modalities, different mathematical FTR formulations, transmission pricing and network congestion, flowgate rights (FGRs), forward and spot auction markets, market power issues in FTR markets, FTR-based merchant mechanisms for transmission expansion, combined merchant-regulatory mechanisms, FTRs in an experimental-economics framework, and even an alternative view on advantages and disadvantages of FTRs. The second part of the book deals with more practical issues such as revenue adequacy in markets using LMPs; real-world aspects in allocating, trading, and bidding of FTRs; financial hedging and risk management strategies; as well as insightful surveys on the most recent status in countries implementing (or discussing the introduction of) FTRs.

International geographical areas covered include diverse markets in North America, Europe, and Oceania.

More specifically, in Chap. 1, William Hogan makes a comprehensive presentation of FTRs. With a standard market design centered on a bid-based, security-constrained, economic dispatch with locational prices, the natural approach is to define FTRs that offer payments based on prices in the actual dispatch. Different models have been proposed for point-to-point and flowgate rights, obligations, and options. Hogan presents a consistent framework that provides a comparison of alternative rights. The comparison addresses issues of modeling approximations, revenue adequacy, auction formulation, and computational requirements.

In Chap. 2, Perez-Arriaga et al. analyze transmission pricing, which is a basis for derivation of FTRs. They argue that the transmission grid significantly affects investment and operation decisions and that network effects result in different nodal prices. However, the application of efficient short-term marginal prices results in prices that are unable to fully recover the regulated cost of the grid, which is necessary to ensure the viability of the regulated transmission service. Therefore, additional, complementary, transmission charges must be applied. Complementary charges should match the shortfall of the grid cost while not interfering with efficient short-term energy prices and provide efficient long-term locational signals. Thus, transmission charges must be calculated according to transmission investor responsibilities, which involve being independent of the commercial transactions taking place. The complementary charges must be then calculated once and for all in advance of construction.

Oren analyzes in Chap. 3 a specific issue on the introduction of FTRs to electricity markets with LMPs, namely, the mechanics and fundamental relationships between point-to-point FTRs and flow-gate rights (FGRs). Then he investigates the issue of revenue adequacy in FTR/FGR markets and the possibility of short positions by transmission owners of FGRs. Such positions allow their holders to capture some of the FTR auction revenues in exchange for assuming liability for the corresponding FTR market revenue shortfall, which can be avoided through improvements in line ratings.

Hobbs et al. present in Chap. 4 another specialized topic: an auction model that implements a sequence of forward and spot auction markets operated by an ISO or a regional transmission organization (RTO) for energy and several types of FTRs, simultaneously. The model includes point-to-point rights as well as options and obligations and FGRs. This nonlinear model has several applications, including forward auctions for FTRs conducted on an alternating current (AC) load flow model. The extension to real power markets includes reactive power (which would also require an AC model) and the modification of auctions for FTRs on a DC (direct current) load flow model to incorporate nonlinear losses for the purpose of loss hedging.

In Chap. 5, Joung et al. study the effects of market power on FTRs. They specifically investigate how generators' ownership of FTRs may influence the effects of the transmission lines on competition. They show that introducing FTRs in an appropriate manner may reduce the physical capacity needed for the full benefits of competition. Among the competitive effects of ownership of FTRs, the focus is on the effects on two possible pure strategy equilibria: the unconstrained Cournot equilibrium and the passive/aggressive equilibrium.

Kristiansen and Rosellón propose in Chap. 6 a merchant mechanism to expand electricity transmission based on long-term FTRs. Due to network loop flows, a change in network capacity might imply negative externalities on existing FTRs. The ISO thus needs a protocol for awarding incremental FTRs that maximize investors' preferences and preserve certain currently unallocated FTRs (or proxy awards) so as to maintain revenue adequacy. Kristiansen and Rosellón define a proxy award as the best use of the current network along the same direction as the incremental awards. They then develop a bi-level programming model for allocation of long-term FTRs (LTFTRs) according to this rule and apply it to different network topologies. They find that simultaneous feasibility for a transmission expansion project crucially depends on the investor-preference and the proxy-preference parameters. Likewise, for a given amount of preexisting FTRs, the larger the current capacity the greater the need to reserve some FTRs for possible negative externalities generated by the expansion changes.

In Chap. 7, Rosellón presents a combined FTR-based merchant-regulatory mechanism to incentivize transmission expansion. There are two main disparate (*non-Bayesian*) analytical approaches to transmission investment: one employs the theory based on LTFTR to transmission (merchant approach) while the other is based on the incentive-regulation hypothesis (performance-based-regulation (PBR) approach). Practical approaches to transmission expansion have then to a large extent been designed according to particular criteria as opposed to being based on general economic theory or on the more specific regulatory economics literature. Rosellón reviews recently developed approaches for PBR and merchant mechanisms. Furthermore, he provides insights so as to build a comprehensive approach that combines both mechanisms in a setting of price-taking electricity generators and loads.

Henze et al. analyze in Chap. 8 FTRs within an experimental-economics framework. They describe the results of an experiment that considers the behavioral properties of LTFTRs, with a focus on market efficiency. The setting is one in which network users can act strategically because they have market power and are better informed than the network operator about future demand growth. They measure spot and LTFTR prices, capacity, and welfare and compare them to a simulated benchmark. They find that, overall, LTFTRs perform well, though they exhibit considerable heterogeneity.

Benjamin discusses in the last chapter of the theory section, Chap. 9, the advantages and disadvantages of FTRs. According to Benjamin, FTRs are claimed to serve four main purposes: transmission price hedging, provision of revenue sufficiency for contracts for differences, distribution of the merchandizing surplus that an RTO accrues in market operations, and provision of price signals for transmission and generation developers. Benjamin argues that FTR allocation has important distributional impacts and related implications for retail rates. The RTO's practices have important implications for the hedging characteristics of FTRs. He argues, via counterexample, that, even in theory, FTRs may not serve as a perfect hedge against congestion charges. Next, he examines the hedging properties of FTRs more carefully, commenting on the effectiveness of FTRs as a tool in hedging profits. Finally, Benjamin looks at the effectiveness of FTRs in hedging congestion costs in practice.

The practical part of the book starts with Chap. 10, where Bautista Alderete discusses revenue adequacy, that is, the condition in which the congestion funds available from the forward energy market are sufficient to cover all FTR payments. Revenue adequacy is one of the metrics closely studied by market operators and is an indication of the overall performance of the FTR process. Attaining revenue adequacy is a challenge due to the inherent changing nature of the variables impacting both the release of FTRs and the funds collected in the energy market (namely, outages and derates of transmission elements). He then covers several practical issues of revenue adequacy in markets using LMPs and FTRs to hedge congestion.

In Chap. 11, Arce, an expert trader of FTRs, carries out a big picture description of the challenges existing in real-life operation of FTRs. The FTR business has evolved substantially in the last 10 years, with more markets to trade and more sophisticated FTR operations. Furthermore, the low correlation between FTRs and other financial products has made FTRs very appealing not only to financial institutions but also to a diverse set of investors. However, there are still challenges to be addressed before realizing the full value of FTRs as a financial product. Accordingly, this chapter describes some of them, from the perspective of a proprietary trading operation, covering three main aspects: building an FTR portfolio and executing its trade, managing risk and the role played by the FTR desk, as well as a potential evolution of the FTR business.

In Chap. 12, Adamson and Parker review the history of the New York Independent System Operator (NYISO)'s implementation of FTR auctions. They explore the evolution of participation in the NYISO FTR auction market over the period 2000–2010 by types of firms including utilities, generators/marketers, investment banks, and specialist funds. Furthermore, they summarize previous analyses of the NYISO FTR market efficiency, finding that the market was relatively inefficient at inception, but quickly reduced the spreads between forward and spot prices.

Read and Jackson discuss in Chap. 13 the implementation of FTRs in Oceanian markets. They discuss the way in which the New Zealand proposal is designed to deal with locational price differentials resulting from losses and ancillary service requirements. In the Australian market, approximate congestion rental rights are available between zones, although they are less firm than the underlying transmission capacity. In both cases, Read and Jackson discuss extensions to the FTR concept that have been proposed to deal with the remaining volatility in price differentials.

Aertrycke and Smeers analyze in Chap. 14 the introduction of FTRs in Europe. The short-term European electricity market is a zonal system organized along two different paradigms. "Market splitting" is the rule in the Nordic market (Denmark, Finland, Norway, and Sweden), while "market coupling" is becoming the reference in the rest of the continental market. The nodal-price model is officially not on the agenda in Europe even when it is mentioned from time to time. Aertrycke and Smeers analyze proposals for FTRs in Europe with reference to market coupling and for the transfer capacity and flow-based models. They conclude that the organization of both the transfer capacity and flow-based models makes it unlikely that the firmness of FTRs can be guaranteed in Europe without artificially restricting the possibilities of the grid.

The book ends with Chap. 15, where Rosellón et al. present an application of a mechanism (also discussed in Chap. 7) that provides incentives to promote transmission network expansion in Pennsylvania, New Jersey, and Maryland (PJM). The applied mechanism combines the merchant and regulatory approaches to attract investment into transmission grids. It is based on rebalancing a two-part tariff in the framework of a wholesale electricity market with locational pricing. The expansion of the network is carried out through the sale of FTRs for the congested lines. Under Laspeyres weights, they show that prices are able to converge to the marginal cost of generation, the congestion rent decreases, and the total social welfare increases. The mechanism is shown to adjust prices effectively given either nonpeak or peak demand.

We really hope that the wide spectrum of issues addressed, the various depth analyses, practical experiences, and techniques in this book make FTR-related concepts accessible to all those academically interested as well as practically working with such financial instruments. We would like to deeply thank all of those who contributed a chapter in this book as well as the Springer publishing team.

## References

Hogan W (1992) Contract networks for electric power transmission. J Regul Econ 4:211–242

Lyons K, Fraser H, Parmesano H (2000) An introduction to financial transmission rights. Electr J 13(10):31–37

# Chapter 1
# Financial Transmission Rights: Point-to Point Formulations

**William W. Hogan**

## 1.1 Introduction

Transmission rights stand at the center of market design in a restructured electricity industry. Beginning with the intuition that electricity markets require some rights to use the transmission system, simple models of transmission rights soon founder after confronting the limited capacity and complex interactions of a transmission grid. The industry searched for many years without success looking for a workable system of physical rights that would support decentralized decisions controlling use of the grid.

The physical interpretation of transmission rights was the principal complaint that buried the Federal Energy Regulatory Commission's (FERC) original Capacity Reservation Tariff (FERC1996). Any attempt to match a large number of scheduled transactions to a set of transmission rights creates a burden that threatens the flexibility of trade needed to support a market or the flexibility of operations needed to maintain reliability. And in a design built on the centerpiece of a coordinated spot market (FERC 2002a), physical transmission rights or any associated scheduling priority would create perverse incentives and conflicts with priorities defined by the bids used in a security-constrained economic dispatch. The idea that a simple physical right can be made to work soon mutates into a complex system of rules intended to force market participants to act against market incentives. In the end, the right becomes not so physical and not much of a right. The idea dies hard, but the physical rights model deserves a decent burial.

If physical rights will not work, then something different is needed to achieve the same objective in providing a compatible definition of transmission rights for a competitive electricity market. As electricity market design developed, the focus

W.W. Hogan (✉)
Mossavar-Rahmani Center for Business and Government, John F. Kennedy School of Government, Harvard University, Cambridge, MA 02138, USA
e-mail: william_hogan@harvard.edu

J. Rosellón and T. Kristiansen (eds.), *Financial Transmission Rights*,
Lecture Notes in Energy 7, DOI 10.1007/978-1-4471-4787-9_1,

turned from so-called physical transmission rights to a redefinition of transmission rights as financial instruments defined with a close connection to both the transmission grid and a spot market organized through a bid-based, security-constrained, economic dispatch (Hogan 1992). The financial approach separates actual use of the grid from ownership of the transmission rights and provides many simplifications that avoid the principal obstacles encountered in the search for physical rights. A coordinated spot market with locational prices complemented by financial transmission rights is a hallmark of market design that works.

There are many possible definitions of financial transmission rights, each with its advantages and disadvantages. Further, the basic building blocks of financial transmission rights could support a secondary market with a wide variety of other trading instruments, just as a forward contract can be decomposed into a variety of elements with different risk properties.

The basic building blocks under different definitions have different properties. The purpose here is to organize a common analysis covering different types of point-to-point financial transmission rights and compare them in regards to four critical aspects of the transmission rights model. The common notation is an eclectic synthesis designed to bridge the electrical engineering and economic market formulations.[1]

The four aspects of the design cover modeling approximations, revenue adequacy, auction formulation, and computational requirements. These do not include important related subjects such as investment incentives. However, an understanding of at least these four aspects of the formulations would be important in choosing among the types of rights to include in a market design. The same would be true of a decision to include all types of rights, where the market participants could ask for any combination (O'Neill et al. 2002).

Approximation refers to the simplifications inherent in the transmission rights model in comparison to the complexity of the real transmission system. To illustrate the point, the simplification that there are no loop flows makes the contract-path transmission model workable in theory. But the simplification deviates from the reality and the contract-path model became recognized as inefficient and unworkable in practice. The different transmission right definitions depend to different degrees on approximations of the reality of the network. The discussion here begins with a simplified but explicit characterization of an alternating current load flow to then specialize it in the market context for an examination of different transmission rights.

Revenue adequacy refers to a financial counterpart of physical "available transmission capacity." A financial transmission right as defined here is a contract for a financial payment that depends on the outcome of the spot market. By definition, the system is revenue adequate whenever the net revenue collected by the system operator for any period of the spot market is at least equal to the payment

[1] This paper is an abridged version of the working paper, Hogan (2002). The working paper includes an elaboration of flowgate financial transmission rights and hybrid models.

obligations under the transmission rights. The analogous physical problem would be to define the available capacity for transmission usage rights such that the transmission schedules could be guaranteed to flow in any given period. A common requirement of both is to maintain the capability of the grid, but the complex interactions make it impossible to guarantee that physical rights could flow no matter what the dispatch conditions. By contrast, we examine here conditions that do ensure revenue adequacy for the financial transmission rights.

A natural approach to allocating some or all transmission rights is through an auction. The auction design also extends to regular and continuing coordinated auctions that could be employed to reconfigure the pattern of transmission rights, supplemented by secondary market trading. The auction formulation interacts with the conditions for revenue adequacy, with different implications for different definitions of financial transmission rights.

The computational requirements for execution of a transmission rights auction differ for the different models. The inherent scale of the security-constrained economic dispatch model takes the discussion into a realm where the ability to solve the problem cannot be taken for granted. In some cases, the auction model is no more complicated than a conventional security-constrained economic dispatch, and commercial software could be and has been adapted successfully for this purpose. In other cases, the ability to solve the formal model is not assured, and new approaches or various restrictions might be required. Hence, proposals for more ambitious financial transmission right formulations have been offered with the caveat that the expanded service beyond point-to-point rights should be offered "as soon as it is technically feasible" (FERC 2002b).[2]

The purpose here is to identify some of the issues raised in the evaluation of technical feasibility. The comparison of transmission rights models involves tradeoffs. Some versions may be impossible to implement. At a minimum, ease of both implementation and use for alternative transmission rights models should not be taken for granted.

## 1.2 Transmission Line Load Flow Model

Every alternating current (AC) electrical network has both real and reactive power flows. The sinusoidal pattern of instantaneous power flow produces a complex power representation with real and imaginary parts that correspond to real and reactive power. The real power flows are measured in Mega-Watts (MWs), and the reactive power flows are measured in Mega-Volt-Amperes-Reactive (MVARs). The VAR is the product of voltage and current, which is the same unit as the watt; the notational difference is maintained to distinguish between real and reactive power. Real power is defined as the average value of the instantaneous

[2] Similar qualifications appear in discussions of an introduction of options or flowgate rights in PJM, New York, New England, the Midwest, and so on.

power and is the "active" or "useful" power. Reactive power is the peak value of the power that "travels back and forth" over the line and has average value of zero and is "capable of no useful work . . . [and] represents a 'nonactive,' or 'reactive,' power (Elgerd 1982)."[3] The combination of real and reactive power flow is the apparent power in Mega-Volt-Amperes (MVA), which is a measure of the magnitude of the total power flow.

The basic model characterizing electricity markets and financial transmission rights (FTR) centers on the description of a network of lines and buses operating in an electrical steady-state. A critical element is the representation of a transmission line. There is a developed literature on this subject. The choices here do not exhaust all that is relevant, but illustrate the basic issues in the treatment of AC networks for purposes of modeling economic dispatch, locational pricing and the related definition of financial transmission rights. In particular, although the focus is on real power flow, the model includes non-linear features of real and reactive power and control devices to illustrate the implications of various simplifications and approximations often suggested for economic dispatch, pricing and definition of financial transmission rights. Further extensions to include other elements of flexible AC transmission systems (FACTS) could be added, with the associated non-linear characterizations of even the effects on real power flows (Ge and Chung 1999).

A generic transmission line as represented here is illustrated in Fig. 1.1. The data include the resistance ($r$), reactance ($x$), and line charging capacitance ($2B_{cap}$). Variable controls include a transformer with winding tap ratio ($t$) and a phase shift angle ($\alpha$). The voltage magnitude at bus $i$ is $V_i$ and the voltage angle is $\delta_i$. The flow of real and reactive power bus from $i$ towards $j$ is the complex variable $Z_{ij}$. Assuming a steady-state flow can be achieved, the conditions relate the flow of complex power on a line to the control parameters including the voltage magnitudes and angles. Due to losses, the flow out of one bus is not the same as the flow into the other. With these sign conventions, positive flow away from a bus adds to net load at the bus.

The sign conventions support an interpretation of an increase in net load as typically adding to economic benefit and associated with a positive price. Correspondingly, an increase in generation reduces net load and typically adds to cost.[4]

The flow of power in an AC electric network can be described by a system of equations known as the AC load flow model.[5]

---

[3] For an excellent summary of the basics for those other than electrical engineers, see Elgerd (1982), pp. 19–32.

[4] Atypical negative prices are allowed, and in the presence of system congestion may not be so atypical.

[5] In anticipation of later simplifications, the notation here follows the development of the "DC" Load Flow model in Schweppe et al. (1988), Appendices A and D. The DC Load flow refers to the real power half of the nonlinear AC load flow model. Under the maintained assumptions, there is a weak link between the reactive power and real power halves of the full problem. And the real power flow equations have the same general form as the direct current flow equations in a purely resistive network; hence the name "DC Load Flow." Similar linear approximations are available for reactive power flow, but the approximation is poor in a heavily loaded system. Hence, if in addition to real power flow, voltage constraints and the associated reactive power are important, then we require the full AC model and spot pricing theory as in Caramanis (1982).

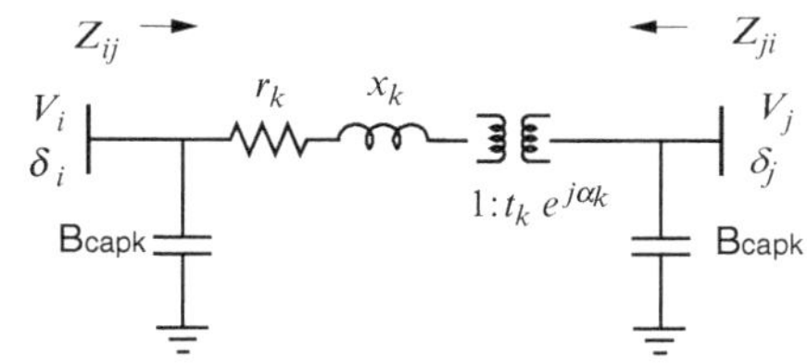


**Fig. 1.1** Generic transmission line representation

Let:

$n_B$ = Number of buses,

$n_L$ = Number of transmission lines, with each line having per unit resistance $r_k$, reactance $x_k$, and shunt capacitance $Bcap_{ij}$ for the Π_equivalent representation of line $k$,[6]

$\tilde{y}_P = d_P - g_P = n_{B-1}$ vector of net real power bus loads, i.e. demand minus generation, $y_P^t = (y_{Ps}, \tilde{y}_P^t)$ where $y_{Ps}$ is at the swing bus,

$\tilde{y}_Q = d_Q - g_Q = n_{B-1}$ vector of reactive power bus loads, i.e. demand minus generation, $y_Q^t = \left(y_{Qs}, \tilde{y}_Q^t\right)$ where $y_{Qs}$ is at the swing bus,

$\delta = n_B$ Vector of voltage angles relative to the swing bus, where by definition $\delta_s = 0$,

$V = n_B$ Vector of voltage magnitudes, where by assumption the voltage at the swing bus, $V_s$, is exogenous,

$t_k$ = ideal transformer tap ratio on line $k$,

$\alpha_k$ = ideal transformer phase angle shift on line $k$,

$A$ = the oriented line-node incidence matrix, the network incidence matrix with elements of 0, 1, −1 corresponding to the network interconnections. If link $k$ originates at bus $i$ and terminates at bus $j$, then $a_{ki} = 1 = -a_{kj}$.

[6] For a development of the Π_equivalent representation of a transmission line, see Bergen (1986), Chap. 4. Here we follow Wood and Wollenberg (1984) in representing *Bcap* as one-half the total line capacitance in the Π_equivalent representation; (Wood and Wollenberg 1984), p.75. A. See also Skilling (1951), pp. 126–133.

Define[7]

$G_k = r_k/(r_k^2 + x_k^2)$,
$\Omega_k = x_k/(r_k^2 + x_k^2)$,
$z_{Pijk}$ = real power (MWs) flowing out of bus $i$ towards bus $j$ along line $k$, and
$z_{Qijk}$ = reactive power (MVARs) flowing out of bus i towards bus j along line k.

Then the complex power flow $Z_{ij}$ includes the real and reactive components[8]:

$$z_{Pijk} = G_k[V_i^2 - (V_iV_j/t_k)\cos(\delta_i - \delta_j + \alpha_k)] + \Omega_k(V_iV_j/t_k)\sin(\delta_i - \delta_j + \alpha_k),$$
$$z_{Pjik} = G_k[(V_j/t_k)^2 - (V_jV_i/t_k)\cos(\delta_j - \delta_i - \alpha_k)] + \Omega_k(V_jV_i/t_k)\sin(\delta_j - \delta_i - \alpha_k). \tag{1.1}$$

and

$$\begin{aligned} z_{Qijk} &= \Omega_k[V_i^2 - (V_iV_j/t_k)\cos(\delta_i - \delta_j + \alpha_k)] - G_k(V_iV_j/t_k)\sin(\delta_i - \delta_j + \alpha_k) \\ &\quad - V_i^2 B_{cap_k}, \\ z_{Qjik} &= \Omega_k[(V_j/t_k)^2 - (V_jV_i/t_k)\cos(\delta_j - \delta_i - \alpha_k)] \\ &\quad - G_k(V_jV_i/a_k)\sin(\delta_j - \delta_i - \alpha_k) - V_j^2 B_{cap_k}. \end{aligned}$$

Real losses on line $k$ are given by

$$l_{Pk} = z_{Pijk} + z_{Pjik}.$$

Hence, in terms of the angles and voltages we have

$$l_{Pk}(\delta, V, t, \alpha) = \mathrm{G}_k\left[V_i^2 + (V_j/t_k)^2 - 2(V_iV_j/t_k)\cos(\delta_i - \delta_j + \alpha_k)\right].$$

Similarly, reactive power losses are

$$l_{Qk} = z_{Qijk} + z_{Qjik},$$

[7] Here the notation follows Schweppe et al. (1988). The purpose is to connect to the discussion of the economics of spot markets and the definition of FTRs. However, the electrical engineering literature follows different notational conventions. For example, Wood and Wollenberg (1984) and others use a different sign convention for $\Omega$. Also note that here $V_i$ is the magnitude of the complex voltage at bus i, not the complex voltage itself as in the appendix. Finally, we use $y$ to denote the net loads at the buses. This should not be confused with the complex admittance matrix, often denoted as Y, which is composed of the elements of $G$ and $\Omega$. See the appendix for further discussion.

[8] For details, see the appendix.

or

$$l_{Qk}(\delta, V, t, \alpha) = \Omega_k\left[V_i^2 + (V_j/t_k)^2 - 2(V_iV_j/t_k)\cos(\delta_i - \delta_j + \alpha_k)\right] \\ - \left(V_i^2 + V_j^2\right)Bcap_k.$$

Given these flows on the lines, conservation of power at each bus requires that the net power loads balance the summation of the flows in and out of each bus. Under our sign conventions and summing over every link connected to bus $i$, we have

$$d_{Pi} + \sum_{k(i,j)} z_{Pijk} = g_{Pi} - \sum_{k(j,i)} z_{Pjik}, \quad and \\ d_{Qi} + \sum_{k(i,j)} z_{Qijk} = g_{Qi} - \sum_{k(j,i)} z_{Qjik}.$$

Here the summation includes each directed line that terminates at $i$ $(k(j,i))$ or originates at $i$ $(k(i,j))$ Hence, the net loads satisfy:

$$y_{Pi} \equiv d_{Pi} - g_{Pi} = -\sum_{k(j,i)} z_{Pjik} - \sum_{k(i,j)} z_{Pijk}, \quad and \\ y_{Qi} \equiv d_{Qi} - g_{Qi} = -\sum_{k(j,i)} z_{Qjik} - \sum_{k(i,j)} z_{Qijk}.$$

Recognizing that the individual flows can be expressed in terms of the several variables, we obtain the relation between net loads, bus angles, voltage magnitudes, transformer ratios, and phase angle changes:

$$\begin{bmatrix} \tilde{y}_P \\ \tilde{y}_Q \end{bmatrix} = \begin{bmatrix} \tilde{y}_P(\delta, V, t, \alpha) \\ \tilde{y}_Q(\delta, V, t, \alpha) \end{bmatrix} = \tilde{Y}(\delta, V, t, \alpha).$$

Assuming that there is convergence to a non-singular solution for the steady-state load flow, this system can be inverted to obtain the relation between the bus angles, voltage magnitudes and the net power loads given the transformer ratios and phase angle changes[9]:

[9] The convention here is that gradients are row vectors. Hence, with

$$f(u,v) = \begin{bmatrix} f_1(u,v) \\ f_2(u,v) \end{bmatrix}, \quad \nabla f = \begin{bmatrix} \partial f_1(u,v)/\partial u & \partial f_1(u,v)/\partial v \\ \partial f_2(u,v)/\partial u & \partial f_2(u,v)/\partial v \end{bmatrix}.$$

$$\begin{bmatrix} \delta \\ V \end{bmatrix} = \begin{bmatrix} J_\delta(\tilde{y}_P, \tilde{y}_Q, t, \alpha) \\ J_V(\tilde{y}_P, \tilde{y}_Q, t, \alpha) \end{bmatrix} = J(\tilde{y}_P, \tilde{y}_Q, t, \alpha), and$$

$$\nabla J = \begin{bmatrix} \nabla J_{\delta P} & \nabla J_{\delta Q} \\ \nabla J_{VP} & \nabla J_{VQ} \end{bmatrix} = \begin{bmatrix} \nabla y_{P\delta} & \nabla y_{PV} \\ \nabla y_{Q\delta} & \nabla y_{QV} \end{bmatrix}^{-1} = \nabla \tilde{Y}^{-1}.$$

This formulation treats all buses, other than the swing bus, as load buses, with given real and reactive power loads. These are sometimes referred to as PQ buses.[10] In practice, many generator buses are operated as PV buses, where $\tilde{y}_P$ and $V$ are given and the required reactive power is determined in order to maintain the voltage (Bergen 1986). There are $4(n_{B-1})$ variables (i.e., $\tilde{y}_P$, $\tilde{y}_Q$, $\delta$, $V$) and $2(n_{B-1})$ independent node balance equations. Hence, half of the variables must be specified and then the solution obtained for the remainder. The corresponding change on the representation of the equations for different treatment of the buses is straightforward. For example, in the DC-Load model discussed below, all buses are treated as PV where the first step is to fix $\tilde{y}_P$ and $V$ to solve for $\delta$ and implicitly $\tilde{y}_Q$.

The power flow entering a line differs from the power leaving the line by the amount of the losses on the line. Typically, but not always, real power losses will be a small fraction of the total flow and it is common to speak of the power flow on the line. In the DC-Load case discussed below, losses are ignored and the real power flow is defined as the same at the source and destination. In the case of an AC line, we could select either or both ends of the line as metered and focus on the flow at that location for purposes of defining transmission constraints.

We can use these relations to define the link between the power flows on the lines and the net loads at the buses:

$$z = \begin{bmatrix} z_P(\delta, V, t, \alpha) \\ z_Q(\delta, V, t, \alpha) \end{bmatrix} = \begin{bmatrix} z_P(J(\tilde{y}_P, \tilde{y}_Q, t, \alpha), t, \alpha) \\ z_Q(J(\tilde{y}_P, \tilde{y}_Q, t, \alpha), t, \alpha) \end{bmatrix} = \begin{bmatrix} \tilde{K}_P(\tilde{y}_P, \tilde{y}_Q, t, \alpha) \\ \tilde{K}_Q(\tilde{y}_P, \tilde{y}_Q, t, \alpha) \end{bmatrix} = \tilde{K}(\tilde{y}_P, \tilde{y}_Q, t, \alpha),$$

*and*

$$\nabla \tilde{K}_y(\tilde{y}_P, \tilde{y}_Q, t, \alpha) = \begin{bmatrix} \nabla \tilde{K}_P \\ \nabla \tilde{K}_Q \end{bmatrix} = \begin{bmatrix} \nabla z_P \nabla J \\ \nabla z_Q \nabla J \end{bmatrix} = \begin{bmatrix} \nabla z_P \nabla \tilde{Y}^{-1} \\ \nabla z_Q \nabla \tilde{Y}^{-1} \end{bmatrix}. \quad (1.2)$$

[10] The swing bus is a $\delta V$ bus for which the angle and the voltage are exogenous.

Summing over all lines gives total losses as:

$$\begin{bmatrix} L_P \\ L_Q \end{bmatrix} = \begin{bmatrix} \sum_k l_{Pk}(\delta, V, t, \alpha) \\ \sum_k l_{Qk}(\delta, V, t, \alpha) \end{bmatrix} = \begin{bmatrix} l_P(J(\tilde{y}_P, \tilde{y}_Q, t, \alpha), t, \alpha) \\ l_Q(J(\tilde{y}_P, \tilde{y}_Q, t, \alpha), t, \alpha) \end{bmatrix} = \begin{bmatrix} L_P(\tilde{y}_P, \tilde{y}_Q, t, \alpha) \\ L_Q(\tilde{y}_P, \tilde{y}_Q, t, \alpha) \end{bmatrix},$$

*and*

$$\nabla L = \nabla l \nabla J = \nabla l \nabla \tilde{Y}^{-1}.$$

Finally, conservation of power determines the required generation at the swing bus, $g_{Ps}$ and $g_{Qs}$, as:

$$\begin{aligned} g_{Ps} &= -y_{Ps} = L_P(\tilde{y}_P, \tilde{y}_Q, t, \alpha) + \iota^t \tilde{y}_P, and \\ g_{Qs} &= -y_{Qs} = L_Q(\tilde{y}_P, \tilde{y}_Q, t, \alpha) + \iota^t \tilde{y}_Q. \end{aligned}$$

where $\iota$ is a unity column vector, $\iota^t = \begin{pmatrix} 1 & 1 & \cdots & 1 \end{pmatrix}$. Equivalently,

$$\begin{aligned} L_P(\tilde{y}_P, \tilde{y}_Q, t, \alpha) + \iota^t y_P &= 0, and \\ L_Q(\tilde{y}_P, \tilde{y}_Q, t, \alpha) + \iota^t y_Q &= 0. \end{aligned}$$

These relationships summarize Kirchoff's Laws that define the AC load flow model in terms convenient for our subsequent characterization of the optimal dispatch problem. Given the configuration of the network consisting of the buses, lines, transformer settings, resistances and reactances, the load flow equations define the relationships among (1) the net inputs at each bus, (2) the voltage magnitudes and angles, and (3) the flows on the individual lines.

## 1.3 Optimal Power Flow

The optimal power flow or economic dispatch problem is to choose the net loads, typically by controlling the dispatch of power plants, in order to achieve maximum net benefits within the limits of the transmission grid. Under its economic interpretation, the solution of the power flow problem produces locational prices in the usual way. For our present purposes we define abstract benefit and cost functions. The model developed here includes three simplifications. First, strictly for notational convenience, we assume that all transmission constraints are defined in terms of the effects of net loads at buses. In reality, transmission constraints may treat loads and generation differently. Incorporating different buses for generation and load connected by a zero impedance line would accommodate different effects of load and generation. This would allow for different prices for load and generation by treating them as at different locations.

The second simplification is to focus on the real power part of the problem, even in the AC case. Here we anticipate a market in which we have FTRs for real power but none are required or available for reactive power and there is no reactive power market. This is not a trivial simplification. It would be appropriate as a model under the assumption that there are no direct costs of producing reactive power and the dispatch of reactive power sources is fully under the control of the system operator. Finally, we abstract from explicit consideration of generation operating reserves.[11]

With these assumptions, we formulate the economic dispatch problem and then extend it to the case of security-constrained economic dispatch.

### 1.3.1 Economic Dispatch

We first specialize the notation to represent the transmission constraints, and then the simplified aggregate benefit function.

The constraints for the economic dispatch problem derive from the characterization of the power flow in transmission lines. Under the simplifying assumptions, we treat the real and reactive power elements differently. Henceforth, we drop the subscript and treat the variable $y = y_P = d_P - g_P$ as the real power bus loads, including for the swing bus ($y^t = (y_s, \tilde{y}^t)$). We further subsume all other parameters above in the generic control vector $u$, with its own constraints as in:

$$u = \begin{pmatrix} y_Q \\ t \\ \alpha \end{pmatrix},$$

$$u \in U.$$

In addition to these control variables, we recognize that system operators may change to topology of the network as summarized in $A$. For simplicity, we limit attention to differentiable elements of $u$. However, in the applications discussed below, the incidence matrix could change. The principal impact of changes in $A$ is to introduce discrete choices with complications for the optimization problem but not for the main results for FTRs.

With this notational adjustment, we restate the transmission flows as the function $\tilde{K}(y, u)$ and the losses as $L(y, u)$. We assume that the flows are constrained. In addition, we incorporate the constraint limits as part of the function and append any other constraints on the real power flows. For example, a constraint on MVA of apparent power flow at a metered end of the line would be:

$$z_{Pijk}^2 + z_{Qijk}^2 - b_{MVA\_MAXk} \leq 0. \tag{1.3}$$

---

[11] Cadwalader et al. (1998) provides an outline of transmission rights and revenue adequacy in the context of explicit reserve markets. The analysis is limited to point-to-point obligations, as discussed below, but could be extended to include other types of financial transmission rights.

We treat this as simply another element of $\tilde{K}(y,u)$. All joint constraints on real power flows and the various control parameters, including interface and other operating limits, appear under $\tilde{K}(y,u)$. The separate limits on the control variables appear in the set $U$. Hence, the summary of the constraints is:

$$
\begin{aligned}
&L(y,u) + \iota^t y = 0,\\
&\tilde{K}(y,u) \leq 0,\\
&u \in U.
\end{aligned}
$$

The objective function for the net loads derives from the benefits of load less the costs of generation. Anticipating a bid-based economic dispatch from a coordinated spot market, we formulate the benefit function for net loads as:

$$
\begin{aligned}
B(y) = \ &\underset{d\in D, g\in G}{Max}\ Benefits(d) - Costs(g)\\
&s.t.\\
&\qquad d - g = y.
\end{aligned}
$$

Under the usual convexity assumptions, the constraint multipliers for this optimization problem define a sub-gradient for this optimal value problem. For simplicity in the discussion here, we treat the sub-gradient as unique so that $B$ is differentiable with gradient $\nabla B$. This gives the right intuition for the resulting prices, with the locational prices of net loads at $p^t = \nabla B$. The more general case would require little more than recognizing that market-clearing prices might not be unique, as for example at a step in a supply function.

Then the economic dispatch problem is:[12]

$$
\begin{aligned}
&\underset{y,u\in U}{Max}\, B(y)\\
&s.t.\\
&L(y,u) + \iota^t y = 0,\\
&\tilde{K}(y,u) \leq 0.
\end{aligned}
\tag{1.4}
$$

In general, this can be a complicated non-linear and typically non-convex problem. In most cases, but not all, the economic dispatch problem is well-behaved in the sense that there is a solution with a corresponding set of Lagrange multipliers and no duality gap. The problem may still be hard to solve, but that is the challenge for software implementation.

[12] This is similar to the formulation in Caramanis et al. (1982); the principal difference is in imposing the thermal limit not just on the real power flow, but on the total MVA flow to account for the total thermal impact. The constraints could also include generator capability tradeoffs. See Feinstein et al. (1988), pp. 22–26, for a discussion of the generator capability curve tradeoffs between real and reactive power.

Cases where there may be no solution present a real challenge to electrical systems, as when there is no convergence to a stable load flow, or for markets, when there may be no price incentives that can support a feasible equilibrium solution. Both pathological circumstances would present difficulties for electricity markets that go beyond the discussion of FTR formulations. Hence, while not claiming that all such economic dispatch problems are well-behaved, we will restrict attention to the case when (1.4) is well-behaved.

There are many conditions that could be imposed to guarantee that the economic dispatch problem in (1.4) meets this condition. For our purposes, it is simple to restrict attention to problems that satisfy the optimality conditions[13]:

$$\begin{gathered}
\textit{There exists}(y^*, u^*, \lambda, \eta),\ \textit{such that} \\
L(y^*, u^*) + \iota^t y^* = 0, \\
\tilde{K}(y^*, u^*) \leq 0, \quad \eta^t \tilde{K}(y^*, u^*) = 0, \\
\eta \geq 0, \quad u^* \in U, \\
(y^*, u^*) \in \arg\max_{y,u \in U} \left[B(y) - \lambda(L(y,u) + \iota^t y) - \eta^t \tilde{K}(y,u)\right].
\end{gathered}$$

Hence, there is no duality gap (Bertsekas 1995). The Lagrange multipliers provide the "shadow prices" for the constraints. The solution for the economic dispatch problem is also a solution for the corresponding dual function for this economic dispatch problem:

$$\mathop{Max}_{y,u \in U} \left[B(y) - \lambda(L(y,u) + \iota^t y) - \eta^t \tilde{K}(y,u)\right].$$

Assuming differentiability, the first order conditions for an optimum $(y^*, u^*)$ include:

$$\nabla B(y^*) - \lambda\left(\nabla L_y(y^*, u^*) + \iota^t\right) - \eta^t \nabla \tilde{K}_y(y^*, u^*) = 0.$$

Hence, we have the locational prices as

$$p^t = \nabla B(y^*) = \lambda \iota^t + \lambda \nabla L_y(y^*, u^*) + \eta^t \nabla \tilde{K}_y(y^*, u^*).$$

The locational prices have the usual interpretation as the price of power at the swing bus ($p_G = \lambda$), the marginal cost of losses $\left(p_L = \lambda \nabla L_y(y^*, u^*)\right)$ and the

[13] As an historical note, apparently the early work on optimality conditions by Kuhn and Tucker was motivated by an inquiry into the theory of electrical networks. Kuhn (2002), p. 132.

marginal cost of congestion $\left(p_C = \eta^t \nabla \tilde{K}_y(y^*, u^*)\right)$.[14] These locational prices play an important role in a coordinated spot market and in the definition of FTRs.

### *1.3.2 Security-Constrained Economic Dispatch*

The optimal power flow formulation in (1.4) ignores the standard procedure of imposing security constraints to protect against contingent events. Although the formulation could be interpreted as including security constraints, it is helpful here to be explicit about the separate security constraints in anticipation of the later discussion of FTR formulations and auctions that include the many contingency limits.

The basic idea of security-constrained dispatch is to identify a set of possible contingencies, such as loss of a line or major facility, and to limit the normal dispatch so that the system would still remain within security limits if the contingency occurs. The modeled loss of the facility leaves the remaining elements in place, suggesting the name of *n−1* contingency analysis.[15]

Hence, a single line may have a normal limit of 100 MW and an emergency limit of 115 MW.[16] The actual flow on the line at a particular moment might be only 90 MW, and the corresponding dispatch might appear to be unconstrained. However, this dispatch may actually be constrained because of the need to protect against a contingency. For example, the binding contingency might be the loss of some other line. In the event of the contingency, the flows for the current pattern of generation and load would redistribute instantly to cause 115 MW to flow on the line in question, hitting the emergency limit. No more power could be dispatched than for the 90 MW flow without potentially violating this emergency limit. The net loads that produced the 90 MW flow, therefore, would be constrained by the dispatch rules in anticipation of the contingency. It would be the contingency constraint and not the 90 MW flow that would set the limit. The corresponding prices would reflect these contingency constraints (Boucher et al 1998).

Depending on conditions, any one of many possible contingencies could determine the current limits on the transmission system. During any given hour, therefore, the actual flow may be, and often is, limited by the impacts that would occur in the event that the contingency came to pass. Hence, the

---

[14] The dispatch and prices are not changed by the arbitrary designation of the swing bus. However, the choice of the reference bus for pricing, which need not be the same as the swing bus, does affect the decomposition of the prices.

[15] A simultaneous loss of multiple facilities would be defined as a single contingency.

[16] Expressing the limits in terms of MW and real power is shorthand for ease of explanation. Line limits in AC models appear in terms of MVA for real and reactive power.

contingencies do not just limit the system when they occur; they are anticipated and can limit the system all the time. In other words, analysis of the power flows during contingencies is not just an exception to the rule; it is the rule. The binding constraints on transmission generally are on the level of flows or voltage in post-contingency conditions, and flows in the actual dispatch are limited to ensure that the system could sustain a contingency.

For instance, suppose that the contingency $\omega$ is the loss of a line. For sake of simplicity in the illustration, assume that the only adjustment in the case of the contingency is to change the net load at the swing bus to rebalance the system. Then there would be a different network, different flows, and different losses, leading to a new set of power flow constraints described as:

$$\begin{aligned} &L^{\omega}\left(y_s^{\omega}, \tilde{y}, u\right) + y_s^{\omega} + \iota^t \tilde{y} = 0, \\ &\tilde{K}^{\omega}\left(y_s^{\omega}, \tilde{y}, u\right) \leq 0, \\ &u \in U. \end{aligned} \tag{1.5}$$

The values of the constraint limits could be different in different contingencies, including changes in monitored elements. Extension of this model to allow other changes in dispatch or control parameters present no problem in principle, but would add to the complexity of the notation. The set of constraints and balancing equations would be different for each contingency.

If we treat normal operations as the contingency $\omega = 0$, then the combined set of constraints on the dispatch would be:

$$\begin{aligned} L^{\omega}\left(y_s^{\omega}, \tilde{y}, u\right) + y_s^{\omega} + \iota^t \tilde{y} = 0, \quad & \omega = 0, 1, 2, \cdots, N, \\ \tilde{K}^{\omega}\left(y_s^{\omega}, \tilde{y}, u\right) \leq 0, \quad & \omega = 0, 1, 2, \cdots, N, \\ u \in U. & \end{aligned}$$

The security-constrained economic dispatch imposes all these constraints on the net loads in advance of the realization of any of the contingencies. However, since the swing bus net load is different in every contingency, we subsume the load balance impacts for $\omega > 0$ in the definition of the constraints, and keep explicit only the loss balance in normal conditions. Then with the appropriate change in notation with ($y^t = \left(y_s^0, \tilde{y}^t\right)$), we arrive at a compact representation of the constraints as:

$$L(y,u)+\iota^t y=0,$$

$$K(y,u)\equiv\begin{pmatrix}\tilde{K}^0\left(y_s^0,\tilde{y},u\right)\\ \tilde{K}^1\left(y_s^1,\tilde{y},u\right)\\ \vdots\\ \tilde{K}^\omega\left(y_s^\omega,\tilde{y},u\right)\\ \vdots\\ \tilde{K}^N\left(y_s^N,\tilde{y},u\right)\end{pmatrix}\leq 0,$$

$$u\in U.$$

With this notational convention, we can then restate the security-constrained economic dispatch problem as:

$$\begin{aligned}&\underset{y,u\in U}{Max}\,B(y)\\ &\quad s.t.\\ L(y,u)+\iota^t y&=0,\\ K(y,u)&\leq 0.\end{aligned}\tag{1.6}$$

However, we now recognize that the single loss balance equation that affects the benefit function is appended by many contingency constraints that limit normal operations. If there are thousands of monitored elements for possible overloads of lines, transformers, or voltage constraints, and there are hundreds of contingencies that enter the protection set, the total number of constraints in $K$ would be on the order of hundreds of thousands. This large scale is inherent in the problem, and a challenge for FTR models.

It is a remarkable fact that system operators solve just such contingency-constrained economic dispatch problems on a regular basis. Below we summarize a basic outline of a solution procedure to capture the elements relevant to the FTR formulations. This method exploits a relaxation strategy and the feature that as we get closer to the actual dispatch, the pattern if loads are better known and the list of plausible contingencies and monitored elements reduces accordingly. Anticipating the discussion of FTRs, however, the larger potential set of constraints would be relevant.

Under the assumed optimality conditions, the corresponding prices obtained from the solution appear as:

$$p^t=\nabla B(y^*)=\lambda\iota^t+\lambda\nabla L_y(y^*,u^*)+\eta^t\nabla K_y(y^*,u^*).$$

Hence, the congestion cost could arise from any of the (many) contingency constraints.

### 1.3.3 Market Equilibrium

The security-constrained economic dispatch problem has the familiar close connection to the competitive partial equilibrium model where market participants act as profit maximizing or welfare maximizing price takers.

Assume that each market participant has an associated benefit function for electricity defined as $B_i(y_i)$, which is concave and continuously differentiable.[17] In FERC terminology, the market participants are the transmission service customers. The customers' benefit functions can arise from a mixture of load or demand benefits and generation or supply costs. In this framework, the producing sector is the electricity transmission provider, with customers injecting power into the grid at some points and drawing power out of the grid at other points. The system operator receives and delivers power, coordinates a spot market, and provides transmission service across locations.

The competitive market equilibrium applied here is based on the conventional partial equilibrium framework that stands behind the typical supply and demand curve analysis.[18] The market consists of the supply and demand of electric energy and transmission service plus an aggregate or numeraire "good" that represents the rest of the economy. Each customer is assumed to have an initial endowment $\tilde{w}_i$ of the numeraire good. In addition, each customer has an ownership share $s_i$ in the profits "$\pi$" of the electricity transmission provider, with $\sum_i s_i = 1$.

An assumption of the competitive model is that all customers are price takers. Hence, given market prices, $p$, customers choose the level of consumption of the aggregate good, $c_i$, and electric energy including the use of the transmission system

[17] A sufficient condition for these to obtain would be that the demand and supply functions at each node are continuous, additively separable and aggregate into a downward sloping net demand curve. The benefit function would be the area under the demand curves minus the area under the supply curves in the usual consumer plus producer surplus interpretation at equilibrium. To avoid notational complexity, the assumption here is that each participant has a continuously differentiable concave benefit function defined across the net loads at every location. Concavity is important for the analysis below of the equivalence of economic dispatch and market equilibrium, if there is a market equilibrium. This would eliminate from this competitive market analysis the related unit commitment problem which includes non-convex start-up conditions. As is well known, in the presence of non-concave benefit functions there may be no competitive market equilibrium. Differentiability can be relaxed, with no more than the possibility of multiple equilibrium prices. Restricting the benefit function to definition at a subset of the locations would be more realistic, but different only in the need to account for the corresponding variable definitions. It would not affect the results presented here. In practice, as is often assumed, the benefits functions may be separable across locations.

[18] The partial equilibrium assumptions are that electricity is a small part of the overall economy with consequent small wealth effects, and prices of other goods and services are approximately unaffected by changes in the electricity market. See Mas-Colell et al. (1995), pp. 311–343. Importantly, we adopt here a relaxed set of assumptions that do not include convexity of the set of feasible net loads.

according to the individual optimization problem maximizing benefits subject to an income constraint:

$$\begin{aligned} &\underset{y_i, c_i}{Max}\, B_i(y_i) + c_i \\ &\qquad s.t. \\ &p^t y_i + c_i \leq \tilde{w}_i + s_i \pi. \end{aligned} \tag{1.7}$$

In this simple partial equilibrium model of the economy, there is only one producing entity, which is the system operator providing transmission service. Under the competitive market assumption, the producer is constrained to operate as a price taker who chooses inputs and outputs ($y_i$) that are feasible and that maximize profits. The profits amount to $\pi = p^t \sum_i y_i$. Hence, the transmission system operator's problem is seen as:

$$\begin{aligned} &\underset{\substack{y, u \in U \\ y_i}}{Max} \quad p^t y \\ &s.t. \\ &y = \sum_i y_i, \\ &L(y, u) + \iota^t y = 0, \\ &K(y, u) \leq 0. \end{aligned} \tag{1.8}$$

Of course, the transmission service provider is a monopoly and would not be expected to follow the competitive assumption in the absence of regulatory oversight. However, the conventional competitive market definition provides the standard for the service that should be required of the system operator.[19]

Given the initial endowment of goods $\tilde{w}_i$, and the ownership shares $s_i$, a competitive market equilibrium is defined as a vector of prices, $p$, profits, $\pi$, controls, $u$, and a set of net loads, $y_i$, for all $i$ that simultaneously solve (1.7) and (1.8).

A competitive equilibrium will have a number of important properties that we can exploit. First, note that $\sum_i c_i = \sum_i \tilde{w}_i$, which is implied and necessary for feasibility. Furthermore, every customer's income constraint is binding and the derivative of each benefit function will equal the common market prices, $p = \nabla B_i^t$. Hence, the equilibrium price at each location is equal to the market clearing

[19] It is the standard formulation to include both the consumption (1.7) and production (1.8) sectors as part of the definition of competitive market equilibrium. Failure to follow this well established convention leads to confusion when the term "market equilibrium" is applied excluding the producing sector in (1.8), as in Wu et al. (1996), pp. 5–24. For a further discussion of equivalence results, see Boucher and Smeers (2001), pp. 821–838.

marginal benefit of net load and the marginal cost of generation and redispatch to meet incremental load.

Finally, a motivation for the connection with economic dispatch is that a market equilibrium $(\{y_i^*\}, u^*)$ must also be a solution to the economic dispatch problem with $B(y) = \sum_i B_i(y_i)$. If not, there would be a set of feasible net loads $\{y_i^1\}$ with $\sum_i B_i(y_i^1) > \sum_i B_i(y_i^*)$. Therefore, by concavity of $B$ we would have:

$$p^t\left(\sum_i (y_i^1 - y_i^*)\right) = \sum_i \nabla B_i(y_i^1 - y_i^*) \geq \sum_i (B_i(y_i^1) - B_i(y_i^*)) > 0.$$

But this would violate the optimality of $(\{y_i^*\}, u^*)$. Hence, a market equilibrium is also a solution to the economic dispatch problem.

Therefore, under the optimality conditions assumed, the market equilibrium would satisfy the same local first-order necessary conditions as an optimal solution to the economic dispatch. In particular, for a market equilibrium we have the pricing condition that:

$$p^t = \nabla B(y^*) = \lambda \iota^t + \lambda \nabla L_y(y^*, u^*) + \eta^t \nabla K_y(y^*, u^*).$$

Another way to look at this problem is to interpret the equilibrium as satisfying the "no arbitrage" condition. At equilibrium, there are no feasible trades of electric loads in (1.8) that would be profitable at the prices $p$. Hence, let $y^1$ be any other feasible set of net loads, such that there is a $u^1$ with:

$$\begin{aligned} &L(y^1, u^1) + \iota^t y^1 = 0, \\ &K(y^1, u^1) \leq 0, \\ &u^1 \in U. \end{aligned}$$

Then by (1.8), we have,

$$p^t(y^* - y^1) \geq 0. \tag{1.9}$$

This no arbitrage condition will be important as part of the analysis of revenue adequacy in the FTR formulations. Importantly, the condition allows for the controls to change from $u^*$. This implies a great degree of flexibility in changing the dispatch while maintaining the no-arbitrage condition for a market equilibrium.

### 1.3.4 Linear Approximation of Constraints

The full AC security-constrained economic dispatch problem is a large optimization problem with very many constraints. Solution procedures for solving this problem often rely on local linearizations of at least the constraints and exploit the condition that in any particular dispatch only relatively few (tens to hundreds) of the many potential constraints might be binding.

One motivation for the linearization follows from the first order conditions for an optimum. Suppose we have a solution to the economic dispatch problem at $(y^*, u^*)$. The usual Taylor approximation gives:

$$L(y, u) \approx L(y^*, u^*) + \nabla L(y^*, u^*) \begin{pmatrix} y - y^* \\ u - u^* \end{pmatrix},$$

$$K(y, u) \approx K(y^*, u^*) + \nabla K(y^*, u^*) \begin{pmatrix} y - y^* \\ u - u^* \end{pmatrix}.$$

Then if we have a solution that satisfies the first order conditions for the security-constrained economic dispatch problem (1.6), this would also satisfy the first order conditions for the linearized constraints as in:

$$\underset{y,u \in U}{Max}\, B(y)$$

$$s.t.$$

$$L(y^*, u^*) + \nabla L(y^*, u^*) \begin{pmatrix} y - y^* \\ u - u^* \end{pmatrix} + \iota^t y = 0,$$

$$K(y^*, u^*) + \nabla K(y^*, u^*) \begin{pmatrix} y - y^* \\ u - u^* \end{pmatrix} \leq 0.$$

If the functions are well behaved, then finding a solution to this approximate problem might also provide a good estimate of the solution to the full problem. Although the functions are not so well behaved as to be everywhere convex, practical computational approaches for solving this problem search for a solution that satisfies the first order conditions. It is not fail safe, and when it fails other approaches would be necessary. However, given a starting point close to the optimum, and some judicious choices, this approximation can work well. Since the actual dispatch involves reoptimization starting with a good solution from the immediate previous period, as well as feedback from metering actual flows and a fair bit of operator judgment, this linearization of the model can be a reasonable approximation. However, as discussed below, the linearization changes with the dispatch.

The local linear approximation suggests an outline for solving this large problem through a familiar relaxation approach by ignoring non-binding constraints (Geoffrion 1970).

*Relaxation Solution Procedure*

**Step 1:** Select an initial candidate solution $(y^0, u^0)$, ignore most (or all) of the constraints in the economic dispatch using only the small subset $K^0(y,u)$, and set the iteration count to $m = 0$.

**Step 2:** Construct the relaxed master problem as:

$$\underset{y,u \in U}{Max}\, B(y)$$

$$s.t.$$

$$L(y^m, u^m) + \nabla L(y^m, u^m)\begin{pmatrix} y - y^m \\ u - u^m \end{pmatrix} + \iota^t y = 0,$$

$$K^m(y^m, u^m) + \nabla K^m(y^m, u^m)\begin{pmatrix} y - y^m \\ u - u^m \end{pmatrix} \leq 0.$$

Let a solution be $(y^{m+1}, u^{m+1})$ and update $m = m + 1$.

**Step 3:** Check to see if the candidate solution $(y^m, u^m)$ violates any of the constraints in (1.6). If so, create a new $K^m(y,u)$ including some or all of these constraints and repeat Step 2. Else done.

The central idea here is that the master problem is much smaller than the full problem and relatively easy to solve. With judicious choices of the initial solution and constraint set, the method works well in practice with relatively few iterations required. In the case that the objective function is represented by a piecewise linearization (as would be true naturally with step-wise representation of supply and demand), the master problem is a linear program for which there are efficient algorithms. Furthermore, in the case of this dispatch problem, evaluation of constraints in Step 3 requires only that a standard load flow be solved for each contingency. Although not trivial, this is well-understood albeit non-linear problem.

One difficulty with this computational approach is the need to calculate $\nabla K^m(y^m, u^m)$.[20] This gradient is the set of "shift factors" summarizing the marginal impact on constraints from changes in the loads and controls. Although it is possible to solve the load flow problem exploiting the sparsity of the network arising from the few links connected to each bus, this sparsity depends on explicit representation of the angles and voltage magnitudes. By contrast, the inverse presentation in $K(y,u)$ is dense. In a sufficiently meshed network, every net load affects every constraint. Hence, virtually every element of $\nabla K^m(y^m, u^m)$ could be non-zero. Part of the art of implementation of this computational outline is in the details of exploiting sparse representations to evaluate load

[20] For more detail on the construction of the gradients, see Weber (1997).

flows, and minimizing the need to calculate or represent $\nabla K^m(y^m, u^m)$. Such commercial dispatch software is well developed and in regular use.[21]

Further note that in general $\nabla K(y^{m_1}, u^{m_1}) \neq \nabla K(y^{m_2}, u^{m_2})$, and this may require frequent updates of the linearization. Finally, in general we have:

$$K(y^{m_1}, u^{m_1}) - \nabla K(y^{m_1}, u^{m_1}) \begin{pmatrix} y^{m_1} \\ u^{m_1} \end{pmatrix} \neq K(y^{m_2}, u^{m_2}) - \nabla K(y^{m_2}, u^{m_2}) \begin{pmatrix} y^{m_2} \\ u^{m_2} \end{pmatrix}.$$

Hence, the "right hand side" of the linearized constraint can be different for each candidate solution. These differences can be quite large, especially for interface constraints in DC-Load approximations.[22]

This presents no difficulty in principle for the dispatch problem. However, these complications are relevant in the discussion of the DC-Load model and in the adaptation of the security-constrained economic dispatch formulation for FTR auctions.

### *1.3.5 DC-Load Approximations*

A common simplification of the load flow model for real power is known as the DC-Load approximation (Schweppe et al. 1988). In terms of the present discussion, the DC-Load model adds further restrictive assumptions that allow us to ignore both real power losses and reactive power loads in determining the real power flows, further specializing the linearization of the constraints.

The key assumptions include:

- There is sufficient reactive power net load at each bus to maintain per unit voltages equal to 1.0 ($V_i \approx 1.0$);
- All phase angle settings are at zero angle change and a fixed tap ratio for transformers ($t = 1.0,\ \alpha = 0$)[23];
- The voltage angle differences across lines are small.

These assumptions imply a choice of controls ($u = u^0$) that yield full decoupling between real and reactive power flow and no transmission losses. The real power flow in (1.1) reduces to:

[21] For example, firms providing such software include ALSTOM ESCA Corporation, Nexant, Inc., Open Access Technology International, Inc.

[22] For examples, see Hogan (2000).

[23] For simplicity, we can assume that the ideal transformers with a fixed tap ratio have been incorporated in a per unit normalization, which results in a simplified Π_equivalent representation of a transmission line. See the appendix for further details

$$z_{Pijk} = G_k[1 - \cos(\delta_i - \delta_j)] + \Omega_k \sin(\delta_i - \delta_j),$$
$$z_{Pjik} = G_k[1 - \cos(\delta_j - \delta_i)] + \Omega_k \sin(\delta_j - \delta_i).$$

Under the small angle difference assumption, we have:

$$\cos(\delta_i - \delta_j) \approx 1,$$
$$\sin(\delta_i - \delta_j) \approx \delta_i - \delta_j.$$

Hence, the real power flow approximation becomes:

$$z_{Pk} = z_{Pijk} = \Omega_k(\delta_i - \delta_j) = -z_{Pjik} = -\Omega_k(\delta_j - \delta_i).$$

This linearity produces a substantial simplification. Let:
$\Omega =$ the diagonal matrix of line transfer factors,
$z =$ the vector of line flows ($z_{Pk}$) in the DC-Load approximation.
Then, with our sign conventions we have:

$$y = -A^t z,$$
$$z = \Omega A \delta.$$

Furthermore, the inversion that eliminates the angles as in (1.2) reduces to another linear equation for the DC-Load formulation with

$$H = \nabla K_y(0, u^0).$$

This is the matrix of shift factors. Under the DC-Load assumptions, $H = \begin{pmatrix} 0 & \tilde{H} \end{pmatrix}$, where $\tilde{H} = -\Omega\tilde{A}\left(\tilde{A}^t\Omega\tilde{A}\right)^{-1}$ with the swing bus dropped in defining $\tilde{A}$.[24] Although $A$ is sparse, the matrix of shift factors is dense, meaning that nearly every net load affects nearly every line. Calculating an element of row of $H$, meaning the shift factors for a particular line in a particular contingency, is about the same amount of work as finding a DC-Load flow for that contingency.

For a given contingency the matrix that links the angles and the net loads, as in

$$y = -A^t \Omega A \delta,$$

is quite sparse, with the only non-zero elements being for the nodes that are directly connected. Furthermore, solving for the angles given the vector of net injections, $y$, involves no more than finding a particular solution for a set of linear equations. In general, this is much less work than solving for the full matrix inverse, and in advanced optimization algorithms this is done quickly and cheaply using sparse

[24] Also the transfer admittance matrix as described in Schweppe et al. (1988), p. 316.

matrix techniques. Once the vector of angles is known for a given set of net loads, it is an easy matter to complete the one matrix multiplication to obtain the complete load flow in $z$ for each contingency. The import of all this is the simplicity of evaluating a particular load flow as compared to calculating the full transfer admittance matrix in $H$.

Note that calculating a particular row of $H$ is about the same order of difficulty as evaluating the load flow for that particular contingency. Let $\varepsilon^i$ be the elementary row vector with all zeros but a $1$ in the $i_{th}$ position. We can obtain any row of $H$, say $h_i$, as the solution to a set of sparse linear equations. By construction:

$$h_i = \varepsilon^i \tilde{H} = -\varepsilon^i \Omega \tilde{A} \left( \tilde{A}^t \Omega \tilde{A} \right)^{-1}.$$

Hence, we have the sparse system:

$$h_i \left( \tilde{A}^t \Omega \tilde{A} \right) = -\varepsilon^i \Omega \tilde{A}. \tag{1.10}$$

In other words, calculating a complete load flow for all the lines is about as much work as calculating the shift factors for one line. Both require solution of a sparse set of linear equations of the dimension equal to the number of nodes. There are specialized sparse matrix techniques for this computation as a part of commercial dispatch software.

With these approximations, the constraints could be restated as:

$$\begin{aligned} \iota^t y &= 0, \\ K\left(0, u^0\right) + Hy &\leq 0. \end{aligned}$$

Letting $b = -K(0, u^0)$, the familiar DC-Load restatement of the security-constrained economic dispatch becomes:

$$\begin{aligned} &\underset{y}{Max}\, B(y) \\ &s.t. \\ &\iota^t y = 0, \\ &Hy \leq b. \end{aligned} \tag{1.11}$$

It is an easy matter to extend the definition of $H$ to include other linear constraints on $y$, including interface constraints expressed as limits on aggregations of flows on lines.

As above, the matrix $H$ for the full security-constrained problem is very large and dense, and successful solution of the security-constrained economic dispatch exploits approaches such as the relaxation algorithm outlined above that avoid unnecessary computation of the elements of $H$ and include only the binding constraints. Furthermore, the DC-Load model is convex and the relaxation algorithm will assure convergence to a global solution.

As discussed below, many models for transmission rights exploit the specialized structure of (1.11) to simplify the problem and guarantee various equivalence conditions between and among different FTRs. In this context, it is important to remember that (1.11) is only a simplified approximation and that key elements of these assumptions are violated by regular operating conditions in the system. The different approximations have different effects on the alternative FTR models and the associated auction problems.

Here we consider the implications of various modifications of these assumptions. Suppose that the phase shifting transformers are set to shift the angles. If we hold the angle shifts fixed, then the approximation under the other DC-Load assumptions becomes:

$$z_{Pk} = \Omega_k\left(\delta_i - \delta_j\right) + \Omega_k \alpha_k.$$

In principle, this changes the inversion in (1.2) to eliminate the bus angles such that even under zero net loads there would be real power flow on all the lines in order to maintain balance at every node. This preserves linearity and a constant $H$, but changes the residual limits for the constraints. Hence, we would have $b = b(\alpha)$, meaning that the limits on the power flow equations would be changing to reflect the phase angle settings. In principle, a phase shift on one line could affect the residual limit on every line.

If the ideal transformer tap ratio ($t$) were to change from 1.0, there would be a modified $\hat{\Omega}$ to reflect the changing impedance.[25] In addition, the inversion depends on the topology of the network as summarized in $A$. This may change from one dispatch to another. In each case, the inversion to eliminate the voltage angles and the associated linearization of the constraints actually depends on the values of $(t, \alpha, A)$. To the extent that these are treated as variables in the economic dispatch, their constraints in $U$ create additional non-linearities. For instance, if a phase-shifting transformer is controlling flow but reaches a limit on the ability to control a line, the representation of the phase angle regulator changes. Although the details depend on the particular case, if there is any possibility of actual changing the topology or settings of phase-shifting transformers, even for the simplified real power only DC-Load approximation we have $H(u^0) = \nabla K_y(0, u^0)$ and $b(u^0) = -K(0, u^0)$. In other words, the linear approximation is not the same across the dispatches.

Therefore, the security-constrained, economic dispatch of the DC-Load approximation could be written as:

[25] For example, see Oliveira et al. (1999), pp. 111–118.

$$
\begin{aligned}
&\underset{y}{Max}\, B(y) \\
&s.t. \\
&\quad \iota^t y = 0, \\
&\quad H(u^0) y \le b(u^0).
\end{aligned}
\tag{1.12}
$$

When this problem is solved at any given hour, for fixed $u^0$ the resulting model takes on the form of the DC-Load approximation. Both constraint limits and shift factors adjust regularly. Hence, it is important below to be explicit about the fact that the linearizations, and therefore, the model itself, changes from dispatch to dispatch, especially for any changes in topology $A$.

Finally, in addition to these changes, other slight modifications of the DC-Load model retain most of the computational simplicity but make the approximation further sensitive to the non-linear properties of the system. For example, consider incorporating line losses:

$$
l_{Pk}(\delta, 1, 1, 0) = \; G_{\mathrm{k}}\left[2 - 2\cos\left(\delta_i - \delta_j\right)\right].
$$

Using the approximation that for small angle differences,

$$
\cos\left(\delta_i - \delta_j\right) \approx 1 - \frac{\left(\delta_i - \delta_j\right)^2}{2},
$$

the approximate line losses are:

$$
l_{Pk}(\delta, 1, 1, 0) = G_{\mathrm{k}}\left[2 - 2\cos\left(\delta_i - \delta_j\right)\right] \approx G_{\mathrm{k}}\left(\delta_i - \delta_j\right)^2 \approx r_k z_{Pk}^2.
$$

Here we have used the condition that $r_k << x_k$.[26]

Define $R$ as the diagonal matrix of line resistances, $|A|$ as the matrix of the absolute values of the incidence matrix, and $z^2$ as the vector of squares of the individual line flows. Then we could include losses in the economic dispatch problem that is almost like the DC-Load model[27]:

[26] This approximation applies to high voltage systems, but is less usable on lower voltage circuits.

[27] This approach is from Transpower in New Zealand.

$$\begin{gathered} Max\, B(y) \\ y, z, \delta \\ s.t. \\ y = -A^t z - \tfrac{1}{2}|A|^t R z^2, \\ z = \Omega A \delta, \\ \delta_s = 0, \\ z \leq b. \end{gathered}$$

Note that this computational form of the problem does not need a separate overall balance equation, as this is accounted for in the individual node equations. Hence, we have net loads (generation) balancing losses as in:

$$\iota^t(g-d) = -\iota^t(d-g) = -\iota^t y = \iota^t A^t z + \tfrac{1}{2}\iota^t|A|^t R z^2 = \iota^t R z^2.$$

This is no longer a linear problem, but the addition of the few quadratic terms in the node balance equations is easier to deal with than a full AC model. However, this simplified formulation would capture some of the interaction between losses and congestion, with the additional power flows needed to account for losses adding to losses and congestion. The inverse linearization of the solution in terms of the net loads would now differ further from the pure DC-Load approximation.[28]

## 1.4 Point-to-Point Financial Transmission Rights

Financial transmission rights are defined in terms of payments related to market prices. Although many years were spent in the search for well-defined and workable physical transmission rights, the complexity of the grid and rapidly changing conditions of the real market outcomes made it impossible to design physical rights that could be used to determine the use of the transmission system.[29] By contrast, financial transmission rights specify payments that are connected to the market outcomes but do not control use of the system. Rather, the actual dispatch or spot market produces a set of market-clearing prices, and these prices in turn define the payments under the FTRs.

The system operator accepts schedules and coordinates the spot market as a bid-based, security-constrained, economic dispatch. The resulting locational prices apply to purchases and sales through the spot market, or the difference in the locational prices defines the price for transmission usage for bilateral schedules.

[28] A version of this DC-Load-Flow implementation with losses appears in a GAMS model available at www.whogan.com.

[29] For further details, see Harvey et al. (1997).

The need for transmission rights to hedge the locational price differences leads to the interest in FTRs.[30]

### 1.4.1 PTP Obligations

The definition of point-to-point (PTP) forward obligations as FTRs follows closely the notion of bilateral transmission schedules. A generic definition includes both balanced and unbalanced rights. Given a vector of inputs and outputs by location, the *kth* PTP forward obligation is defined by:

$$PTP_k^f = \begin{pmatrix} 0 \\ -g_i \\ 0 \\ d_j \\ 0 \end{pmatrix}.$$

With a corresponding vector of market clearing prices, this FTR is a contract to receive

$$p^t PTP_k^f = p^t \begin{pmatrix} 0 \\ -g_i \\ 0 \\ d_j \\ 0 \end{pmatrix} = p_j d_j - p_i g_i.$$

Although any such vector could be allowed, it is clear that any such FTR could be restated as a mix of balanced and unbalanced rights:

$$PTP_k^f = \begin{pmatrix} 0 \\ -d_j \\ 0 \\ d_j \\ 0 \end{pmatrix} - \begin{pmatrix} 0 \\ g_i - d_j \\ 0 \\ 0 \\ 0 \end{pmatrix}.$$

Motivated by the discussion of options below, it is convenient to define two types of forward obligations, balanced ($\tau_k^f$) and unbalanced ($\bar{g}_k^f$), such as

[30] For further discussion of market structure, see Chandley and Hogan (2002).

$$\tau_k^f = \begin{pmatrix} 0 \\ -x \\ 0 \\ x \\ 0 \end{pmatrix}, \quad \bar{g}_k^f = \begin{pmatrix} 0 \\ 0 \\ g \\ 0 \\ 0 \end{pmatrix}.$$

We can think of the balanced PTP-FTRs providing for the same input and output at different locations. More generally, all that is required of a balanced PTP-FTR is that the inputs and outputs sum to zero, $\iota^t \tau_k^f = 0$. The unbalanced FTRs can be thought of as forward energy sales at any location and would be a contribution towards losses to balance the system. The notation suggests that individuals could hold either or both types of PTP-FTR forward obligations, and there is no need that the locations be the same.

The intended role of the PTP-FTR is to provide a hedge against variable transmission costs. If a market participant has a balanced FTR between two locations and schedules a corresponding bilateral transaction with the same inputs and outputs ($x$), then the charge for using the system would be $(p_j - p_i)x$, which is exactly the payment that would be received under the FTR. Hence, the balanced FTR provides a perfect hedge of the variable transmission charge for the bilateral transaction.

The holder of an unbalanced forward obligation FTR has an obligation to make the payment equal to the value of the energy at the relevant location. If the holder also sells an equal amount of energy at the same location in the actual dispatch, the payment received for the energy is $p_i g$, equal to the payment required under the FTR. Hence, we can think of the unbalanced FTR as a forward sale of energy. Although in principle there would be no difficulty in allowing negative unbalanced PTP-FTRs, equivalent to forward purchases of energy, it is convenient to interpret unbalanced PTP-FTR obligations as forward sales of energy.

In this case of obligations, the PTP-FTRs are easily decomposable. For example, an FTR from bus 1 to bus 2 can be decomposed into two PTP-FTR obligations from 1 to a Hub and the Hub to 2. The total payment is $(p_2 - p_{HUB}) + (p_{HUB} - p_1) = (p_2 - p_1)$. This provides support for trading at market hubs and the associated trading flexibility. Periodic FTR auctions provide other opportunities to obtain other reconfigurations of the pattern of FTRs

An attraction of the FTR is that the spot market can operate to set the actual use of the transmission system and the FTRs operate in parallel through the settlements system to administer financial hedges. Importantly, the system of payments will be consistent as long as the set of PTP-FTRs satisfies a simultaneous feasibility condition.

#### 1.4.1.1 Revenue Adequacy

Suppose that we have a set of balanced ($\tau^f_k, k = 1, \ldots, N$) and unbalanced ($\bar{g}^f_k$, $k = 1, \ldots, N$) PTP-FTRs obligations for any possible locations. Consider the constraints from the security-constrained dispatch in (1.6) or equivalently in (1.8). We say that the set of FTRs is simultaneously feasible if there is a $u \in U$ such that:

$$\begin{aligned} y &= \sum_k \tau^f_k - \sum_k \bar{g}^f_k, \\ L(y, u) &+ \iota^t y = 0, \\ K(y, u) &\leq 0. \end{aligned} \tag{1.13}$$

Assume the set of PTP-FTR forward obligations is simultaneously feasible. If we have a market equilibrium ($p$, $y^*$, $u^*$) in the spot market, then from (1.9) it follows immediately that we meet the revenue adequacy condition,

$$p^t y^* - p^t\left(\sum_k \tau^f_k - \sum_k \bar{g}^f_k\right) = p^t\left(y^* - \sum_k \tau^f_k + \sum_k \bar{g}^f_k\right) = p^t(y^* - y) \geq 0.$$

In other words, at the market equilibrium prices the net payments collected by the system operator through the actual dispatch ($p^t y^*$) would be greater than or equal to the payments required under the PTP-FTR forward obligations $\left(p^t\left(\sum_k \tau^f_k - \sum_k \bar{g}^f_k\right)\right)$. This revenue adequacy condition is general enough to accommodate a great deal of flexibility.

Note that the simultaneous feasibility condition does not require that the set of PTP-FTRs be feasible at the current set of controls ($u^*$) associated with the market equilibrium. All that is required is that the system operator could choose a set of controls that would make the PTP-FTRs feasible. There could be a very different set of actual operating conditions, including changes in the configuration of the grid, but as long as the controls and configuration could be set to make the PTP-FTRs feasible, the simultaneous feasibility condition holds and revenue adequacy follows. This is true even though actual physical delivery to match the FTRs would be impossible at the current settings of the grid controls at $u^*$. This is an important simplification compared to physical rights and a primary attraction of using financial rights.

The intuition of revenue adequacy is clear. If the dispatch of PTP-FTRs were more valuable than the market equilibrium, in violation of the revenue adequacy condition, the system operator could have selected this dispatch outcome. Since we have by assumption a market equilibrium that differs from the PTP-FTRs, and the PTP-FTRs are simultaneously feasible, the market equilibrium from (1.8) must be at least as valuable as the payment obligation under the PTP-FTRs.[31]

#### 1.4.1.2 PTP-FTR Auction

Allocation rules for FTRs follow different procedures. For example, in PJM Load Serving Entities (LSE) are required to purchase network service and meet installed capacity requirements. As part of this process, LSEs acquire FTRs. Grandfathering rules under existing contracts might be another source of allocation, and so on.

A natural way to allocate PTP-FTR forward obligations would be to conduct an auction. Suppose that we represent bids for balanced forward-obligations by $(t^f_k, \tau^f_k)$ and for unbalanced forward obligations by $(\rho^f_k, \bar{g}^f_k)$. Here the first element is the scalar amount of the FTR and the second element is the vector pattern of inputs and outputs. For simplicity, we subsume any upper bounds on the awards are part of the concave and differentiable bid function $\beta_k\left(t^f_k, \rho^f_k\right)$. With these notational conventions, a formulation of the PTP-FTR forward obligation auction would be:

$$\begin{aligned}
&\underset{y,u\in U, t^f_k \geq 0, \rho^f_k \geq 0}{Max} \sum_k \beta_k\left(t^f_k, \rho^f_k\right) \\
&s.t. \\
&y = \sum_k t^f_k \tau^f_k - \sum_k \rho^f_k \bar{g}^f_k, \\
&L(y,u) + \iota^t y = 0, \\
&K(y,u) \leq 0.
\end{aligned} \tag{1.14}$$

A solution of this problem would determine the award of FTRs and the associated market clearing prices for the awards. The locational price $\widehat{p}$ would be of the same form as in the market equilibrium model, with

$$\widehat{p}^t = \widehat{\lambda}\iota^t + \widehat{\lambda}\nabla L_y(y^*, u^*) + \widehat{\eta}^t \nabla K_y(y^*, u^*).$$

However, the prices here would be based on the expected value of the hedge over the many dispatches to which it applies. The corresponding market clearing prices

[31] The definition of FTRs could be extended to include the sharing rule for allocation of any difference between the collections and payments. This is formalized in the market equilibrium model as $s_i$. In practice, the FTR implementations for existing system redistribute any excess collection to reduce access charges or some similar purpose. Although this is a more important issue for defining incentives for system expansion, it does not affect the analysis here.

for the auction awards would be the difference in the locational prices for the balanced obligations and the locational price for loss contributions. Hence,

$$\widehat{p}_{t_k^f} = \frac{\partial \beta_k\left(t_k^f, \rho_k^f\right)}{\partial t_k^f} = \widehat{p}^t \tau_k^f, \quad \widehat{p}_{\bar{g}_k^f} = -\frac{\partial \beta_k\left(t_k^f, \rho_k^f\right)}{\partial \rho_k^f} = \widehat{p}^t \bar{g}_k^f.$$

By construction, the FTRs would be simultaneously feasible. In addition to an initial sale to allocate FTRs for the existing grid, this same format accommodates offers to sell existing FTRs. By this means, regular auctions of this form also provide opportunities to reconfigure the pattern of FTRs.

It is obvious that the PTP-FTR auction problem in (1.14) is essentially of the same form as the security-constrained economic dispatch problem in (1.6) or the market equilibrium problem in (1.8), with the addition of a set of simple linear constraints on the net loads as dictated by the bids. Furthermore, the addition of the linear constraints on the awards could be included in the master problem of the relaxation solution procedure described above, allowing for a direct adaptation of familiar optimal dispatch software to solve the auction problem. This is the essence of the AC-formulation of the PTP-FTR obligation auction conducted by the New York Independent System Operator (NYISO), where the computational feasibility of the solution procedure has been verified in practice.[32]

In the case of a dispatch that prices losses and includes losses in the PTP-FTRs, the consistent model anticipates that market participants will take on the forward commitment to meet the financial requirements for losses. Various approximations might be considered where this is a requirement is modified.[33] In the early implementations, the focus of PTP-FTRs was on congestion costs.

#### 1.4.1.3 PTP-FTR for Congestion

The initial PJM implementation employed a DC-Load dispatch model similar to (1.12).[34] The dispatch and the resulting market prices do not explicitly treat marginal losses. Hence, the prices differ across locations only due to the effects of congestion. The PTP-FTRs are defined for payments on congestion cost, and in this case are the full hedge for the difference in locational prices. Under this system, the payments for losses are treated as part of an uplift charge, and not covered by the FTRs. Since the congestion costs define the only locational price differences

[32] For results of New York auctions, see: http://www.nyiso.com/markets/tcc_auctions/2001_2002_winter.html.

[33] For a further discussion see Harvey and Hogan (2002).

[34] In PJM, financial transmission rights are called fixed transmission rights (FTR). http://www.pjm.com/energy/ftr/ftrauc.html.

charged or hedged, revenue adequacy follows from the simultaneous feasibility condition for the PTP-FTRs.

The implementation in New York differs in its treatment of losses. Losses are included in the dispatch model and the associated market prices. However, the PTP-FTRs are defined as balanced rights only and provide for payment of congestion costs but not the cost of losses. The auction for FTRs uses an AC formulation as in (1.14). Market participants obtain balanced FTRs and the NYISO includes provisions for losses in the auction, in order to obtain a feasible solution in the auction. However, the NYISO does not assume financial responsibility for loss hedges. In New York, the FTRs provide a hedge only for congestion costs.[35] This New York type implementation leads to a different version of the revenue adequacy condition.

Let the allocation of balanced FTRs in the auction be $\tau^f = \sum_k t^f_k \tau^f_k$. Choose an arbitrary unbalanced vector of loss contributions $\bar{g}^f$ such that $\left(\tau^f, \bar{g}^f\right)$ is simultaneously feasible. Let there be a market equilibrium $(p, y^*, u^*)$ from the actual dispatch. The prices decompose into the price of generation ($p_G = \lambda$), the marginal contribution to losses ($p^t_L = \lambda \nabla L_y(y^*, u^*)$), and the cost of congestion ($p^t_C = \eta^t \nabla K_y(y^*, u^*)$). By the simultaneous feasibility of the PTP-FTRs, we have

$$p^t y^* \geq p^t\left(\tau^f - \bar{g}^f\right) = p^t_C \tau^f + \left(p_G \iota^t + p^t_L\right)\tau^f - p^t \bar{g}^f.$$

Define the loss rentals on the FTRs as the difference between the payment for losses at the marginal cost and the average cost of the losses. Hence,

$$\pi_L \equiv p^t_L \tau^f - p^t \bar{g}^f = \left(p_G \iota^t + p^t_L\right)\tau^f - p^t \bar{g}^f.$$

If we have these loss rentals as non-negative, $\pi_L \geq 0$, then the simultaneous feasibility test coupled with this condition is enough to ensure that the total net payments from the dispatch are at least as large as the congestion payments under the PTP-FTRs, as in:

$$p^t y^* \geq p^t_C \tau^f. \tag{1.15}$$

Since $\bar{g}^f$ is arbitrary but feasible, we could have chosen $\bar{g}^f$ to maximize the loss rentals for the FTRs given the prices for this hour. In other words, if we have sufficiently inexpensive locations at which to deem the unbalanced FTR loss contribution, the loss rentals would be non-negative and along with the

[35] In New York, financial transmission rights are called Transmission Congestion Contracts (TCC).

simultaneous feasibility condition would be sufficient to ensure revenue adequacy in the sense of (1.15) for congestion hedges only.[36]

In the case of New York, the loss prices and loss rentals may be small, and the typical situation would be that losses would be costly with the maximum loss rentals implied for the FTRs being positive. Under typical conditions, therefore, simultaneous feasibility would guarantee revenue adequacy for the congestion payments under the FTRs.

### *1.4.2 PTP Options*

A PTP-FTR obligation is a financial contract for the payment of the locational price difference. When matched with a corresponding delivery of power, the charge for transmission usage just balances the FTR payment, and there is a perfect hedge. This is true whether or not the price difference is positive or negative. If the price difference is negative, the schedule provides valuable counterflow for which the provider is paid, and the payment from the spot market dispatch just balances the obligation under the FTR. There is a perfect match either way.

A natural complement to the PTP-FTR obligation would be a PTP-FTR option that did not require payment when the price difference was negative. Hence for the balanced PTP-FTR option $\tau_k^o$ the payment would be $\max\left(0, p^t \tau_k^o\right)$. This financial contract might be more attractive as a tool for hedging purposes, and it is typically the first suggestion from market participants because of the perception that there is a closer analogy to the presumed option not to schedule under a physical right. The option might also be more valuable for speculators who want to trade rights but don't plan to match the FTR with a schedule.

Unlike obligations, PTP-FTRs are not decomposable in the sense of to and from a hub. The difficulty is inherent in the option. For example, an FTR option from bus 1 to bus 2 cannot be decomposed into two PTP-FTR options from 1 to a Hub and the Hub to 2. The total payment under the two options would be $\max(0, p_2 - p_{HUB}) + \max(0, p_{HUB} - p_1) \neq \max(0, p_2 - p_1)$. Hence, reconfiguration of options would require coordination in a formal auction.

Whatever the merits of the PTP-FTR option, it presents complications that do not arise in the case of obligations. The difficulty flows from the simple fact that the dispatch formulation (1.6) does not include options; in the real dispatch everything is an obligation. Hence the auction model for options does not follow directly from the formulation for economic dispatch. Further, the associated settlement rules for options do not follow immediately from the analysis for obligations.

---

[36] It is a conjecture, but not proven, that this "optimized" FTR-loss rental is always non-negative, and that simultaneous feasibility alone is sufficient for revenue adequacy in this congestion-only case.

The analytical problem for options is similar to the problem for physical rights. Without knowing all the other flows on the system, it is not possible in general to know if any particular transaction will be feasible. Hence, to guarantee feasibility it is necessary to consider all possible combinations of the exercise of options. For example, if too few of the other options are exercised, there may be insufficient counterflow to support a particular transaction; or if all the options are exercised, some other constraint might be limiting. This ambiguity does not arise with obligations, which by definition are always exercised.

#### 1.4.2.1 Revenue Adequacy for Options

As with PTP-FTR obligations, simultaneous feasibility of the exercised options is a necessary condition to guarantee revenue adequacy.[37] To demonstrate that simultaneous feasibility is also sufficient requires an expansion of the definition and test for simultaneous feasibility. Once we know which options are exercised, we can treat them like obligations for settlement purposes, so if the exercised rights are simultaneously feasible, we will have revenue adequacy. But the test of feasibility of all possible combinations of exercise of options requires an expansion of the model.

Here we consider only the possibility of balanced PTP-FTR options, combined with both balanced and unbalanced forward obligations. As above, we have a set of balanced $(\tau^f_k, k = 1, \ldots, N)$ and unbalanced $(\bar{g}^f_k, k = 1, \ldots, N)$ PTP-FTR obligations for any possible locations. In addition, define the balanced options as $(\tau^0_k, k = 1, \ldots, N)$. Let $x_k$ be the fraction of each option exercised. Since different exercise patterns produce different losses, we need some flexibility in the total loss provision. As with contingency constraints, we impose this balancing adjustment at the swing bus. For the moment, assume the unbalanced obligations are large enough to ensure that this adjustment is non-negative. Then for feasibility we require by analogy to (1.13) that there is a $u \in U$ and a scalar balancing adjustment at the swing bus with $\varepsilon^0_s \geq 0$ such that:

$$y = \sum_k \tau^f_k + \sum_k x_k \tau^o_k - \sum_k \bar{g}^f_k + \begin{pmatrix} 1 \\ 0 \end{pmatrix} \varepsilon^0_s,$$
$$L(y, u) + \iota^t y = 0,$$
$$K(y, u) \leq 0.$$

Since this must be true for an arbitrary exercise of options and applies to all constraints collectively, it must be true for each contingency and constraint

[37] The FTRs may be revenue adequate under some dispatch cases without simultaneous feasibility, but not under all dispatch cases. For instance, if the FTRs follow the same pattern as the dispatch, but imply even more of the valuable flows than is feasible, the FTRs would not be revenue adequate.

combination. A formulation that allowed for a different $u \in U$ for each exercise of the options would be the weakest condition. A somewhat simpler test that provides a sufficient condition for simultaneous feasibility is to require that any exercise of the options be feasible for the same $u \in U$.[38]

Consider first the constraints in $K(y,u)$. The constraints do not depend on the value of $y$ at the swing bus that is merely a balancing adjustment. Hence, the constraints would be satisfied if there is a $u \in U$ such that

$$\underset{\substack{y=\sum_k \tau_k^f+\sum_k x_k\tau_k^o-\sum_k g_k^f,\\ 0\le x_k\le 1}}{Max}\ \underset{i,\omega}{Max}\, K_i^\omega(y,u) = \underset{i,\omega}{Max}\ \underset{\substack{y=\sum_k \tau_k^f+\sum_k x_k\tau_k^o-\sum_k g_k^f,\\ 0\le x_k\le 1}}{Max} K_i^\omega(y,u) \le 0. \tag{1.16}$$

Recall from (1.5) that there is a loss function for each contingency, and many constraints. Here we represent these loss functions and constraints explicitly to make clear the nature of the constraints induced by the options. Hence, define a new function $w_i^\omega$, meaning constraint $i$ in contingency $\omega$:

$$\begin{aligned} w_i^\omega\left(\tau^f,\{t_k^o\},\bar{g}^f,u\right) = \ & \underset{\substack{\varepsilon_s,y\\ 0\le x_k\le 1}}{Max}\ K_i^\omega(y,u) \\ s.t.\ & \\ y = \tau^f + \sum_k x_k t_k^o \tau_k^o - \bar{g}^f + & \begin{pmatrix}1\\0\end{pmatrix}\varepsilon_s, \\ & L^\omega(y,u) + \iota^t y = 0. \end{aligned} \tag{1.17}$$

The notation $\{t_k^o\}$ refers to the vector of award levels of the options. Here $\varepsilon_s$ is the load adjustment at the swing bus to achieve balanced loads in the contingency. This notation allows and anticipates a different solution $y$ for every constraint and contingency combination. Apparently the condition that the constraint $K_i^\omega$ is satisfied for all possible exercise of options is equivalent to:

$$w_i^\omega\left(\tau^f,\{t_k^o\},\bar{g}^f,u\right) \le 0.$$

This $w_i^\omega$ is an optimal-value function, the result itself of an optimization problem (Shimuzu et al. 1997). However, it is a well-defined function that would allow restatement of the auction problem in terms of the variables defining the auction awards.

For the contingency we define:

---

[38] These two definitions would be the same if there is a saddle point for the function $f(y,u) = \underset{i,\omega}{Max}\, K_i^\omega(y,u)$. However, the usual convexity arguments would not apply to guarantee a saddle point as it seems unlikely that $f$ would be concave in $y$, Ponstein (1965), pp. 181–188. In any event, the former computational problem appears more difficult.

$$w^{\omega}\left(\tau^{f},\left\{t_{k}^{o}\right\},\bar{g}^{f},u\right)\equiv\begin{pmatrix} w_{1}^{\omega}\left(\tau^{f},\left\{t_{k}^{o}\right\},\bar{g}^{f},u\right)\\ w_{2}^{\omega}\left(\tau^{f},\left\{t_{k}^{o}\right\},\bar{g}^{f},u\right)\\ \vdots\\ w_{n}^{\omega}\left(\tau^{f},\left\{t_{k}^{o}\right\},\bar{g}^{f},u\right)\end{pmatrix}.$$

Hence, the sufficient condition in (1.16) for simultaneous feasibility of PTP-FTRs with options requires:

$$w\left(\tau^{f},\left\{t_{k}^{o}\right\},\bar{g}^{f},u\right)\equiv\begin{pmatrix} w^{0}\left(\tau^{f},\left\{t_{k}^{o}\right\},\bar{g}^{f},u\right)\\ w^{1}\left(\tau^{f},\left\{t_{k}^{o}\right\},\bar{g}^{f},u\right)\\ \vdots\\ w^{\omega}\left(\tau^{f},\left\{t_{k}^{o}\right\},\bar{g}^{f},u\right)\\ \vdots\\ w^{m}\left(\tau^{f},\left\{t_{k}^{o}\right\},\bar{g}^{f},u\right)\end{pmatrix}\leq 0.$$

Finally, to treat losses and ensure that $\varepsilon_{s}^{0}\geq 0$, define the worst case for the contribution of losses and the unbalanced obligations:

$$L_{O}^{0}\left(\tau^{f},\left\{t_{k}^{o}\right\},\bar{g}^{f},u\right)=\underset{\substack{y\\ 0\leq x_{k}\leq 1}}{Max}\ L^{0}(y,u)$$

$$\begin{aligned} &s.t.\\ &y=\tau^{f}+\sum_{k}x_{k}t_{k}^{o}\tau_{k}^{o}-\bar{g}^{f},\\ &L^{0}(y,u)+\iota^{t}y=0.\end{aligned}$$

If we have enough loss obligations to meet this maximized exercise of FTR losses, then we have enough total forward unbalanced obligations to meet or exceed the exercised FTR losses and ensure that we meet the assumption above that $\varepsilon_{s}^{0}\geq 0$. Therefore, we set the simultaneous feasibility condition with PTP-FTR obligations and options as:

$$\begin{aligned} &L_{O}^{0}\left(\tau^{f},\left\{t_{k}^{o}\right\},\bar{g}^{f},u\right)-\iota^{t}\bar{g}^{f}=0,\\ &w\left(\tau^{f},\left\{t_{k}^{o}\right\},\bar{g}^{f},u\right)\leq 0,\\ &u\in U.\end{aligned}$$

Consider a market equilibrium $(p,y^{*},u^{*})$. Let $\tau^{o^{*}}$ be the corresponding aggregate of exercised options from the simultaneously feasible combination, $\left(\tau^{f},\left\{t_{k}^{o}\right\},\bar{g}^{f},u\right)$.

In other words, $\tau^{o^*}$ is the aggregate of all the options with $p^t\tau_k^o \geq 0$. Then let $\varepsilon_s^*$ be the difference in the net load at the swing bus required to achieve balance of the FTR in the pre-contingency case $\omega = 0$, i.e.,

$$y = \tau^f + \tau^{o^*} - \bar{g}^f + \begin{pmatrix} 1 \\ 0 \end{pmatrix} \varepsilon_s^*,$$

$$L^0(y, u) + \iota^t y = 0.$$

By construction it must be that $\varepsilon_s^* \geq 0$. Further, let

$$\bar{g}^{f*} = \bar{g}^f - \begin{pmatrix} 1 \\ 0 \end{pmatrix} \varepsilon_s^*.$$

Then since $\bar{g}^{f*}$ differs from $\bar{g}^f$ only for the swing bus, which is allowed to adjust freely for each contingency in the definition of $w$, we have a $u \in U$ with

$$w\left(\tau^f, \left\{t_k^o\right\}, \bar{g}^{f*}, u\right) \leq 0.$$

Therefore, the exercise of the options must be feasible. Hence, we have a balanced load that satisfies every constraint, or

$$y = \tau^f + \tau^{o*} - \bar{g}^{f*},$$
$$L(y, u) + \iota^t y = 0,$$
$$K(y, u) \leq 0,$$
$$u \in U.$$

Following (1.9) we must have:

$$p^t(y^* - y) \geq 0.$$

The payments under the PTP-FTRs equal $p^t\left(\tau^f + \tau^{o*} - \bar{g}^f\right) = p^t y - p_s \varepsilon_s^*$. By construction, $\varepsilon_s^* \geq 0$. Hence, if the swing bus price $p_s \geq 0$, the net revenue from the dispatch will be adequate to pay out the obligations and exercise of options for the PTP-FTRs. Typically, $\varepsilon_s^*$ should be small so that even with a negative price at the swing bus, any revenue inadequacy would be bounded by the small value of the difference in losses.

#### 1.4.2.2 PTP-FTR Auction with Options

With this background, the natural extension of the auction for PTP-FTRs in (1.14) becomes:

$$\begin{aligned}
&\underset{u \in U, t_k^f \geq 0, t_k^o \geq 0, \rho_k^f \geq 0}{Max} \quad \sum_k \beta_k\left(t_k^f, t_k^o, \rho_k^f\right) \\
&s.t. \\
&L_O^0\left(\sum_k t_k^f \tau_k^f, \{t_k^o\}, \sum_k \rho_k^f \bar{g}_k^f, u\right) - \iota^t \sum_k \rho_k^f \bar{g}_k^f = 0, \\
&w\left(\sum_k t_k^f \tau_k^f, \{t_k^o\}, \sum_k \rho_k^f \bar{g}_k^f, u\right) \leq 0.
\end{aligned} \tag{1.18}$$

This is a well-defined model and the objective function is well-behaved.[39] The major change from the AC auction model with obligations only is that the conventional constraint functions $K$ have been replaced with the more complicated constraint functions $w$. Evaluating any element of the function $K$ requires solving an AC load flow problem, one for each contingency. Evaluating any element of $w$ requires solution of an AC optimal power flow problem, one for each contingency and constraint combination. This is a significant increase in computational burden.

In a relaxation and sequential approximation approach for solving the AC auction model with obligations only, the corresponding model from (1.14) is:

$$\begin{aligned}
&\underset{y, u \in U, 0 \leq t_k^f, 0 \leq \rho_k^f}{Max} \quad \sum_k \beta_k\left(t_k^f, \rho_k^f\right) \\
&s.t. \\
&y = \sum_k t_k^f \tau_k^f - \sum_k \rho_k^f \bar{g}_k^f, \\
&L(y, u) + \iota^t y = 0, \\
&K(y, u) \leq 0.
\end{aligned}$$

A computational approach to this problem would exploit the close similarity with security-constrained optimal dispatch problem. The sequential approximation approach begins with a simplified version of the problem that ignores many of the constraints and is solved via a sequential linearization. Then a candidate solution $(\hat{y}, \hat{u}) = \left(\hat{t}^f - \hat{\bar{g}}^f, \hat{u}\right)$ is tested for feasibility by solving a load flow to evaluate $K_i^\omega(\hat{y}, \hat{u})$. If the constraint is violated, determine the gradient of the function and impose the new constraint:

$$K_i^\omega(\hat{y}, \hat{u}) + \nabla K_i^\omega(\hat{y}, \hat{u})^t \begin{pmatrix} y - \hat{y} \\ u - \hat{u} \end{pmatrix} \leq 0.$$

[39] This is a parametric satisfaction problem in the terminology of Shimuzu et al. (1997), p. 285.

This linearized constraint would be appended to the auction model, and there would be further iteration until a solution is found that optimizes the bid function and satisfies all the constraints. Typically we are limited to search algorithms that find solutions to the first-order Karush-Kuhn-Tucker (KKT) conditions and, therefore, to a guarantee only of local optimal solutions.

Applying this same idea to the AC auction with options would require a method for (1) evaluating $w$ and (2) finding a linear approximation whenever the constraint is violated.

Consider first the question of evaluating a constraint. For each contingency constraint, a good guess as to the solution of the unconstrained optimal power flow in (1.17) would be to use the DC-Load approximation above to determine the value for $x$, the pattern of the exercise of the option. For each option, if $H_i^{\omega}\tau_k^o > 0$ set the corresponding $kth$ element $x$ to 1, otherwise set the element to zero. Let the result be the vector $\tilde{x}_i^{\omega}$ that achieves this value for the $ith$ constraint in contingency $\omega$. This is the same solution for $x$ that would be obtained in the DC-Load case.

Then compute $\nabla K^{\omega}_{i\,\tilde{x}_i^{\omega}}\left(\hat{t}^f + \sum_k \tilde{x}^{\omega}_{ik} t^o_k \tau^o_k - \hat{\bar{g}}^f, \hat{u}\right)$, the change in the constraint as we change the exercise of the options. If the solution satisfies the condition that the elements of this gradient vector have positive signs when and only when the corresponding elements of $\tilde{x}_i^{\omega}$ are at the upper bound, then we can show that $\tilde{x}_i^{\omega}$ satisfies the first-order conditions for achieving the maximum for the optimal value function. If so, then we would expect that this is the optimal solution for $w_i^{\omega}$, at least for a well-behaved network. If the first order condition is satisfied at a local optimum that is not a global optimum, then an ordinary local search algorithm may not be able to find a global solution.

In practice, we accept approximate solutions of the first-order conditions as optimal solutions. If the problem is well-behaved, then the simple solution based on the DC-Load model should define the worst-case exercise of options for each constraint without the necessity to conduct a further search. (Note that this is not the same thing as saying that the DC-Load estimate of $K$ is acceptable. We use the DC-Load guess for the solution $x$, but use a full AC load flow to evaluate the constraint).

If the first order condition is not satisfied, then this should be a good starting point for a search to find an acceptable solution to maximize $K_i^{\omega}(y, u)$. This case would require iterative solution of an optimal power flow problem for the applicable contingency. This is easier than finding the full security-constrained solution for the auction model.

In any event, let the end result of evaluating the optimal value function $w_i^{\omega}$ be $\hat{x}_i^{\omega}$, with corresponding solution $(\hat{y}, \hat{u})$ where $w_i^{\omega}\left(\sum_k \hat{t}_k^f \tau^f_k, \{\hat{t}^o_k\}, \sum_k \hat{\rho}^f_k \bar{g}_k^f, \hat{u}\right) = K_i^{\omega}$ $(\hat{y}, \hat{u})$. This gives us an evaluation of the constraint. If the value is greater than zero, the constraint is violated.

Recognize that there will be different value of $\hat{x}_i^{\omega}$, the implied exercise of the options, for each constraint $i$ and contingency $\omega$. This is not an obstacle in principle because in using the optimal-value function we are interested only in the value of the violated constraint and its linear approximation relative to the option awards, not to the exercised awards. Hence we need only use the exercised awards temporarily, at each constraint, to evaluate the function and calculate the linear approximation.[40]

In the case of a violated constraint, the optimal-value function is not in general differentiable or even convex. However, it does have a generalized gradient $\partial^o w$ that serves a similar purpose (Clarke 1990).[41] In the present application the generalized gradient of the optimal value function $w_i^\omega$ has a simple form that limits the domain where it is nondifferentiable to those points where some of the elements of the options awards are zero. These are important points, since not all options will have positive awards. Hence, the lack of a regular gradient is relevant.

The following vector will always be an element of the generalized gradient:

$$\phi_i^\omega\left(\hat{\tau}^f,\left\{\hat{t}_k^o\right\},\hat{\bar{g}}^f,\hat{u}\right)^t \equiv \begin{bmatrix} \nabla_y K_i^\omega(\hat{y},\hat{u})^t \\ \left\{Max\left(0,\nabla_y K_i^\omega(\hat{y},\hat{u})\tau_k^o\right)\right\}^t \\ -\nabla_y K_i^\omega(\hat{y},\hat{u})^t \\ \nabla_u K_i^\omega(\hat{y},\hat{u})^t \end{bmatrix}^t \in \partial^o w_i^\omega\left(\hat{\tau}^f,\left\{\hat{t}_k^o\right\},\hat{\bar{g}}^f,\hat{u}\right). \tag{1.19}$$

To see this, note that the special nature of the problem in (1.17), where the swing bus net load is determined freely to meet the condition, could be restated as:

$$w_i^\omega\left(\tau^f,\left\{t_k^o\right\},\bar{g}^f,u\right) = \underset{\substack{\varepsilon_s, y \\ 0\le x_k\le 1}}{Max}\ \tilde{K}_i^\omega(\tilde{y},u)$$

$$s.t.$$

$$\begin{pmatrix} y_s \\ \tilde{y} \end{pmatrix} = \tau^f + \sum_k x_k t_k^o \tau_k^o - \bar{g}^f + \begin{pmatrix} 1 \\ 0 \end{pmatrix}\varepsilon_s.$$

[40] Note: in the early stages of the computation, we might accept both the DC-Load solution and the associated DC-Load shift factors as the estimates of the linearized constraint. However, when close to the solution, the assumption that the DC-Load model is inadequate means that we need an exact evaluation of both the function and the linearized representation of any violated constraint.

[41] Here we follow the applications Shimuzu et al. (1997), p. 28. A generalized gradient of a function *f(x)* at the point $\bar{x}$ is defined as $\partial^o f(\bar{x})$ in terms of the generalized directional derivative as the set of vectors

$$\partial^o f(\bar{x}) = \left\{\gamma \in R^n \middle| f^o(\bar{x}:s) \ge \gamma^t s, \forall s \in R^n\right\},$$

*where*

$$f^o(\bar{x}:s) = \limsup_{\substack{x\to\bar{x} \\ \tau\downarrow 0}} \frac{f(x+\tau s)-f(x)}{\tau}.$$

In other words, $y_s$ does not enter the objective function and the resulting gradients depend only on the objective function derivatives. At most points, $w$ is differentiable. But at points where it is not differentiable, the generalized gradient exists and equals the convex hull of the limit points of the gradients, including (1.19), (Shimuzu et al. 1997).

When the option award is zero, any element in the interval $[Max\left(0, \nabla_y K_i^\omega(\hat{y}, \hat{u})\tau_k^o\right), +\infty)$ would also give rise to a generalized gradient. Thus the vector $\phi_i^\omega\left(\hat{\tau}^f, \{\hat{t}_k^o\}, \hat{\bar{g}}^f, \hat{u}\right)$ is an extreme point of the generalized gradient. It should give an adequate linear representation of the constraint function in the range of interest over the non-negative allocations.

For a violated constraint, therefore, the idea is to introduce the linearized constraint:

$$w_i^\omega\left(\hat{\tau}^f, \{\hat{t}_k^o\}, \hat{\bar{g}}^f, \hat{u}\right) + \varphi_i^\omega\left(\hat{\tau}^f, \{\hat{t}_k^o\}, \hat{\bar{g}}^f, \hat{u}\right)\begin{pmatrix} \sum_k t_k^f \tau_k^f - \hat{t}^f \\ \{t_k^o\} - \{\hat{t}_k^o\} \\ \sum_k \rho_k^f \bar{g}_k^f - \hat{\bar{g}}^f \\ u - \hat{u} \end{pmatrix} \leq 0.$$

This would then serve as a constraint in the sequential approximation of the nonlinear AC auction problem in the corresponding way that the constraint would enter in the case of obligations only. For the linear approximation, the usual first order KKT conditions would generalize to finding zero as an element of the generalized gradient.

As a technical point, this application would depend on a slightly stronger set of assumptions to guarantee that $w$ is Lipschitz near the solution. These conditions would apply for a slightly modified version of the problem where for a sufficiently large value of the penalty $M$ we redefine the value function as:

$$w_i^\omega\left(\tau^f, \{t_k^o\}, \bar{g}^f, u\right) = \underset{\substack{\varepsilon_s, y, 0 \leq x_k, 0 \leq m_k, \\ x_k - m_k \leq 1}}{Max} \quad K_i^\omega(y, u) - M \sum_k m_k$$

$s.t.$

$$y = \tau^f + \sum_k x_k t_k^o \tau_k^o - \bar{g}^f + \begin{pmatrix} 1 \\ 0 \end{pmatrix} \varepsilon_s,$$

$$L^\omega(y, u) + \iota^t y = 0.$$

This allows the function to be finite for all $\left(\tau^f, \{t_k^o\}, \bar{g}^f, u\right)$ and locally Lipschitz everywhere (Shimuzu et al. 1997). The generalized gradient at a non-differentiable point would be bounded by $M$, but the same lower extreme point should define the appropriate local linearization to use in the large optimization problem. The sequential linear approximations would use these function evaluations and selections from the generalized gradient to search for the optimal solution that satisfied the generalized Karush-Kuhn-Tucker conditions for the master problem.

*Relaxation Solution Procedure with PTP-FTR Options*

**Step 1:** Select an initial candidate solution $\left(\hat{\tau}^f, \{\hat{t}_k^o\}, \hat{\bar{g}}^f, \hat{u}\right)^0$, ignore most (or all) of the constraints in the economic dispatch using only the small subset $w_i^{\omega}\left(\hat{\tau}^f, \{\hat{\tau}_k^o\}, \hat{\bar{g}}^f, \hat{u}\right)^0$, and set the iteration count to $m = 0$.

**Step 2:** Construct the relaxed master problem as:

$$\underset{u \in U, t_k^f \geq 0, t_k^o \geq 0, \rho_k^f \geq 0}{Max} \quad \sum_k \beta_k\left(t_k^f, t_k^o, \rho_k^f\right)$$

$$s.t.$$

$$L_O^0\left(\sum_k t_k^f \tau_k^f, \{t_k^o\}, \sum_k \rho_k^f \bar{g}_k^f, u\right) - t^t \sum_k \rho_k^f \bar{g}_k^f = 0,$$

$$w_i^{\omega}\left(\hat{\tau}^f, \{\hat{t}_k^o\}, \hat{\bar{g}}^f, \hat{u}\right)^m + \phi_i^{\omega}\left(\hat{\tau}^f, \{\hat{t}_k^o\}, \hat{\bar{g}}^f, \hat{u}\right)^{mt} \begin{pmatrix} \sum_k t_k^f \tau_k^f - \hat{t}^{fm} \\ \{t_k^o\} - \{\hat{t}_k^o\}^m \\ \sum_k \rho_k^f \bar{g}_k^f - \hat{\bar{g}}^{fm} \\ u - \hat{u}^m \end{pmatrix} \leq 0.$$

Let a solution be $\left(\hat{\tau}^f, \{\hat{t}_k^o\}, \hat{\bar{g}}^f, \hat{u}\right)^{m+1}$ and update $m = m + 1$.

**Step 3:** Check to see if the candidate solution $\left(\hat{\tau}^f, \{\hat{t}_k^o\}, \hat{\bar{g}}^f, \hat{u}\right)^m$ violates any of the constraints. If so, create a new $w_i^{\omega}\left(\hat{\tau}^f, \{\hat{t}_k^o\}, \hat{\bar{g}}^f, \hat{u}\right)^m$ including some or all of these constraints and repeat Step 2. Else done.

Success with this proposed relaxation procedure for solving the auction problem with PTP-FTR options depends on the expectation that relatively few of the (very) many contingency constraints will be binding. This is a well-established condition in the dispatch model and the associated PTP-FTR obligation-only auction model that is of the same form as the dispatch. By contrast, if the introduction of options produces many more bids and many more binding constraints, then the scale of the problem may overwhelm current computational capabilities.

A concern with the potential number of binding constraints applies as well to the case of a DC-Load model for PTP-FTR obligations and options. However, the DC-Load formulation would have the computational advantage that evaluation of the constraints and the associated generalized gradient would be a relatively simple calculation that reduces to calculating the associated shift factors in $H(u^0)$ and evaluating the positive elements to construct the generalized gradient. In the DC-Load formulation ignoring losses, we would have:

$$\begin{aligned} w_{DC}{}_i^{\omega}\left(\tau^f, \{t_k^o\}, u\right) &= \underset{0 \le x_k \le 1}{Max} \; H_i^{\omega}(u)\left(\tau^f + \sum_k x_k t_k^o \tau_k^o\right) - b(u), \\ &= H_i^{\omega}(u)\tau^f + \sum_k t_k^o \max\left(0, H_i^{\omega}(u)\tau_k^o\right) - b(u). \end{aligned}$$

Combining all the constraints and contingencies, we have

$$w_{DC}\left(\tau^f, \{t_k^o\}, u\right) \le 0.$$

The corresponding auction model with bids for balanced forward-obligations by $(t_k^f, \tau_k^f)$ and balanced forward-option by $(t_k^o, \tau_k^o)$ would be

$$\begin{aligned} &\underset{u \in U, t_k^f \ge 0, t_k^o \ge 0}{Max} \quad \sum_k \beta_k\left(t_k^f, t_k^o\right) \\ &s.t. \\ &w_{DC}\left(\sum_k t_k^f \tau_k^f, \{t_k^o\}, u\right) \le 0. \end{aligned} \tag{1.20}$$

Even in the DC-Load case, therefore, this computation is not trivial. For obligations we need to evaluate only the load flow for each contingency given $\tau^f$, the aggregate of the obligations. Following the discussion of (1.10), this amounts to solving a system of linear equations for each contingency but evaluates all constraints in that contingency at once. But in order to evaluate the constraint in (1.20), we need to calculate the shift factors for every constraint in the contingency, each of which involves a similar system of linear equations. In other words, in the relaxation algorithm the need to calculate shift factors expands from the violated constraints only to every constraint when options are included.

Although this does require more computation, the evaluation of the constraints is separable and efficient means should be available to do the many evaluations, at least in the DC-Load case. Furthermore, not every constraint needs to be included in the relaxed master problem. As long as the number of binding constraints is small, meaning hundreds and not hundreds of thousands, this auction model might accommodate PTP-FTR options and obligations and be computationally feasible.

By construction of the constraints, exercise of the options along with the obligations would be simultaneously feasible under the condition that the system operator could select the set of controls needed to satisfy the constraints for the obligations and exercised options. Hence, the revenue collected in the final spot market dispatch would always be sufficient to pay the amounts required by the various PTP-FTR contracts.

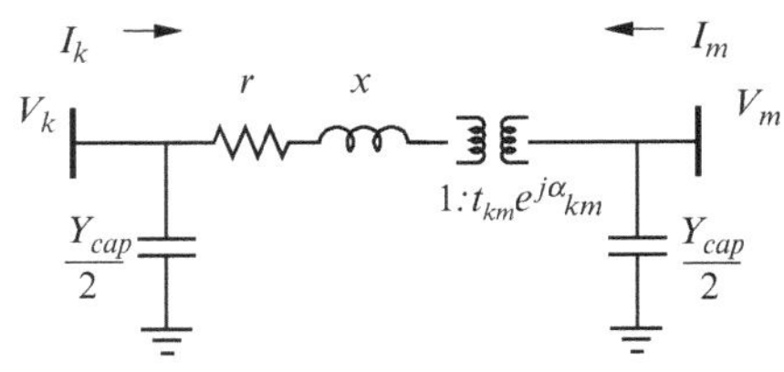


**Fig. 1.2** Transmission line and transformer

## 1.5 Conclusion

So-called physical transmission rights present so many complications for a restructured electricity market that some other approach is required to provide property rights for the grid. Under a standard market design built on a bid-based, security-constrained, economic dispatch with locational prices, the natural approach is to define financial transmission rights that offer payments based on prices in the actual dispatch. Different models have been proposed for point-to-point, including obligations and options. With consistent definitions, the rights can be shown to be simultaneously feasible and revenue adequate in various AC formulations or approximations. The conditions for simultaneous feasibility also define the form of auctions that would award or reconfigure the rights. In the case of point-to-point obligations, the practical feasibility of the approach has been demonstrated with adaptations of commercial dispatch software. In the case of point-to-point options, the computational strategies are more demanding but have been implemented in a limited way.

## Appendix: Generic Transmission Line Representation

The generic transmission line analysis employs complex variables. To avoid confusion here, the indexes for the two terminals of the line are $k$ and $m$. For a development of the model transmission line and transformer model, see Grainger and Stevenson (1994). By choice of parameters, this generic transmission line representation allows for a $\Pi$-equivalent representation of a line with no transformer, an ideal transformer, or a combination of both.

Here we follow Weber (1997)'s notation and conventions. This is useful in that Weber also provides an extensive detail on the characterization of the Jacobian of the power flow equations to provide further insight into the implications of the AC power flow model, including calculation of the derivatives with respect to the transformer parameters. As shown in Fig. 1.2 let $V_k$ represent the complex voltage with magnitude $|V_k|$ and angle $\theta_k$. The data include the line resistance ($r$), reactance ($x$). The transformer includes turns ratio ($t_{km}$) and angle change ($\alpha_{km}$). The line charging capacitance is the complex $Y_{cap}$.

The line admittance ($y$) is the inverse of the line impedance ($z$) formed from the resistance and reactance.

$$y = \frac{1}{z} = \frac{1}{r+jx} = \frac{1}{r+jx}\frac{(r+jx)^*}{(r+jx)^*} = \frac{1}{r+jx}\frac{r-jx}{r-jx} = \frac{r-jx}{r^2+x^2} = g+jb.$$

With $P$ as the real power and $Q$ as the reactive power, the general rules for complex power ($S$) have:

$$S = P + jQ = VI^* = zII^* = z|I|^2 = (P-jQ)^* = (V^*I)^*.$$

The line capacitance is represented here as:

$$\frac{Y_{cap}}{2} = 0 + jB_{cap}.$$

Following Weber, for the generic representation in Fig. 1.2, complex current ($I_k$) from $k$ towards $m$ satisfies:

$$I_k = V_k\left(y + \frac{Y_{cap}}{2}\right) - V_m\frac{e^{-j\alpha_{km}}}{t_{km}}y.$$

Therefore, the complex power flow from $k$ to $m$ is:

$$\begin{aligned}
S_k &= V_k I_k^* = V_k V_k^*\left(y+\frac{Y_{cap}}{2}\right)^* - V_k V_m^*\frac{e^{j\alpha_{km}}}{t_{km}}y^* \\
&= |V_k|^2\left(y+\frac{Y_{cap}}{2}\right)^* - \frac{|V_k||V_m|e^{j(\theta_k-\theta_m+\alpha_{km})}}{t_{km}}y^*, \\
&= |V_k|^2(g-j(b+B_{cap})) - \frac{|V_k||V_m|}{t_{km}}(\cos(\theta_k-\theta_m+\alpha_{km})+j\sin(\theta_k-\theta_m+\alpha_{km}))(g-jb), \\
&= |V_k|^2 g - \frac{|V_k||V_m|}{t_{km}}(g\cos(\theta_k-\theta_m+\alpha_{km})+b\sin(\theta_k-\theta_m+\alpha_{km})) \\
&\quad + j\left(-\frac{|V_k||V_m|}{t_{km}}\left(g\sin(\theta_k-\theta_m+\alpha_{km})-b\cos(\theta_k-\theta_m+\alpha_{km})\right)-|V_k|^2(b+B_{cap})\right).
\end{aligned}$$

The complex current ($I_m$) from $m$ towards $k$ is

$$I_m = -V_k\frac{e^{j\alpha_{km}}}{t_{km}}y + V_m\left(\frac{1}{t_{km}^2}y + \frac{Y_{cap}}{2}\right).$$

Hence,

$$
\begin{aligned}
S_m &= -V_m V_k^* \frac{e^{-j\alpha_{km}}}{t_{km}} y^* + V_m V_m^* \left(\frac{1}{t_{km}^2} y + \frac{Y_{cap}}{2}\right)^*, \\
&= -\frac{|V_m||V_k| e^{j(\theta_m - \theta_k - \alpha_{km})}}{t_{km}} (g - jb) + |V_m|^2 \left(\frac{g}{t_{km}^2} - j\left(\frac{b}{t_{km}^2} + B_{cap}\right)\right), \\
&= |V_m|^2 \frac{g}{t_{km}^2} - \frac{|V_m||V_k|}{t_{km}} (g\cos(\theta_m - \theta_k - \alpha_{km}) + b\sin(\theta_m - \theta_k - \alpha_{km})) \\
&\quad + j\left(-\frac{|V_m||V_k|}{t_{km}} (g\sin(\theta_m - \theta_k - \alpha_{km}) - b\cos(\theta_m - \theta_k - \alpha_{km})) - |V_m|^2 \left(\frac{b}{t_{km}^2} + B_{cap}\right)\right).
\end{aligned}
$$

If the system is normal and the angle change is fixed, then the angle change can be included in the line admittance. Similarly for normal systems, if the transformer tap setting is fixed, the turns ratio can be included in the per unit normalization of the voltages, which would produce appropriately modified values of $y$ but with the elimination of the separate transformer parameters $(t,\ \alpha)$.[42] Ignoring the line capacitance, this simplified representation would be

$$
\begin{aligned}
S_k = {} & |V_k|^2 \hat{g} - |V_k||V_m| \left(\hat{g}\cos(\theta_k - \theta_m) + \hat{b}\sin(\theta_k - \theta_m)\right) \\
& + j\left(-|V_k||V_m| \left(\hat{g}\sin(\theta_k - \theta_m) - \hat{b}\cos(\theta_k - \theta_m)\right) - |V_k|^2 \hat{b}\right).
\end{aligned}
$$

and

$$
\begin{aligned}
S_m = {} & |V_m|^2 \hat{g} - |V_m||V_k| \left(\hat{g}\cos(\theta_m - \theta_k) + \hat{b}\sin(\theta_m - \theta_k)\right) \\
& + j\left(-|V_m||V_k| \left(\hat{g}\sin(\theta_m - \theta_k) - \hat{b}\cos(\theta_m - \theta_k)\right) - |V_m|^2 \hat{b}\right).
\end{aligned}
$$

This is a familiar simplification often seen in the electrical engineering literature. However, if the system is not normal, tap ratios are variable, or phase angle adjustments are variable, it will be necessary to use the more general representation as shown above.

The notation translation to the discussion in the main text has:

$$
G_k = g,\ \Omega_k = -b,\ \delta_i = \theta_k,\ Z_{ij} = S_k,\ \alpha_k = \alpha_{km},\ t_k = t_{km}.
$$

[42] Normal is a term of art, not necessarily intended to mean "usual." A system is normal if for each parallel path the product of ideal transformer gain magnitudes is equal and the sum of ideal transformer phase shifts is the same. See Bergen and Vittal (2000), pp. 154–175.

## References


Bergen AR (1986) Power systems analysis. Prentice Hall, Englewood Cliffs, NJ

Bergen AR, Vittal V (2000) Power systems analysis, 2nd edn. Prentice Hall, Upper Saddle River, pp 154–175

Bertsekas DP (1995) Nonlinear programming. Athena Scientific, Belmont, MA, p 427

Boucher J, Smeers Y (2001) Alternative models of restructured electricity systems, part 1: no market power. Oper Res 9(6):821–838

Boucher J, Ghilain B, Smeers Y (1998) Security-constrained dispatch gives financially and economically significant nodal prices. Elec J 11:53–59

Cadwalader MD, Harvey SM, Hogan WW, Pope SL (1998) Reliability, scheduling markets, and electricity pricing. Center for Business and Government, Harvard University, May 1998

Caramanis MC, Bohn RE, Schweppe FC (1982) Investment decisions and long-term planning under electricity spot pricing. IEEE Trans Power Ap Syst PAS-101(12):3234–3245

Chandley JD, Hogan WW (2002) Independent transmission companies in a regional transmission organization. Center for Business and Government, Harvard University, Cambridge, MA, 8 Jan 2002

Clarke FH (1990) Optimization and nonsmooth analysis. SIAM Reprints, Philadelphia, p 10

Elgerd OI (1982) Electric energy systems and theory, 2nd edn. McGraw Hill, New York, p 23

Federal Energy Regulatory Commission (FERC) (1996) Capacity reservation open access transmission tariffs. Notice of Proposed Rulemaking, RM96-11-000, Washington DC, 24 April 1996

Federal Energy Regulatory Commission (FERC) (2002a) Working paper on standardized transmission service and wholesale electricity market design. Washington, DC, 15 March 2002

Federal Energy Regulatory Commission (FERC) (2002b) Working paper on standardized transmission service and wholesale electricity market design. Washington, DC, 15 March 2002, p 11

Feinstein J, Tscherne J, Koenig M (1988) Reactive load and reserve calculation in real-time computer control system. IEEE Comput Appl Power 1(3):22–26

Ge SY, Chung TS (1999) Optimal active power flow incorporating power flow control needs in flexible AC transmission systems. IEEE Trans Power Syst 14(2):738–744

Geoffrion AM (1970) Elements of large-scale mathematical programming, parts I and II. Manag Sci 16(11):652–691

Grainger JD, Stevenson WD (1994) Power systems analysis. McGraw-Hill, New York, pp 361–367

Harvey SM, Hogan WW (2002) Loss hedging financial transmission rights. Center for Business and Government, Harvard University, 15 Jan 2002

Harvey SM, Hogan WW, Pope SL (1997) Transmission capacity reservations and transmission congestion contracts, Center for Business and Government, Harvard University, 6 June 1996, (Revised 8 March 8 1997)

Hogan H (1992) Contract networks for electric power transmission. J Regul Econ 4:211–242

Hogan WW (2000) Flowgate rights and wrongs. Center for Business and Government, Harvard University, Aug 2000

Kuhn HW (2002) Being in the right place at the right time. Oper Res 50(1):132

Mas-Colell A, Whinston MD, Green JR (1995) Microeconomic theory. Oxford University Press, New York, pp 311–343

Oliveira EJ, Marangon JW, Lima JL, Pereira R (1999) Flexible AC transmission system devices: allocation and transmission pricing. Elec Power Energy Syst 21:111–118

O'Neill RP, Helman U, Hobbs B, Stewart WR, Rothkopf MH (2002) A joint energy and transmission rights auction: proposal and properties. Federal Energy Regulatory Commission, Working Paper, Feb 2002

Ponstein J (1965) An extension of the min-max theorem. SIAM Rev 7(2):181–188

Schweppe FC, Caramanis MC, Tabors RD, Bohn RE (1988) Spot pricing of electricity. Kluwer Academic, Norwell, MA

Shimuzu K, Ishizuka Y, Bard JF (1997) Nondifferentiable and two-level mathematical programming. Kluwer Academic, Boston, pp 188–228

Skilling HH (1951) Electric transmission lines. McGraw Hill, New York, pp 126–133

Weber JD (1997) Implementation of a Newton-based optimal power flow into a power system simulation environment. MS Thesis, University of Illinois at Urbana-Champaign, Urbana, Illinois

Wood AJ, Wollenberg BF (1984) Power generation, control, and operation. Wiley, New York, p 75

Wu F, Varaiya P, Spiller P, Oren S (1996) Folk theorems on transmission access. J Regul Econ 10(1):5–24

# Chapter 2
# Transmission Pricing

Ignacio J. Pérez-Arriaga, Luis Olmos, and Michel Rivier

## 2.1 Introduction

The transmission grid has a major impact on the operation and investment decisions in electric power systems. This impact is more noticeable when the electricity sector is organized around a wholesale market, where the transmission network becomes the meeting point of producers and consumers. The relevance of transmission is presently increasing with the growing penetration of intermittent renewable energy sources, frequently distant from the main load centres and significantly adding to the variability of flow patterns.

This chapter examines the economic impact of the transmission network on its users. This impact is twofold. On the one hand the network modifies the bulk prices of electrical energy, due to the presence of network losses and congestions. On the other hand, the costs of investment and operation of the transmission network have to be allocated to its users, according to some reasonable criterion. In principle both impacts should have a locational component. Injections or withdrawals of power in the grid affect losses and constraints differently depending on the node where they occur. Besides, the responsibility of network users in the reinforcements to the network generally depends on the location of these generators and loads. Thus, the allocation of the cost of the grid to its users should be guided by the location of the latter.

The chapter starts by discussing in Sect. 2.2 the effect of the transmission grid on system operation costs: how network constraints modify the economic dispatch of generation plants, and the costs of transmission losses. Section 2.3 presents the concept of nodal prices (locational marginal prices) and how to compute them.

I.J. Pérez-Arriaga (✉)
Comillas Pontifical University, Alberto aguilera 23, 28015 Madrid, Spain

Massachusetts Institute of Technology, 77 Massachusetts Avenue, Cambridge, MA 02139, USA
e-mail: ignacio@iit.upcomillas.es; ipa@mit.edu

L. Olmos • M. Rivier
Comillas Pontifical University, Alberto aguilera 23, 28015 Madrid, Spain

J. Rosellón and T. Kristiansen (eds.), *Financial Transmission Rights*,
Lecture Notes in Energy 7, DOI 10.1007/978-1-4471-4787-9_2,

The main properties of nodal prices are explained in Sect. 2.4. Section 2.5 describes how the impact of the transmission network on electricity energy prices is accounted for in practice in different power systems. Finally, Sect. 2.6 examines the allocation of the transmission network costs among the network users in the form of regulated network tariffs Sect. 2.7 concludes.

## 2.2 The Effect of Transmission on System Operation Costs

The unavoidable physical limitations of transmission networks when connecting producers to consumers have three main undesirable effects on the operation of the system. First, part of the energy transmitted over the lines and other grid facilities is transformed in heat and, therefore, never reaches the consumption centres. The difference between the amount of power injected at one end of a line and that withdrawn at the other end is called the loss of power in the line, or line **loss**. Second, the transmission grid imposes **constraints** due to a variety of technical reasons on any given set of power transactions that the network users want to make happen. Third, problems affecting the integrity and well functioning of the grid may result in the interruption of power supply to certain (or all) loads or the deterioration of the quality of electricity supplied. Thus, the **quality of the electricity supply service** may be affected by the grid.

Therefore the transmission grid may affect both the operation costs and the set of power injections and withdrawals that are allowed to take place. Conversely, the specific location of generators and loads in the grid is the driving factor behind the need for network expansion, which tries to improve the reliability of electricity supply and to reduce the operation costs derived from losses and network constraints. The locational differentiation of energy prices and network charges sends locational signals to prospective network users to be located so that these adverse effects are minimized.

Understanding the origin of network related operation costs, as well as the main drivers behind these costs and their impact on the system economic dispatch is of essence. Next, each of the three main effects of the grid on the system operation and its costs is discussed separately.

### *2.2.1 Network Losses*

Most of the energy losses in electric power grids are due to the resistance of conductors to the circulation of electric current flows. These are known as **ohmic losses**. Other losses are due to the **corona effect** whereby electrical discharges take place in the air surrounding high voltage line conductors. Losses also occur within network devices like **transformers, reactors, capacitors**. Due to existing losses, consumers receive less energy than generators produce.

Transmission network losses result in additional system costs. More energy has to be produced than is consumed, because part is lost while being transported. These costs correspond to additional production costs, i.e. they are not network costs per se, though they are a consequence of the need to transport power over the transmission grid. The cost of losses is affected by transmission expansion and operation decisions. It is therefore advisable to set efficiency incentives encouraging the System Operator and network users to reduce these costs.

Ohmic losses in a line are nearly proportional to the square of the power flow over the line (more precisely, they are proportional to the square of the current in the wires). This means that the increase in losses per unit increase in the system load (marginal increase in losses) is approximately twice as large as the average amount of losses per unit of load (total amount of losses/total system load). Consequently, the marginal cost of transmission losses (transmission losses cost increase/increase in system load) exceeds their average cost (total cost of losses/total system load).

The increase in transmission losses in the system due to a marginal increase in the load at a certain node depends on the location of this node in the grid, since the resulting changes in line flows depend on the latter as well. Therefore, transmission losses create geographic differences in the marginal cost of supplying electric energy. This implies that the marginal cost of meeting a marginal increase in demand can only be correctly assessed if the exact node where demand is increased is specified. Other factors contributing to these differences are described in the next subsection.

Due to transmission losses, some power plants may take precedence in the merit order of the economic dispatch over other plants whose production costs are lower. The merit order of power plants in the dispatch must be affected by the loss factor corresponding to each plant according to its location in the grid.[1]

### 2.2.2 Network Constraints

Networks restrict in many ways the power transactions that can take place in the system. Most typically, transactions cannot result in a current intensity (roughly proportional to the power flow, for a given voltage level) over any line that exceeds the maximum one that can be handled by this line. The underlying reason to limit the current intensity over a transmission line may be thermal – and therefore dependent upon the physical characteristics of the facility – or related to the conditions of system operation as a whole, like the provisions to guarantee an appropriate system dynamic response to disturbances or to avoid stability related problems that usually increase with the length of lines. Another typical grid constraint is the need to keep **voltages** within certain limits at all nodes, which may call for having some generation unit connected near the node experiencing problems. The maximum allowable **short-circuit power** may also limit grid configuration. Generally speaking, the chief effect of grid constraints is to condition system operation, leading to deviations from the most

[1] The loss factor at a certain node represents the increase in transmission losses in the system resulting from a unit increase in the power injected at this node. Loss factors depend on the existing system operation conditions.

efficient one from an economic point of view. Most common constraints in distribution grids are related to voltage limits and maximum line capacities.

Just as in the case of network losses, the mere existence of the transmission network adds to system costs by requiring the dispatch of more costly generation units to surmount the physical limitations imposed by the grid. This does not imply that network design or development is flawed, since network investments required to ensure the total absence of constraints in the system would probably not be economically justified. Some network constraints may therefore be justified from an economic point of view (provided that they do not systematically prevent the coverage of demand).

The cost of grid constraints, like that of losses, corresponds to additional generation costs that are associated with the characteristics of the network. Therefore, these costs are not part of the cost of the network itself. Operation and expansion decisions may affect the cost of grid constraints, which advises sending economic signals encouraging parties in the system to reduce this cost.

Both losses and grid constraints result in changes in the economic dispatch. The merit order of generation units depends not only on their production costs but also on their location in the grid and their impact on losses and grid constraints. The marginal cost of supplying load depends on the location in the grid of the former and therefore, may vary from one node to another. Additional costs associated with losses and constraints must be assigned to network users.

As explained below, nodal prices applied to the electric energy sold or purchased are economic signals that efficiently internalize all the short-term effects of the network on electricity supply costs. Due to their relevance, next Sects. 2.3 and 2.4 are devoted to discussing nodal prices and their properties.

### 2.2.3 *Quality of Service*

Transmission networks have also an impact on the **quality of the electricity supply** service. In countries where the electricity system is well developed, generation outages or lack of total generation capacity are hardly ever responsible for electricity supply interruptions. In a small percentage of cases, the origin of interruptions lies in joint generation and transmission security failures (although the consequences of such events are usually very severe, since they affect large areas in the system). Supply disruptions are in fact practically always due to local distribution grid failures. Distribution business regulation should strike a balance between the cost of developing the grid and the resulting enhancement of end consumer quality of service. The effect of the transmission grid on the quality of service is not so notorious and will not be discussed further in this section.

## 2.3 Nodal Prices: Definition and Computation

Losses and grid constraints result in differences in the local marginal value of energy among transmission nodes.[2] Locational energy prices affect the short and long term efficiency of the functioning of the system by driving market agent decisions on how much power to produce or consume at each time, as well as where to site the new generation or load they plan to install, which may in turn affect the development of the transmission network.

Short-term locational energy prices also vary over time. Separate prices are computed for each hour in day-ahead markets and in some power systems they are also computed as close as several minutes ahead of real time. Signals sent through these prices are needed to achieve maximum system efficiency. They aim to ensure that the generators with the lowest variable costs are the ones dispatched and demand can respond to the actual costs of supplying energy at each location. Besides, these signals also drive the expansion of the system, since expectations about future values of energy prices at the different locations affect market agents' long-term decisions on the siting of new generation and demand facilities.

### *2.3.1 Concept of Nodal Prices*

Nodal pricing represents the most sophisticated and efficient expression of locational energy prices. The marginal cost of electricity in a system corresponds to the extra cost incurred to serve a differential increase in the system load. It can be demonstrated that pricing the electricity produced or consumed in each node at the local marginal cost leads generators and loads in the system to make efficient operation decisions.

As a result of the existence of the grid, the marginal cost of electricity varies from one node to another. The nodal electricity price, also called locational marginal electricity price, in each node k is the short term cost of supplying most economically a marginal increase in demand in this node while complying with grid constraints. Nodal energy prices can be computed both for active and reactive power, as discussed in (Schweppe et al. 1988). However, nodal prices of reactive power have not been used in any real life system.[3]

When taking into account the actual features of electricity systems, which obviously must include the transmission network, any computed marginal system costs must be node specific. The uniform marginal system cost considered in several electricity markets results from disregarding the effect of the transmission network on the generation economic dispatch. Both short and long term marginal costs can be

---

[2] Nodal prices are also called locational marginal prices. In the pioneering work on this subject, see (Schweppe et al. 1988), the most general term "spot prices" is used.

[3] In some systems, like UK, energy and capacity payments associated to the production of reactive power have been paid to agents located in specific areas of the system where voltage problems may occur. However, no systematic nodal or zonal reactive power pricing scheme has been applied.

computed at system level and for each node. Long term marginal costs consider the option to marginally increase transmission or generation capacity to meet an increment of the load at a certain node.

### 2.3.2 Computation of Nodal Prices

Nodal prices can be readily obtained as by-products of the models widely available to compute the economic dispatch in the short-term taking into account the transmission grid. Models used may be as complex as needed. Using a very simple model, we aim to illustrate the process of computation of nodal energy prices within a centralized economic dispatch where network effects are considered.

In model (2.1) of the system economic dispatch, we make use of linear equations representing the flow of power over the grid according to Kirchhoff laws[4] (DC model). For the sake of simplicity, ohmic losses in each line have been represented as a function of the flow over this line and assigned to the extreme nodes of the line, thus being equivalent to an extra demand in each of the two nodes (half of the losses would be assigned to each node). For other representations of line losses, see, for instance, (Rivier et al. 1990). Besides, in order to make the formulation simpler, the only grid constraints considered are maximum line capacities.

$$
\begin{aligned}
&\max \textstyle\sum_{\mathrm{i}} \{B_{\mathrm{i}}(d_{\mathrm{i}}) - C_{\mathrm{i}}(g_{\mathrm{i}})\} \\
&s.t. \\
&d_i - g_i + \textstyle\sum_m \{\tau_{im} \cdot \phi_{im} - L_{i,m}(\phi_{im}, R_{im})\} = 0 && \forall i && \pi_i \\
&\tau_{im} \frac{\theta_i - \theta_m}{x_{im}} = \phi_{im} && \forall i, m && \xi_{im} \\
&\phi_{im} \le \overline{\phi_{im}} && \forall i, m && \mu_{im} \qquad (2.1) \\
&\theta_{ref} = 0 \\
&g_i \le \overline{g_i} && \forall i && \beta_i \\
&d_i \le \overline{d_i} && \forall i && \alpha_i
\end{aligned}
$$

[4] Kirchhoff laws are two. First one states that at each node, power injections must equal power withdrawals. Second one states that, when flowing among two nodes, power is split among the different parallel paths between these nodes in inverse proportion to the electrical distances along these paths.

In the formulation in (2.1), i is an index representing the set of nodes; m is an alias of i; $B_i(d_i)$ is either the benefit obtained by agents at node i from the power $d_i$ they consume or the offer by agents at node for the power $d_i$ they consume. Then, $B_i(d_i)$ is equal to the cost of electricity for consumers plus the consumer surplus; $C_i(g_i)$ is the either the cost incurred by agents at node i when producing $g_i$ units of power or the bid by agents at node i to produce $g_i$ units of power; $\tau_{im}$ is a binary variable whose value is 1 when nodes i and m are connected through a line and is 0 otherwise; $\phi_{im}$ is the flow over the line between nodes i and m in the direction from i to m; $L_{i,m}(\phi_{im}, R_{i,m})$ is the fraction (half) of transmission losses in the line between nodes *i* and *m* that has been assigned to node i and has therefore been represented as an extra load in this node: losses over a line depend on the flow and resistance of the line; $\theta_i$ is the phase angle at node i; $x_{im}$ is the reactance of the line between nodes i and m; $\overline{\phi_{im}}$ is the maximum flow allowed over the line from i to m in the relevant direction; $\theta_{ref}$ is the reference phase angle; $\overline{g_i}$ is the maximum power production in node i and $\overline{d_i}$ is the maximum demand for power in the same node. Apart from that, $\pi_i$, $\xi_{im}$, $\mu_{im}$, $\beta_i$ and $\alpha_i$ are the dual variables of the corresponding constraints, which are obtained, together with the primal variables, when solving the optimization problem.

The nodal price at node k, $\rho_k$, is, in this simple case, the dual variable of the k$^{th}$ nodal balance equation, $\pi_k$.[5]

$$\rho_k = \pi_k \tag{2.2}$$

The system economic dispatch can be modeled using an alternative formulation, see (2.3). In model (2.3), ohmic losses in the transmission grid are easily represented.

$$\max \sum_i \{B_i(d_i) - C_i(g_i)\}$$

$$s.t.$$

$$\begin{aligned}
&\sum_i (d_i - g_i) + L(d,\ g) = 0 && && \gamma \\
&\sum_i \{PTDF_{i,l} \cdot (g_i - d_i)\} = \phi_l && \forall l && \xi_l \\
&\phi_l \le \overline{\phi_l} && \forall l && \mu_l \\
&g_i \le \overline{g_i} && \forall i && \beta_i \\
&d_i \le \overline{d_i} && \forall i && \alpha_i
\end{aligned} \tag{2.3}$$

[5] Strictly speaking, the nodal price expression will be $\rho_k = \pi_k + \alpha_k$, although $\alpha_k$ will be non-zero only at those nodes where all the demand is fully unserved.

Most symbols in model (2.3) have already been used in this section. New ones are described next. $L(d,\ g)$ represents transmission losses in the system expressed as a function of power injections and withdrawals; $PTDF_{i,l}$, is the Power Transfer Distribution Factor of the flow over line l with respect to the power injection at node i, i.e. it is the sensitivity of the flow over this line with respect to the power injected at this node; $\phi_l$ is the flow over line l and $\overline{\phi_l}$ represents the maximum amount of power allowed over line l in the direction in which the flow actually goes in the scenario considered. Finally, when used as an index, l refers to the set of lines in the system.

Nodal energy prices can be computed from the solution of the economic dispatch in (2.3) as a linear combination of several of the dual variables of constraints in this problem.

$$\rho_k = \gamma + \eta_k = \gamma + \gamma \cdot \frac{\delta L}{\delta d_k} - \sum\nolimits_l (\mu_l \cdot PTDF_{k,l}) \tag{2.4}$$

where $\frac{\delta L}{\delta d_k}$ is the loss factor corresponding to node k; and $\eta_k$ is a variable representing the difference between the energy system price and the nodal price at each node. Then, generally speaking, this variable should be different from zero for all nodes but that taken as a reference for computing the system price. It reflects the impact of the grid on the value of energy at each node and depends on the reference node chosen.

The formulation of the optimization problem in (2.3) should depend on the identity of the node chosen as a reference nodes. The derivative of the systems losses with respect to demand at node $k$ and PTDFs depend on the choice of reference node. The amount of power being dispatched at each node; the overall value of accepted bids; and nodal prices are not affected by the choice of the reference node but dual variable $\gamma$ is. Therefore, (2.4) should instead read as in (2.5).

$$\begin{aligned}\rho_k = \gamma_s + \eta_{k,s} = \gamma_s + \gamma_s \cdot LF_{k,s} - \sum\nolimits_l (\mu_l \cdot PTDF_{k,l,s}) = \rho_s \cdot (1 + LF_{k,s}) \\ - \sum\nolimits_l (\mu_l \cdot PTDF_{k,l,s})\end{aligned} \tag{2.5}$$

## 2.4 Main Properties of Nodal prices

Nodal electricity prices consider the impact of the transmission network on the short term marginal value of energy both from a technical and an economic point of view. The level of these prices depends, at any time and node of the network, on system operation conditions including the following: set of available generation and transmission facilities and their technical features (capacities, line impedances); load level at each node; and variable costs of generators. Amongst nodal electricity prices main properties, there are their ability to send efficient short term signals; the efficient allocation among parties of the cost of losses and network constraints; their ability to recover only part of the cost of the grid; and the option to decompose them into a system (or energy) a loss and a congestion part. The remainder of this subsection discusses these properties in detail.

### *2.4.1 Property 1: Efficient Short-Term Energy Prices*

It can be easily demonstrated that nodal prices are optimal short term economic signals that internalise all the grid effects in a single value – the price to buy or sell energy in €/$/£ per kWh – separately computed for each node. In other words, when the energy produced or consumed at a certain node i is priced at the corresponding nodal price $\pi_i$, market agents located at this node are encouraged to behave most efficiently in order to maximize the social benefit of the system. The proof of this statement can be found in (Schweppe et al. 1988).

Consumers decisions will only be optimal if they exhibit some elasticity to the price of energy (the larger the amount of power purchased, the smaller value they place on, or the price they are willing to pay for, an extra unit of power). However, most consumers do not decide how much power to purchase at any given moment in time based on the price they will have to pay for it. Hence, the amount of power retrieved by consumers at the majority of nodes can be considered to be an input to the dispatch problem where prices are determined. Unless the dispatcher has access to the true utility function of consumers ($B_i$ for consumer at node i), nodal prices will not maximize social welfare in the short term.

Besides, achieving an optimal operation of the system requires bids from generators corresponding to their true production cost function (we have worked under the hypothesis that the cost function $C_i(g_i)$ used in the dispatch and the agent problems is the same). However, generators may bid strategically deviating from their true cost function since, in reality, some degree of market power always exists.

### *2.4.2 Property 2: Efficient Allocation of Network Losses Associated Costs and Redispatch Costs due to Network Constraints*

Nodal prices result in those network users located in areas where the power they produce or consume cause significant losses or network congestion facing less favorable prices (higher for consumers, lower for producers) than those network users that, due to their location in the grid, contribute to reducing network losses or alleviate congestion in the grid. Therefore, besides producing optimal short-term signals, nodal prices are locational signals encouraging agents to install new load or generation in places where the resulting ohmic losses and network congestion are as small as possible.

Note however that, while nodal prices send economic signals in the direction of reducing losses and congestion costs, they are not assigning to agents the social cost of losses and network congestion. This is remarkably clear for losses, since, due to the fact that losses increase with the square of power flows, nodal price differences due to losses result in larger net revenues for the system than the cost of system losses.

### 2.4.3 Property 3: Contribution to the Recovery of Network Investment and Maintenance Costs

The application of nodal prices to the power injected and withdrawn in each node gives rise to a net revenue $NR_t$ at each time *t*, whose expression is provided in (2.6). The overall net revenue for the whole system over a certain period of time, normally a year, is widely known as the Variable Transmission Revenues (*VTR*) of the system, whose mathematical expression is provided in (2.7):

$$NR_t = \sum_n (\pi_{n,t} \cdot d_{n,t} - \pi_{n,t} \cdot g_{n,t}) \tag{2.6}$$

$$VTR = \sum_t NR_t \tag{2.7}$$

where *n* represents the set of nodes in the system and t the time.

As shown in (Olmos 2006), *VTR* can also be computed line by line according to (2.8). Each line *l* between nodes *in* and *out*, where power flows from node *in* to node *out*, can be considered an arbitrageur buying energy $P_{l,in,t}$ injected in the line at node *in* and time *t* and selling energy $P_{l,out,t}$ retrieved from the line at node *out* and the same time. Given that the amounts of power injected into and withdrawn from line *l* differ by the amount of ohmic losses in this line, and nodal prices at time *t* at both line nodes $\pi_{\text{in,t}}$ and $\pi_{\text{out,t}}$ also differ, the commercial exploitation of line *l* will result in a net revenue at time *t* represented in (2.8) by $NR_{l,t}$.[6]

$$VTR = \sum_{l,t} NR_{l,t} = \sum_t \sum_l (\pi_{out,t} \cdot P_{l,out,t} - \pi_{in,t} \cdot P_{l,in,t}) \tag{2.8}$$

Variable Transmission Revenues computed according to (2.7) and (2.8) are the same. These revenues are associated with differences among nodal prices and powers injected into and withdrawn from the grid due to transmission losses and congestion. Network revenues associated with congestion are also known as congestion rents.

---

[6] Exceptionally, "**network revenues**" may be negative when line losses are very large due to corona discharge. Note that network revenue is the profit that the transmission network would earn if energy were purchased from generators at their **nodal price** and sold to consumers at theirs. However, the transmission network should not be allowed to conduct free market transactions, but must rather be treated like a regulated monopoly with pre-established remuneration. Exceptions, namely merchant lines, may be justified for individual lines under special circumstances.

*VTR* critically depend on the level of development of the grid. Overdeveloped grids will result in small losses and congestion, thus leading to small differences among nodal prices. These, in turn, will result in small *VTR*. On the other hand, underdeveloped grids will result in large differences among nodal prices probably leading to large *VTR* (although losses will probably be large as well).

Pérez-Arriaga et al. (1995) demonstrate that, under ideal conditions affecting the planning of the grid, *VTR* in an optimally developed network would amount to exactly 100 % of the network investment costs. Ideal conditions affecting the development of the network to be met for network variable revenues to amount to 100 % of the network costs are investigated in Rubio and Pérez-Arriaga (2000), and mainly include the following:

- Static and dynamic network expansion plans are the same and planning errors do not occur.
- Investments in transmission are continuous.
- Economies of scale do not exist in the transmission activity.
- Reliability constraints considered in system development planning are also considered in system operation.

In real life systems, *VTR* fall short of total transmission costs. The former only manage to recover about 20 % of the costs of the grid, according to estimates in Pérez-Arriaga et al. 1995. Main reasons for revenues from the application of nodal prices being so low are briefly discussed next.

First, economies of scale and the discrete nature of network investments result in an overdevelopment of networks in practice (see Dismukes et al. 1998). In effect, building lines with a large capacity is generally preferable over building a larger number of small lines even when the former are not going to be fully used during the first years of their economic life. As we have just explained, overinvesting in the development of the grid results in small nodal price differences and a small *VTR*.

Second, certain reliability constraints and a wide range of scenarios shall be considered when planning the expansion of the grid, This is due to the fact that there is a high level of uncertainty about the operation conditions that may occur in the system throughout the economic life of investments being decided. However, some of these restrictions and all these scenarios but one will not be considered when computing the operation of the system. Due to the fact that the set of constraints considered when computing the expansion of the system tends to be larger than that considered for operation planning, long term nodal prices computed assuming grid investments are continuous would also differ substantially from short term nodal prices. Specifically, differences among long term nodal prices, and therefore also revenues from their application (which should amount to the exact cost of the grid assuming continuous investments), would be much larger than those computed for short term prices.

Therefore, even if (short term) nodal prices are applied, revenues from their application will not suffice to recover the cost of the grid. Additional transmission charges will need to be levied on network users to complete the recovery of this cost. This is discussed in Sect. 2.6.

Given that Financial Transmission Rights entitle owners to receive the difference between the energy prices at the nodes that these rights refer to, the aggregate value for market agents of all the simultaneously feasible transmission rights (defined as obligations) that can be issued in the system would equal the expected overall net revenues from the application of energy prices. Due to the fact that, as just mentioned, these revenues tend to be much smaller than the total cost of an optimally developed transmission grid, it is highly unlikely that the financing of investments in the transmission grid through the issuance of FTRs would result in an appropriate development of the grid. Most of the required reinforcements could not be financed through this scheme.

Authorities must bear in mind that revenues of transmission companies or the System Operator should generally not depend on revenues resulting from the application of nodal prices. Otherwise, they will have a perverse incentive not to invest in the further development and maintenance of the grid so as to increase nodal price differences and therefore their revenues. Revenues of transmission service providers should generally be regulated (not dependent on nodal price revenues), though *VTR* should probably be devoted to finance part of the payments to these companies.

### 2.4.4 Property 4: Decomposition of Nodal Prices in Their Energy, Losses and Congestion Components

As already pointed out when discussing the computation of nodal prices in Sect. 2.3.2, the nodal energy price in each node can be decomposed into three components: one associated with the marginal cost of producing electricity in the system; another one associated with the effect that increasing the demand in this node has on ohmic losses and the marginal cost of electricity; and a third one related to the effect of marginally increasing the demand in the node on transmission constraints and the cost of these constraints. The decomposition of nodal electricity prices is investigated by Rivier and Pérez-Arriaga (1993), where the mathematical expression of the nodal price in node *k* provided in (2.9) is derived.

$$\rho_k = \gamma_s + \eta_{k,s} = \gamma_s + \gamma_s \cdot LF_{k,s} + \sum_j \mu_j \cdot NC_{j,ks} \tag{2.9}$$

where, $\gamma_s$ can be deemed the cost of producing electricity in the system, which is common to all nodes whose prices are to be computed; and $\eta_{k,s}$ is the part that can be deemed specific to each node k, which comprises the cost of losses caused by an increase in the node demand, $\gamma_s \cdot LF_{k,s}$, and the cost of restrictions affected by this demand increase, $\sum_j \mu_j \cdot NC_{j,ks}$. $LF_{k,s}$ is the loss factor of node k; $\mu_j$ is the cost of each restriction j; and $NC_{j,ks}$ is the impact of an increase in demand in node k on the system variable constrained in restriction j. $LF_{k,s}$ and $NC_{j,ks}$ are therefore sensitivity factors measuring changes of losses and any constrained parameter of the system, respectively, for an increase in demand at node k.

However, defining a one-to-one relationship between each nodal price $\rho$ and its energy, losses and constraint components is not possible. As highlighted in (2.9), components of the price at a node k must be defined taking as a reference the nodal price at another node s, which we shall call reference node from now on. Thus, the energy component of price $\rho_k$, $\gamma_s$, corresponds to the nodal price at node s; the losses component is defined in terms of $LF_{k,s}$, which is the loss factor at node k taking as a reference node s, meaning the increase in the ohmic losses in the system resulting from an increase in power injected in node s to supply a marginal increase in electricity demand in node k; finally, the constraint component of price $\rho_k$ is defined in terms of the impact $NC_{j,ks}$ on the system variable constrained by grid constrain j of an increase in power injected in node s to supply a marginal increase in electricity demand in node k. Changing node s taken as a reference for the computation of nodal price $\rho_k$ would result in a change of the value of its energy, losses and constraint components, while the nodal price itself would not change.

The reference node s may be chosen to be the one(-s) where the marginal generator(-s) in the system economic dispatch is(are) connected. Then, the energy component of the nodal price at node k would refer to the cost of producing electricity with the most efficient generation unit(-s) available, while the losses and constraint components would correspond to the cost for the system of transporting electricity produced by the marginal generator(-s) in the dispatch to node k. This, in any case, must be deemed an arbitrary decomposition of nodal price $\rho_k$, since the system marginal generator may change depending on the set of active constraints and existing losses, and therefore the production cost of this generator cannot be deemed independent of constraints and losses in the system. Therefore, decomposing nodal prices into its energy, losses and constraint components may have practical applications but one should be aware of the limitations of such a composition.

An interesting corollary of the decomposition of nodal prices just discussed is the existing relationship between the prices in any two nodes k1 and k2 in the system, which is provided in (2.10).

$$\rho_{k1} = \rho_{k2} \cdot (1 + LF_{k1,k2}) + \sum_j \mu_j \cdot NC_{j,k1,k2} \tag{2.10}$$

Equation (2.10) results from deriving the expression of nodal price $\rho_{k1}$ according to (2.9) when taking node k2 as the reference one. Rivier and Pérez-Arriaga (1993), discuss other less-relevant properties of nodal prices. Other algorithms have been proposed more recently to overcome the dependence of the decomposition of prices on the reference bus chosen (see Cheng and Overbye 2006). This and other research works try to get around this challenge by imposing constraints on the decomposition problem that determine the identity of the reference bus.

### 2.4.5 Dependence of the Sensitivity of Line Flows with Respect to Nodal Power Injections on the Choice of Reference Node

Factor $PTDF_{k,l}$ refers to the sensitivity with respect to the power injection in node $k$ of a specific type of constrained variable: the flow over line $l$.[7] Sensitivity factors of line flows are commonly used in regulatory approaches normally related to the allocation of the costs of transmission lines. Factor $PTDF_{k,l,s}$ is commonly claimed to represent the marginal use of line $l$ by agents located in node $k$.

As already mentioned, the value of PTDFs depends on the reference node considered when computing them. Then, the sensitivity of the flow in line $l$ with respect to the power injection in node $k$ must be denoted $PTDF_{k,l,s}$, thus refering to the specific node $s$ where an increase in the power withdrawn balances the aforementioned increase in the power injected in node $k$ (neglecting losses, the extra power withdrawn in node s must be the same as that injected in node k[8]).

Given the role that PTDFs have in the allocation of the cost of transmission lines according to some of the methods proposed for this (namely the so-called Marginal Participations method), discussing the effect of the selection of the reference node on the value of these factors is relevant. If losses are neglected and line flows are assumed to be a linear function of power injections and withdrawals, applying the superposition principle it can be easily proved that the PTDFs of line $l$ with respect to the power injected at node $k$ computed using reference nodes $s1$ and $s2$ are related by the expression in (2.11).

$$PTDF_{k,l,s2} = PTDF_{k,l,s1} + PTDF_{s1,l,s2} \tag{2.11}$$

Note that $PTDF_{s1,l,s2}$ does not depend on the reference node chosen. This involves that changing the slack node results in a uniform increase (either positive or negative) of the sensitivities of the flow in each line with respect to the power injected in all nodes of the system. Therefore, absolute differences among the sensitivities of the flow in a line with respect to power injections in different nodes of the system do not depend on the reference node used to compute these sensitivities.

---

[7] Power Transfer Distribution Factors are normally defined as the sensitivities of flows with respect to power injections, while sensitivity factors of constrained variables in general, NC, are normally defined with respect to power withdrawals. Therefore, changing the sign of factors NC corresponding to line flows is necessary to compute PTDFs. Besides, it must be noted that PTDFs are defined by some authors as the sensitivity of line flows with respect to point to point transactions rather than power injections. Thus, for example, authors in Galiana et al. (2003) compute the sensitivity of line flows with respect to equivalent bilateral power exchanges (whereby each demand is assigned a fraction of each generation and each generator is assigned a fraction of each demand in a uniform manner) to allocate the cost of these lines to their users.

[8] If losses are considered, the amount of power withdrawn in the reference node should not be 1 MW (a unit increase) but an amount slightly larger or smaller depending on the effect on transmission losses in the system of the considered power transaction between node k and reference node s.

Then, if part of the cost of transmission lines is allocated to agents according to the sensitivities of the flows in the former with respect to power injections by the latter, differences among the transmission charges to be paid by different agents would not depend on the reference node chosen to compute line flow sensitivities. However, this does not mean that charges computed using any reference node make engineering and economic sense. As explained in Olmos and Pérez-Arriaga (2009), only those cost allocation methods whose underlying principles are sound can be deemed to produce sound transmission charges.

If transmission losses are taken into account, the choice of the reference node has a small, albeit nonzero, influence on differences among the sensitivities of a line flow with respect to power injections in different nodes, as shown in (2.12) and (2.13), which have been derived from the discussion on the subject in Rivier and Pérez-Arriaga (1993):

$$(1 + LF_{k,s2}) = (1 + LF_{k,s1}) \cdot (1 + LF_{s1,s2}) \tag{2.12}$$

$$PTDF_{k,l,s2} = PTDF_{k,l,s1} + PTDF_{s1,l,s2} \cdot (1 + LF_{k,s1}) \tag{2.13}$$

Differences among line flow sensitivities with respect to different injection nodes are dependent on the choice of the reference node because the change in the sensitivity factor for a certain injection node resulting from a change of the reference node is a function of the loss factor of this injection node. However, differences among loss factors computed for different injection nodes are likely to be very small. Hence, generally speaking, differences among line flow sensitivity factors can be deemed slack node independent.

## 2.5 Main Locational Energy Pricing Schemes: Alternatives to Nodal Pricing

The management and pricing of the effect that the transmission network has on the energy dispatch is one of the areas where the power system academic community has been more prolific recently (see Chao and Peck 1996; Stoft 1998; Ruff 1999; Chao and Peck 2000; Tabors and Caramanis 2000; Boucher and Smeers 2001; ETSO 2001; Henney 2002; Hogan 2002; O'Neill et al. 2002; ETSO 2004; ETSO/EuroPEX 2004 as a sample of relevant works on the subject). The choice of the transmission pricing scheme to be applied should condition the definition of Financial Transmission Rights, as we shall explain below for each of the main types of schemes. Any transmission pricing scheme to be implemented must comply with sound engineering and economic principles but it must also be politically acceptable. This section describes and critically analyses the most relevant options for the pricing of the effects of transmission on power system dispatch. We discuss only market based methods, i.e. those which aim to maximize the economic value of energy and transmission capacity bids accepted.

Pricing schemes can be classified according to different criteria. Probably the two most relevant ones are (1) the type of interface involved in these schemes between energy and transmission pricing and (2) the level of locational differentiation (granularity) in final energy prices that result from them. According to the first criterion, pricing schemes can be classified into implicit schemes, where energy prices computed include the effect of the transmission grid on the economic value of energy, and explicit ones, where the effect of the network on the value of energy at each location is priced separately from energy itself. According to the second criterion, one may distinguish among nodal pricing, where a separate energy price is computed for each transmission node; zonal pricing, whereby the system is divided into pricing areas and a separate price is computed for each of them; and single pricing, where a single energy price is applied at all nodes in the system.

We shall here review main pricing alternatives according to the location differentiation in final energy prices they create. Within each main option corresponding to a level of disaggregation of prices, a distinction may be made between implicit and explicit schemes if appropriate.

### 2.5.1 Nodal Differentiation of Energy Prices

By far, the most relevant scheme within this category is nodal energy pricing (also called Locational Marginal Pricing), which produces a separate price for the energy consumed and generated at each transmission grid node. Energy prices computed through nodal pricing implicitly include the effect of grid losses and transmission congestions, internalising both effects in a single value (€/$/£ per kWh) that varies at each system node. Therefore, nodal pricing is an implicit transmission pricing scheme that produces perfectly efficient signals for decisions concerning the (short-term) economic operation of generation and demand, since nodal prices correctly convey the economic impact of losses and constraints at the different producer and consumer locations.

When focusing on the effect of grid congestion on the dispatch, nodal pricing may be seen as the outcome of a joint competitive auction of energy and physical rights to use the transmission capacity. O'Neill et al. (2002), provide an example of implementation of a contingency constrained auction for both energy and transmission rights where the authors consider both options and obligations. Auctions proposed in O'Neill et al. (2002), are different from other designs of implicit auctions in the sense that authors propose using them both in the short and the long term.

The academic community has come up with several designs to run implicit auctions in a decentralized manner. Thus, Aguado et al. (2004), decomposes the original problem into several simpler ones. The optimal outcome at regional level is found through an iterative process. The concept, properties and way to compute nodal energy prices have already been extensively discussed in the preceding sections within this chapter.

Instead of integrating the effect of transmission on the energy dispatch, one may think of separately pricing the effects that network congestion or losses should have on the final price of energy. However, if we are not able to define areas of uniform energy prices, which result from the application of a zonal, instead of a nodal, pricing scheme, separating the allocation of energy and capacity is not possible (or feasible from a practical point of view). When zonal prices cannot be defined, any power transaction significantly affects the flow through the congested lines and has to participate in the transmission capacity allocation process. Then, the unconstrained energy dispatch taking place after the allocation of transmission capacity (where limits to power flows imposed by the network are not considered) has to replicate exactly the outcome of the capacity allocation process (either the capacity auction or the outcome of the bilateral trading process taking place among agents to buy and sell transmission capacity rights).

However, the effect of transmission losses on efficient energy prices can effectively be computed separately from the energy system price (the so called lambda in nodal pricing nomenclature) through the application of loss factors. Therefore, there is no need to forgo the short-term loss signals that contribute to the economically efficient system operation. The losses attributable to each player, either computed as a marginal or average value, can be applied in the form of corrective factors to determine the prices to be paid or earned by this player or, rather preferably, the net amount of energy produced or consumed by the former. This should lead players to internalise the losses they are responsible for in their offers.

When energy prices differ by node, Financial Transmission Rights can be used to hedge against possible financial losses from the volatility in the differences among prices at two or more nodes (ETSO 2006). FTRs hedging a certain power transaction may be issued by any party. However, leaving their issuance in the hands of the TSO responsible for transmission among the nodes in the transaction would ensure revenue adequacy (Hogan 1992). According to this criterion, the issuing party should in this case be the corresponding national or State TSO for local transactions and the regional TSO for cross-border transactions.

Examples of **nodal pricing** can be found in electricity markets in Chile, Argentina, New Zealand and several Regional Transmission Organizations (RTOs) in the USA, such as the PJM system (Pennsylvania, New Jersey, Maryland), the Electric Reliability Council of Texas (ERCOT) system, or the California system. Loss factors are used for instance in the Irish Single Electricity Market.

Revenues from the application of nodal prices correspond to the economic value produced by the transmission grid by transporting power from nodes where it has a lower value (price) to those where its value is higher. Then, these revenues should be devoted to pay the regulated revenues to be earned by grid owner(-s).

### 2.5.2 Zonal Differentiation of Prices

Zonal price differentiation schemes involve applying the same final energy price within each of a set of areas while allowing price differences to take place among these areas. Normally, under zonal price schemes, a single market price is applied to all agents in the system unless significant network congestion occurs restricting the energy flows among pre-defined areas. In the latter case, prices differ among areas but the same price is applied to all nodes within any of these areas. Therefore, zonal price differences are normally caused by grid congestion, though a system of zonal loss factors is applied in some systems.

Energy price differences among electrical zones can result from the application of both implicit and explicit schemes. Zonal type implicit pricing schemes are normally referred to as zonal pricing or market splitting. Explicit mechanisms normally take the form of a coordinated auction of the capacity of the corridors linking price zones.

Zonal pricing normally involves the computation of a single, centralized, energy dispatch in the whole national or regional system where network effects within each uniform price area are neglected. It is therefore a simplification of nodal pricing. Market splitting, which can be considered a particular case of zonal pricing, involves the consideration of only one offer curve and one demand curve for the whole system in a first step. If the resulting pattern of flows causes significant congestion on the corridors linking the predefined areas, separate offer and demand curves are considered for each price area and, according to these curves, power is transacted among areas so that existing congestion is solved. This implementation of market splitting agrees with that of many others in the academic literature and the industry (see ETSO 1999; Newbery et al. 2003). Market splitting is applied within the Nordel region and in Italy. Zonal pricing has been also used in California.

Alternatively, the network capacity of likely-to-be-congested corridors linking uniform price areas may be explicitly allocated prior to running an only-energy market within each area. Market agents must acquire the right to use the inter-area transmission capacity they need to carry out the commercial transactions they want to get involved in, i.e. physical transmission rights over this capacity. Agents may buy this capacity (the right to use it) in a centralized explicit auction where the right to use the transmission network is allocated to those agents who value it most. Alternatively, agents may negotiate bilaterally the acquisition of those rights previously issued by the corresponding TSO.

Chao and Peck were the first ones to propose the utilization of rights over the capacity of likely-to-be-congested flow-gates (corridors) (see Chao and Peck 1996), where authors demonstrate that, under ideal conditions, this system would converge towards efficient energy prices. Similarly, Oren and Ross 2002, propose in an auction for flow-gate rights prior to the energy dispatch. Authors propose a system whereby SOs responsible for the energy dispatch in the different control areas would coordinate to manage the flow on the congested lines that is the responsibility of transactions taking place within different areas. There are other works on the use of flow-gate rights in combination with unconstrained energy markets (see Tabors and Caramanis 2000, for an example).

Once transmission capacity rights have been assigned in one way or the other, the energy auction takes place. Only those transactions that have acquired capacity rights to access the congested transmission they use can participate in the energy market. Auctioning transmission capacity at regional level requires some centralized coordination (see ETSO 2001). If flow patterns due to the different transactions were not considered jointly they might result in unexpected violations of network constraints unless significant security margins were applied. But employing security margins would most likely result in an underutilization of the transmission grid.

In those systems where explicit auctions are used, local authorities are in charge of the dispatch of energy within their corresponding areas. Thus, areas or countries enjoy a high level of independence. For this reason, capacity auctions have been widely applied in real life power systems. Up till recently, this was the method used to manage congestion on the borders between Austria and the Czech Republic, Belgium and the Netherlands, Denmark and Germany or France and the United Kingdom, among others (see Consentec/Frontier 2004).

The implementation of both zonal pricing schemes and mechanisms for the explicit allocation of transmission capacity on congested corridors implies the definition of internally well-meshed areas which can be considered as super nodes for congestion management purposes. Nodal energy prices computed within any of these predefined areas should be very similar if losses are ignored and serious congestion is limited to the interconnections between areas. Then, these areas can be regarded as "Single Price Areas" (SPAs) as far as congestion management is concerned (Christie and Wangensteen 1998; Stoft 1998; Chao and Peck 2000).

Balanced transactions within a SPA should not significantly affect the flow over inter-area links. In other words, any bilateral transaction within a SPA should not create loop flows outside this area which may significantly contribute to congestions inter Single Price Areas. The definition of Single Price Areas, whenever applicable, is not a trivial matter without practical consequences, see (Boucher and Smeers 2001). In zonal pricing schemes it will affect the validity of the energy dispatch and energy prices computed. What is more, as explained when discussing nodal pricing schemes, if we were not able to define SPAs, separating the allocation of energy and capacity, and therefore applying explicit capacity pricing mechanisms, would not be possible.

Borders among Single Price Areas may probably not coincide with political ones. Thus, assuming SPAs that are the same as existing control areas or countries may result in an inefficient dispatch or, even worse, in one that is far from being feasible. Thus, the plans to implement an implicit auction among Power Exchanges in Europe should be reconsidered carefully (see ETSO/EuroPEX 2004).

Financial Transmission Rights to be defined in this case should refer to two or more of the pricing zones whose definition has just been discussed, as price differences to hedge within each of these zones would be zero.

Revenues from the application of pricing schemes with zonal differentiation should be devoted to the coverage of network allowed regulated revenues, as with nodal prices, since they are just a simplified version of the nodal pricing scheme.

### *2.5.3 Single Pricing*

In densely meshed transmission grids with no systematic or structural congestions, applying pricing mechanisms introducing nodal or zonal energy price differentiation is often regarded to be an unnecessary sophistication. Then, a single energy price is computed for the whole system. Once the outcome of the only energy market is known, one can check whether the pattern of commercial transactions violates any network constraint. Only when a constraint is violated does the System Operator need to re-dispatch some generation. Several implementations of re-dispatch are possible. According to some of them, the cost of the re-dispatch carried out to solve any violation of the network constraints should be minimum (see Rau 2000; Tao and Gross 2002). In these cases, market-based mechanisms must be used to modify the pattern of generation in the system. In other words, changes to the dispatch must be based on the bids sent by market agents indicating how much they ask for in order to change their market positions. Other re-dispatch algorithms aim to minimize the number and size of the adjustments to the original dispatch (see Galiana and Ilic 1998; Alomoush and Shahidehpour 2000). Fang and David 1999, describe other possible schemes.

Alomoush and Shahidehpour (2000) and Biskas and Bakirtzis (2002), are aimed at re-dispatching generation and load in the context of regional markets. These algorithms must achieve coordination among the different zones. Thus, Biskas and Bakirtzis (2002), decomposes the original problem using Lagrangian relaxation techniques. The coordination variables are the prices of the power exchanges between zones.

Counter-trading is a specific implementation of the method of re-dispatch. In counter-trading, the System Operator nominates pairs of generators that modify their outputs to create a power flow that goes in the opposite direction to the one causing network congestion in the unconstrained energy dispatch. Obviously, one could generalize and say that re-dispatch is nothing but counter-trade, since any increase in the output of a generator has to be matched by a corresponding and identical (except for losses) reduction in the output of another generator.

Typically, the extra cost of re-dispatch or counter-trade is socialized to all consumers thus leading to uniform energy prices in the whole system (single pricing). In this case, any economic signals resulting from the management of congestion, which could have been used to emulate nodal or zonal pricing, are lost. Conceptually speaking, assigning the cost of re-dispatch to those market agents that "create" the network constraint is possible. Economic signals would thus not be completely lost. This is a technically complex task, nevertheless Rivier and Pérez-Arriaga (1993), and others. Tao and Gross (2002), allocate the cost of re-dispatch taking into account the participations of agents (injections and withdrawals considered separately) in the flow over the congested lines. In order to do this, they express the flow over the congested lines as a function of power injections and withdrawals. Similarly, Baran et al. (2000), determines the participation of each transaction in the flow over congested lines. Afterwards, the total cost of re-dispatch is allocated among congested lines taking into account both the marginal cost of the restriction on the flow through each congested line and the incremental cost of the re-dispatch necessary to avoid violating this restriction.

Experience with counter-trade in California shows that those schemes based on re-dispatch may be subject to gaming by market agents who artificially create congestion in the grid in order to be paid afterwards to remove it. In any case, nodal pricing or implicit auctions seem to be superior to congestion management mechanisms based on re-dispatch. Singh et al. (1998), compare nodal pricing to a mechanism based on decentralized bilateral trade among market agents, followed by the minimum cost re-dispatch necessary to solve infeasibilities. They conclude that price signals resulting from nodal pricing are more efficient, unless the cost of re-dispatch is efficiently allocated to the agents responsible for congestion in the grid. However, as we have explained before, efficiently allocating the cost of re-dispatch is not an easy task nor is there an indisputable way to do it.

However, nodal and zonal pricing schemes may also result in extra incentives to exercise market power when, due to the reduction in the size of the relevant market under these schemes in the presence of congestion, market agents gain power to unilaterally affect the energy price in any of the pricing zones the system is divided into. Auctioning Financial transmission Rights may aggravate this problem when market agents enjoying market power in an importing area are allowed to buy transmission rights into this area, see Olmos and Neuhoff 2006.

Applying a single energy pricing scheme does not result in net revenues (congestion rents) to be devoted to partially covering the cost of regulated transmission grid lines. On the other hand, as already mentioned, if redispatching generation and or load is necessary, a net cost will be incurred. Many national power systems apply single energy pricing schemes internally (within their borders). These include almost all European countries and Colombia.

Obviously, implementing single pricing within a system would render FTRs useless at local level, since there would not be energy price differences to hedge. Market agents would only need to be hedged against potential differences among single energy prices applied in different local (national, State) systems. For the most part, this is the case within the Internal Electricity Market of the European Union.

## 2.6 Completing the Recovery of the Network Cost: Computation of the Complementary Transmission Charges

### *2.6.1 Fundamentals*

Electric power transmission is nearly regarded a natural monopoly. Therefore, transmission should be a regulated business. Both under traditional cost of service regulation and under incentive regulation, the allowed annual revenues of the regulated transmission company, which are set by the regulator, must be paid by transmission network users. We discuss here how network related economic signals should be designed to achieve the recovery of the allowed transmission revenues while promoting efficiency in the short-term (encouraging agents to make optimal operation decisions) and in the long-term (driving agents' decisions on the location of new generators and loads).

As already shown, energy prices applied may have different levels of spatial differentiation due to the existence of losses and constraints in the grid. Energy price differences among nodes give raise to location-related economic signals to network users and result in some partial recovery of the total allowed revenues of the regulated transmission company. As already explained, revenues from the application of nodal prices comprise those obtained well ahead of real time through the sale of Financial (or Physical) Transmission Rights over the capacity of likely to be congested corridors, or hedging differences in prices among different nodes, and those obtained in the day-ahead and real time markets through the application of these prices to power injections and withdrawals. However, as Rubio and Pérez-Arriaga (2000), show, the net revenue resulting from the application of nodal prices amounts only to a small fraction of the total regulated cost of the grid. Revenues resulting from the application of alternative energy pricing schemes are expected to be lower. The fraction of regulated transmission revenues recovered from the application of energy prices is normally referred to as Variable Transmission Revenues (VNR).

Therefore, completing the recovery of the cost of the grid requires applying additional charges, normally called complementary charges, that relate to the fraction of the grid cost not recovered through energy prices. Complementary charges should also send economic signals to agents encouraging them to reduce the cost of expansion of the grid. Therefore, these charges should encourage agents to install new generation or load in those locations where reinforcements needed for the grid to cope with the resulting incremental flows are least costly.

Additionally, complementary charges should be compatible with the application of efficient short-term economic signals. Complementary charges refer to all transmission business costs associated with network infrastructure including investment costs (asset depreciation as well as a return on net fixed assets), operation and maintenance costs, and other administrative and corporate costs. On the other hand, line losses and generation costs due to grid constraints, System Operator costs and those costs related to the provision of Ancillary Services should be levied on system users through other charges. Then, complementary charges are related to the allocation of long term costs not to be affected by short-term decisions by agents (the cost of lines already existing is not conditioned by how much power each generator or load is transacting at each time). As a consequence of this, complementary charges should interfere as little as possible with short-term economic signals, so as not to compromise the efficiency of system operation.

Transmission charges can be divided into Connection charges and Use of the System (UoS) charges. Connection charges are employed to allocate the cost of transmission facilities directly connecting a network user, or group of users, to the rest of the grid. UoS charges are related to the costs of the rest of transmission facilities. Economic principles advocate allocating the cost of each transmission line according to the responsibility of grid users on the construction of that line. Applying this principle is easy when it is about allocating the cost of connection facilities: those responsible for their construction are the users connecting through them to the rest of the system. On the other hand, determining the responsibility of generators and loads in the construction of the bulk of the transmission grid is much more difficult, especially when the grid is meshed. The remainder of this section is devoted to the discussion of the design of UoS charges. Both the allocation method employed to determine which fraction of the grid should be paid by each agent and the design of transmission charges are discussed next.

## 2.6.2 *Allocation of the Cost of the Main Grid to Its Users*

Determining those generators and loads that were responsible for the construction of some lines has proven to be a very difficult task. Then, it is most sensible to use some proxy of cost causality, such as the level of network utilization of each line by each agent, as the basic criterion for the allocation of the cost of this line. This involves assuming that the responsibility of each agent in the construction of a line is proportional to the amount of use of the line by the agent.

However, the cost of those expensive lines that only benefit a subset of network users, in non-well-meshed networks, should be allocated according to the responsibility of network users in the construction of the former. The fraction of the cost of each line that each network user is responsible for can be computed based on the a priori estimation of the benefits produced by this line for this user.

Unfortunately, computing the electrical utilization of lines by agents is not a simple task either, since there is no indisputable method to do it. Several methods to determine the use of the network by agents have been proposed and applied, with results that vary significantly from one another. It is important to keep in mind that the final objective is not computing the use of the network by each agent, but determining the responsibility of this agent in the construction of the line.

Transmission tariffs in most countries do not contain any locational signal. They disregard the need to allocate efficiently line costs (see for instance ETSO 2008; Lusztig et al. 2006). Regulators have settled for simple transmission charges that socialize the cost of the network to its users. However, in our view, as time passes and all kinds of new generation compete to enter into the system, sending clear locational signals – including transmission tariffs – will become more relevant.

### 2.6.2.1 Computing the Responsibility of Agents in Network Costs

Whenever computing the benefits that network users obtain from transmission lines is not possible, the responsibility of these users in network costs should be determined taking as a reference the best estimate possible of their use of the grid. Olmos and Pérez-Arriaga (2007) point out that methods to be used to compute the use of the grid by generators and loads shall be in agreement with the underlying technical and economic principles of the functioning of power systems. Even when there is no indisputable method to compute the utilization of lines by agents, some proposed in the literature, like the method of Average Participations (AP) described first in Bialek (1996) and Kirschen et al. (1997), or the Aumann-Shapley method, whose application for the computation of transmission tariffs is analyzed in Junqueira et al. (2007), seem to be sensible options.

Most usage based network cost allocation methods providing sensible results (like AP or Aumann-Shapley) aim at determining the "average" use of the grid by each generator or load as if the latter had always been in place. However, the responsibility of agents in network reinforcements is directly related to the incremental flows produced by the decisions of these agents to install new generators or loads in specific places. Hence, usage based cost allocation factors produced by methods like AP or Aummann-Shapley should be modified to take account of the different possible patterns of change of the flows in the system caused the installation of each generator or load and the time when this generators or load and the lines in the system were built. The application of these principles to the process of computation of transmission charges is discussed in detail in Olmos and Pérez-Arriaga (2009).

Olmos and Pérez-Arriaga (2009) point out, the loading rate of each transmission line and the desired split of total transmission costs between generation and load in the system should also condition the level of transmission tariffs (complementary charges) paid by each network user. The fraction of the total cost of a line to be allocated to agents according to their responsibility in the construction of the line should probably be limited to the ratio of the loading rate of the line to that of other similar lines in the system. The remainder of the cost of this line should probably be socialized, since current users of the grid cannot be deemed responsible for the construction of the fraction of the capacity of a line that is expected not to be used until long time in the future (for lines that are underutilized in the present).

As already mentioned, the split of total transmission charges between generation and load should probably take place according to the total benefits that generation on the one hand, and load, on the other, will obtain from the grid. However, given that estimating these benefits may turn out to be very difficult in most cases, a 50/50 split of costs between the two groups may be adopted unless system authorities have sound arguments to set a lower limit to the overall fraction of costs to be paid by generators (operation decisions by generators may be more sensitive to the level of transmission charges than those by loads).

### *2.6.3 Designing UoS Charges*

Designing transmission charges involves not only developing the methodology for computing the responsibility of agents in the cost of the transmission grid, but also providing adequate answers to many implementation issues. We now focus on the most relevant aspects of the implementation of locational transmission grid charges that are not directly related to the cost allocation algorithm applied. These include computing the number of operation scenarios to be considered; defining the structure of charges and their updating procedure; and deciding the way to deal with grandfathering issues arising in the process of implementation of these charges.

As Olmos and Pérez-Arriaga (2009) point out, tariffs should be published based on the expected future operation of the system over a set of scenarios that are representative of the different set of situations that may exist in the future once the considered generator or load has entered into operation. The relative weight given to each scenario in the computation of the allocation of the cost of a line should be in accordance with the reasons justifying the construction of this line. The total cost of the line should be apportioned into two parts: one representative of the weight that the reduction of transmission losses had on the decision to build the line and another one representative of the weight of the decrease in congestion costs. Then, the relative weight given to each scenario in the process of allocation of the cost of the fraction of the line deemed to be built to reduce losses should be proportional to the system losses in this scenario. The relative weight given to each scenario in the process of allocation of the cost of the fraction of the line attributable to the reduction of congestion costs should be proportional to the level of congestion costs in this scenario, which, as a proxy, can be deemed proportional to the load level.

As aforementioned, operation decisions by network users, which are short-term decisions, should not be conditioned by the level of the transmission charge paid by these agents to recover the total network costs, which should be a long term signal. Short-term locational signals can be sent via nodal energy prices (locational marginal prices, LMP in the US terminology). If transmission tariffs are applied in the form of energy charges (€/MWh), i.e. a charge that depends on the amount of energy produced or consumed by the corresponding agent, network users will internalize these charges in their energy bids to the Power Exchange or in their bilateral contracts, therefore causing a distortion in the original market behaviour of these agents and the outcome of the wholesale market. It is then concluded that the transmission charge should have the format of a capacity charge (€/MW. year) or of just an annual charge (€/year). The first option runs into the problem of applying the same charge to all generation units with the same maximum capacity, which may have quite differing operation profiles. (the same occurs with demands that have widely different utilization factors and the same contracted capacity). The transmission charge should therefore be an annual charge (€/year) or a capacity charge computed separately for each type of generator or demand in each type of area in the system (see Olmos and Pérez-Arriaga 2009).

Olmos and Pérez-Arriaga also argue that the transmission tariff to be applied to each generator or load must be computed once and for all before its installation, since the level of this tariff should be based on the expected incremental contribution of this generator or load to the use of the grid (this is the driver of transmission investments). This means that the transmission charge to be paid by a network user should not be modified after its installation. Otherwise, the locational signal sent through this charge would be severely weakened.

Lastly, the process of implementation of new tariffs must be thought carefully. In order to avoid making big changes to the level of tariffs paid by already existing network users when introducing a new tariff scheme, the application of charges computed according to the new scheme could be limited to new network users. Alternatively, charges paid by already existing users could gradually evolve from the old tariff regime to the new one. In any case, the difference between the total cost of the grid and revenues from the application of tariffs should be socialized (preferably to demand).

## 2.7 Conclusions

Chapter 2 has analysed the effect that the grid should have on prices paid and earned by network users. Prices set should send both efficient short term signals driving operation decisions and long term ones driving the development of the system. Additionally, prices should provide an adequate remuneration of the transmission service guaranteeing its economic viability. Therefore, prices applied should be able to recover 100% of the regulated cost of the grid. No single set of prices seems to be able to meet all the aforementioned requirements, nor the sale of FTRs aimed at hedging the corresponding energy price differences. Thus, at least two set of transmission related prices must be applied.

Energy prices are aimed at driving operation decisions. Nodal prices, also called locational marginal prices, are deemed to be optimal energy prices because, assuming perfect information and competition, they encourage market agents to make socially optimal short-term decisions. Nodal prices internalize the effect of network losses and congestion on operation costs. However, in many real life systems, differences among nodal prices are small. Then, applying a single energy price (Single Pricing) or a price common to all the nodes within each of a set areas (zonal pricing) is considered to be preferable.

Net revenues resulting from the application of locationally differentiated energy prices, or from the sale of FTRs corresponding to commercial power transactions taking place, fall short of those needed to recover the whole cost of the grid. Then, additional charges, normally called transmission charges, or complementary charges, must be applied to complete the recovery of the grid cost. Complementary charges applied should allocate the cost of lines to those network users responsible for their construction. The electrical usage of lines by agents may be used as a proxy to network cost causality. However, it is the incremental usage made of new lines by new agents what determines the network reinforcements to be made. Therefore, network usage factors produced by most network cost allocation methods are useless, while average network usage factors produced by other methods like Average Participations or the Aumann-Shapley method must be modified to reflect the incremental nature of flows driving the development of the grid. Last but not least, in order for transmission charges not to interfere with the short term decisions by network users (to be driven by energy prices), they should be computed, once and for all, before the corresponding generators or loads are installed, and should take into account the expected increase in network flows that may result from the installation of the latter over all the set of possible operation situations that may occur along the economic life of these generators or loads. Besides, network tariffs should be applied as a fixed annual charge or a capacity charge computed separately for each type of generator or demand in each area in the system.

## References

Aguado JA, Quintana VH, Madrigal M, Rosehart WD (2004) Coordinated spot market for congestion management of inter-regional electricity markets. IEEE Trans Power Syst 19(1): 180–187

Alomoush MI, Shahidehpour SM (2000) Contingency-constrained congestion management with a minimum number of adjustments in preferred schedules. Electr Power Energy Syst 22: 277–290

Baran ME, Babunarayanan V, Garren KE (2000) Equitable alloation of congestion relief cost to transactions. IEEE Trans Power Syst 15(2):579–585

Bialek J (1996) Tracing the flow of electricity. IEE Proc Gener Trans Distrib 143(4):313–320

Biskas PN, Bakirtzis AG (2002) Decentralised congestion management of interconnected power systems. In: IEE Proc Gener Trans Distrib, pp 432–438

Boucher J, Smeers Y (2001) Towards a common European electricity market-paths in the right direction... still far from an effective design. Harvard Electricity Policy Group. http://www.ksg.harvard.edu/hepg/Standard_Mkt_dsgn/Smeers_Interconnections1_4jni_3.do1.pdf

Chao HP, Peck S (1996) A market mechanism for electric power transmission. J Regul Econ 10: 25–29

Chao HP, Peck S (2000) Flow-based transmission rights and congestion management. Electr J 13(8):38–59

Cheng X, Overbye TJ (2006) An energy reference bus independent LMP decomposition algorithm. IEEE Trans Power Syst 21(3):1041–1049

Christie RD, Wangensteen I (1998) The energy market in Norway and Sweden: congestion management. IEEE Power Eng Rev 18:61–63

Consentec/Frontier (2004) Analysis of cross-border congestion management methods for the EU internal electricity market, consentec and frontier economics. Prepared for the European Commission. Directorate-General Energy and Transport. Web page: http://europa.eu.int/comm/energy/electricity/publications/index_en.htm

Dismukes DE, Cope RF III, Mesyanzhinov D (1998) Capacity and economies of scale in electric power transmission. Utilities Policy 7(3):155–162

ETSO (2001) Co-ordinated auctioning: a market-based method for transmission capacity allocation in meshed networks. Association of European Transmission System Operators (ETSO). Web address: http://www.etso-net.org/upload/documents/Coordinated%20Auctioning.pdf, p 22

ETSO (2004) An overview of current cross-border congestion management methods in Europe. European association of Transmission System Operators (ETSO). Web address: http://www.etso-net.org/upload/documents/Current_CM_methods_final_20040908.pdf, p 31

ETSO (2006) Transmission risk hedging products. Solutions for the market and consequences for the TSOs. ETSO Background Paper, 20 April 2006. http://www.entsoe.eu

ETSO/EuroPEX (2004) Flow-based market coupling: a joint ETSO-EuroPEX proposal for cross-bcongestion management and integration of electricity markets in Europe. Association of European Transmission System Operators (ETSO) and association of European Power Exchanges (EuroPEX)address: http://www.etso-net.org/upload/documents/ETSO-EuroPEX_Interimreport_Sept-2004-.pdf, p 26

European Transmission System Operators (ETSO) (1999) Evaluation of congestion management methods for cross-border transmission. http://www.etso-net.org/, p 22

European Transmission System Operators (ETSO) (2008) ETSO overview of transmission tariffs in Europe: synthesis 2007. ETSO tariffs task force. Web page: http://www.etso-net.org/upload/documents/11.a.%20Final_Synthesis_2007_18-06-08.pdf, p 27

Fang RS, David AK (1999) Optimal dispatch under transmission contracts. IEEE Trans Power Syst 14(2):732–737

Galiana FD, Ilic M (1998) A mathematical framework for the analysis and management of power transactions under open access. IEEE Trans Power Syst 13(2):681–687

Galiana FD, Conejo AJ, et al. (2003) Transmission network cost allocation based on equivalent bilateral exchanges. IEEE Trans Power Syst 18(4):1425–1431

Henney A (2002) What the US could learn from Western Europe and elsewhere. Electr J 15(10): 53–64
Hogan WW (1992) Contract networks for electric power transmission. J Regul Econ 4:211–242
Hogan WW (2002) Financial transmission right formulations. Cambridge. Center for Business and Government, John F. Kennedy School of Government, Harvard University. http://www.ksg.harvard.edu/hepg
Junqueira M, da Costa LC, Barroso OGC, Thomé LM, Pereira MW (2007) An aumann-shapley approach to allocate transmission service cost among network users in electricity markets. IEEE Trans Power Syst 22(4):1532–1546
Kirschen D, Allan R, Strbac G (1997) Contributions of individual generators to loads and flows. IEEE Trans Power Syst 12(1):52–60
Lusztig C, Feldberg P, Orans R, Olsonet A (2006) A survey of transmission tariffs in North America. Energy 31(6–7):1017–1039
Newbery DV, Damme E, von der Fehr NM (2003) Benelux market integration: market power concerns. Dutch energy regulator (DTe). Web page: http://www.dte.nl/nl/Images/12_13216.pdf
O'Neill R, Helman PU, Hobbs BF, Stewart WR Jr, Rothkopf MH (2002) A joint energy and transmission rights auction: proposal and properties. IEEE Trans Power Syst 17(4):1058–1067
Olmos (2006) Regulatory design of the transmission activity in regional electricity markets. Ph.D. dissertation, Universidad Pontificia Comillas
Olmos L, Neuhoff K (2006) Identifying a balancing point for electricity transmission contracts. IEEE Trans Power Syst 21(1):91–98
Olmos L, Pérez-Arriaga IJ (2007) Evaluation of three methods proposed for the computation of inter-TSO payments in the internal electricity market of the European Union. IEEE Trans Power Syst 22(4):1507–1522
Olmos L, Pérez-Arriaga IJ (2009) A comprehensive approach for computation and implementation of efficient electricity transmission network charges. Energy Policy 37(12):5285–5295
Oren SS, Ross AM (2002) Economic congestion relief across multiple regions requires tradable physical flow-gate rights. IEEE Trans Power Syst 17(1):159–165
Pérez-Arriaga IJ, Rubio FJ, Puerta JF, Arceluz J, Marín J (1995) Marginal pricing of transmission services: an analysis of cost recovery. IEEE Trans Power Syst 10(1):546–553
Rau NS (2000) Transmission loss and congestion cost allocation: an approach based on responsibility. IEEE Trans Power Syst 15(4):1401–1409
Rivier M, Pérez-Arriaga IJ (1993) Computation and decomposition of spot prices for transmission pricing. In: Proceedings of the power systems computation conference (PSCC), Avignon
Rivier M, Pérez-Arriaga IJ, Luengo G (1990) JUANAC: a model for computation of spot prices in interconnected power systems. In: Proceedings of the 10th PSCC conference, Graz, 19–24 Aug 1990
Rubio FJ, Pérez-Arriaga IJ (2000) Marginal pricing of transmission services: a comparative analysis of network cost allocation methods. IEEE Trans Power Syst 15(1):448–454
Ruff LE (1999) Competitive Electricity markets: why they are working and how to improve them. http://www.ksg.harvard.edu/hepg,n/e/r/a
Schweppe FC, Caramanis M, Tabors RD, Bohn RE (1988) Spot pricing of electricity. Kluwer, Boston
Singh H, Hao S et al (1998) Transmission congestion management in competitive electricity markets. IEEE Trans Power Syst 13(2):672–680
Stoft S (1998) Congestion pricing with fewer prices than zones. Electr J 11(4):23–31
Tabors R, Caramanis M (2000) Real flow, a preliminary proposal for a flow-based congestion management system. Cambridge, MA. http://www.ksg.harvard.edu/hepg/flowgate/Real%20Flow.pdf
Tao S, Gross G (2002) A congestion management allocation mechanism for multiple transaction networks. IEEE Trans Power Syst 17(3):826–833

# Chapter 3
# Point to Point and Flow-Based Financial Transmission Rights: Revenue Adequacy and Performance Incentives

**Shmuel S. Oren**

## 3.1 Introduction

The prevalent market mechanism for defining transmission rights in North American restructured electricity markets is through financial instruments that enable energy traders to hedge congestion risk. The underlying quantities for such instruments are either Locational Marginal Prices (LMP) or shadow prices on transmission flowgates which are determined as part of an Optimal Power Flow (OPF) calculation. There are three prevalent forms of financial transmission rights whose settlements are based on the above underlying quantities:

FTR Obligations – These are LMP SWAPS defined over specific time intervals and between specific nodes, whose holder is entitled to receive, or obligated to pay, the nodal price difference between designated locations per MW denomination.

FTR Options – These are one sided LMP SWAPS defined over specific time intervals and between specific nodes, whose holder is entitled to receive the nodal price difference between designated locations per MW denomination if that difference is positive (but can walk away if it is negative.)

FGR – These are directional rights defined over specific time intervals and specific links, entitling their holder to the shadow price on the links capacity constraint in the designated direction per MW denomination.

Alternative forms of entitlements to the transmission infrastructure which have been used in the past or are still used in parts of the world include contract path rights which are based on a fictional "commercial path" between designated locations, or physical capacity rights between designated locations or on specific network interfaces. One major shortcoming of such physical rights is that they require coordination between the dispatch and transmission rights ownership. Furthermore, when the

S.S. Oren (✉)
Department of IEOR, University of California at Berkeley, Berkeley, USA
e-mail: oren@ieor.berkel0065y.edu

J. Rosellón and T. Kristiansen (eds.), *Financial Transmission Rights*, Lecture Notes in Energy 7, DOI 10.1007/978-1-4471-4787-9_3,

rights definition is not consistent with the physical flows induced by specific point to point energy transactions (as is the case for contract path), then the available transmission capacity between points (ATC) varies depending on overall dispatch patterns, making it difficult to issue entitlements that extend over long time periods. By contrast, financial rights have the advantage of enabling complete decoupling between the actual dispatch and the settlement of congestion charges. The system operator can dispatch generation resources in the most efficient way with no regard to how transmission rights ownership, and impose congestion charges based on actual use of the network. The congestion revenues are then distributed to the rights holders so that a network user whose transmission rights holdings match its network use breaks even. Any discrepancies between use and financial rights holdings will result in financial shortfalls or surpluses but will not impact dispatch efficiency. Furthermore, insuring that the amounts of FTRs and FGRs issued conform with physical feasibility enables the issuance of long term rights with minimal financial risk to the underwriters.

FTRs defining point to point financial transmission rights have been first introduced within a general framework of contract networks by Hogan (1992) and have been widely adopted in the US as an integral part of the nodal market designs implemented by the various independent system operators. Flow based transmission rights (financial or physical) have been first introduced in a seminal paper by Chao and Peck (1996). The potential use of FGRs, which are financial flow based rights, as substitutes or complements to FTRs has been discussed by Chao and Peck (1996), Chao et al. (2000), Ruff (2001), O'Neill et al. (2002) (and in numerous follow-up papers). However, FGRs, are rarely used in today's markets since energy traders prefer FTRs that are more suitable for hedging point to point congestion risk. Specifically, a bilateral energy transaction of X MW from node A to another node B in the network is exposed to congestion risk between the two location and is liable for a congestion charge that equals to the difference of LMPs between the two node. That charge is equivalent to the net cost resulting from selling the power at node A and buying it back at node B at the respective nodal prices. A trader can offset such a congestion charge by holding an FTR from node A to B for X MW which entitles him to the nodal price difference between node B and node A time X. Hence the FTR payoff exactly equals the congestion charge. Conceptually, however, FTRs and FGRs are equivalent due to a fundamental relationship between nodal price differences and flowgate shadow prices which is explained in the next section (see Chao et al. 2000). To understand the relationship between FTRs, FGRs and how they relate to optimal dispatch and locational marginal pricing we begin with a brief tutorial explaining these basic concepts in the following section.

## 3.2 A Primer to LMPs, FTRs and FGRs

The objective of Optimal Power Flow (OPF) is to find the output levels for a set of generation resources that are distributed over a transmission network (and are already running and synchronized), so as to minimize total cost of serving specified loads (or maximize social welfare if loads are characterized by price sensitive loads), while accounting for losses and without violating transmission flow constraints. In general

 

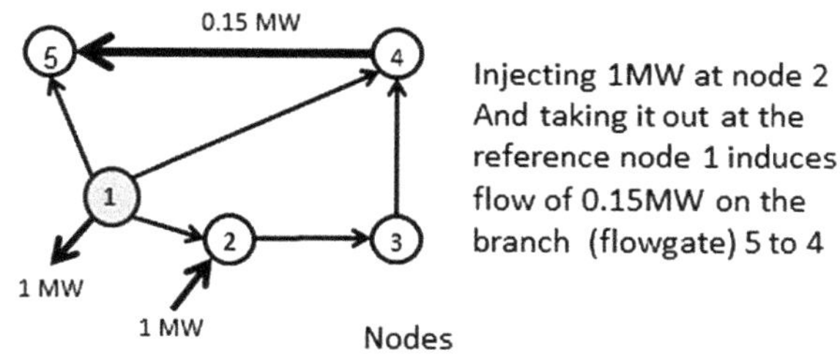


Nodes

| Branches | 1 | 2 | 3 | 4 | 5 |
|---|---|---|---|---|---|
| 1to2 | 0.00 | 0.67 | 0.54 | 0.19 | 0.03 |
| 2to3 | 0.00 | -0.33 | 0.54 | 0.19 | 0.03 |
| 3to4 | 0.00 | -0.33 | -0.46 | 0.19 | 0.03 |
| 4to5 | 0.00 | **-0.15** | -0.29 | -0.37 | 0.11 |
| 1to4 | 0.00 | 0.18 | 0.25 | 0.44 | 0.09 |
| 1to5 | 0.00 | 0.15 | 0.21 | 0.37 | 0.89 |

**Fig. 3.1** PTDF matrix for five node example

flows on transmission links are determined by Kirchhoff laws for Alternating Current (AC) and they must satisfy thermal and voltage limits. For the purpose of this exposition, however, we will ignore losses and assume a Direct Current (DC) approximation of Kirchhoff's laws in which case flows are only constrained by thermal limits specified for each transmission line.

Under such simplifications the flow pattern in a network can be characterized in terms of a matrix of Power Transfer Distribution Factors (PTDF) whose *ij* element specifies the incremental flow induced on each transmission link *j* by injecting one incremental MW at node *i* and withdrawing it at some designated reference node. The transmission links are specified as directional so negative flow indicate flow in the opposite direction. In the following, for clarity, we will denote the transmission links by pairs of indices representing the adjacent from/to nodes so that *hk* represents the directional link from node *h* to node *k*. The PTDF matrix can be easily computed through simulation or directly from the electrical properties (susceptances) of the transmission lines. As an illustrative example Fig. 3.1 gives the PTDFmatrix corresponding to the 5 node network shown, with node 1 as reference node. This example due to Fernando Alvarado (2000, personal communication) portrays a stylized representation of the PJM system.

According to the PTDF matrix in Fig. 3.1, $PTDF_{45,2} = -0.15$, indicating that injecting 1 MW at node 2 and withdrawing it at the reference node 1 results in 0.15 MW flow on the line connecting nodes 4 and 5 in the direction from 5 to 4 (opposite to the designated *45* direction). The PTDF matrix can be used to determine the impact of injections and withdrawals at any pair of nodes on any transmission line using superposition. For instance, the flow on the line 1–4 resulting from injecting 1 MW at node 2 and withdrawing it at node 5 is given by $PTDF_{14,2} - PTDF_{14,5} = 0.18 - 0.09 = 0.09$. This calculation is invariant to the choice of reference node since, the PDTF matrix for any reference node *i* can be obtained from the given matrix by subtracting the column corresponding to the reference node in the given PTDF matrix from each of the columns.

As indicated above the underlying quantities for financial transmission rights are locational marginal prices (LMPs) or line shadow prices (SP). These quantities are meaningful in the context of optimal power flow or optimal dispatch. A well-known property of optimal dispatch is that if no transmission constraint is binding, then the marginal cost of serving one incremental unit of energy at any node is identical and there is at least one marginal generation unit that can be moved to produce such an incremental unit at that cost. A less obvious result is that if one transmission line is congested and the system is dispatched optimally, then supplying an incremental unit of energy at any node without violating the binding constraint can be achieved by adjusting the output of up to two generation units, so called, marginal generators which can be moved up or down. This principle can be generalized in the sense that when the OPF results in m binding constraints then supplying an incremental unit of energy at a specific node without violating the constraints may require change in output levels of up to m + 1 marginal generators. Solving an OPF problem determines the output levels of all operating generators and identifies the marginal units which implicitly determines the LMPs and transmission line shadow prices. Following are intuitive definitions of these two concepts.

**Locational Marginal Price (LMP):** *The least cost of providing an incremental unit of energy at a node under optimal dispatch, without violating the binding transmission constraints.*

**Line Shadow Price (SP):** *The maximum dispatch cost savings, under optimal dispatch that can be achieved due to an incremental unit increase in the lines' flow capacity constraint without violating any of the binding transmission constraints.*

Given the set of marginal generators corresponding to an OPF solution and the PTDF matrix we can calculate the LMPs and Shadow prices according to the above definitions. Clearly only lines operating at the limit have positive shadow prices and LMPs at nodes with generators that are free to move up or down will equal that generator's marginal cost. However, at nodes with no generation or with generators operating at their capacity limits (up or down), the LMP can be positive or negative. To illustrate the LMP calculation consider the example in Fig. 3.1 and assume that the line connecting nodes 4 and 5 is operating at its limit in the 5–4 direction under optimal dispatch where the two marginal generation units are at node 1 with marginal cost of \$15/MWh and at node 4 with a marginal cost of \$30/MWh. To determine the LMP at node 2 we must calculate the incremental outputs $Q_1, Q_4$ of the marginal units at nodes 1 and 4 so as to deliver 1 MWh to node 2 without increasing the flow on the congested line. From the PTDF matrix in Fig. 3.1 we can determine that 1 MW injected at node 1 and withdrawn at node 2 will increase flow on line 4–5 by 0.15 MW. Likewise injecting 1 MW at node 4 and withdrawing it at node 2 will increase flow on line 4–5 by $-0.37 + 0.15 = -0.22$ MW. Thus the quantities $Q_1, Q_4$ must satisfy the system of equations:

$$\begin{bmatrix} 0.15 & -0.22 \\ 1 & 1 \end{bmatrix} \begin{bmatrix} Q_1 \\ Q_4 \end{bmatrix} = \begin{bmatrix} 0 \\ 1 \end{bmatrix} \Rightarrow Q_1 = 0.59 \quad Q_4 = 0.41$$

Hence the cost of supplying a marginal 1 MW at node 2 which is the LMP at node 2 is given by:

$$\mathrm{LMP}_2 = 15 \cdot \mathrm{Q}_1 + 30 \cdot \mathrm{Q}_4 = 15 \times 0.59 + 30 \times 0.41 = \$21.15/\mathrm{MWh}$$

A similar calculation can be performed to determine the shadow price on the line connecting nodes 4 and 5 in the congested direction 4–5. Now the objective is to perturb the outputs of the marginal units by incremental amounts $Q_1$, $Q_4$ so as to increase the flow on the congested line 5–4 while maintaining the energy balance. The resulting quantities can be determined by solving the system of equations:

$$\begin{bmatrix} 0.15 & -0.22 \\ 1 & 1 \end{bmatrix} \begin{bmatrix} Q_1 \\ Q_4 \end{bmatrix} = \begin{bmatrix} 1 \\ 0 \end{bmatrix} \Rightarrow Q_1 = 2.7 \quad Q_4 = -2.7$$

Which tell us that the increased capacity enables us to increase output from the cheap marginal unit at node 1 by 2.7 MW while reducing the output of the expansive marginal unit at node 4 by the same amount. Thus, the incremental change in dispatch cost due to a unit increase in capacity of the congested line (flowgate) which is the flowgate shadow price is given by:

$$\mathrm{SP}_{54} = 15 \cdot \mathrm{Q}_1 - 30 \cdot \mathrm{Q}_4 = (15 - 30) \times 2.7 = \$40.5/\mathrm{MW/h}$$

It should be noted that shadow prices are direction specific and have non zero values only if the line flow is at capacity. So in the above example $\mathrm{SP}_{45} = 0$, since the line flow capacity constraint in the direction 4–5 is not binding.

Clearly there is a close relationship between LMPs and flowgate shadow prices both of which are calculated from the same data. In general it can be shown that for any pair of nodes $i,j$ the following fundamental relationship holds.

$$LMP_j - LMP_i = \sum_{\text{all flowgates hk}} SP_{hk} \cdot (PTDF_{hk,j} - PTDF_{hk,i}) \tag{3.1}$$

As explained earlier a 1 MW point to point FTR obligation is a forward contract entitling (or obligating) its holder to receive or pay the stream of LMP differences between two specific nodes over a designated time period. Likewise a 1 MW FGR is a forward contract entitling its holder to receive the stream of shadow prices on a specific flowgate over a designated time period. Hence, the above fundamental relationship can be extended to relate point to point FTR obligations and FGRs implying that a point to point FTR obligation may be viewed as a portfolio of FGRs weighted by the corresponding PTDF differences. This relationship, however, becomes more complicated with respect to point to point FTR options. A simplistic approximation, suggested by O'Neill et al. (2002), is to calculate the payoff (or price) of a point to point FTR option as the partial summation of the weighted FGR payoffs (or prices) over flowgates for which the PTDF difference in the above formula is positive. Since shadow prices and hence FGR payoffs are nonnegative such an approximation ensure a nonnegative payoff for the point to point FTR option. Such a calculation, however, overcompensates point to point FTR options in cases where the payoff is positive but reduced by the presence of "couterflow" branches. Unfortunately, the decomposition of point to point FTR options into FGRs enabled by the above approximation is essential for a joint auction that offers the different instruments simultaneously.

## 3.3 Managing Congestion Risk

In LMP based markets, energy transactions in the Day Ahead market are exposed to congestion rents that are determined as the LMP difference between the injection and withdrawal nodes. A trader buying energy at one location to be delivered at another location, incurs such congestion rents as the difference between the selling price of energy at the source and the buying price at the delivery point when the transactions are cleared through the ISO market. Alternatively, if the delivery is scheduled as a firm bilateral transaction then it is subject to a congestion charge imposed by the ISO that equals to the LMP difference between the injection and withdrawal locations. In either case a trader can hedge its exposure to the congestion charges by acquiring financial transmission rights.

In view of the fundamental relationship between point to point FTRs and FGRs explained above, a trader could achieve the same protection against congestion charges provided by a point to point FTR obligation by buying the equivalent portfolio of FGRs. To illustrate this equivalence consider the three node network in Fig. 3.2 with identical susceptances for all three lines and flow limits as indicated on the respective lines.

Injecting 1 MW at node 1 and withdrawing it at node 3 produces (2/3)MW flow on the line 1–3 and (1/3)MW flow on the lines 1–2 and 2–3. Suppose that G1 has a bilateral contract with L3 to deliver 150 MW and wishes to hedge the contract against congestion charges. This can be done by procuring 150 MW FTR obligation from node 1 to 3. In real time the congestion rent charged to the bilateral transaction is the nodal price difference between the two nodes times the 150 MW transacted. That amounts is also the settlement payment for the 150 MW FTR from node 1 to 3. Thus the FTR settlement exactly offsets the congestion charge. Alternatively, the bilateral transaction can be hedged against congestion by procuring a portfolio of FGRs as follows: 100 MW FGR on line 1–3, 50 MW FGR on line 1–2 and 50 MW FGR on line 2–3. Each FGR is paid in real time the corresponding shadow price per MW.

Assume that only the line 2–3 is congested then the shadow price on the other two lines is zero and the settlement payment for the above FGR portfolio is $50 \cdot SP_{23}$ but from the fundamental relationship between nodal prices and shadow prices on transmission lines we know that $LMP_3 - LMP_1 = \frac{1}{3}SP_{23}$ . Hence the settlement payment for the FGR portfolio is $50 \cdot 3 \cdot (LMP_3 - LMP_1)$ which is identical to the settlement for the FTR from node 1 to 3 both of which equal the congestion charge for the bilateral transaction.

**Fig. 3.2** Three node example

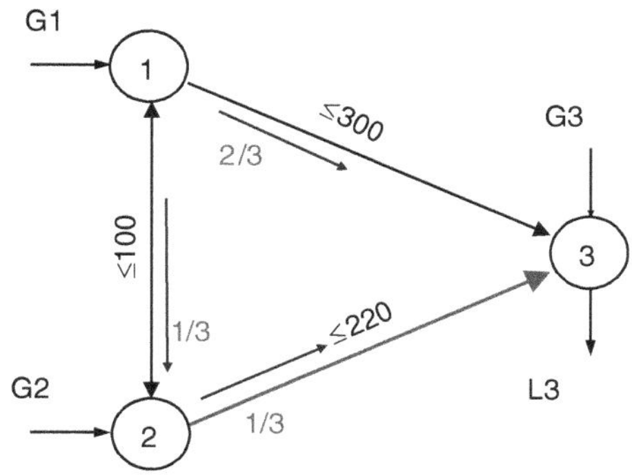


The difference between using FTRs or FGRs in hedging congestion risk arises when considering changes in the network topology which will produce changes in the PTDFs such changes may result from contingencies or deliberate control actions switching lines in or out. Whether FTRs or FGRs are used to define property rights and hedging mechanisms has also implication regarding the extend to which the physical capacity of the network can be fully subscribed and the ability of market participants to fully hedged their energy transactions. Since the payoff of a point to point FTR obligation is based on the actual LMP difference which is also used in computing the congestion rents, a 1 MW FTR obligation between two nodes provides a perfect hedge against the congestion charges imposed on a 1 MW energy transaction between the same nodes. Such a hedge provides insurance against congestion risk resulting from changes in dispatch patterns and LMPs as well as changes in the network topology, including line capacity ratings and the PTDFs.

The availability of perfect hedging instruments does not imply, however that all transactions that can be accommodated in real time by the physical system can be hedged while assuring that the real time congestion revenues suffice to pay off the settlements to all outstanding FTRs (i.e., revenue adequacy). As discussed below, the conditions that will guarantee revenue adequacy result in unsubscribed flowgate capacity which in turn can lead to congestion revenue surplus. Such surplus indicates that some energy transactions could not be fully hedged. When FGRs portfolios are used to hedge congestion risk associated with energy transactions, it is the responsibility of the FGRs holder to assemble a portfolio that synthesizes the LMP differences that are used to compute congestion charges, such a portfolio protects the holder against fluctuations in shadow prices on the flowgates and against changes in the flowgate capacity ratings but does not provide insurance against variation in the PTDFs. So it is the responsibility of the insured to track such variations to ensure that the FGR portfolio produces sufficient settlement revenue to cover the congestion charges that are based on the LMP differences.

On the other hand, FGR allocations are based on the full flowgate capacity as opposed to FTR allocation that only subscribes the flowgate capacity corresponding to the allocated FTRs. Thus, the entire wire capacity can be subscribed through FGRs and as long as flowgate capacities are not reduced, the congestion revenues (which can be assigned to flowgates based on the real time PTDFs), will match the FGR settlements (i.e., revenue adequacy is automatically guaranteed). Another consideration is the need for centralized coordination in issuing and secondary trading of the various forms of financial transmission rights. As will be discussed below, Assuring revenue adequacy for point to point FTRs requires central coordination since available transmission capacity between any two nodes depends on the entire constellation of other point to point FTRs issued.

Consequently, the issue of point to point FTRs is always done by a central authority such as an ISO and any secondary trading takes place through centrally coordinated reconfiguration auctions. By contrast, since FGRs are only tied to specific flowgate capacities, they can be issued by multiple entities owning specific flowgate assets or producing counterflow and can be traded independently in secondary markets. This issue has come up, for instance, in the European Union where congestion revenues on international interconnect flowgates are collected by the interconnected countries which are also vested with the right to issue long term contracts for the use of such facilities.

Arguments in favor and against employing FGRs in practice as hedges against congestion risk can be found in Chao et al. (2000) and Ruff (2001). In our subsequent discussion we will not duel any further on this debate and only exploit the conceptual interpretation of FTRs as an FGR portfolio.

## 3.4 Revenue Adequacy and Simultaneous Feasibility

Hogan (1992) has shown that if the outstanding FTRs satisfy a "simultaneous feasibility test" (SFT) and the network topology is fixed then the FTR market is "revenue adequate". Revenue adequacy means that congestion revenues and merchandising surplus (i.e., the difference between the buying cost and the sales revenues for energy traded through the pool) collected by the system operator from bilateral transactions and local sales and purchases at the LMPs, will cover the FTR settlements. The SFT requires that if all the FTRs were exercised simultaneously as physical bilateral transactions then the transmission flow constraints would not be violated.

In FTR auctions bidders submit bids for specific FTRs and the ISO selects winning bids by treating FTR bids as proposed schedules using a security constrained OPF that maximizes the FTR auction bid value. These constraints are also imposed if any portion of the FTRs is being allocated based on historical use or other allocation criteria. As mention above, the hypothetical dispatch (referred to as the FTR point) corresponding to simultaneous bilateral schedules replicating all outstanding FTRs must meet all security and flow constraints i.e. the grid must be able to support all the bilateral transactions covered by the FTRs. The auction produces a set of winning bids and uniform clearing prices for each pair of nodes that equal to the LMP differences of the auction OPF.

Clearly, the FTR point characterizing the mix of awarded FTRs, may differ from real time dispatch. However, but if the topology hasn't changed the FTR point represents a feasible but not necessarily optimal dispatch. Hence, if the nomogram is convex, then the congestion revenues will be sufficient to cover the FTR settlements.

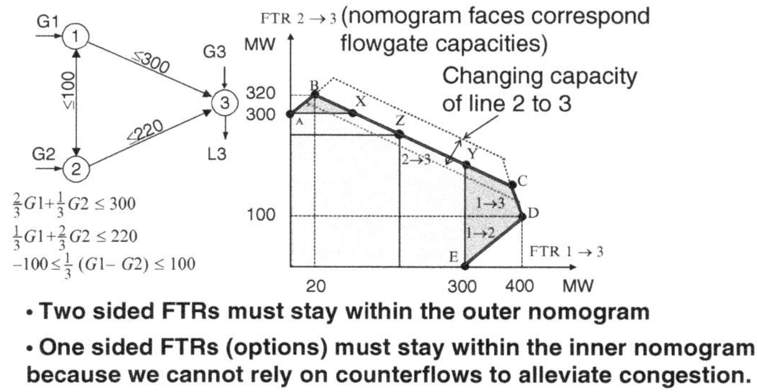


**Fig. 3.3** Feasibility region of FTR options and obligations and the effect of flowgate capacity rating

This follows from a theoretical argument based on duality of linear programs, showing that minimum cost dispatch is equivalent to maximizing congestion revenues.

Figure 3.3 below illustrates the nomogram representing feasible dispatch for the three node DC system introduced earlier with identical susceptances for all lines but different flow limits as shown. The vertical axis of the graph represents injection at node 2 and withdrawal at node 3 while the horizontal axis represent injection at node 1 and withdrawal at node 3. The feasible region given the flow constraints is characterized by a convex polyhedron defined by the system of linear inequality constraints implied by Kirchhoff's law and the flow limits on the lines. The same constraints also characterize the feasible set of FTRs from node 1 to node 3 and from node 2 to node 3 that will meet the SFT described above. The facets of the polyhedron correspond to the flow capacity constraints and adjusting these capacities is represented by a parallel shift of these facets as shown for line 2–3.

We note that the system can accommodate up to 400 MW transaction from node 1 to 3 if there is a 100 MW transaction from node 2 to node 3 which produces counterflow on the congested link from node 1 to node 2. In the absence of such counterflow, the system can only accommodate a 300 MW transaction from node 1 to node 3. In the context of the SFT, reliance on conterflow translates to reliance on an FTR obligation with a negative real time settlement which will supplement the congestion revenues to produce sufficient income for FTR payoffs.

FTRs with an expected negative real time settlement have negative value and those who are willing to assume such an obligation would expect to be paid upfront and will submit negative bids (i.e. offers) in the FTR auction to undertake the obligation. If the holder of such an FTR obligation from node 2 to node 3 actually executes the corresponding transaction in real time, by injecting power at node 2 and taking it out at node 3, it produces counterflow for which it will collect negative real time congestion charges (i.e., counterflow payments) that will exactly offset the negative settlement of the FTR obligation from node 2 to node 3. In such a case, the auction income from taking on an FTR obligation with negative payoff is a net gain to the FTR holder which can be used, to subsidize a forward contract at a price below marginal production cost if executing the transaction produces couterflow that will offset the negative FTR settlement.

However, undertaking such an FTR obligation entails exposure to performance risk in case that the FTR holder cannot execute the transaction due to a generator outage, for instance. To avoid such exposure, market participant would prefer (assuming all else being equal) FTR options that protects them from potential liability that comes with an FTR obligation. Issuing FTR options rather than obligations implies, however, that the ISO cannot rely in the SFT on counterflows and cannot rely on the supplemental revenue produced by FTR obligations with negative settlemet. Hence, the feasible region for FTR options in the case depicted by Fig. 3.3 is the chopped off light portion of the nomogram. While FTR options are attractive from a risk management perspective their use is limited since they severely limit the simultaneously feasible FTRs that can be issued and they turn out to be expensive as compared to the two sided FTR obligations. One of the important uses of FTR options is to convert historical entitlements to physical transmission rights held by MUNIs, for instance, (which are inherently options) to financial transmission rights.

To illustrate how FTRs can facilitate efficient forward energy trading, lets assume that the marginal cost of G1 is \$30/MWh the marginal cost of G2 is \$45/MWh and of G3 is \$100/MWh. The load at L3 is 500 MW and the capacities of all three generators exceeds 500 MW. The optimal dispatch for this case is at point D of the nomogram in Fig. 3.3, which corresponds to supplying the load at L3 with 400 MW from G1 and 100 MW from G2. The corresponding LMPs at nodes 1,2,3 are \$30/MWh, \$45/MWh and \$40/MWh respectively. Both, line 1–3 and line 1–2 are operating at the flow limit with corresponding shadow prices of \$5/MW/h and \$20/MW/h, respectively. If the optimal dispatch and LMPs are forecasted correctly, the FTR auction will clear with 100 MW FTR obligations from node 1 to 3 awarded at \$10/MW/h and 400 MW FTR obligation from 2 to 3 awarded at −\$5/MW/h (i.e. the bidder gets paid for assuming the obligation).

Both G1 and G2 can enter into forward contracts to deliver energy to L3 at \$40/MWh which for G1 would result in a gain of \$10/MWh and for G2 in a loss of \$5/MWh. G1 can then hedge its exposure to real time congestion charges by using its forward contract surplus to buy FTR obligations from node 1 to 3 in an amount matching the forward energy contract. Likewise, G2 can offset the forward contract deficit with expected real time counterflow payments or lock in these payments by taking on FTR obligations from node 2 to 3 so as to match the forward contract quantity. The system operator collects from G1 congestion rents for 400 MW from node 1 to 3 in the amount of \$10/MWh (based on the LMP difference) and pays to G2 \$5/MWh for 100 MW of counterflow totaling \$3,500/h. The FTR settlement amount to \$10/MWh times 400 MW for FTRs from node 1 to 3 less the amount collected from the FTRs from node 2 to 3 of \$5/MWh times 100 MW, adding up to \$3,500/h. So in this case the ISO breaks even.

Suppose, however, that the real time LMPs were not forecasted correctly in the FTR auction and the bids resulted in an FTR point other than point D on the nomogram. Specifically, assume that the FTR auction awards corresponded to point E on the nomogram with 300 MW FTRs from node 1 to 3 and no FTRs from node 2 to 3. Then, the FTR settlement amounts to $300 \times 10 = \$3{,}000$/h resulting in a congestion revenue surplus of \$500/h. In general the real time settlement for any feasible FTR award combination will be less than or equal to the congestion revenue corresponding to the optimal dispatch point D.

FGRs can be used in a similar way to the above although achieving proper hedging places more burden on energy traders. In an FGR auction all the FGRs corresponding to the lines capacities rating (in both directions) are being allocated. However, if the dispatch is correctly forecasted in the FGR auction, only the FGRs on the line from node 1 to 3 and from node 1 to 2 have positive clearing prices which in our example equal to \$5/MW/h and \$20/MW/h respectively. The total auction revenue will be the same as in the corresponding FTR auction totaling $5 \times 300 + 20 \times 100 = 10 \times 400 - 5 \times 100 = \$3{,}500$/h. as in the case of FTRs, G1 and G2 can hedge their forward energy contracts to deliver energy to L3 at \$40/MWh. In this case G2 would buy (100/3)MW FGRs on line 1–3 (backed by wire capacity) and sell (100/3)MW FGRs on line 1–2 (backed by counterflow it expects to produce) at a total gain of $(100/3) \times (20-5) = 500$/h which exactly offsets its forward energy contract deficit. G1 could buy (800/3)MW FGRs on line 1–3 (backed by wire capacity) and (400/3)MW FGRs on line 1–2 of which 100 MW is backed by wire capacity and (100/3)MW is backed by counterflow.

The total FGR cost to G1 is $(800/3) \times 5 + (400/3) \times 20 = \$4{,}000$/h which exactly matches its forward energy contract surplus. These FGR procurements match the expected flows induced by the transactions corresponding to the forward contracts entered into by G1 and G2 (based on the system PTDFs). In real time G1 pays as before congestion charges of \$10/MWh for 400 MW it delivers to L3 and receives FGR settlements (based on shadow prices) of $(800/3) \times 5 + (400/3) \times 20 = \$4{,}000$/h which exactly offsets the congestion charges. Likewise G2 collects \$500 in counterflow payments from the ISO which cover its net FGR settlement liability resulting from \$5/MW/h income for (100/3)MW FGRs it sold on line 1–3 less its \$20/MW/h payout for its short position on (100/3)MW FGR on line 2–3, totaling \$500/h. In the above setting the ISO always breaks even since the wires capacity is fully sold while both the congestion rents and the FGR settlements are based on the same flows and shadow prices.

The revenue surplus we have identified when FTRs are being used results from the fact that an FTR auction only allocates flowgate capacity corresponding to the FTRs that are sold leaving the remaining flowgate capacity in the hands of the ISO. Hence when the real time dispatch differs from the FTR point, unsold flowgate capacity may become valuable and the congestion revenue corresponding to that unsold capacity translates into a revenue surplus for the ISO. For instance in Fig. 3.3, if FTRs awarded in the auction correspond to point E, then the constraint on line 1–3 is not binding and 100 MW of flowgate capacity on line 1–3 remains unsold. Then when the real time dispatch moves to point D on the nomogram and the shadow price on line 1–3 goes to \$5/MW/h, the congestion rents on that unsold flowgate capacity retained by the ISO produce a revenue surplus of \$500/h.

## 3.5 Line Derating and Topology Changes

Flowgate capacity ratings will affect the feasible SFT nomogram as illustrated in Fig. 3.3 for a three node DC network. Consequently, if in real time operation, a flowgate rating is decreased from what was assumed in the SFT or if the flowgate failed due to a contingency, then, the FTR operating point may not be feasible in the real dispatch topology as shown in Fig. 3.4.

Such line derating may result in revenue shortfall, i.e., the congestion rents that are based on the real time LMP differences may not suffice to cover the settlements to all outstanding FTRs. To illustrate such revenue shortfall more explicitly consider a three node example introduced by Hedman et al. (2011) and shown in Fig. 3.5 . In this example FTRs are allocated based on an SFT which assumes the depicted topology. In particular 60 MW FTR obligations from node A to B and 30 MW FTR obligation from node A to C have been sold through an auction (or allocated by any other means).

The feasible region for the SFT is characterized by the set of linear inequalities:

$$\begin{aligned}
&-50 \leq \frac{2}{3}AB + \frac{1}{3}AC \leq 50 \\
&-100 \leq \frac{1}{3}AB + \frac{2}{3}AC \leq 100 \\
&-100 \leq \frac{1}{3}AB - \frac{1}{3}AC \leq 100 \\
&-100 \leq AB + AC \leq 100 \\
&-100 \leq AB \leq 100
\end{aligned} \tag{3.2}$$

This region is illustrated in Fig. 3.6 as the triangle consisting of areas 1, 2, and 4. The outstanding FTRs represent a point on the boundary of the feasible region (depicted by the gray square) and hence they satisfy the SFT for this topology.

If the topology doesn't change then the optimal dispatch coincides with the FTR allocation and hence the corresponding congestion revenues exactly cover thepayments to FTR holders. Suppose, however, that in operation one of the lines between node A and B fails. Such a contingency will shrink the feasible region to area 4 in Fig. 3.5 which is represented by the inequalities:

$$\begin{aligned}
&-25 \leq \frac{1}{2}AB + \frac{1}{4}AC \leq 25 \\
&-100 \leq \frac{1}{2}AB + \frac{3}{4}AC \leq 100 \\
&-100 \leq \frac{1}{2}AB - \frac{1}{4}AC \leq 100
\end{aligned} \tag{3.3}$$

Thus, the outstanding FTRs are no longer simultaneously feasible under the new topology.

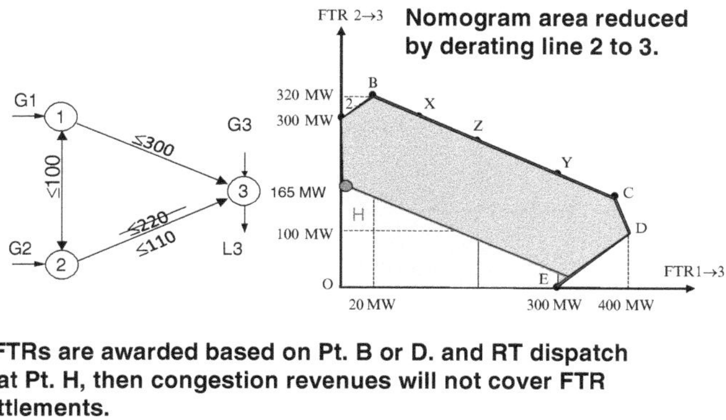


**Fig. 3.4** The effect of derating flowgate capacity

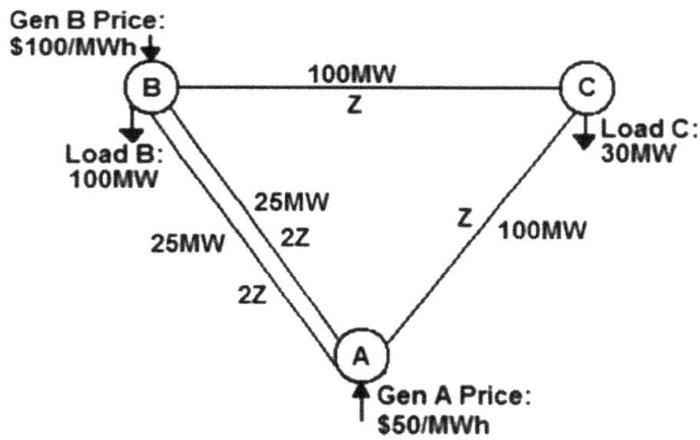


**Fig. 3.5** Revenue adequacy example

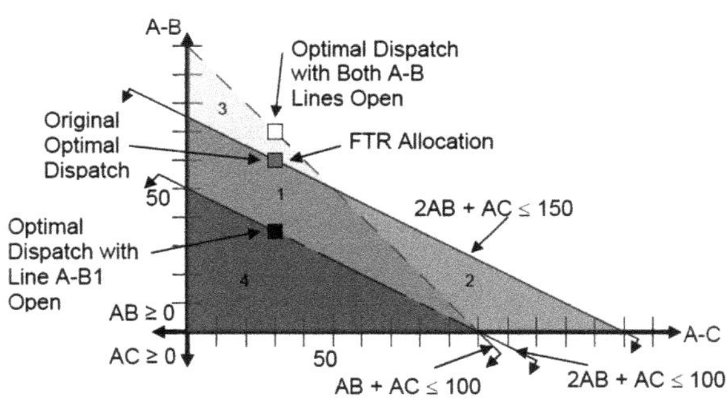


**Fig. 3.6** Feasible region for different topologies

The optimal dispatch under the above contingency is represented by the black square in Fig. 3.6. Tables 3.1, 3.2, and 3.3 below show that the congestion revenues corresponding to this dispatch fall short of covering the settlement payments to the

**Table 3.1** Optimal dispatch results with all lines in

| Node | Gen output | LMP | Gen cost | Trans-action | MW | Cong. rent |
|---|---|---|---|---|---|---|
| A | 90 MW | $50/MWh | $4,500 | A–B | 60 MW | $3,000 |
| B | 40 MW | $100/MWh | $4,000 | A–C | 30 MW | $750 |
| C | 0 MW | $75/MWh | $0 | **Congestion rent:** | | **$3,750** |
| **Total generation cost:** | | | **$8,500** | | | |

**Table 3.2** Optimal dispatch results with one line A–B out

| Node | Gen output | LMP | Gen cost | Trans-action | MW | Cong. rent |
|---|---|---|---|---|---|---|
| A | 65 MW | $50/MWh | $3,250 | A–B | 35 MW | $1,750 |
| B | 65 MW | $100/MWh | $6,500 | A–C | 30 MW | $750 |
| C | 0 MW | $75/MWh | $0 | **Congestion rent:** | | **$2,500** |
| **Total generation cost:** | | | **$9,750** | | | |

**Table 3.3** FTR settlements

| Source to sink: | FTR quantity: | FTR settlements (all lines in) | FTR settlements (one line A–B out): |
|---|---|---|---|
| A to B | 60 MW | $3,000 (LMP gap: $50/MWh) | $3,000 (LMP gap: $50/MWh) |
| A to C | 30 MW | $750 (LMP gap: $25/MWh) | $750 (LMP gap: $50/MWh) |
| **Total FTR settlements:** | | **$3,750** | **$3,750** |

FTR holders. In this case the contingency affected the generators' output and flows but did not affect the LMPs and hence the FTR payments. Specifically, the congestion revenues dropped from $3,750 to $2,500 while the FTR settlement remains $3,750 resulting in a shortfall of $1,250.

Surprisingly, revenue adequacy can be restored and generation cost reduced in this case by switching off the other line between nodes A and B. The feasible region corresponding to the topology with both lines between node A and B out is defined by the constraint:

$$AB + AC \leq 100$$

Since both A to B and A to C transactions must share the line between A and C. Hence, the feasible region is now represented by the triangle consisting of areas 1, 3 and 4 in Fig. 3.6 whereas the optimal dispatch moved from the black rectangle to the white rectangle. Furthermore, the gray rectangle representing the outstanding FTRs is now within the feasible region and can, therefore, be interpreted as a suboptimal feasible dispatch. Since an optimal dispatch solution also maximizes congestion rents (by duality theory of linear programming), it follows that the congestion rents exceed the FTR settlements which equal to the congestion rents corresponding to a feasible suboptimal dispatch. The above observations are verified numerically by the results in Tables 3.4 and 3.5. The optimal dispatch results with both lines between node A and B out are summarized in Table 3.4 and the corresponding FTR settlements are given in Table 3.5. We note that generation cost dropped to $8,000 which is below the

**Table 3.4** Optimal dispatch results with two lines A–B out

| Node | Gen output | LMP | Gen cost | Trans-action | MW | Cong. rent |
|---|---|---|---|---|---|---|
| A | 100 MW | $50/MWh | $5,000 | A–B | 70 MW | $3,500 |
| B | 30 MW | $100/MWh | $3,000 | A–C | 30 MW | $1,500 |
| C | 0 MW | $100/MWh | $0 | **Congestion rent:** | | **$5,000** |
| **Total generation cost:** | | | **$8,000** | | | |

**Table 3.5** FTR settlements with the two lines A–B out

| Source to sink: | FTR quantity: | FTR settlements (both lines A–B open): |
|---|---|---|
| A to B | 60 MW | $3,000 (LMP gap: $50/MWh) |
| A to C | 30 MW | $1,500 (LMP gap: $50/MWh) |
| Total FTR settlements: | | **$4,500** |

optimal dispatch with all lines in, while congestion revenues increased to $5,000 which is sufficient to cover the $4,500 FTR settlement payments.

## 3.6 Allocating Revenue Shortfalls

When a revenue shortfall occurs, i.e. congestion revenues cannot cover the settlement payments to FTR holders, the system operators must make up the difference. The various approaches adopted by system operators in the US for addressing such revenue shortfalls include:

- Full payment to FTRs based on nodal prices and uplift of the shortfall to sellers or buyers of energy (full funding approach)
- Prorate settlement to all FTRs to cover shortfall ("haircut" approach)
- Intertemporal smoothing of congestion revenue accounting by carrying over revenue surpluses and shortfall over an extended time period.
- Prorate settlement to FTRs based on impact of derated flowgates
- Full funding of FTRs and assignment of shortfall to owners of derated flowgates.

The first three alternatives socialize the cost of derated lines to energy sellers or buyersor to the FTR holders or across time periods. In the extreme case when a derated line is radial such socialization is vulnerable to gaming. An FTR holder on a derated but underutilized radial line has the incentive to congest that line though fictitious transactions in order to capture FTR revenues. The last two alternatives, which we advocate in this paper, directly assigns shortfalls to users or owners of derated flowgates. An important motivation for such an approach is to prevent potential gaming through overscheduling intended to induce congestion that will increase the payoff on certain FTRs. To illustrate such direct assignment consider the three node example in Fig. 3.2. In that example 1 MW FTR from node 1 to 3 contains 1/3 MW flow on line 2–3, whereas 1 MW FTR from node 2 to 3 contains 2/3 MW flow on line 2–3. Thus, if line 2–3 is derated by 50 % the congestion revenue shortfall will be 110 times the shadow price $SP_{23}$ on line 2–3.

The aforementioned shortfall can be assigned to the line owner while preserving full funding of the outstanding FTRs. Alternatively it can be assigned to the FTRs by reducing their settlement payment in accordance to the proportion of the derated line flow that they contain. Specifically since the capacity of line 2–3 was reduced by 50 %, The payment to a 1 MW FTR from node 1 to 3 is reduced by $0.5 \times (1/3) \times SP_{23}$ and the payment to a 1 MW FTR from node 2 to 3 is reduced by $0.5 \times (2/3) \times SP_{23}$. The SFT requires that the number of FTRs from node 1 to 3 times 1/3 plus the number of FTRs from node 2 to 3 times 2/3 does exceed the thermal limit of line 2–3 which is 220 MW (and it equals to that limit when the shadow price $SP_{23}$ is positive.) Hence, the reductions of FTR settlement payments above adds up exactly to $110 \times SP_{23}$ which is the revenue shortfall due to the derating of line 2–3.

Consider now the case when more than one line is derated. Suppose that line 2–3 is derated by 50 % and line 1–3 is derated by 20 %. Direct assignment the of revenue shortfall will again reduce the settlement payments to each FTR based on its flow share on each derated line. Thus payments to 1 MW FTR from node 1 to 3 is reduced by $0.5 \times (1/3) \times SP_{23} + 0.2 \times (2/3) \times SP_{13}$. Likewise payments to 1 MW of FTR from node 2 to 3 is reduced by $0.5 \times (2/3) \times SP_{23} + 0.2 \times (1/3) \times SP_{13}$. An intuitive analogy to the above approach is to think of FGRs as stocks and of FTRs as mutual funds which contain the various FGRs in proportions reflecting the corresponding PTDFs. When a line is derated by 50 % it is equivalent in our analogy to a stock loosing half its value. In the financial analogy it is natural that when a stock loses part of its value then the different mutual funds containing that stock will be impacted in proportion to their holdings of that stock. It would seem unreasonable to suggest that the loss of a stock would be born equally by all mutual funds offered by a brokerage house regardless of the holdings of the stock in each fund. Likewise it is natural and fair to allocate the revenue shortfall due to derating of a line according to the flow impact of each FTR on the derated line.

## 3.7 Expanding the FTR Feasible Region via Short FGRs

While derating line capacities reduces the feasible set of FTRs that the network could support without revenue shortfalls, increasing line capacity ratings will increase the set of FTRs that can be awarded in the auction as shown in Fig. 3.7 below. Such an increase could result from a physical change in line capacity due to an upgrade of a line or improved maintenance. Alternatively, an increase in line capacity used for the purpose of the SFT can be "virtual" and supported by short positions on FGRs, just as an increased number of available FTRs between two points can be underwritten by counterflow commitments. A short position on an FGR amounts to an obligation to either increase the flowgate capacity or underwrite the settlement cost of the added FTRs. The holder of a 1 MW short FGR position on a particular line is paid the shadow price on that line in the SFT power flow calculation and is liable for the shadow price on that line in real time. The payment received by such a short position holder in the FTR auction is financed by the revenue from the additional FTRs that can be sold due to the increase in the SFT feasible nomogram.

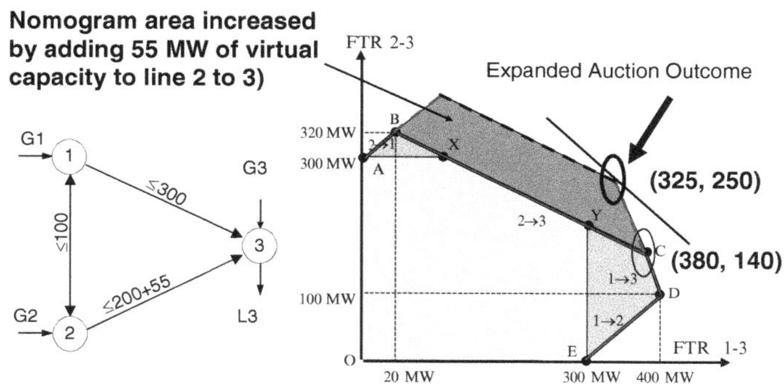


**Fig. 3.7** Expanding FTR feasibility with short FGR positions

The real time settlement paid by the short FGR holder supplements the congestion revenues and will cover any FTR revenues shortfall resulting from the oversold FTRs. If the line for which the short FGR position was issued is not congested in real time then the holder of that position gets to pocket the auction revenue for underwriting that position. To illustrate, suppose that the auction clearing price on both FTRs depicted along the axis in Fig. 3.6 (Node 2–3 and node 1–3) is $10/MW/h, then the corresponding shadow price on line 2–3 is also $10/MW/h. A short position of 55 MW on line 2–3 will earn its underwriter $550/h. Such a short position expands the feasible region in the SFT as shown in Fig. 3.7 and changes the results of the FTR auction clearing so that the number of FTRs awarded from node 2 to 3 increase from 140 MW to 250 MW while the number of FTRs awarded from node 1 to node 3 is reduced from 380 to 325. In this particular case the expansion of the feasible region did not change the FTR clearing prices only their awarded quantities. Thus the net gain in FTR auction revenue is $10 \times (250-140) + 10 \times (325-380) = \$550$/h which is exactly the amount paid by the auctioneer for the 55 MW short FGRs. In real time the underwriter of the short FGRs is liable for $55 \times SP_{23}$ which should cover any revenue shortfall resulting for the incremental FTRs awarded against the short FGR position. However, if the line 2–3 turns not to be congested $SP_{23}$ is zero and no revenue shortfall occurs so that the short FGR underwriter got to pocket the short position income.

Short FGR positions can be assumed by any entity that wishes to bet against certain lines being congested. However, such instruments are ideally suited for transmission owners (TOs) who are in a position to upgrade the line or maintain it so as to increase its real time rating. Thus, short flowgate positions provide incentives for incremental improvements and maintenance (e.g. vegetation control) that can enhance real time transmission capacity. If a line is not binding in real time then the TO retains the auction income for the short position taken. Similarly, short positions on long term flowgate rights can finance planned upgrades and investments that will alleviate congestion on the shorted flowgates while enabling the ISO to issue long term FTRs against such upgrades.

Like in every performance based incentive scheme, performance must be measured and verified against a credible and stable yardstick (e.g. PBR scheme for NGC in the UK). TOs should get assurances that they will not face a moving target and improvements they make will not change the nominal line rating used in subsequent FTR auctions. Furthermore, active participation by TOs in FTR trading must be regulated to insure correct incentives (e.g. long positions by TOs should not be allowed since they create incentives to restrict flow).

## 3.8 Conclusion

Just as point to point FTRs provide a convenient hedge against congestion charge risk for point to point energy transactions, FGRs are convenient instruments for managing flowgate capacity risk and reward investment in such capacity. When a revenue shortfall occurs allocating the losses based on the imbedded FGR content of various FTRs or directly to the TO of the affected flowgate, eliminates socialization that can cause inefficiencies and gaming. Conversely FGR short position that expand possible FTR awards provide a useful means for financing investment and reward performance that improves flowgate ratings. These positions also allow private parties to underwrite FTR revenue shortfalls due to flowgate capacity risk. Such activities, however, must be carefully regulated and monitored to avoid perverse incentives and abuses.

**Acknowledgement** This chapter is intended as a tutorial and review of previous work. Much of the text and most of the figures used are adopted from a joint conference paper with Kory Hedman, published online in the proceeding for the IREP 2010 symposium Oren and Hedman (2010). I also adopted material, especially the example in Fig. 3.1, that was developed by Fernando Alvarado as part of a tutorial we jointly presented on financial transmission rights in the year 2000. This work was supported by the National Science Foundation Grant IIP-0969016 and by the Power Systems Engineering Research Center.

## References

Chao H-P, Peck S (1996) A market mechanism for electric power transmission. J Regul Econ 10(1):25–59

Chao H-P, Peck S, Oren SS, Wilson RB (2000) Flow-based transmission rights and congestion management. Electr J 13(8):38–58

Hedman KW, Oren SS, O'Neill RP (2011) Optimal transmission switching: economic efficiency and market implications. J Regul Econ 40:111–140

Hogan WW (1992) Contract networks for electric power transmission. J Regul Econ 4:211–242

O'Neill RP, Helman U, Hobbs BF, Stewart WR, Rothkopf MH (2002) A joint energy and transmission rights auction: proposal and properties. IEEE Trans Power Syst 1(4):1058–1067

Oren SS, Hedman KW (2010) Revenue adequacy, shortfall allocation and transmission performance incentives in FTR/FGR markets. In: Proceedings of the IREP 2010 symposium, Bouzios, 1–6 Aug 2010

Ruff LE (2001) Flowgates, contingency-constrained dispatch, and transmission rights. Electr J 14(1):34–55

# Chapter 4
# A Joint Energy and Transmission Rights Auction on a Network with Nonlinear Constraints: Design, Pricing and Revenue Adequacy

**Richard P. O'Neill, Udi Helman, Benjamin F. Hobbs, Michael H. Rothkopf, and William R. Stewart**

## 4.1 Introduction

The forward and real-time (spot) auction markets operated by independent system operators (ISOs) allow for trade in multiple wholesale electricity products, differentiated by time and location on the transmission network.[1] This chapter

[1] In the United States, there are two types of independent system operators established under federal jurisdiction – Regional Transmission Organizations (RTOs) and Independent System Operators (ISOs). RTOs have additional geographical requirements compared to the original ISOs, such as encompassing a larger multi-state region, as well as some functional differences, such as regional transmission planning. However, wholesale market design is not differentiated between the two types of organizations. Since ISO is a more generic term, we will use this term to refer to both types of organization in the remainder of the chapter. In the U.S., ISOs and RTOs include the California ISO, ERCOT (encompassing most of Texas, and not subject to federal jurisdiction), PJM RTO, the Midwest ISO (MISO), New York ISO, ISO New England, and the Southwest Power Pool (SPP). For a survey of the designs of some of these markets in the United States, see O'Neill et al. (2006). Each of the U.S. ISOs and RTOs also has a website with extensive documentation of market rules and procedures as well as data on market outcomes. We refer to some of these below.

R.P. O'Neill (✉)
Federal Energy Regulatory Commission, Washington, DC 20426, USA
e-mail: richard.oneill@ferc.gov

U. Helman
BrightSource Energy, Oakland, CA 94612, USA

B.F. Hobbs
Johns Hopkins University, Baltimore, MD 21218, USA
e-mail: bhobbs@jhu.edu

M.H. Rothkopf
Pennsylvania State University, University Park, USA

W.R. Stewart
College of William and Mary, Williamsburg, VA 23187, USA
e-mail: william.stewart@business.wm.edu

J. Rosellón and T. Kristiansen (eds.), *Financial Transmission Rights*,
Lecture Notes in Energy 7, DOI 10.1007/978-1-4471-4787-9_4,

presents a general auction model that implements key features of the ISO markets, including definition of several market products, the rules for joint auctioning of the products in a sequence of forward and spot markets, the rules for financial settlement of those products, and the requirements to ensure revenue adequacy of the auctioneer. The model formulation is focused on a joint energy and transmission rights auction (JETRA; henceforth, the 'auction model' or 'auction'), along with a non-linear representation of the transmission network constraints. However, the formulation can be extended, in some cases with modification, to other market products. Our earlier paper (O'Neill et al. 2002) explored properties of this auction with linear transmission constraints.

At its inception, this auction model informed deliberations at the U.S. Federal Energy Regulatory Commission (FERC) in the early 2000s over a possible standard market design tariff for the wholesale power markets under its jurisdiction. A key objective at the time was to establish a framework for introducing a more complete set of financial transmission rights for the ISOs, including both point-to-point rights and "flowgate" rights, then considered to be mutually exclusive designs (see, e.g., Chao et al. 2000; Hogan 2000). Subsequently, political factors made it impossible for FERC to require implementation of a standardized wholesale market design.[2] Nevertheless, individual U.S. ISO market designs have since converged on certain products and pricing rules represented in our model formulation, such as point-to-point financial transmission rights and day-ahead and real-time markets with locational marginal pricing (LMP) of energy incorporating marginal congestion and loss charges. Other products discussed below have, however, not yet been introduced, such as forward locational energy sales integrated with the transmission rights auctions, and flowgate rights.

Despite this progress, the wholesale market design process has not been completed in the U.S., and there are almost continuous efforts at each ISO to introduce new products and pricing rules – some standardized across the ISOs, some not. This process advances market completeness by expanding the set of products and prices to a fuller range of the services provided by generation, non-generation,[3] and transmission assets, as is required for economic efficiency, especially under changing market and system conditions (such as integration of variable renewable generation). As some of these possible new market products, such as a reactive power product, require representation of non-linear transmission network constraints, whether for forward sales or real-time settlement purposes, our model continues to be applicable to the evolution of U.S. ISO market designs as well as regulatory reforms in other countries. At the same time, our illustrative extension to new products does not necessarily reflect an endorsement: as the history of market design in the U.S. has shown, for any specific ISO, the

[2] The standard market design tariff was proposed by FERC in 2002, but failed to achieve sufficient political support in certain regions to be implemented in its original form.

[3] "Non-generation resources" is the term adopted by FERC to refer to demand response, storage and other non-generation resources that may provide market services.

determination of the products for inclusion should be left to the market participants based on their needs and preferences as well as the physical characteristics of the regional power system (as well as being subject to approval by FERC or state regulators).

In deciding whether or not to adopt a more elaborate market design, such as that proposed in this chapter, the market operator, stakeholders, and regulators have to balance several criteria. Two of them are emphasized in this chapter, and motivate our design: efficiency in the allocation and trading off of various market products (in our case here, forward energy and transmission rights), given what bidders say they are willing to pay for them; and revenue adequacy for the market operator. Others that are relevant include: incentive compatibility (the extent to which an auction design encourages bidders to reveal their true valuations and costs in their bids); complexity and cost of implementation relative to anticipated benefits to the market; transparency; and perceived fairness (definable in several ways).

### 4.1.1 General Features of the Forward and Spot Market Designs

We refer to forward markets[4] as any ISO market that clears prior to the ISO's physical dispatch, or real-time, market. As a general matter, offers and bids that clear forward markets are financially but not physically binding,[5] whereas those that clear the real-time auctions are treated as physical commitments that typically must follow the system operator's instructions or be subject to warnings or financial penalties.

In practice, ISOs hold forward markets on a variety of time-frames that reflect operating requirements and constraints, market needs or simply utility/regulatory conventions. The basic market sequence is characterized in Table 4.1. The types of market products shown are not offered uniformly in all ISOs (for example, only one ISO provides pre-day-ahead forward reserves); we provide further detail on product definition in the next section, but focus in this section on a general description of the market sequence and the features reflected in our auction formulation.

The number and timing of forward markets in JETRA is a market design decision that needs to reflect the conditions that pertain in the market and stakeholder preferences. The minimal requirement of the ISO is that it run a real-time market; it is possible to provide all forward products through formal or informal markets operated by other parties. However, non-ISO operated markets that do not clear using

[4] We only consider ISO forward auction markets here, not any off-ISO bilateral power exchanges that can also operate in forward time-frames and in the same geographical territory. The existence of ISO auctions does not preclude operation of secondary non-ISO forward markets for transmission rights or bilateral energy transactions. In the U.S., ISO and non-ISO markets are generally regulated under a just and reasonable standard originating and under a fraud and abuse standard in the Federal Power Act.

[5] The exception to this rule is sales of forward capacity that create performance obligations in real-time.

**Table 4.1** Characterization of existing forward and spot U.S. ISO markets

| | Time-frame | Auction periodicity | Financial settlement interval | Types of market products |
|---|---|---|---|---|
| Spot markets (physical) | Real-time | Hourly | 5–60 min | Energy (physical only) |
| Forward markets (financial) | Hour-ahead | Hourly | 1 h | Energy, operating reserves |
| | Day-ahead | Daily (24 h) | 1 h for energy<br>Daily for 'make whole' payments | Energy, operating reserves, residual capacity |
| | Pre-day-ahead | Semi-annually or annually | Months, possibly differentiated by time of day | Operating reserves, financial transmission rights, capacity |

a good representation of the network and the full dispatch run a significant risk of infeasible trades. This has happened in the CalPX in the early days of the California market and several European exchanges. For revenue adequacy, FTRs require assumptions about the network configuration, but if non-ISO markets only trade simple flowgates, then they can avoid the need to make such assumptions. However, the downside of only selling simple flowgates is that the rights holder is not guaranteed a perfect hedge for a bilateral power contract between two points. A distinct advantage of a central forward market operated by the ISO is that it is in the best position to incorporate network constraints along with the rest of the generation, load, and net imports. The other advantage is that the ISO can back FTR payments with congestion revenues, which an independent party cannot.

**Pre-day-ahead markets.** The pre-day-ahead ISO markets have conventionally been used to transact products denominated in time-periods of months or multiple months, such as financial transmission rights and capacity. Some ISOs have used such markets to procure forward operating reserves. In the auction design we propose in Sect. 4.3, the mathematical formulation explicitly represents only energy and financial transmission rights for pre-day-ahead auctions. A key generalization of the model has been to accommodate the joint auction of products that were previously advocated as mutually exclusive market designs, intended to support different visions of how forward market institutions should develop. Specifically, the auction model in (O'Neill et al. 2002) – and the analogous one presented here for the nonlinear case – synthesize and extend several prior auction models to allow for the simultaneous auction of flowgate, or flow-based, transmission rights and point-to-point transmission rights specified as options or obligations (Chao and Peck 1996; Harvey et al. 1997; Hogan 2000, 2002), in addition to real energy and possibly other products.[6]

[6] The debate over the implementation of alternative transmission rights formulations is recounted in Hogan (2000, 2002) and O'Neill et al. (2002), among other sources, and will not be repeated here.

Pre-day-ahead energy transactions, whether in separate energy-only auctions[7] or in joint auctions with financial transmission rights, have not been introduced explicitly by any of the U.S. ISOs. However, conceptually, energy commitments could enter the pre-day-ahead auctions for transmission rights where the energy may be needed to make up losses on the transmission network, or to supply inertia or reactive power that are not explicitly modelled to create additional transmission capacity into transmission constrained areas. An example of the latter is the "San Francisco nomogram" constraint discussed in (O'Neill et al. 2002). Our introduction of energy into the forward transmission rights auction is a generalization of these applications.

**Day-ahead and real-time markets.** In day-ahead and real-time auction markets, products include real energy priced and settled using LMPs, and regulation and operating reserves settled at system-wide or zonal prices. With respect to day-ahead energy markets, ISOs typically use a two-phase day-ahead market clearing, in which first both physical[8] and financial (or virtual) bids are accepted and day-ahead prices determined, and second, a reliability unit commitment is conducted using bids associated with physical generation only and forecast load. Financial demand and supply bids[9] have some unique properties in that they are not associated with physical energy supply or demand, or physical transmission capacity. They can be used for financial hedging and are permitted, in part, in the forward markets to counter market power and to aid in producing better price convergence (on average) between and among the forward markets and the real-time market.

Simultaneously, transmission users are charged for marginal transmission costs (congestion and possibly losses) and congestion revenues are used to settle the financial transmission rights awarded in the pre-day-ahead auctions. Generally, these settlements take place using day-ahead market LMPs, unless the ISO only operates a real-time market, in which case they are settled against real-time LMPs. Point-to-point transmission rights are settled based on the differences in the LMP congestion components between their injection and withdrawal points, while if the ISO offered them, the flowgate rights would be settled using transmission shadow prices (called flowgate marginal prices in FERC 2002).[10] Settlement rules are defined precisely in the next section.

---

[7] For example, some ISOs have evaluated additional energy auctions prior to the day-ahead auction, but not integrated with other products.

[8] That is, bids backed by physical assets. Selection in the day-ahead auction market does not require that the seller of the physical asset deliver in real-time; the seller still has the option to not perform and sell or buy back its position in real-time. The incentive to perform is thus primarily financial. In contrast, in real-time, failure to perform as instructed may result in administrative penalties.

[9] In this chapter we will use the term 'bid' at times to include either a bid or offer.

[10] While ISOs do not offer flowgate rights through auctions, there are a number of applications of flow-based capacity reservations that are used by the ISOs and affect energy prices in real-time. For example, currently, ISOs exchange flowgate capacity with their neighbors through Joint Operating Agreements to feasibly and optimally allocate loop flow.

Finally, the real-time markets begin at midnight of the operating day, clear every 5–10 min, and settle every 5–60 min.[11] In these markets, only physical bids are allowed, subject to performance requirements, and all financial positions are re-settled. Forward markets close in time to real-time, such as markets held one or more hours before the operating hour, are more "physical" in nature than financial, although the ISO has less time to recover from failure to perform than in the day-ahead market, where it has time to conduct reliability commitments and procure additional reserves.

In all these markets, various additional rules have been established to prevent market power and market manipulation by entities that also hold other property rights (including physical transmission scheduling), and appropriate creditworthiness rules are required for all cleared bids.

The actual timing of the sequence of ISO market clearing for the various market products is due to a mix of factors, including scheduling conventions inherited from predecessor utilities, regional system operators (e.g., power pools) and reliability organizations, market design decisions and computational constraints at the ISOs, and the interests of the market participants as new market designs were developed. Unfortunately, the timing of the sequence has tended to differ among ISOs, including contiguous ones, resulting in "seams" issues, some of which have been resolved over time through improved coordination (see, e.g., O'Neill et al. 2006).

### 4.1.2 *Auctions with Non-linear Transmission Network Constraints*

To formalize and generalize the design of these forward and real-time markets, the authors first introduced a multi-settlement, joint energy and transmission rights auction on a network characterized by an approximate linearized 'dc' load flow model (O'Neill et al. 2002, 2003) (for a derivation of the dc load flow approximation, see, e.g., Schweppe et al. 1988). In order to simplify auction clearing and financial settlements, linear network constraints are used in all U.S. ISO markets. For example, forward auctions for obligation and option point-to-point Financial Transmission Rights (FTRs) in PJM employ a dc load flow model.[12] As noted, some market operators create additional linear 'nomogram' constraints or 'cuts', often proxies for voltage limits, to ensure feasibility of the underlying physical system. According to our communications with software developers, the more general linear model in O'Neill et al. (2002) has been a basis of the development of the recently implemented transmission rights markets for the ISOs in ERCOT (Texas) and California in the U.S..

---

[11] That is, some ISOs financially settle on a 5–10 min basis, while others settle on the basis of an hourly integrated price.

[12] www.pjm.com/markets/ftr

But other ISOs have implemented auctions with non-linear transmission constraints. In the New York ISO, the obligation point-to-point financial transmission rights are called Transmission Congestion Contracts (TCCs). In contrast to PJM, the auction is conducted using an approximate AC optimal power flow model that respects thermal, voltage and stability constraints within the New York control area.[13] There are other market products that would require consideration of non-linear constraints. The inclusion of reactive power in the auction market would also require the AC load flow model (FERC 2005; Hogan 1993; Kahn and Baldick 1994; O'Neill et al. 2008) or a linear or quadratic approximation to the model. Moreover, proposals for forward hedging of marginal losses through unbalanced point-to-point transmission rights would require auctions with a dc load flow model and quadratic losses (Harvey and Hogan 2002). This chapter thus generalizes the linear auction model in O'Neill et al. (2002) to the case with nonlinear constraints.

Whatever the final set of products, a key goal of the market design is to ensure the revenue adequacy of the auctions, which means that the ISO collects sufficient revenues to cover payment obligations. A theoretical result presented here is that for the auction with nonlinear transmission constraints that define a convex feasible region, the forward and spot auction sequence can be revenue adequate (the analogous proof for the linear case is shown by O'Neill et al. 2002). However, as with any transmission rights auction, additional rules are needed to account for revenue inadequacy due to changes in system topology. While we show the formal conditions for revenue adequacy, we do not explore in detail how market participants are affected financially when there is a shortfall. There are currently different rules for dealing with shortfalls. For example, in PJM, revenue inadequacy of FTRs is addressed by prorating the shortfall among the FTR holders. In NYISO, revenue inadequacy of TCC holders is covered by the transmission owners to provide incentives for efficient timing of transmission maintenance.

### *4.1.3 Additional Extensions of the General Auction Model*

In each step of the sequence of auctions, our general model framework can be extended to include additional products,[14] pricing rules, settlements, or linkages with auctions for other wholesale market products. Some of these extensions are discussed in the subsequent sections, but we summarize several others here.

For example, some ISOs have established forward capacity (MW) auction markets to satisfy annual or multi-year local area and system-wide planning reserve margins (or resource adequacy requirements). These forward markets pay a locational clearing price for capacity, which in some designs is set by an administrative

[13] www.nyiso.com

[14] Including those, such as generator start-up, that requires mixed integer programming formulations, as discussed in Sect. 4.4.

demand curve. The network models are also zonal rather than nodal, another difference with our model formulation as presented here. One linkage between the capacity auctions and the model of this chapter is that, as a general rule, offers that clear the capacity auction then have an offer obligation in the ISO day-ahead markets, making the capacity payment equivalent to the ISO buying a call option (on behalf of the load-serving entities that have the capacity obligation) on energy that pays the LMP when exercised. Hence, the model presented here can be viewed as a framework for final settlement of the energy call option associated with capacity rights.

Closer to actual operations, the sequential auction market design can be implemented with additional settlements between day-ahead and real-time energy markets in order to better accommodate variable energy generation by renewable sources, such as wind and solar generation, whose production forecast uncertainty decreases as the real-time market approaches. A sequence of auction markets, for example, occurring every six hours with rolling horizons, might allow for more efficient adjustments as the uncertainty decreases.

The remainder of the chapter is organized as follows. Section 4.2 offers a description of the types of energy and transmission right bids in the auction. Section 4.3 presents the mathematical statement of the auction model with nonlinear transmission constraints, and provides more mathematical detail on how transmission rights are specified for the auction. Section 4.4 discusses the settlement system and conditions for maintaining revenue adequacy. Section 4.5 provides an example based on a dc load flow with quadratic losses. Section 4.6 offers conclusions. An appendix presents the proof of revenue adequacy for a sequence of forward and real-time market auctions with 'expanding' transmission constraints that define a convex feasible region.

## 4.2 Auction Products

We now turn to the set of energy and transmission rights products modeled in the auction design, a subset of those discussed above. The types of electricity products that can be traded in the auction mechanism proposed in this chapter have been described by Baldick et al. (2005), Chao and Peck (1996), Chao et al. (2000), Harvey et al. (1997), and O'Neill et al. (2002, 2003, 2006). This section provides further qualitative description of these products, while the next section introduces our model's notation.

**Energy.** Several types of bids are typically allowed in energy and transmission auctions: supply offers, demand bids, financial bids, and transmission bids. Point-to-point transmission bids represent what a bilateral energy transaction is willing to pay for marginal congestion charges (and possibly losses) associated with its transmission schedule. If both the points are inside the ISO, the product is financial. Physical point-to-point bids are typically used on the boundaries of ISO systems where there is no fully arbitraged LMP on the "other side" of the boundary, which is often called a proxy bus or interface. The model presented here can accommodate each of these types of bids. For purposes of this discussion, some important aspects of energy auctions are not considered, such as the inclusion of unit commitment start-up and no-load costs, restrictions on bids to control the exercise of market power[15] and changes in network topology.[16]

Currently, energy offers (to sell) and bids (to buy) have only been allowed in the day-ahead and real-time markets. In a pre-day-ahead ISO auction market for energy and transmission, as discussed above, energy transactions could be used also to balance point-to-point transmission rights in a lossy system, or to increase transmission capacity for forward sale. These one-sided or unbalanced "rights" (actually, obligations) can be called "nodal revenue rights."

**Simple Transmission Capacity Rights and Portfolio Combinations.** As noted in the flowgate or flow-based rights literature (e.g., Chao and Peck 1996; Chao et al. 2000), there are two types of elementary transmission rights, which we call here the "simple rent collection right" and the "simple rent payment right." The simple rights are defined over single transmission elements, which include lines, transformers, other transmission elements or collections of transmission elements whose capacity is limited by exogenous thermal, stability, or contingency considerations. Such rights are often generically called "flowgate" rights (FERC 2002). For each element, the direction of the flows covered by the simple rights is defined separately and arbitrarily, in either a positive or negative direction. The simple rent collection right on a transmission element confers to the buyer the right to collect the rents that would occur when that element is congested, for the capacity specified in the right. Because the flow-based right is directional, the holder of a rent collection right only collects non-negative rents.

The simple rent payment right obliges the seller to pay any rents on a transmission element, for the capacity specified in the right. The rent payment right allows a market participant to create or consume financial capacity on a specific transmission element. Moreover, if the ISO did not itself allocate rights, but simply facilitated an auction of buyers and sellers (see Sect. 4.3), then all transmission owners could offer physical transmission rights. The simple rights can be aggregated into more complex rights through linear combinations or portfolios, for example, covering several transmission lines, nomograms, or constructing "point-to-point" rights on the basis of power flow distribution factors (O'Neill et al. 2002).

The combination of buying a rent collection rights on some transmission element and selling rent payment rights on other transmission element creates portfolio of

[15] Bid restrictions for market power reasons can include a uniform, "safety net" bid cap for all generators, bid thresholds on generators that trigger market power mitigation, a requirement to bid approximate marginal costs, and other measures.

[16] Network topology changes can be either purposeful, to increase market surplus, or due to planned outages, such as maintenance, or to unplanned outages. Topology changes to increase market surplus, called optimal transmission switching, can ironically cause revenue inadequacy in the point-to-point transmission rights settlements. Corrective switching to stabilize or re-optimize the system can follow unplanned outages.

flowgate rights. For a set of simple rights that constructs a point-to-point right, holding this portfolio on each transmission element in the set is analogous in the linear dc JETRA model to the point-to-point obligation rights with a constant topology. In general, however, the individual rights and the portfolios are more likely to offer an imperfect rather than a perfect hedge against congestion charges associated with an energy transaction. Since an exact match between a particular point-to-point transaction and a portfolio of the rights would be difficult to create and maintain (although some authors propose that the ISO provide subsidies to maintain particular portfolios as complete hedges, for example, Chao et al. 2000). A transmission right that offers a perfect, or complete, congestion hedge is defined as one in which the congestion charges associated with real-time market transactions are equal to the congestion revenues obtained by the rights holder. An imperfect hedge is one in which the congestion charges are not equal to the revenues to transmission rights holders. For many holders, then, the flowgate right will be used to collect rents on heavily congested transmission elements rather than to hedge any particular power transaction.

Flowgate rights can be made available or withdrawn in the real-time market due to forced outages, the use of short-term ratings instead of steady-state ratings or unanticipated changes in weather. For example, changes in ambient temperature and wind speed can change the transmission line's carrying capacity.

**Point-to-Point Transmission Rights.** There are two types of point-to-point rights, the obligation right and the option right. An obligation right is more accurately described as a "contract" (Harvey et al. 1997), since it embodies an obligation to pay congestion revenues, but is now conventionally termed a financial transmission right. A point-to-point obligation transmission right is defined as the right to receive a payment or the obligation to pay the congestion charge rents that result from the physical flows associated with putting power into the system at a point of injection (POI) and taking power out of the system at a point of withdrawal (POW) (Harvey et al. 1997). Note that for a point-to-point obligation, flow in one direction adds an equivalent amount of "counterflow capacity" in the other direction. This can be generalized to multiple point-to-multiple point rights, which we will call network rights. These rights may simply aggregate point-to-point rights or may be "contingent" rights, when they hedge multiple possible POIs and POWs (discussed in O'Neill et al. 2002). The point-to-point obligation transmission right is equivalent to the forward transmission congestion contracts (TCCs) described in Harvey et al. (1997). The network rights were described in FERC's proposed capacity reservation tariff (FERC 1996).

The amount that is received (or paid, if negative) by the holder of the obligation right is the nodal price at the POW minus the nodal price at the POI multiplied by the quantity specified in the right. (A variant implemented at some ISOs pays only the difference in the congestion portion of the LMP price and not the loss component.) If the injections and withdrawals of power specified in the right are scheduled in the market in which the right is settled (and then executed in the real-time market, if different from the settlement market), then the right provides a complete congestion hedge, i.e., no additional payment for congestion will be necessary.

The point-to-point option transmission right is defined as the option to put power into the system at one or more POIs and take power out of the system at one or more POWs. The option TCCs discussed by Harvey et al. (1997) are similar to these point-to-point option rights in the linearized dc load flow model (O'Neill et al. 2002). It can be interpreted as the right to collect congestion rents if they exceed zero, without the obligation to pay that amount if negative. (I.e., they are options with a strike price of zero.) This option faces considerable computational challenges in an auction model with nonlinear transmission constraints, in that a separate load flow has to be calculated for each combination of possible exercised options (Hogan 2002). However, using a linearized dc load flow approximation model, the computation can be reduced sufficiently, thus facilitating the implementation of point-to-point options (alternatively, portfolios of flowgate rights could be used to approximate a point-to-point option right). In an auction with linear constraints, the point-to-point option is shown to be equivalent to setting aside capacity in each transmission constraint for positive increments of flow associated with the right but ignoring negative flows ("counterflow") in the opposite direction (e.g., O'Neill et al. 2002). This allows the auction to be run using a single set of power flow distribution factors (PTDFs), but no analogous reduction has been developed for the nonlinear case. Moreover, as we showed previously (O'Neill et al. 2002), the reduction in the linear case implies that an appropriately defined bundle of flowgate rights dominates the point-to-point option in the sense that there exists such a bundle whose cost is the same as the option right but which will pay off at least as much as an option right and, under some possible outcomes, it will pay strictly more. Although a point-to-point option has been included in some ISO markets, it has been excluded in others for various reasons. These include the fact that such rights would excessively diminish the available rights in locations where there are physical set-asides to honor prior physical transmission scheduling rights; a lack of stakeholder interest in such options as a hedging instrument; and to the software development costs and computational requirements of its implementation.

Point-to-point rights can be balanced or not balanced. A balanced right is one in which the quantity injected is equal to the quantity withdrawn. An unbalanced right does not have this requirement, so that an entity can approximate losses (average or marginal) by specifying a higher quantity injected than withdrawn.

Finally, as with the flowgate right, point-to-point rights can be bought from or sold into the auction.

## 4.3 The Auction with Nonlinear Constraints

### *4.3.1 Mathematical Statement*

The types of energy bids and transmission rights described in Sect. 4.2 are represented in the mathematical statement of the auction model with non-linear constraints, JETRA-NL, below with more detail in Sect. 4.3.2. For ease of recognition, the notation used in the model borrows and extends from standard references, such as Chao and Peck (1996) and Harvey et al. (1997). All variables are assumed to be real power; however, the framework allows for the inclusion of reactive power (VARs). Units of the decision variables and right hand sides (RHS) of the constraints are in megawatts (MW or MWh/hour), while the objective function coefficients are in \$/MWh.

The JETRA-NL model is formally stated below. In brief, the formulation maximizes the net economic value (4.1) of accepted energy and transmission bids subject to definition of the net injection at each bus (4.2), inequality constraints upon injections and flows (4.3, 4.4, and 4.5) (whose capacity can be sold as rights), load flow constraints (4.6), upper bounds on transmission and energy rights (4.7, 4.8, 4.9, and 4.10), and nonnegativity restrictions.

$$\textit{JETRA-NL: max } v(t^F, t^P, g, x, y, f^+, f^-) = b^F t^F + b^P t^P + c^+ g^+ + c^- g^- \quad (4.1)$$

$$A^P t^P + A^+ g^+ + A^- g^- - y = 0 \qquad (\pi) \qquad (4.2)$$

$$B^N t^F + K'(\mathrm{x}, y, f) \le F^N \qquad (\mu^N) \qquad (4.3)$$

$$B^+ t^F + f^+ \le F^+ \qquad (\mu^+) \qquad (4.4)$$

$$B^- t^F + f^- \le F^- \qquad (\mu^-) \qquad (4.5)$$

$$K''(x, y) - f^+ + f^- = 0 \qquad (\gamma) \qquad (4.6)$$

$$t^F \le T^F \qquad (\psi^F) \qquad (4.7)$$

$$t^P \le T^P \qquad (\psi^P) \qquad (4.8)$$

$$g^+ \le G^+ \qquad (\rho^+) \qquad (4.9)$$

$$g^- \le G^- \qquad (\rho^-) \qquad (4.10)$$

$$t^F, t^P, g^+, g^-, f^+, f^- \ge 0$$

To avoid unnecessary notation, the bids are shown as having a lower bound of zero; more generally, quantity bids could have nonzero lower bounds. This generalization is a simple transformation in the linear parts of models. We assume a feasible solution exists; for instance, zero for all decision variables will be feasible if $K'(0,0,0) = 0$. The notation is defined as follows:

#### 4.3.1.1 Index Sets

$I$ is the set of nodes, $i = 1, \ldots, n^I$, in the system. $F$ is the set of transmission (or flowgate) bids to buy or sell rights on individual transmission elements (e.g., a line, capacitor, transformer, or other transmission equipment) or a set of transmission elements, and is indexed by $k = 1, \ldots, n^F$. $P$ is the set of transmission bids to buy or sell point-to-point rights, with index $k = 1, \ldots, n^P$. $M^+$ is the set of bids to sell (inject) energy, indexed by $m = 1, \ldots, n^{M+}$. $M^-$ is the set of bids to buy (withdraw) energy, with index $m = 1, \ldots, n^{M-}$ . $H$ is the set of transmission elements in the system on which rights are purchased and sold, and the associated constraints are (4.4) and (4.5). It uses index $h = 1, \ldots, n^H$, where $n^H$ defines the cardinality of $H$. $H'$ is the set of additional interaction constraints that result from analysis of voltage, angle, stability, and contingency constraints, sometimes called nomogram or cut set constraints. On these constraints, rights can be purchased and sold. These constraints are indexed by $h = 1, \ldots, n^{H'}$, where $n^{H'}$ defines the cardinality of $H'$. The set $H'$ is associated with the mapping $K'$ in (4.3).

#### 4.3.1.2 Variables

$f = f^+ - f^-$ is a vector-valued variable describing flows on the transmission elements. $f^+{}_h$ and $f^-{}_h$, $h \in H'$, representing the flow induced by $x$ and $y$ on transmission element $h$ in the positive and negative direction respectively (defined arbitrarily).

$g = g^+ - g^-$ is a vector, where $g^+{}_m$, $m \in M^+$ represents the quantity of energy sold by the *mth* energy bid and $g^-{}_m$, $m \in M^-$ represents the quantity of energy purchased by the *mth* energy bid

$t^F$, $\{t^F{}_k,\ k \in F\}$, and $t^P$, $\{t^P{}_k,\ k \in P\}$, are vectors where $t^F{}_k$ represents the quantity of rights awarded to (bought by or sold to) the *kth* bid for flowgate ($F$) transmission type rights and $t^P{}_k$ represents the quantity of rights awarded to the *kth* bid for point-to-point ($P$) transmission type rights.

$x$ is the set of variables that affect the topology and performance of the network, e.g., phase shifter settings, dc line settings, reactive power compensation and contingency set-asides on transmission elements for locational reserves. In today's practice, these variables are typically determined either exogenously or as a part of an iterative procedure, but the auction can accommodate bidding for these settings in the auction; see, e.g., O'Neill et al. (2002).

$y$ is a vector, $\{y_i, i \in I\}$, where $y_i > 0$ is the amount of real power injected at node $i$, and $y_i < 0$ is the amount withdrawn at node $i$ that is induced by the $t^P$, $g^+$ and $g^-$ bids.

$\pi$, $\mu^N$, $\mu^+$, $\mu^-$, $\gamma$, $\psi^F$, $\psi^P$, $\rho^+$, $\rho^-$ are vectors of Lagrange multipliers associated with sets of primal constraints in the auction.

#### 4.3.1.3 Parameters and Functions

$b^F$, $\{b^F{}_k,\ k \in F\}$, and $b^P$, $\{b^P{}_k,\ k \in P\}$, are vectors. $b^F{}_k$, $k \in F$ and $b^P{}_k$, $k \in P$ represents the \$/MWh value that the bidder associates with a transmission bid. Bids to buy are positive and bids to sell are negative.

$F^+, \{F^+{}_h,\ h \in H'\}, F^-, \{F^-{}_h, h \in H'\}$, and $F^N, \{F^N{}_h,\ h \in H''\}$, are transmission capacity constraints including thermal, stability or contingency limits associated with one or more transmission elements (e.g., several transmission elements grouped as a flowgate). Each individual constraint in the third category of capacity constraints (condition (4.3)) involve two or more flows simultaneously and so we refer them to interaction constraints. In practice, they are often called nomogram constraints.

$B^+$, $B^-$ are matrices, $\{B^+{}_{hk},\ h \in H',\ k \in F\}$, $\{B^-{}_{hk},\ h \in H',\ k \in P\}$, where $B^+{}_{hk}$ represents the quantity in the positive direction on transmission element $h$ that is requested in bid $k$ and $B^-{}_{hk}$ represents the quantity in the negative direction on transmission element $h$ that is requested in bid $k$.

$B^N$ is a matrix, $\{B^N{}_{hk}, h \in H'', k \in F\}$, where $B^N{}_{hk}$ defines the quantity of the *hth* transmission network interaction constraint that the *kth* bid for a $F$ right requires. An 'interaction' constraint is any constraint that is not simply a lower or upper bound on some variables (especially flows) or otherwise associated with a single transmission element. Examples include voltage and stability constraints. The set of network constraints $H''$ includes these constraints.

$c^+$, $\{c^+{}_m,\ m \in M\}$, and $c^-$, $\{c^-{}_m, m \in M^-\}$ are vectors where $c^+{}_m < 0$ represents the unit \$/MWh value to sell energy bid $m$ and $c^-{}_m > 0$ represents the unit value to buy energy bid $m$.

$A^+$, $\{a^+{}_{im},\ i \in I,\ m \in M\}$, and $A^-$, $\{a^-{}_{im},\ i \in I,\ m \in M^-\}$, is a matrix where $a^+{}_{im} = 1$, if there is an injection of energy at node $i$ associated with energy bid $m$; $a^-{}_{im} = -1$, if there a withdrawal at node $i$ associated with energy bid $m$; and zero otherwise for simple trades. The formulation also permits energy portfolio bids where the matrix entries are not restricted to 0, 1 or $-1$.

$K'(x, y, f)$ is the mapping that defines additional inequality constraints upon flows resulting from off-line studies of contingencies, stability, voltage and angle constraints.

$K''(x, y)$ is the mapping from $x$ and $y$ to flows $f$. These are the basic load flow constraints, expressing flows as a function of injections. Consequently, $\partial K''(x,y)/\partial y$ can be viewed as a matrix of the power transfer distribution factors (PTDFs).

The set of optimal bids accepted by the auction is denoted as $\{t^{F*},\ t^{P*},\ g^{+*}, g^{-*}\}$ and the set of Lagrange multipliers that satisfy the Karush-Kuhn-Tucker (KKT) conditions for the auction is denoted $\{\pi^*,\ \mu^{N*},\ \mu^{+*},\ \mu^{-*},\ \gamma^*,\ \psi^{F*},\ \psi^{P*},\ \rho^{+*},\ \rho^{-*}\}$. If there are no losses, then the congestion rents (i.e., opportunity costs) resulting from flows are $\mu^N F^N + \mu^+ F^+ + \mu^- F^-$.

Constraint (4.2) includes the net injections from the energy part of the auction along with net injections implied by the point(s)-to-point(s) transmission auction; their sum yields the overall net injections at each node, $y$. Constraints (4.4 and 4.5) require that the flowgate $F$ rights plus flows induced by $y$ and $x$ are subject to the bounds on each transmission element. Constraints (4.3) further require that the $F$ rights and flows induced by $y$ and $x$ are subject to the interaction constraints on the system (i.e., represent a feasible physical dispatch with respect to those constraints). Constraint (4.6) calculates the flows induced by $x$ and $y$. For instance, (4.2) and (4.6) together could be based on the linearized dc load flow analogues of Kirchhoff's current and voltage laws, respectively (as in the example at the end of this chapter). Constraints (4.7, 4.8, 4.9 and 4.10) enforce the upper bounds on each type of bid.

In general, the underlying physical constraints of a reliable AC system yield a nonconvex set. Let it be called $C$. Let $\underline{C}$ be the set that satisfies (4.2, 4.3, 4.4, 4.5, 4.6, 4.7, 4.8, 4.9 and 4.10). $\underline{C}$ is often represented by an energy management system combined with judgment of experienced operators, various approximations and the results of contingency analyses. The set $C$ includes relationships between power, reactive power, Kirchhoff's law, losses, voltage, phase angle regulators, dc lines and all specified contingencies. These constraints ensure the reliability/feasibility of the implied dispatch. Here we assume $\underline{C} \subset C$, that is, JETRA-NL is a restriction of the AC problem. In general, a full AC model would include a doubling of the size of $y$ to include reactive power. More generally, we could define $g^{+}{}_{m} \in G^{+}{}_{m}$, $g^{-}{}_{m} \in G^{-}{}_{m}$ could define additional constraints on generators and load such as ramp rate constraints or total energy limits over a series of hours (e.g., hydro energy constraints).

Several further generalizations are worth mentioning. First, the model could allow "all or nothing" or binary bids for rights. This can be accomplished by adding integer variables and replacing the upper bound constraints such as the following for $g_m$: If transmission switching was considered, it would also affect $K''$ (due to KVL); this complication is not considered in this chapter.

$$g_m - G_m z_m \leq 0$$

where $z_m$ are 0/1 variables. Lower bounds could be similarly specified as follows:

$$g_m - \underline{G}_m z_m \geq 0$$

where the underlining denotes a lower bound.

Furthermore, the introduction of integer variables allows for unit commitment (i.e., dynamic optimization) of generation (e.g., Hobbs et al. 2001) and transmission switching (FERC 2005; O'Neill et al. 2005a), as well as for consideration by longer-term auction markets of entry by technologies with investment costs, as is characteristic of generation and transmission projects. Elsewhere, we have shown that efficient market-clearing prices in auction markets with non-convexities in technology and production exist using a two-part pricing scheme in which the integral activity (e.g., start-up) is offered a specific ("non-anonymous" or discriminatory) price while the associated commodity (e.g., energy) is cleared through a single or uniform market clearing ("anonymous" or non-discriminatory) price

(Elmaghraby et al. 2004; O'Neill et al. 2005b). Most ISOs have adopted such a two-part pricing regime (often called a revenue sufficiency or bid-cost recovery guarantee) for generator offers accepted in the day-ahead market and real-time market. The omission of these binary variables yields suboptimal solutions with lower market surplus and possibly an infeasible dispatch, but their inclusion threatens revenue adequacy and may induce changes in the settlement rules.

Finally, to this point, we have assumed that the ISO is defining and selling transmission rights. An initial allocation of rights can be done through an auction or by other methods. For example, in most U.S. ISO markets, the ISO first allocates transmission rights or the rights to a portion of transmission auction revenues. Next, the ISO conducts the transmission auction as if it owns the transmission rights under its control, but then returns auction revenues to transmission holders. In this approach, the capacity held by the ISO, $F^+$ and $F^-$, is the unallocated capacity. If $F^+ = F^- = 0$, the ISO offers no transmission rights and trading takes place among the rights holders.

### 4.3.2 *Specifying the Bids for Energy and Transmission Rights*

Because in some cases our notification diverges from familiar notation from prior transmission rights models (e.g., Chao and Peck 1996; Harvey et al. 1997), this section elaborates on the product definitions and characteristics introduced in Sect. 4.2, reviewing the mathematical formulation of the products as required by the auction model.

**Energy.** An simple energy bid (real or financial) to sell is defined by scalars, $G^+{}_m$ and $c^+{}_m$, and the vector $a^+{}_m$; $c^+{}_m$ (usually $c^+{}_m < 0$) is the cost (e.g., in \$ per MWh) for a step $m$, and $G^+{}_m$ (e.g., in MWh) is the maximum quantity for sale in step $m$ ($g^+{}_m \leq G^+{}_m$). Adding the locational aspect, $a^+{}_m$ is a vector of 0 s and a single $a^+{}_{im} = 1$ defining the injection node $i$. Symmetrically, an energy bid (real or financial) to buy is defined by scalars, $G^-{}_m$, and $c^-{}_m$, and the vector $a^-{}_m$; $c^-{}_m$ specify the bid value (e.g., in \$ per MWh) for step $m$ up to $G^-{}_m$, the maximum quantity for the step ($g^-{}_m \leq G^-{}_m$). Adding the locational aspect, $a^-{}_m$ is a vector of 0 s and a single $a^-{}_{im} = -1$ defining the withdrawal node $i$. For example, to define a simple bid to sell one unit of energy at node 6 in a network, $\mathrm{a^+{}_{6m}} = 1$ and $\mathrm{a^+{}_{im}} = 0$ for $i \neq 6$. If $\mathrm{a^-{}_{6m}} = -1$, then it would be a bid to buy one unit of energy at node 6. An individual bid can be part of a step-wise function with each step a separate value of the index $m$.

**Simple Transmission Capacity Rights and Portfolio Combinations.** A bid for a transmission right of either the flowgate ($F$) or the point-to-point ($P$) type is defined by $b$ and $T$. What differentiates the bids for $F$ and $P$ rights is that flowgate rights are directly associated with a transmission element and/or combination of transmission elements while point-to-point rights are associated with injections and withdrawals independent of the topology.

To sell the simple rent payment transmission right, $b^F{}_k < 0$ is the lowest amount a bidder is willing to accept to sell up to $T^F{}_k$ units. A bid, $k \in F$, for this right on transmission element $j$ in the positive direction is defined by inserting $B^+{}_{hk} = -1$ in the flow constraint (4.4) for $h = j$ and 0 for $h \neq j$. Similarly, a bid on transmission element $j$ in the negative direction is defined by $B^-{}_{hk} = -1$ in flow constraint (4.5) for $h = j$ and 0 for $h \neq j$.

To buy the simple rent collection transmission right, $b^F{}_k > 0$ is interpreted as the highest amount a bidder is willing to pay to buy up to $T^F{}_k$ units. A bid, $k \in F$, for this right on transmission element $j$ in the positive direction is defined as $B^+{}_{hk} = 1$ for $h = j$ in (4.4) and 0 for $h \neq j$. Similarly, a bid on transmission element $j$ in the negative direction is defined by $B^-{}_{hk} = 1$ in (4.5) for $h = j$ and 0 for $h \neq j$.

Those parameters (extending notation introduced by Chao and Peck 1996) indicate how much capacity on transmission element $h$ is taken up by a unit of this type of right. In fact, a portfolio of flowgates $k$ is defined by, $B^+{}_{hk}$, $B^-{}_{hk}$, $B^N{}_{hk}$, the proportions of each flowgate in the portfolio.

**Point-to-Point Transmission Rights.** As noted, the point-to-point transmission bids, $l \in P$, are defined over one or more POIs and one or more POWs at the $n^I$ nodes in the system (more than one POI or POW defines a so-called network right). In the auction, the bidder would further have to specify whether the right is desired as an option or obligation; if options are allowed, this would result in different and more complicated computations (Hogan 2002). For the buyer of the $P$ right, $b^P{}_l$ (usually $b^P{}_l > 0$) represents the highest amount bidder $l$ is willing to pay to buy up to $T^P{}_l$ units. For sellers of the rights, $b^P{}_l$ (usually $b^P{}_l < 0$) is the lowest amount a bidder $l$ is willing to accept to sell up to $T^P{}_l$ units. $A^P{}_l$ is a vector of net injection coefficients defining the net injection at each node $i$ in each $l \in P$, with elements $a^P{}_{il}$. For a POI (conversely, POW), $a^P{}_{il} > 0$ (conversely, $a^P{}_{il} < 0$). Hence, for balanced rights in a lossless transmission system, $\sum_i a^P{}_{il} = 0$.

The portfolio of flowgate rights can be constructed that provides the same payoffs as a specified set of point-to-point rights if the topology is known and unchanging. However, if the network topology changes, then, in general, the flow patterns associated with a given point-to-point right will change. Generally, the point-to-point rights are independent of the topology, but flowgate rights depend specifically on the topology.

## 4.4 Forward and Dispatch Markets: Financial Settlement and Revenue Adequacy

ISO auction markets operate in a sequence of forward and real-time market auctions, with products such as transmission rights and generation capacity being traded pre-day-ahead, while energy and bid-based ancillary services are typically traded day-ahead and in real-time. As noted above, the exact timing and content of these product auctions are a matter of market design based on the history and characteristics of specific ISO markets. This section provides the general mathematical procedure for financial settlement and its link to revenue adequacy, focused on the two types of transmission rights and energy. A few brief simple examples are also given.

There are alternative sets of market rules that could be used for selling all or part of a set of transmission rights and/or forward energy commitments. Here, our formulation mathematically liquidates all rights in each auction. Carrying the rights to the next stage could be accomplished by bidding an equal specification to the current rights with a corresponding large bid value (although this rule could conflict with market power mitigation rules) or submitting a fixed bid, that is, a bid with an upper and lower bound equal to current holdings. Holdings are liquidated by simply not submitting a bid. By convention, in ISO markets, point-to-point transmission rights are formally settled in the day-ahead market, while financial energy trades through the ISO auctions can be transacted day-ahead but cashed out at the real-time physical dispatch prices. We do not require any financial bid to be cashed out until the real-time market. Energy sales and purchases are settled financially in each forward market.

The notation, $s$, is introduced to designate the sequence of energy and transmission auctions, where $s = S, S - 1, \ldots, 1, 0$, and the *sth* auction is defined as JETRA$^s$. JETRA$^0$ is the final, real-time dispatch auction. The optimal values for energy and transmission rights resulting from the *sth* auction are designated $t^{Fs*}$, $t^{Ps*}$, $g^{+s*}$ and $g^{-s*}$. The optimal dual values will be similarly superscripted.

### 4.4.1 Multi-settlement System

Table 4.2 summarizes the multi-settlement system for the auction model using a uniform clearing price rule. The table shows the market design in which transmission rights contracts and nodal revenue rights contracts are settled finally in the real-time dispatch market (s = 0). In essence, for each auction s∈S, the ISO settles the rights contracts acquired in auction s + 1.

Row one of Table 4.2 shows that in each auction, $s$, transmission and energy rights contracts from auction $s + 1$ are settled (or liquidated) at the auction price times their contract holdings from the $s + 1$ auction (note again that incrementing by 1 is moving the auction backwards in time). Row two shows the contracts established in auction $s$ will pay or are paid the auction price times the quantity of transmission rights and forward energy contracts that clear the market.

The real-time dispatch market, $s = 0$, settlements shown in rows three and four follow the same logic as the forward markets with respect to holders of transmission rights or forward energy contracts, who are paid the auction price times their holdings from the prior auction iteration, $s = 1$. Only physical injections and withdrawals are traded in auction 0, but the forward rights from $s = 1$ are settled.

**Table 4.2** Calculation of settlement payments in auction $s$ for rights allocated in auctions $s + 1$ and $s$ using uniform clearing price rule

| | Flowgate rights (F) | Point-to-point rights (P) | Energy supply and demand (g) |
|---|---|---|---|
| JETRA$^s$ ($s \geq 1$) (forward market): Payment to holders of contracts from previous auction $s+1$ | $\mu^{N,s^*} B^{N,s+1} t^{F,s+1^*}$ (interaction constraints) | $\pi^{s^*} A^{P,s+1} t^{P,s+1^*}$ | $\pi^{s^*} A^{+,s+1} g^{+,s+1^*}, \pi^{s^*} A^{-,s+1} g^{-,s+1^*}$ |
| | $\mu^{+s^*} B^{+,s+1} t^{F,s+1}$ (flowgates in + direction) | | |
| | $\mu^{-s^*} B^{-,s+1} t^{F,s+1}$ (flowgates in − direction) | | |
| JETRA$^s$ ($s \geq 1$) (forward market): Payment by purchasers of contracts in auction $s$ | $\mu^{N,s^*} B^{N,s} t^{F,s^*}$ (interaction constraints) | $\pi^{s^*} A^{P,s} t^{P,s^*}$ | $\pi^{s^*} A^{+,s} g^{+,s^*}, \pi^{s^*} A^{-,s} g^{-,s^*}$ |
| | $\mu^{+,s^*} B^{+,s} t^{F,s^*}$ (flowgates in + direction) | | |
| | $\mu^{-s^*} B^{-,s} t^{F,s^*}$ (flowgates in − direction) | | |
| JETRA$^0$ (real-time market): Payment to holders of contracts from previous auction 1 | $\mu^{N,0^*} B^{N,1} t^{F,1^*}$ (interaction constraints) | $\pi^{0^*} A^{P,1} t^{P,1^*}$ | $\pi^{0^*} A^{+,1} g^{+,1^*}, \pi^{0^*} A^{-,1} g^{-1^*}$ |
| | $\mu^{+0^*} B^{+,1} t^{F,0^*}$ (flowgates in + direction) | | |
| | $\mu^{-0^*} B^{-,1} t^{F,1^*}$ (flowgates in − direction) | | |
| JETRA$^0$ (real-time market): Payment by purchasers of physical energy in auction 0 | $\mu^{N,0^*} B^{N,0} t^{F,0}$ (interaction constraints) | $\pi^{0^*} A^{P,0} t^{P,0^*}$ | $\pi^{0^*} A^{+,0} g^{+,0^*}, \pi^{0^*} A^{-,0} g^{-,0^*}$ |
| | $\mu^{+0^*} B^{+,0} t^{F,0^*}$ (flowgates in + direction) | | |
| | $\mu^{-0^*} B^{-,0} t^{F,0^*}$ (flowgates in − direction) | | |

The implicit congestion charge associated with any pair of injections and withdrawals at different nodes is the difference in the auction LMPs at those two nodes. For instance, using our notation, if two awards for bids $m$ and $m'$ result in $g^*{}_m = g^*{}_{m'}$, where $g^*{}_m$ is an injection at node 1 ($a_{1m} = 1$) and $g^*{}_{m'}$ is a withdrawal at node 2 ($a_{2m} = -1$), then the total congestion charge associated with these two transactions is $\pi^*{}_2 g^*{}_m - \pi^*{}_1 g^*{}_{m'} = (\pi^*{}_2 - \pi^*{}_1) g^*{}_m$.

A property of this settlement system that follows from convexity of the JETRA model and the optimality of its solution is that the prices in auction $s$ are such that there remain no arbitrage opportunities among the rights awarded in that auction. As an example, a pair of energy rights, one involving injection of 1 MW at one node $i$ and the other involving withdrawal at another node $i'$ would result in exactly the same settlement as an equivalent point-to-point right from $i'$ to $i$, so that no profitable arbitrage can be undertaken between those two types of rights. In a sense, the numerical process of finding an optimal solution can be viewed as consisting of searching for and taking advantage of all profitable arbitrage among the bids; if there remained profitable arbitrage opportunities at as solution, then the solution by definition could not be optimal.

Pre-day-ahead forward energy transactions, or nodal revenue rights, are not yet offered in ISO auctions. Hence their financial settlement deserves some further explanation. Settlement would take place, as with other transmission rights, in the day-ahead market (or in the real-time market if there is no day-ahead market). The holder of the injection right gets paid the nodal price for the energy it produces but is obligated to pay the nodal price to the ISO for energy represented in its nodal energy right, while the holder of the withdrawal right is obligated to pay the nodal price for the energy it actually consumes but is paid the nodal price for the energy quantity specified in its forward right. As with the two-sided, point-to-point right, executing the physical transaction specified in the right results in a net zero financial position in settlement. There are practical issues to implementing such a forward energy auction, most notably creditworthiness.

### *4.4.2 Revenue Adequacy of the Auction Sequence*

Revenue adequacy could pertain to each pair of auctions in the sequence. Also, revenue adequacy could pertain to the entire sequence. If all pairs are revenue adequate, the full sequence is revenue adequate. A set of sufficient conditions for revenue adequacy is that the constraint sets are convex and the constraint set does not contract over the auction sequence. The proof is in the appendix. Even if the constraint set is not convex, if it is not contracting (i.e., if all feasible solutions in previous iterations remain feasible in subsequent iterations), then even if the prices do not result in revenue adequacy, each and every market participant can in theory be made better off by re-allocating the surplus. This is because the objective function (total surplus) can only improve if the feasible region is non-contracting.

This re-allocation may require a deviation from the uniform clearing price settlement, for example using two-part tariffs or fixed monetary transfers.

Beginning with each auction clearing, a requirement for revenue adequacy is that the auction result respects the set of transmission constraints. For point-to-point rights, this is commonly known as "simultaneous feasibility," meaning that the power flow induced by the injections and withdrawals associated with the rights awarded is feasible (Harvey et al. 1997). Here "simultaneous feasibility" applies to all rights in each auction.

Turning next to the conditions on the auction sequence, we have assumed heretofore that each auction in the sequence, $\mathrm{JETRA}^s$, is conducted with the same set of transmission constraints. However, an important feature of actual electricity markets is that in the forward markets for transmission rights, the transmission constraints modeled may be either more or less restrictive than the set operative in the real-time market.

The further ahead a forward market is of the physical dispatch auction $s = 0$, the greater the uncertainty about the network topology that will apply in the dispatch. This could justify a conservative transmission constraint set in the further forward auctions. For forward auctions closer in time to the dispatch, some uncertainty will be resolved and this will justify increased offerings by relaxing the constraints. For example, equipment may need to be derated if it is extremely hot, but temperature is not known until a time closer to the dispatch. The uncertainty can be captured in auction models through either multi-state or chance-constrained models, but these models are large and harder to solve and may require different settlement rules.

In general, the recursion of the auction markets is revenue adequate as long as the transmission capacity constraints form a nested, expanding sequence, a restriction which is stated more formally in the proof in the appendix. If $K''$ is linear and $K'$ is convex, the constraint set is convex. For $s' > s$, if the constraint set defines a feasible region that is convex and non-contracting, that is, $F^{+,s'} \leq F^{+,s}$, $F^{-,s'} \leq F^{-,s}$ and $F^{N,s} \leq F^{N,s}$, then the auction sequence is revenue adequate. Non-contracting means that in each auction in the sequence, the transmission constraint set must be no more restrictive than the prior auction. This is an obvious requirement to prevent overselling of flowgate transmission rights.

A expanding constraint set can be thought of as the ISO holding back some of the rights until it is reasonably sure they will be available. Therefore, it is not unusual for the auction sequence to start with a conservative estimate of the availability of the network topology. Some ISOs have adopted simple rules to accommodate this requirement; for example, the California ISO sells forward transmission rights to only a small percentage of its transmission capacity (Bautista-Alderete 2010). Long-term point-to-point transmission rights are usually made available on conservative basis to account for the long-term uncertainty. Operational experience will be required to determine what quantity of alternative types of transmission rights can be made available in each forward market (annual, monthly, weekly, etc.).

As noted above, if the auction sequence is not revenue adequate in actual market operations, for example due to unplanned transmission outages affecting day-ahead and real-time market settlements, then each ISO has rules for how revenue shortfalls to rights holders are allocated.

## 4.5 Auction Example with Quadratic Losses

This section presents a numerical example of the auction model in a simplified network based upon a linearized dc load flow with quadratic losses (e.g., Hobbs et al. 2008; Schweppe et al. 1988). The only transmission elements considered are lines. Constraint (4.3) is omitted and (4.6) is modified to represent the dc analogues to Kirchhoff's Current and Voltage Laws:

$$\text{Current Law: } -y + D(f^+ - f^-) + f^{-T}L^-f^- + f^{+T}L^+f^+ \leq 0 \qquad (4.11)$$

$$\text{Voltage Law: } R(f^+ - f^-) = 0, \qquad (4.12)$$

where the new notation is as follows:

$D$ is a matrix that maps flow variables to the associated current law (enerrgy balance) constraints. The rows of the vector correspond to buses, and the columns correspond to lines of the network.

$L^+$, $L^-$ are tensors of rank 3, where the only nonzero elements in $L^+$ ($L^-$) are $l^+{}_{ikk}$ ($l^-{}_{ikk}$), representing the resistance loss coefficients (decrease in imports to bus $i$) due to a positive (negative) flow through transmission line $k$.

$R = \{r_{vk}\}$ are line reactances used in the voltage law analogues. Each element is the value of reactance for transmission line $k$ that appears in voltage loop $v$. $r_{vk} = +R_k$ or $-R_k$ if line $k$ occurs in loop $v$, depending on whether a positive flow $(f^+ - f^-)$ is in the same or opposite sense of flow around $v$. On the other hand, $r_{vk} = 0$ if link $k$ does not occur in loop $v$. Consistent with the dc model, the number of independent loops $v$ must be equal to $K - N + 1$, where $K$ is the number of lines considered and $N$ is the number of buses.

Note that (4.11) is a relaxation of the Kirchoff's Current Law (energy balance) equality constraint that results in a convex feasible region (Chao and Peck 1998). An example is given below to illustrate (4.11) and (4.12).

An important property, noted in Harvey and Hogan (2002), is that if $l_{ikk} > 0$, for some $k$, then in general no set of balanced P (point-to-point) rights will be feasible (revenue adequate) by themselves (except in the degenerate case of $t_P = 0$). This is because of losses. Revenue adequacy is thus possible only if sufficient energy rights are also sold (in particular, "rights" that oblige the rights holder to make payments to the ISO; i.e., rights g whose coefficients in $Ag$ are positive). A combination of such energy and balanced point-to-point rights $g$ and $T_F$ can also be viewed as a set of imbalanced point-to-point rights.

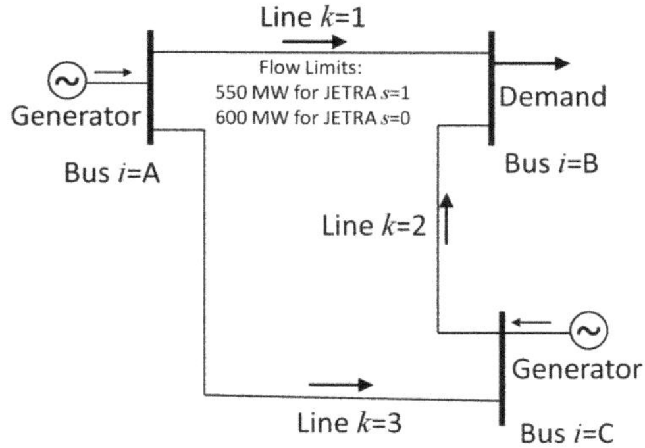


**Fig. 4.1** Network for three bus JETRA-NL example

The numerical example takes place on the three node network in Fig. 4.1, in which the arrows show the direction of flow on lines $k = 1, 2, 3$ for injections at buses A and C (with a larger one at A) and a withdrawal at bus B. These directions also coincide with the positive directions of flows associated with those lines. All loss factors on all lines equal 0.0001 [MW/MW$^2$], and all lines have a physical flow limit of 600 MW (only the limit for $k = 1$ is shown because that is the only one that binds in the solutions below). All line reactances, $R_k = 1$. Then for this network, (4.11) and (4.12) become:

$$KCL_A: -y_A + ({f^+}_1 - {f^-}_1) + ({f^+}_3 - {f^-}_3) + 0.0001\left(({f^-}_1)^2 + ({f^-}_3)^2\right) \leq 0$$

$$KCL_B: -y_B - ({f^+}_1 - {f^-}_1) - ({f^+}_2 - {f^-}_2) + 0.0001\left(({f^+}_1)^2 + ({f^+}_2)^2\right) \leq 0$$

$$KCL_C: -y_C + ({f^+}_2 - {f^-}_2) - ({f^+}_3 - {f^-}_3) + 0.0001\left(({f^-}_2)^2 + ({f^+}_3)^2\right) \leq 0$$

$$KVL: ({f^+}_1 - {f^-}_1) - ({f^+}_2 - {f^-}_2) - ({f^+}_3 - {f^-}_3) = 0$$

Notice that if the only existing transmission or energy right is, say, a balanced $t_P$ involving an injection of 1,000 MW at $A$ ($y_A = +1{,}000$) and a withdrawal of 1,000 MW at $B$ ($y_B = -1{,}000$), this would be infeasible. There are two reasons for this. First, such an injection-withdrawal pair would induce more than 600 MW of flow on line $k = 1$ (in the lossless case, 667 MW would flow). Second, because of line losses, there is no set of nonnegative flows $\{{f^+}_1, {f^-}_1, {f^+}_2, {f^-}_2, {f^+}_3, {f^-}_3\}$ that would simultaneously satisfy all four of the above constraints. Thus, there would either need to be some additional energy injected to make up for the loss, or the point-to-point right would need to be imbalanced, with more injected at A than withdrawn at B. The infeasibility of this right implies that the ISO might be revenue deficient if it settled that right at nodal prices from an optimal dispatches subject to the above constraints; this is indeed the case, as we see below

We illustrate a sequence of JETRA-NL with this network. In auction $s = 1$, due to ISO caution, only 550 MW of rights are released on each line $k$ rather than the full 600 MW. As mentioned, this is the current policy of certain ISOs in order to lessen the likelihood of revenue inadequacy. We assume that in this auction there are the following bidders for transmission and energy rights:

- Bidder 1 is willing to pay up to $60 per MWh per hour for up to 700 MW of point-to-point obligation transmission rights from node $i$ = A to $i$ = B;
- Bidder 2 is willing to pay up to $30 per MWh per hour for up to 300 MW of point-to-point rights from C to B;
- Bidder 3 bids is willing to pay up to $80 per MW per hour for up to 100 MW of flowgate rights on line $k = 1$ in the direction from bus A to bus B; and
- Bidder 4 offers to sell up to 100 MW of forward energy rights at node B at a price of $90/MWh.

The resulting formulation of JETRA-NL is as follows:

$$JETRA - NL, s = 1 : max\ 60{t^P}_1 + 30{t^P}_2 + 80\,{t^F}_3 - 90g_4$$

$${t^P}_1 - y_A = 0$$

$$-{t^P}_1 - {t^P}_2 + g_4 - y_B = 0$$

$${t^P}_2 - y_C = 0$$

$${t^F}_3 + ({f^+}_1 - {f^-}_1) \leq 550$$

$$-({f^+}_1 - {f^-}_1) \leq 550$$

$$({f^+}_2 - {f^-}_2) \leq 550$$

$$-({f^+}_2 - {f^-}_2) \leq 550$$

$$({f^+}_3 - {f^-}_3) \leq 550$$

$$-({f^+}_3 - {f^-}_3) \leq 550$$

$$-y_A + ({f^+}_1 - {f^-}_1) + ({f^+}_3 - {f^-}_3) + 0.0001\left(({f^-}_1)^2 + ({f^-}_3)^2\right) \leq 0$$

$$-y_B - ({f^+}_1 - {f^-}_1) - ({f^+}_2 - {f^-}_2) + 0.0001\left(({f^+}_1)^2 + ({f^+}_2)^2\right) \leq 0$$

$$-y_C + ({f^+}_2 - {f^-}_2) - ({f^+}_3 - {f^-}_3) + 0.0001\left(({f^-}_2)^2 + ({f^+}_3)^2\right) \leq 0$$

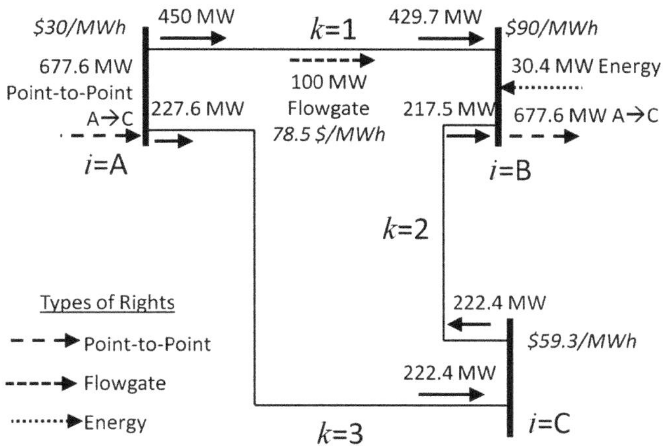


**Fig. 4.2** Awarded financial transmission and energy rights and LMPs for dispatch round $s = 1$ of the JETRA-NL example

$$(f^{+}{}_{1} - f^{-}{}_{1}) - (f^{+}{}_{2} - f^{-}{}_{2}) - (f^{+}{}_{3} - f^{-}{}_{3}) = 0$$

$$t^{P}{}_{1} \leq 700$$

$$t^{P}{}_{2} \leq 300$$

$$t^{F}{}_{3} \leq 100$$

$$g_4 \leq 100$$

$$t^{P}{}_{1},\ t^{P}{}_{2},\ t^{F}{}_{3},\ g_4 \geq 0$$

The ISO runs JETRA-NL for this auction, and makes the following awards of rights:

- 677.6 MW of point-to-point rights to Bidder 1, who pays the ISO $40,655 for these rights (equal to the nodal price difference between A and C times the awarded
- 0 MW of point-to-point rights to Bidder 2
- 100 MW of flowgate rights to Bidder 3, who pays $7856 for those rights (flowgate 1's shadow price times the award)
- 30.4 MW of energy rights from Bidder 4, who the ISO pays $2,734 (B's nodal price times the energy right sold)

Figure 4.2 shows this solution to the auction, along with the nodal and flowgate prices. The ISO's net receipts from the auction are $45,776, which is less than the

\$45,922 objective function for the auction model.[17] This discrepancy arises because Bidder 3's upper bound is binding, meaning that she pays less for the rights than they are worth to her.

We now move to the next (and final) JETRA-NL iteration, $s = 0$, which is the physical dispatch. We assume that there are two power plants, neither with capacity limits. The plant at A offers to sell energy at \$20/MWh (variable $g_A$), while C's plant offers at \$50/MWh (variable $g_C$). There is a 1,000 MW load at B (variable $g_B$). The ISO makes available the full 600 MW of flow capacity in each line for this iteration. The resulting JETRA-NL is:

$$JETRA\text{-}NL, s = 0 : max - 20g_A - 50g_C$$

$$g_A - y_A = 0$$

$$- g_B - y_B = 0$$

$$g_C - y_C = 0$$

$$(f^+{}_1 - f^-{}_1) \leq 600$$

$$- (f^+{}_1 - f^-{}_1) \leq 600$$

$$(f^+{}_2 - f^-{}_2) \leq 600$$

$$- (f^+{}_2 - f^-{}_2) \leq 600$$

$$(f^+{}_3 - f^-{}_3) \leq 600$$

$$- (f^+{}_3 - f^-{}_3) \leq 600$$

$$- y_A + (f^+{}_1 - f^-{}_1) + (f^+{}_3 - f^-{}_3) + 0.0001\left((f^-{}_1)^2 + (f^-{}_3)^2\right) \leq 0$$

$$- y_B - (f^+{}_1 - f^-{}_1) - (f^+{}_2 - f^-{}_2) + 0.0001\left((f^+{}_1)^2 + (f^+{}_2)^2\right) \leq 0$$

$$- y_C + (f^+{}_2 - f^-{}_2) - (f^+{}_3 - f^-{}_3) + 0.0001\left((f^-{}_2)^2 + (f^+{}_3)^2\right) \leq 0$$

$$(f^+{}_1 - f^-{}_1) - (f^+{}_2 - f^-{}_2) - (f^+{}_3 - f^-{}_3) = 0$$

[17] Round off errors result in slight discrepancies in results. For instance, \$45,922 is the exact objective function value resulting from the exact decision variable values, while the values of the decision variables presented here, which are rounded off, yield \$45,920 instead.

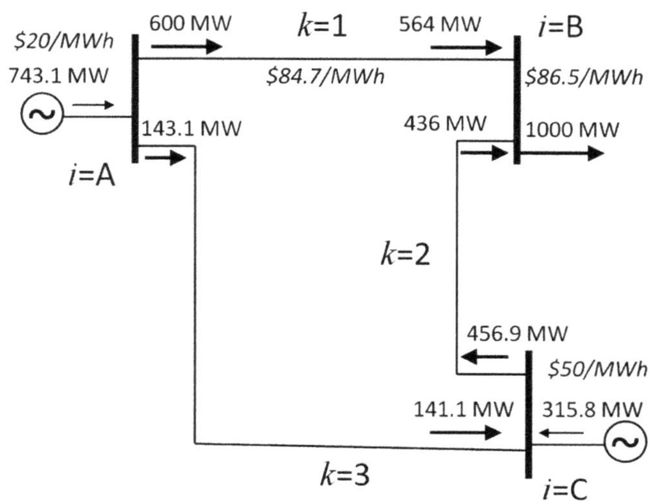


**Fig. 4.3** Nodal injections and withdrawals and LMPs for dispatch round $s = 0$ of the JETRA-NL example

$$g_B = 1000$$

$$g_A, g_C \geq 0$$

The resulting dispatch is shown in Fig. 4.3, along with the nodal prices. The ISO pays a total of $30,652 to the two generators for their energy, while receiving $86,470 from the load at B. The resulting total surplus gained by the ISO is $55,818. If congestion portion of this surplus is calculated as the sum of the flowgate shadow prices times the flows, the congestion surplus is $50,797, while the loss surplus is the remaining $5020. The loss surplus arises because of the quadratic nature of losses, which means that marginal losses are roughly double the average loss. Consumers pay for marginal losses. In this example, the ISO essentially gets to keep the difference between marginal and average losses. In practice, the U.S. ISOs are required to refund excess revenues to market participants.

From its surplus, the ISO must pay the holders of financial transmission and energy rights awarded in the earlier JETRA-NL $s = 1$. The following awards are made to financial rights holders:

- Bidder 1, who holds 677.6 MW of point-to-point rights from A to B that were awarded in $s = 1$, is paid the nodal price difference ($86.5–$20) times those rights, or $45,039.
- Bidder 2 owns no rights, and so receives no payment
- Bidder 3 is paid 100 MW times the flowgate shadow price for $k = 1$, or $8466.
- Bidder 4 has to pay the ISO $2,627 for its 30.4 MW of energy injection rights at node B.

The net payments to financial rights holders by the ISO is \$50,879. Note that each bidder happens to make money on their financial rights. Bidders 1 and 3 get paid more in $s = 0$ for their transmission rights than they paid in $s = 1$, while Bidder 4 pays less to settle her energy right in $s = 0$ than she got paid in $s = 1$. Note that since Bidder 4 has no physical asset, her energy right is what is known in U.S. markets as a virtual energy right, in which energy is bought in one market, and then the same amount is sold back in the next, arbitraging the difference in prices. Bidder 4 is what is known as a virtual supplier, since she supplied power in the first auction $s = 1$. Because the energy price in $s = 1$ was greater than $s = 0$, she makes money on that energy transaction.

The fact that the financial rights holders made money on their rights has no implications for revenue neutrality of auction $s = 0$. In fact, the ISO's surplus in the final dispatch round $s = 0$ of JETRA-NL (\$55,818) exceeds its net payments to owners of financial rights awarded in $s = 1$ (\$50,879, as just noted)., This is necessarily the case because the dispatch model is convex (the feasible region defined by the load flow constraints (4.11 and 4.12) plus capacity constraints is convex, while the objective function is linear), and the transmission flows that would induced by the financial rights awarded in $s = 1$ are feasible in the dispatch model. In particular, note that the $s = 1$ flows in Fig. 4.2 are feasible if the transmission limits were the 600 MW values assumed in the $s = 0$ dispatch optimization.

As an example of financial rights that would not be revenue adequate, return again to the simple example mentioned before in which the only transmission or energy rights held after $s = 1$ are 1,000 MW of point-to-point rights from A to B. This set of rights would violate the load flow and capacity constraints of the network in Fig. 4.1. The settlement in that case, based on $s = 0$'s nodal prices, would be (\$86.5 − \$20) × 1,000 MW, or \$66,500; this would exceed the ISO's surplus of \$55,818 in $s = 0$, violating revenue adequacy.

## 4.6 Conclusion

The nonlinear auction model presented here provides a general framework for representing and implementing a more complete version of combined energy and transmission rights auctions that have been proposed and discussed in the United States. With all types of energy and transmission capacity bids allowed, the auction framework can be extended to most types of forward hedging. Frequent auctions increase liquidity by providing additional opportunities to trade while considering the network constraints that bilateral markets have difficulty factoring in. In addition, this framework could facilitate the efficient operation of off-ISO forward bilateral markets, which should benefit from more liquid transmission rights, such as the rights on commonly congested flowgates or possibly hub-to-hub rights. The proof of revenue adequacy that we previously provided for the auction with linear constraints (O'Neill et al. 2002) has been extended to the auction with convex constraints. However, the introduction of non-convex constraints invalidates the proof of revenue adequacy, assuming that a uniform pricing rule is used to determine the prices of rights and to settle them.

The practical obstacles to implementation of the model are computational requirements, implementation costs and transactions costs. For these reasons, while there is now broad consensus on many elements of market design, such as locational marginal prices for energy and financial transmission rights, market design proposals have allowed for phased implementation of different types of transmission rights and different auction products to allow for development of software and resolution of cost allocation issues.

The nonlinear auction model provides an analytic framework for exploration of additional market design features. For example, more frequent auctions add liquidity to the market. Future research to be conducted by the authors within this framework includes the modeling and pricing of locational reserves, pricing of reactive power (e.g., FERC 2005), property right awards for transmission expansion, pricing under optimal topologies, and unit commitment of transmission elements (e.g., O'Neill et al. 2005a).

**Acknowledgments** The authors would like to thank R. Baldick, H.-P. Chao, R. Entriken, W. Hogan, D. Mead, and S. Oren for helpful comments on this and previous descriptions of the JETRA proposal, as well as editors of this volume and anonymous reviewers of a previous version.

## Appendix: Proof of Revenue Adequacy for the Auction Sequence

This appendix provides a set of sufficient conditions and a proof of revenue adequacy of the auction sequence. This proof extends the revenue adequacy proofs for transmission in Harvey et al. (1997) and O'Neill et al. (2002), both of which considered the case of linear transmission constraints, to an auction with both flowgate or flow-based and point-to-point rights together with nonlinear transmission constraints that define a convex feasible region. To simplify the presentation, the auction model is mapped into a more compact and general non-linear program (NLP) representing an auction in the following way:

As before, the rights bid for and awarded in the $s$-th auction in a sequence of auctions determine the distribution of revenues from the subsequent auction, $s - 1$. Meanwhile, the prices obtained in the $s$-th auction determine how the rights awarded in the previous auction $s + 1$ are financially settled, as well as how much winning bidders in auction $s$ pay for the rights they win.

Define $g^s$ as the vector of quantities awarded to $P$- and $G$-type bids (encompassing $t^P$, $g^+$ and $g^-$ in the JETRA-NL model) with upper bound $G^s$ in the $s$-th auction in the sequence. Define a general benefit function $c(g^s)$ (based on the bids by those seeking rights) for the bid award level, $g^s$. The vector $y^s$ represents net injections in the $s$-th auction associated with rights $g^s$. $K^s(y)$ represents the flows induced by $y^s$ as a result of the applicable load flow equations. Define $t^s$ as the vector of $F$ transmission rights ($t^F$ in the JETRA-NL model) with upper bound $T^s$ in the $s$-th auction. $F^s$ is the vector of bounds in auction $s$ for transmission elements and network flow constraints. Define $\pi$ as the vector of dual values for the nodal energy balance constraint, which can be interpreted as the shadow or clearing prices for energy. Finally, define $\mu$ as the vector of dual values associated with transmission constraints, which can be interpreted as the shadow prices for transmission rights.

Using the resulting model NLP, the *sth* auction in the auction sequence $s + 1, s, s - 1, \ldots, 0$, termed NLP$^s$, is:

$$\text{NLP}^s : \max\ b^s t + c^s(g)$$

$$Ag - y = 0 \qquad (\pi)$$

$$B^s t + K^s(y) \leq F^s \qquad (\mu)$$

$$t \leq T^s \qquad (\psi)$$

$$g \leq G^s \qquad (\rho)$$

Note that all constraint and objective function parameters can depend on $s$.

The optimal solution to NLP$^s$ is defined as $\{y^s, t^s, g^s\}$ and the corresponding optimal dual variables are $\{\pi^s, \mu^s, \psi^s, \rho^s\}$. To demonstrate revenue adequacy of the auction sequence, prices and payments must be defined for the bids for $g$ and $t$ that are accepted. Duals $\pi^s$ are the market prices for $g^s$, and $\mu^s$ are the market prices for $F^s$, and are treated as row vectors in the below. The rights held as a result of the $s + 1^{\text{st}}$ auction in the sequence are $g^{s+1}$ and $t^{s+1}$. Financial settlements (payments by the auctioneer) in NLP$^s$ for rights to its revenues, analogous to those defined above for the full auction model, are $\pi^s A g^{s+1}$ and $\mu^s B^{s+1} t^{s+1}$ for the two types of rights awarded in the previous auction NLP$^{s+1}$, where the superscript T is the transpose operator Meanwhile, the winning bidders for the two types of rights awarded by NLP$^s$ pay $\pi^{s\,\mathrm{T}} A g^s$ and $\mu^{s\mathrm{T}} B^s t^s$, respectively.

The following theorem concerns the revenue adequacy of this sequence of auctions, and is a generalization of our earlier results for the linear JETRA (O'Neill et al. 2002):

**Theorem 1** *If $B^s(g)$ is concave, $K^s(y)$ is convex, $K^s(y) \leq K^{s+1}(y)$ for all $y$, and* $F^{s+1} \leq F^s$, then each auction in the sequence of auctions $\{S - 1, \ldots, s, \ldots, 1, 0\}$, is revenue adequate; that is:

$$\pi^{s\mathrm{T}}(A^s g^s - A^{s+1} g^{s+1}) + \mu^{s\mathrm{T}}(B^s t^s - B^{s+1} t^{s+1}) \geq 0.$$

*Proof* By convexity of $K^s$,

$$K^s(y^{s+1}) \geq K^s(y^s) + \nabla K^s(y^s)(y^{s+1} - y^s).$$

Rearranging, we obtain,

$$\nabla K^s(y^s)y^s \geq \nabla K^s(y^s)y^{s+1} + K^s(y^s) - K^s(y^{s+1}).$$

Premultiplying by the row vector of transmission capacity shadow prices $\mu^s \geq 0$,

$$\mu^s \nabla K^s(y^s)y^s \geq \mu^s \nabla K^s(y^s)y^{s+1} + \mu^s K^s(y^s) - \mu^s K^s(y^{s+1}) \tag{4.13}$$

From the KKTs to NLP$^s$,

$$\mu^s(B^s t^s + K^s(y^s)) = \mu^s F^s \tag{4.14}$$

Since $K^s(y) \leq K^{s+1}(y)$ and $F^s \geq F^{s+1}$ and because $(B^{s+1}t^{s+1}, y^{s+1})$ is a feasible solution to NLP$^{s+1}$,

$$B^{s+1}t^{s+1} + K^s(y^{s+1}) \leq F^s \tag{4.15}$$

Multiplying both sides by $\mu^s \geq 0$,

$$\mu^s(B^{s+1}t^{s+1} + K^s(y^{s+1})) \leq \mu^s F^s \tag{4.16}$$

Combining (4.14) and (4.16) and multiplying both sides by $-1$ (which requires reversing the inequality),

$$-\mu^s(B^{s+1}t^{s+1} + K^s(y^{s+1})) \geq -\mu^s(B^s t^s + K^s(y^s)) \tag{4.17}$$

Adding (4.13) and (4.17), eliminating terms that cancel, and finally rearranging,

$$\mu^s \nabla K^s(y^s)y^s - \mu^s(B^{s+1}t^{s+1}) \geq \mu^s \nabla K^s(y^s)y^{s+1} - \mu^s(B^s t^s)$$

Substituting $\pi^s = \mu^s \nabla K^s(y^s)$ from the KKT condition for $y^s$ for problem NLP$^s$ and rearranging,

$$\pi^s(y^s - y^{s+1}) + \mu^s(B^s t^s - B^{s+1}t^{s+1}) \geq 0$$

Finally, in NLP$^s$, $A^s g^s = y^s$ while in NLP$^{s+1}$, $A^{s+1}g^{s+1} = y^{s+1}$; substitution of these constraints establishes the desired result:

$$\pi^s(A^s g^s - A^{s+1}g^{s+1}) + \mu^s(B^s t^s - B^{s+1}t^{s+1}) \geq 0.$$

Note that this result does not explicitly depend on the form of the objective function NLP$^s$. The objective can be linear or nonlinear, as long as it is concave so that the KKT conditions describe an optimal solution, then the use of the KKT conditions in the above proof remains valid.

## References

Baldick R, Helman U, Hobbs BF, O'Neill RP (2005) Design of efficient generation markets. Proc IEEE (Special Issue on Power Technology & Policy) 93(11):1998–2012

Bautista-Alderete G (2010) Implementation and evolution of the congestion revenue right market in California. In: Proceedings, IEEE Power Engineering Society general meeting, Minneapolis, MN., USA, 25–29 July

Chao H-P, Peck S (1996) A market mechanism for electric power transmission. J Regul Econ 10(1):25–59

Chao H-P, Peck S (1998) Reliability management in competitive electricity markets. J Regul Econ 14(2):189–200

Chao H-P, Peck S, Oren S, Wilson R (2000) Flow-based transmission rights and congestion management. Electr J 13(8):38–58

Elmaghraby W, O'Neill RP, Rothkopf MH, Stewart WR Jr (2004) Pricing and efficiency in 'lumpy' energy markets. Electr J 17:54–64

Federal Energy Regulatory Commission (2002) Remedying undue discrimination through open access transmission service and standard electricity market design, notice of proposed rulemaking. Docket No. RM01-12-000 (July 31). [Online] Available: www.ferc.gov

Federal Energy Regulatory Commission (2005) Principles for efficient and reliable reactive power supply and consumption. Staff report, Docket No. AD-05-1-000, Washington, DC (February 4). [Online] Available: www.ferc.gov

Federal Energy Regulatory Commission (FERC) (1996) Capacity reservation open access transmission tariffs: notice of proposed rulemaking. Docket No. RM96-11-000 (April 24). [Online] Available: www.ferc.gov

Harvey SM, Hogan WW (2002) Loss hedging financial transmission rights. Technical report, John F. Kennedy School of Government, Harvard University, Cambridge, MA (January 15). [Online] Available: ksghome.harvard.edu/.whogan.cbg.Ksg

Harvey SM, Hogan WW, Pope SL (1997) Transmission capacity reservations and transmission congestion contracts. Technical report, John F. Kennedy School of Government, Harvard University, Cambridge, MA (Revised March 8). [Online] Available: ksghome.harvard.edu/.whogan.cbg.Ksg

Hobbs BF, Rothkopf MH, O'Neill RP, Chao H-P (eds) (2001) The next generation of electric power unit commitment models. Kluwer, Boston

Hobbs BF, Drayton G, Fisher EB, Lise W (2008) Improved transmission representations in oligopolistic market models: quadratic losses, phase shifters, and dc lines. IEEE Trans Power Syst 23(3):1018–1029

Hogan WW (1993) Markets in real electric networks require reactive prices. Energy J 14(3):171–200

Hogan WW (2000) Flowgate rights and wrongs. Technical report, John F. Kennedy School of Government, Harvard University, Cambridge, MA (August 20). [Online] Available: ksghome.harvard.edu/.whogan.cbg.Ksg

Hogan WW (2002) Financial transmission right formulations. Technical report, John F. Kennedy School of Government, Harvard University, Cambridge, MA (January 15). [Online] Available: ksghome.harvard.edu/.whogan.cbg.Ksg

Kahn E, Baldick R (1994) Reactive power is a cheap constraint. Energy J 15(4):191

O'Neill RP, Helman U, Hobbs BF, Stewart WR Jr, Rothkopf M (2002) A joint energy and transmission rights auction: proposal and properties. IEEE Trans Power Syst 17(4):1058–1067

O'Neill RP, Helman U, Baldick R, Stewart WR Jr, Rothkopf M (2003) Contingent transmission rights in the standard market design. IEEE Trans Power Syst 18(4):1331–1337

O'Neill RP, Baldick R, Helman U, Rothkopf M, Stewart WR Jr (2005a) Dispatchable transmission in RTO markets. IEEE Trans Power Syst 20(1):171–179

O'Neill RP, Sotkiewicz PM, Hobbs BF, Rothkopf M, Stewart WR Jr (2005b) Efficient market-clearing prices in markets with nonconvexities. Eur J Operat Res 164:269–285

O'Neill RP, Helman U, Hobbs BF, Baldick R (2006) Independent system operators in the USA: history, lessons learned, and prospects. In: Sioshansi FP, Pfaffenberger W (eds) International experience in restructured electricity markets: what works, what does not, and why? Elsevier, London

O'Neill RP, Fisher EB, Hobbs BF, Baldick R (2008) Towards a complete real-time electricity market design. J Regul Econ 34(3):220–250

Schweppe FC, Caramanis MC, Tabors RD, Bohn RE (1988) Spot pricing of electricity. Kluwer, Boston

# Chapter 5
# Generator Ownership of Financial Transmission Rights and Market Power

**Manho Joung, Ross Baldick, and Tarjei Kristiansen**

## 5.1 Introduction

Game theory is well suited to analyze a situation with strategic interdependence of multiple decision makers. Electricity markets include both physical and operational attributes. Likewise, electricity markets are characterized by a relatively small number of large market players, limited competitiveness and strategic behavior. Cournot models compete in quantities while Bertrand models compete in prices. Supply function equilibrium function models assume market players compete both in quantity and price. These are realistic assumptions for electricity markets where market players submit a price-quantity schedule. However these models are complex to solve and may not incorporate all technical attributes of electricity markets. Cournot models are easily solvable and yield under reasonable conditions a unique Nash equilibrium. They are also more suitable for short term analysis.

Competition is introduced in most electricity markets around the world. Markets are also increasingly coupled with interconnectors and thus may exhibit stronger price convergence. To supply more power to a region a decision maker has three choices: build power plant assets, reduce local consumption or build transmission assets. Transmission assets may bring increased competitive benefits to a market. The main objective of transmission rights is to hedge against locational price differences. But an FTR is also a transmission property right. Such a right brings the benefits associated

M. Joung
Zigi Solutions, LLC. 9085 Judicial Dr. #2127, San Diego, CA 92122, USA
e-mail: manhojoung@gmail.com

R. Baldick (✉)
Department of ECE, The University of Texas at Austin, Austin, TX 78712, USA
e-mail: baldick@ece.utexas.edu

T. Kristiansen
Åsegårdsvegen 65, N-6017, Ålesund
e-mail: tarjeikr@yahoo.com

J. Rosellón and T. Kristiansen (eds.), *Financial Transmission Rights*,
Lecture Notes in Energy 7, DOI 10.1007/978-1-4471-4787-9_5,

with transmission capacity and facilitates efficient use of scarce resources. Property rights are also a mechanism to reward transmission investments.

Among researchers (Joskow and Tirole 2000; Léautier 2000; Gilbert et al. 2004) there is consensus about the need to mitigate market power for any FTR auction to be efficient. Joskow and Tirole (2000) study a radial line network under different market structures for both generation and FTRs. They demonstrate that FTR market power by a producer in the importing region (or a consumer in the exporting region) aggravates their monopoly (monopsony) power, because dominance in the FTR market creates an incentive to curtail generation (demand) to increase the value of the FTRs. Allocation of FTRs to a monopoly generator depends on the structure of the market (Joskow and Tirole 2000). When the FTRs are allocated initially to a single owner that is neither a generator nor a load, the monopoly generator will want to acquire all FTRs. When all FTRs initially are distributed to market players without market power, the generator will buy no FTRs. When the FTRs are auctioned to the highest bidders, the generator will buy a random number of FTRs. Extending this analysis, Gilbert et al.(2004) analyze ways of preventing perverse incentives by identifying conditions where different FTR allocation mechanisms can mitigate generator market power during transmission congestion. In an arbitraged uniform price auction, generators will buy FTRs that mitigate their market power, while in a pay-as-bid auction FTRs might enhance their market power. Specifically, in the radial line case, market power might be mitigated by not allowing generators to hold FTRs related to their own energy delivery. In the three-node case, mitigation of market power implies defining FTRs according to the reference node with the price least influenced by the generation decision of the generator. In practical implementations of the FTR model, market power mitigating rules are designed (Rosellón 2003). The Federal Energy Regulatory Commission (FERC) has included market power mitigation rules in the standard market design (FERC 2002). FERC indicates that insufficient demand-side response and transmission constraints are the two main sources for market power. FERC differentiates between high prices because of scarcity and high prices resulting from exercising market power. Using a merit-order spot market mechanism FERC proposes to use a bid cap for generators with market power in a constrained region and a "safety net" for demand side response. Regulated generators are also subject to a resource adequacy requirement. Chandley and Hogan (2002) claim that this mechanism is inefficient because the use of penalties for under-contracting (with respect to the resource adequacy requirement) would not permit prices to clear the energy and reserve markets. Moreover, long-term contracting should be voluntary, and based on financial hedging, not on capacity.

Borenstein et al. (2000) studied the economic benefits of linking markets with a transmission line. Their work demonstrated that there may be no direct relationship between the level of competition and the actual physical line utilization. For a sufficiently large transmission line capacity, a competitive outcome may be achieved even if the flow is zero. A market outcome similar to two merged markets would be replicated. Borenstein et al. (2000) applied their duopoly model to the California electricity market. Willems (2002) conducted a similar study but included the role of the network operator to enhance the competition level. Leautier (2000) studied regulatory contracts for transmission system operators and introduced a contract that incentivizes the operators to optimally expand the transmission network. Stoft (1999) studied market power arising from generation shipped to consumers over congested transmission lines. He included FTRs and the congestion rent distribution. Joskow and Tirole (2000) conducted a more general study of market power and transmission rights and suggested possible regulatory mechanisms. Cho (2003) researched the competitive equilibrium in a network with limited transmission capacity and a developed a tool to identify an efficient equilibrium. He included transmission right markets for specific electricity market structures. Gilbert et al. (2004) studied the market power effects of transmission rights. Their initial analysis focused on a simple two-node network and was extended to meshed networks.

This chapter analyzes the FTR ownership effects on the strategic behavior of electricity generators in a Cournot framework developed by Joung (2008). We follow Borenstein et al. (2000) with two identical but geographical distinct markets. Each market has an identical monopoly supplier and cost function. The model setup is similar to Borenstein et al. but also includes FTRs. Various FTR models are introduced and market efficiency is studied under FTR ownership. Joskow and Tirole studied FTRs in a two node market model and Pritchard and Philpott (2005) considered a similar model. However the market structure was simplified by assuming that only one market had consumers while the other only had producers. The market power structures were limited to monopolistic and oligopolistic competition in one market while the other market remained competitive. Thus the competitive effects of FTRs were not considered in the more typical case where producers in both markets are imperfectly competing. Likewise Cho (2003) analyzed FTRs in a two stage model where stage 1 included the transmission market with strategic behavior and stage 2 the energy market with price taking behavior. He demonstrated that inefficient equilibria may exist. However real world electricity markets differ from the proposed model and the results are thus not directly applicable. Gilbert et al. (2004) proposed a three stage model with transmission right allocation, trading and energy market output allocation. The model is solved backwards starting with the energy market. However the two node model has limitations since there is competition only among generators located in one market while the other market is perfectly competitive. Similar to Joskow and Tirole (2000) this does not consider the case when generators in different markets are imperfectly competing. Likewise the transmission line is always assumed to be congested. These limitations influence the results of each stage of the game and therefore limit the analysis of the energy market.

Game-theoretic studies focused on the competitive effects of FTRs have tried to consider mixed strategy equilibria. Borenstein et al. (2000) utilized a numerical method, Gilbert et al. (2004) presented analytic results for mixed strategy equilibria but the model limitations excluded the competitive effects of FTRs when generators in different markets are imperfectly competing.

This chapter studies the interactions between two incompletely competitive markets. In particular, this chapter investigates the competitive effects of FTR ownership of generating firms for their market strategy formulation. A Cournot framework is applied and the best response curves provide implications for FTR ownership effects in

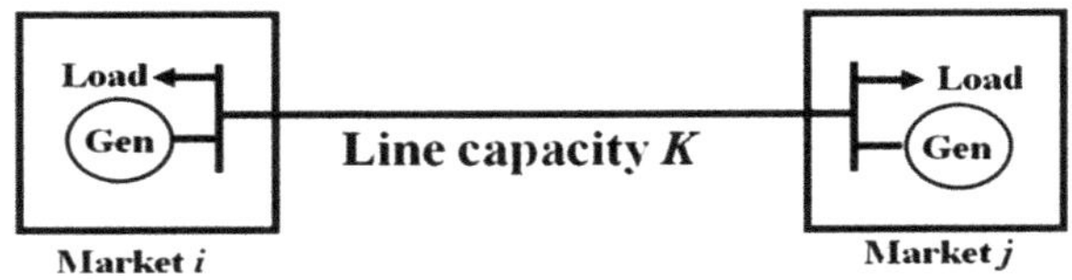


**Fig. 5.1** Two market model

an unconstrained Cournot equilibrium. Allocation of outward directional FTRs to generators could result in a lower needed transmission line capacity than in Borenstein et al.'s work (2000) to achieve full competition. These FTRs which are directed from the generator to other markets hedge the generator's exposure to prices in its own home market and therefore mitigate its market power. If the generator possesses an FTR from another market to its own, the FTR causes a negative effect on competition. The FTR increases exposure to prices in the generator's home market and increases its market power. The model is also extended to include the analysis of asymmetric markets and where one market is competitive.

## 5.2 Two Market Model

We consider a model of two markets. Demand in each market is assumed to be characterized by an affine inverse-demand function. In each market there is a single generating firm. These two markets are linked by a single transmission line whose capacity is $K$. The transmission line is operated by a third entity and the electricity pricing follows the nodal pricing rules (Schweppe et al. 1988). Both generating firms try to maximize their profits by employing quantity strategies (Cournot competition). Figure 5.1 conceptually depicts the market model.

To make competitive analysis more tractable, we assume two markets are identical. That is, demand in each market is assumed to be identical and to be characterized by the same inverse-demand function denoted by $P$: $\Re_+ \to \Re_+$., and we also assume that both generating firms have an identical cost function $C$: $\Re_+ \to \Re_+$. Asymmetric markets will be discussed as an extension of the symmetric market model.

In order for the model to be more concrete, we make the following assumptions:

- The inverse demand $P(q)$ in each market is represented by an affine curve with a negative slope:

$$P(q) = -\alpha q + \beta, \text{ where } a, \beta \in \Re_+, \tag{5.1}$$

- Generating firms' generating costs $C(q)$ are represented by a convex quadratic function:

$$C(q) = \frac{a}{2} q^2 + bq + c, \text{ where } a,\ c \in \Re_+, \text{ and } b \in \Re. \tag{5.2}$$

**Fig. 5.2** FTR directions

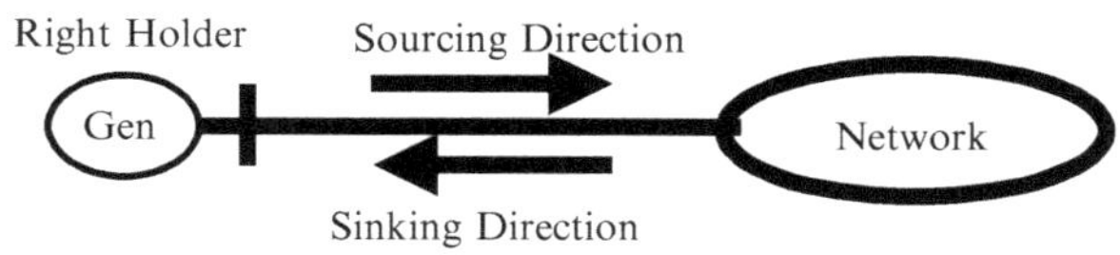


## 5.3 FTR Models

In this chapter, three FTR models are defined: the reference model, the FTR option model, and the FTR obligation model. The reference model considers the case in which neither firm has any FTRs.

The FTR option model is the market model with generators' owning FTR options. An FTR option is a financial contract for collecting the amount of money determined by the locational price difference and the share of the right. This option gives the owner the right to collect a portion of the congestion rents when the price difference is positive, but does not require payment when the price difference is negative.

The FTR obligation model is the market model with generators' owning FTR obligations. An FTR obligation is a similar financial contract to an FTR option, but it has negative payoff if the nodal prices reverse. That is, if the price difference is positive, a holder collects the congestion rents of the transmission line, while for the negative price difference, the holder makes a payment. Obligation-type rights also have two possible directions.

We define the "direction" of FTRs from the point of view of the generating firm that holds the transmission rights. We say the "sourcing" direction for FTRs that are in the direction from the market where the right holding generating firm is located to the other market. That is, the payoff of sourcing FTRs is defined by the nodal price in the other market minus the nodal price at the generator. The opposite direction is called the "sinking" direction. That is, the payoff of sinking FTRs is defined by the nodal price at the generator minus price in the other market. These two directions are illustrated in Fig. 5.2.

## 5.4 Competitive Effects of FTRs

In this section, we derive analytical expressions for the best response of each firm for different FTR models: the reference model without financial transmission rights (in Sect. 5.4.1), the FTR option model (in Sect. 5.4.2), and the FTR obligation model (in Sect. 5.4.3). We also analyze the competitive effects of the corresponding financial transmission rights for each model using best response analysis.

Following Borenstein et al. (2000), for the best response analysis, we define two categories of optimal responses: optimal aggressive output and optimal passive output. First, suppose that firm $i$ is in the situation such that the opponent, firm $j$, is producing nothing (more generally, that firm $j$ is producing so little energy that there is transmission congestion on the line in the direction from market $i$ to market $j$). In this case, the best response of firm $i$ is to produce its optimal quantity given that the line is congested from $i$ to $j$. Under the nodal pricing scheme, this quantity will be the same as the monopoly output for firm $i$ when the market is isolated but with the demand shifted to the right by $K$. This is called the optimal aggressive output for $i$ and denote it with a superscript $+$ .

Now, suppose that firm $i$ is in the situation such that the opponent, firm $j$, is producing a great amount of electric power (more generally, firm $j$ is producing enough energy to cause line congestion from market $j$ to market $i$). In this case, the best response of firm $i$ is to produce its optimal quantity given that the line is congested in the direction from market $j$ to market $i$. Under the nodal pricing scheme, this quantity will be the monopoly output for firm $i$ when the market is isolated with the demand shifted to the left by $K$. This monopoly quantity is called the optimal passive output for $i$ and will be denoted with a superscript $-$.

Besides the optimal aggressive and passive outputs, one more category of best response behavior is needed to cover the uncongested case. Since the resulting quantity is equivalent to the unconstrained Cournot best response output for the merged markets, this output is called the Cournot best response output and is denoted with a superscript $C$.

### 5.4.1 Reference Model

If the electricity market is perfectly competitive and there is no market power, the introduction of FTRs into the market has no effect on the prices for energy or the dispatch of generators. As a reference model, the case is considered such that neither firm has any rights on the transmission line. The reference case will be denoted with a superscript $r$. In this case, the optimal aggressive and passive outputs, and the Cournot best response output, which are denoted by $q^{r+}(K)$, $q^{r-}(K)$, and $q_i^{rC}(q_j)$ respectively, are expressed by (5.3), (5.4), and (5.5).[1]

$$q^{r+}(K) = \frac{\beta + \alpha K - b}{2\alpha + a}, \tag{5.3}$$

$$q^{r-}(K) = \frac{\beta - \alpha K - b}{2\alpha + a}, \tag{5.4}$$

$$q_i^{rC}(q_j) = -\frac{\alpha}{2(\alpha + a)} q_j + \frac{\beta - b}{\alpha + a}. \tag{5.5}$$

Here, it can be observed that the function $q^{r+}$ is increasing in its argument while the function $q^{r-}$ and the function $q_i^{rC}$ are both decreasing in their argument (Note that $q^{r+}$

[1] There is no case where (5.3) and (5.4) are achieved in an equilibrium in a symmetric model; however, in an asymmetric model, passive/aggressive equilibria are possible, in which case a pair of (5.3) and (5.4) will be an equilibrium output pair.

and $q^{r-}$. are functions of line capacity $K$, while $q_i^{rC}$ is a function of production by the other firm, $q_j$.).

This reference model is equivalent to the symmetric two-firm model of Borenstein et al. (2000). This section serves to review their results. The line will be congested only when the difference between the outputs of two firms is greater than $2K$, since otherwise, by transferring a smaller amount of electricity than the line capacity $K$, the two markets' prices would be equalized.

Let us consider the best response of firm $i$ with respect to the other firm $j$'s strategy, $q_j$. When firm $j$ is producing any amount up to $q^{r+}(K) - 2K$, firm $i$ can maximize its profit by producing the fixed amount $q^{r+}(K)$. As firm $j$'s output increases above $q^{r+}(K) - 2K$, however, firm $i$ can maximize its profit and still export $K$ by producing $2K$ more than firm $j$. That is, firm $i$ maximizes its profit by producing $q_j + 2K$, accounting for the segment of slope 1 in the best responses shown in Fig. 5.2. Note that as $q_j$ keeps increasing, firm $i$'s resulting payoff from maintaining an aggressive response is decreasing. As firm $j$'s output continues to increase, two situations can be thought of.

On the one hand, if $K$ is small, then producing the optimal passive output $q^{r-}(K)$ becomes more profitable for firm $i$ before the value of $q_i = q_j + 2K$ reaches the unconstrained Cournot best response $q_i^{rC}(q_j)$. This is shown by the dashed curve in Fig. 5.3.

On the other hand, if the line capacity is large enough, say, $K'$ as shown in Fig. 1.3 as the solid curve, then firm $i$'s best response will change from $q_j + 2K$ to $q_i^{rC}(q_j)$. However, even in this situation, as $q_j$ keeps increasing, producing $q^{r-}(K)$ will eventually be more profitable for firm $i$ than producing $q_i^{rC}(q_j)$. This accounts for the transition in the best responses to $q^{r-}(K')$ and $q^{r-}(K)$, respectively, for high enough $q_j$.

To summarize, the situations for the two values of line capacity are illustrated in Fig. 5.3. The solid curve shows the case of relatively large capacity $K'$ where firm $i$'s optimal response includes some values equal to the Cournot unconstrained best response. The dashed curve shows the case of relatively small capacity $K$ where the best response never includes values equal to the Cournot unconstrained best response.

As shown in Fig. 5.3, the best responses of both firms will have different characteristics according to the transmission line capacity $K$. Specifically, increase of physical line capacity implies both increase of the optimal aggressive output $q^{r+}(K)$ and decrease of the optimal passive output $q^{r-}(K)$. Borenstein et al. (2000) shows that this, in turn, implies an increase in the competition-promoting effects of the transmission line:

- Decrease in the equilibrium price of the mixed strategy equilibrium, and
- Increase in the range of market demand conditions that result in the pure strategy Cournot equilibrium.

The results of Borenstein et al. (2000) also shows that if $K$ is very small, then there is no pure strategy equilibrium, while if $K$ is large enough, the Cournot duopoly equilibrium will be reached as the unique equilibrium. That is, the equilibrium is

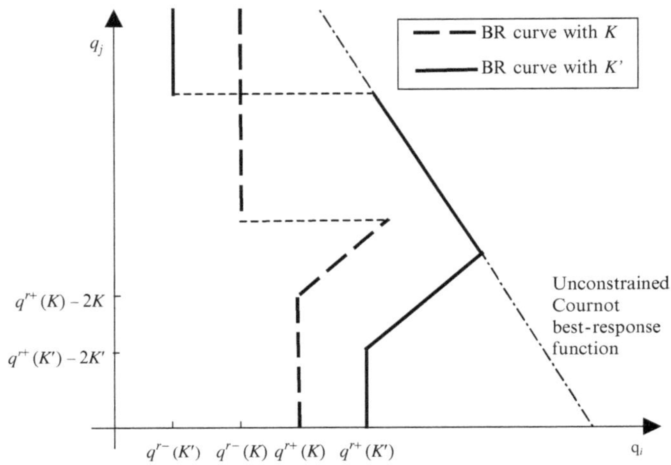


**Fig. 5.3** Best response curves for firm $i$ ($K < K'$)

specified by (5.5), with zero flow along the line but with the line providing the full competitive benefits of merged markets.

## 5.4.2 FTR Option Model

An FTR option is a financial contract for collecting the amount of money determined by the locational price difference and the share of the right. This option gives the owner the right to collect a portion of the congestion rents when the price difference is positive, but does not require payment when the price difference is negative. FTR options have been implemented in PJM first (PJM 2011) and are being introduced in several other markets in the United States, recently including the Electric Reliability Council of Texas (ERCOT) "nodal" market in 2010 (ERCOT 2011).

An FTR option has a specified exercise direction and if the nodal price difference is positive in this direction, then the FTR provides a positive payoff. There is zero payoff for price differences in the other direction. This means that each firm $i$ has two possible directions for his FTR option in this two market model; that is, a direction from market $i$ to $j$ (the sourcing direction) and one from $j$ to $i$ (the sinking direction).

Let $\eta_i^{ij}$ and $\eta_i^{ji}$ denote generating firm $i$'s FTR option share from market $i$ to $j$ and from market $j$ to $i$, respectively, such that $\eta_i^{ij}, \eta_i^{ij} \in [0, 1]$. That is, $\eta_i^{ij}$ describes the share of sourcing FTR, while $\eta_i^{ji}$ describes the share of sinking FTRs. We use superscript *uo* to denote options.

We have:

**Lemma 1.** *Let $q_i^{uoij+}$, $q_i^{uoij-}$, and $q_i^{uoijC}$ be the optimal aggressive, passive, and Cournot responses for firm i holding share $\eta_i^{ij}$. Let $q_i^{uoji+}$, $q_i^{uoji-}$, and $q_i^{uojiC}$ be the* *optimal aggressive, passive, and Cournot responses for firm i holding share* $\eta_i^{ji}$. *Then:*

$$q_i^{uoij+}\left(K,\eta_i^{ij}\right) = q^{r+}\left(\left(1+\eta_i^{ij}\right)K\right), \tag{5.6}$$

$$q_i^{uoij-}(K) = q^{r-}(K), \tag{5.7}$$

$$q_i^{uoijC}(q_j) = q_i^{rC}(q_j), \tag{5.8}$$

$$q_i^{uoji+}(K) = q^{r+}(K), \tag{5.9}$$

$$q_i^{uoji-}\left(K,\eta_i^{ji}\right) = q^{r-}\left(\left(1+\eta_i^{ji}\right)K\right), \tag{5.10}$$

$$q_i^{uojiC}(q_j) = q_i^{rC}(q_j). \tag{5.11}$$

*Proof.* See Appendix.

Lemma 1 suggests that the ownership of an FTR option is equivalent to expanding the capacity of the line in one direction. This specific relationship is mainly from the linearity of demand. When the demand linearity is relaxed, this relationship would change, but a similar qualitative effect would be expected.

To summarize, an FTR option results in the change of either the optimal aggressive output (see (5.6)) or optimal passive output (see (5.10)) compared to the reference model. By possessing an $\eta_i^{ij}$ FTR option, firm *i*'s optimal aggressive output increases as indicated by (5.6), observing that by (5.3), $q^{r+}$ is increasing in its argument. By possessing an $\eta_i^{ji}$ FTR option, firm *i*'s optimal passive output decreases as indicated by (5.10), observing that by (5.4), $q^{r-}$ is decreasing in its argument. The change of the best response due to an FTR option is illustrated in Fig. 5.4. Note that, in order to differentiate two different response curves in Fig. 1.4, there are some line segments that are illustrated as being close together although they are in fact coincident.

As shown in Fig. 5.4, according to its direction, each FTR option has one of two different effects: either increase of the optimal aggressive output as shown in Fig. 5.4a or decrease of the optimal passive output as shown in Fig. 5.4b. This, in turn, affects the range of conditions for realization of the pure strategy equilibrium. Here, we focus on the effect on the occurrence of three forms of equilibrium: the unconstrained Cournot equilibrium, passive/aggressive equilibrium, and mixed strategy equilibrium Borenstein et al. (2000). We do not consider overlapping equilibria as described in the work of Borenstein et al. (2000).

Increase of the optimal aggressive output has no effect on achieving the unconstrained Cournot equilibrium since the unconstrained Cournot best response region is the same as that in the reference case and the range of conditions for the unconstrained Cournot equilibrium will be also the same as shown in Fig. 5.4a. On the other hand,

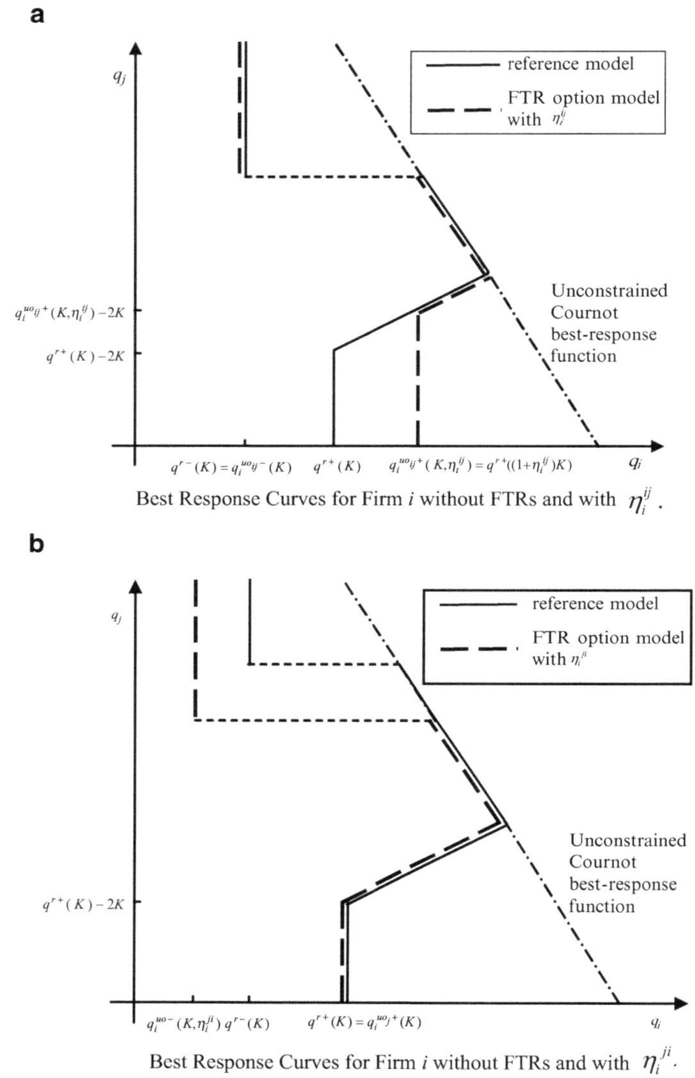


**Fig. 5.4** Comparison of best response curves. (**a**) Best response curves for firm $i$ without FTRs and with $\eta_i^{ij}$. (**b**) Best response curves for firm $i$ without FTRs and with $\eta_i^{ji}$

decrease of the optimal passive output reduces the unconstrained Cournot best response region since the right holder becomes more inclined to the optimal passive output. That is, the transition of its best response from the unconstrained Cournot response to the optimal passive output occurs at a smaller value of the other firm's output as shown in Fig. 5.4b.

Consider a case where, without FTRs, the capacity of the transmission line is enough to achieve the unconstrained Cournot equilibrium. Figure 5.5a illustrates this case. From the previous argument, if firm $i$ possesses an $\eta_i^{ij}$ FTR option and/or firm $j$ possesses an $\eta_j^{ji}$ FTR option, then the resulting equilibrium will be the same as the unconstrained Cournot equilibrium in the reference case as shown in Fig. 5.5b.

In contrast, suppose that firm $i$ possesses an $\eta_i^{ji}$ FTR option. In this case, the resulting equilibrium may change from the unconstrained Cournot equilibrium to a mixed strategy equilibrium. This is illustrated in Fig. 5.5c. Figure 5.5c shows that by $i$ possessing an $\eta_i^{ji}$ FTR option, the change of best response curve of firm $i$ may result in a mixed strategy equilibrium instead of the unconstrained Cournot equilibrium that is achieved without FTRs (Fig. 5.5a). A similar effect can occur if firm $j$ possesses an $\eta_j^{ij}$ FTR option.

However, for the range of $\eta_i^{ji} \in [0, 1]$, the introduction of FTR options cannot create enough asymmetry to yield a passive/aggressive equilibrium.

**Lemma 2.** *Suppose that, without FTRs, the capacity of the transmission line is enough to achieve the unconstrained Cournot equilibrium. In this case, by firm i's possessing an* $\eta_i^{ji}$ *FTR option, the resulting equilibrium cannot change to a passive/aggressive equilibrium.*

*Proof.* Suppose that, with firm $i$'s possessing an $\eta_i^{ji}$ FTR option, a passive/aggressive equilibrium is achieved. Then, the price difference $P_{ij}(q_i, q_j)$, between two markets is obtained as:

$$\begin{aligned} P_{ij}\left(q_i^{uoji-}, q_j^{r+}\right) &= P\left(q_i^{uoji-} + K\right) - P\left(q_j^{r+} - K\right) \\ &= \left(\frac{\left(2 + \eta_i^{ji}\right)\alpha}{2\alpha + a} - 2\right)\alpha K < 0 \end{aligned} \tag{5.12}$$

This contradicts the assumption of achieving a passive/aggressive equilibrium since, with negative price difference, FTR options will not generate any additional payoffs and, therefore, firm $i$'s best response will not become the optimal passive output.

**Q.E.D.**

### *5.4.3 FTR Obligation Model*

An FTR obligation is a similar financial contract to an FTR option, but it has negative payoff if the nodal prices reverse. That is, if the price difference is positive, a holder collects the congestion rents of the transmission line, while for the negative price difference, the holder makes a payment. Obligation-type rights also have two possible directions. FTR obligations are implemented in several markets in the Eastern US, including PJM (PJM 2011), New York ISO (New York ISO 2011), and

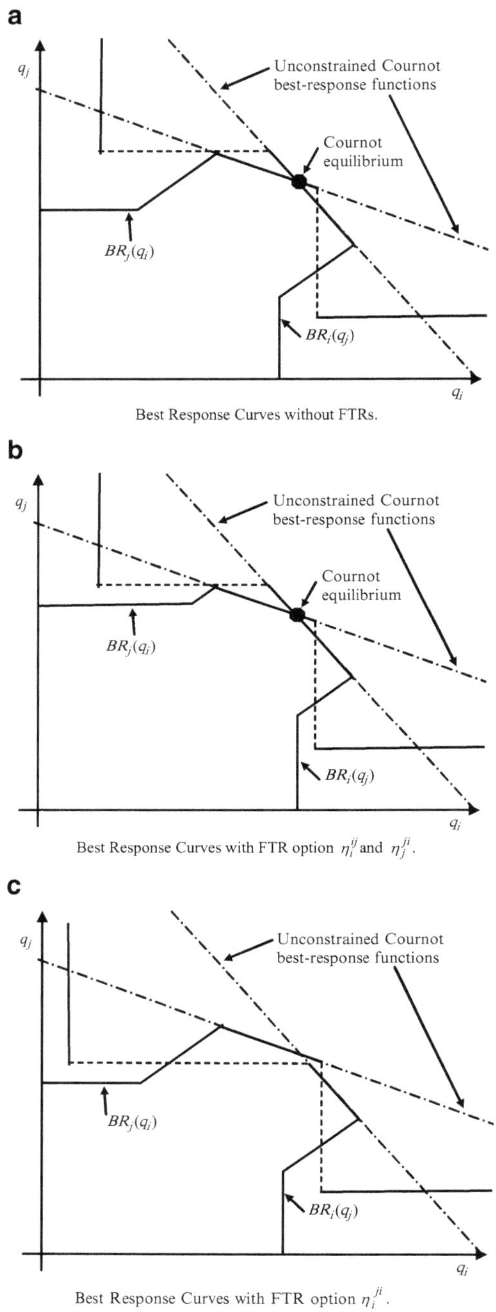


**Fig. 5.5** Illustration of the effects of FTR options on the Cournot equilibrium. (**a**) Best response curves without FTRs. (**b**) Best response curves with FTR option $\eta_i^{ij}$ and $\eta_j^{ji}$. (**c**) Best response curves with FTR option $\eta_i^{ji}$

New England ISO (New England ISO 2011), and are available in California ISO (California ISO 2011), Midwest ISO (Midwest ISO 2011) and the ERCOT nodal market (ERCOT 2011).

Let the FTR obligation share of firm $i$ be denoted by $\gamma_i \in [-1, 1]$, where the sourcing direction is assumed positive. That is, firm $i$ collects or pays $\gamma_i$ portion of the total congestion rents. We use superscript *ob* to denote FTR obligations.

We have:

**Lemma 3.**

$$q_i^{ob+}(k, \gamma_i) = \frac{\beta + (1 + \gamma_i)\alpha k - b}{2\alpha + a} = q^{r+}((1 + \gamma_i)k), \tag{5.13}$$

$$q_i^{ob-}(k, \gamma_i) = \frac{\beta - (1 - \gamma_i)\alpha k - b}{2\alpha + a} = q^{r-}((1 - \gamma_i)k), \tag{5.14}$$

$$q_i^{obC}(q_j) = -\frac{1}{2(\alpha + a)} q_j + \frac{\beta - b}{\alpha + a} = q_i^{rC}(q_j). \tag{5.15}$$

*Proof.* See Appendix.

These results imply that firm $i$ possessing $\gamma_i$ FTR obligation has the same effects on the firms' strategic behaviors as having two directional transmission lines with different capacities: capacity $(1 + \gamma_i)K$ MW from market $i$ to $j$ and capacity $(1 - \gamma_i)$ $K$ MW from market $j$ to $i$. The resulting competitive effects are different depending on the sign of $\gamma_i$.

First, suppose that $\gamma_i \geq 0$. This means that, in terms of its effect on competitive behavior, the effective line capacity increases by the amount of $\gamma_i K$ MW in the $i$ to $j$ direction, while the effective line capacity decreases by $\gamma_i K$ MW in the opposite direction. This results in an increase of both the optimal aggressive output and the optimal passive output compared to those in the reference model.[2]

On the other hand, if $\gamma_i < 0$, the opposite results are obtained; that is, the optimal aggressive and passive outputs decrease. Figure 5.6 shows these effects of an FTR obligation.

As shown in Fig. 5.6a, a positive FTR obligation will have positive effect on achieving the unconstrained Cournot equilibrium by increasing the unconstrained Cournot best response region. A negative FTR obligation will have a negative effect on achieving the unconstrained Cournot equilibrium as shown in Fig. 5.6b. Suppose that, without FTRs, the unconstrained Cournot equilibrium is achieved as illustrated in Fig. 5.7a. Here, we consider only the effect of firm $i$'s possession of FTR obligations. In this case, if firm $i$ possesses a positive FTR obligation then the only possible pure strategy equilibrium will be the same unconstrained Cournot equilibrium as shown in Fig. 5.7b.

Now consider a case where, without FTRs, the capacity of the transmission line is not enough to achieve the unconstrained Cournot equilibrium. Figure 5.7c illustrates

[2] Of course, the amount of power transferred over the line remains limited to $K$.

a

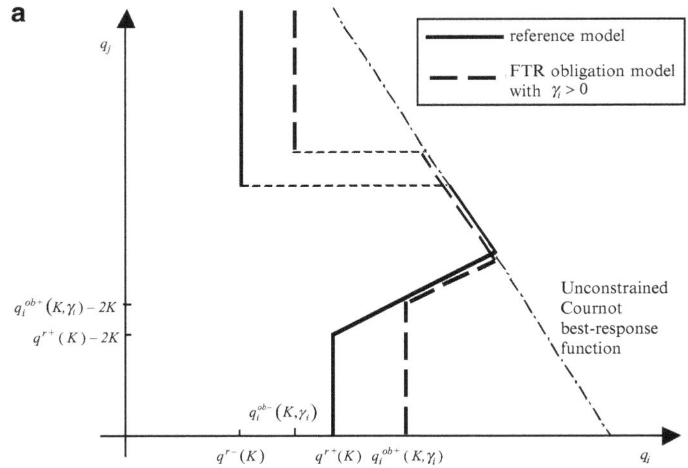


Best Response Curves For Firm $i$ without FTR and with $\gamma_i > 0$.

b

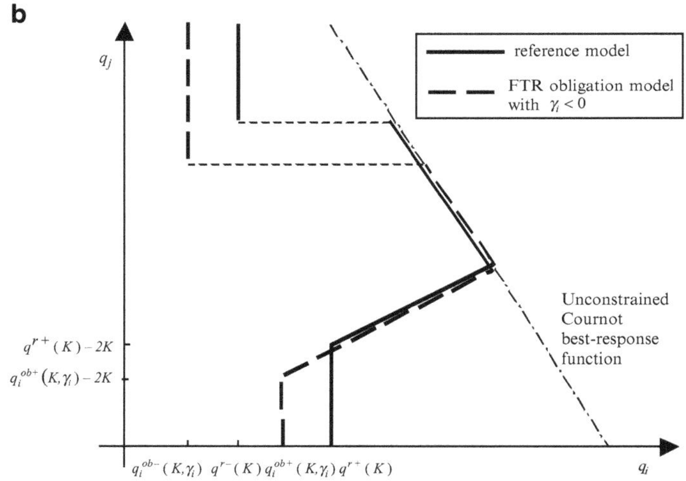


Best Response Curves For Firm $i$ without FTR and with $\gamma_i > 0$.

**Fig. 5.6** Comparison of best response curves. (**a**) Best response curves for firm $i$ without FTR and with $\gamma_i > 0$. (**b**) Best response curves for firm $i$ without FTR and with $\gamma_i < 0$

this case. As shown in the figure, due to the insufficient line capacity, two best response curves do not intersect at the unconstrained Cournot equilibrium. Without FTRs, only a mixed equilibrium can occur. By Borenstein et al. (2000), the expected price will be higher than in the Cournot equilibrium. Figure 5.7d shows that by $i$ and $j$ each possessing a positive FTR obligation, two best response curves intersects at the unconstrained Cournot equilibrium due to both firms' changed Cournot best response regions. As illustrated in Fig. 5.7d, a positive FTR obligation may result in the unconstrained Cournot equilibrium when it was impossible without FTRs.

**a**

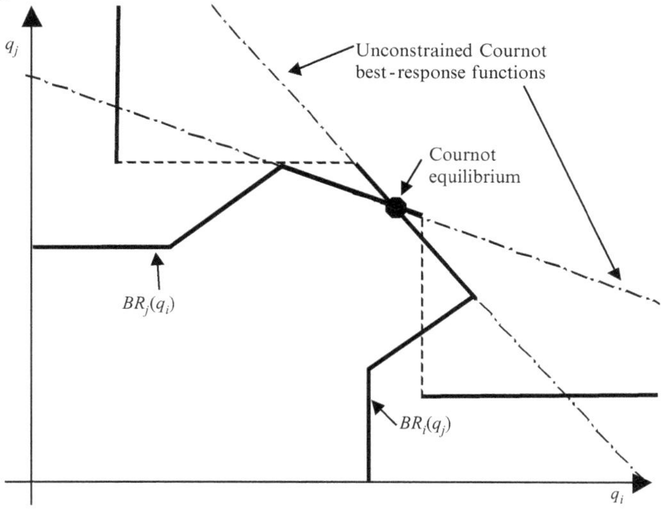


Best Response Curves without FTR.

**b**

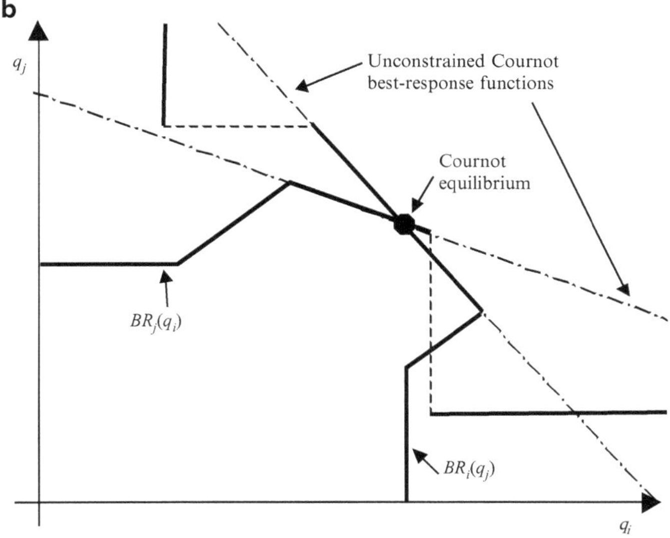


Best Response Curves with Positive FTR Obligations.

**Fig. 5.7** (continued)

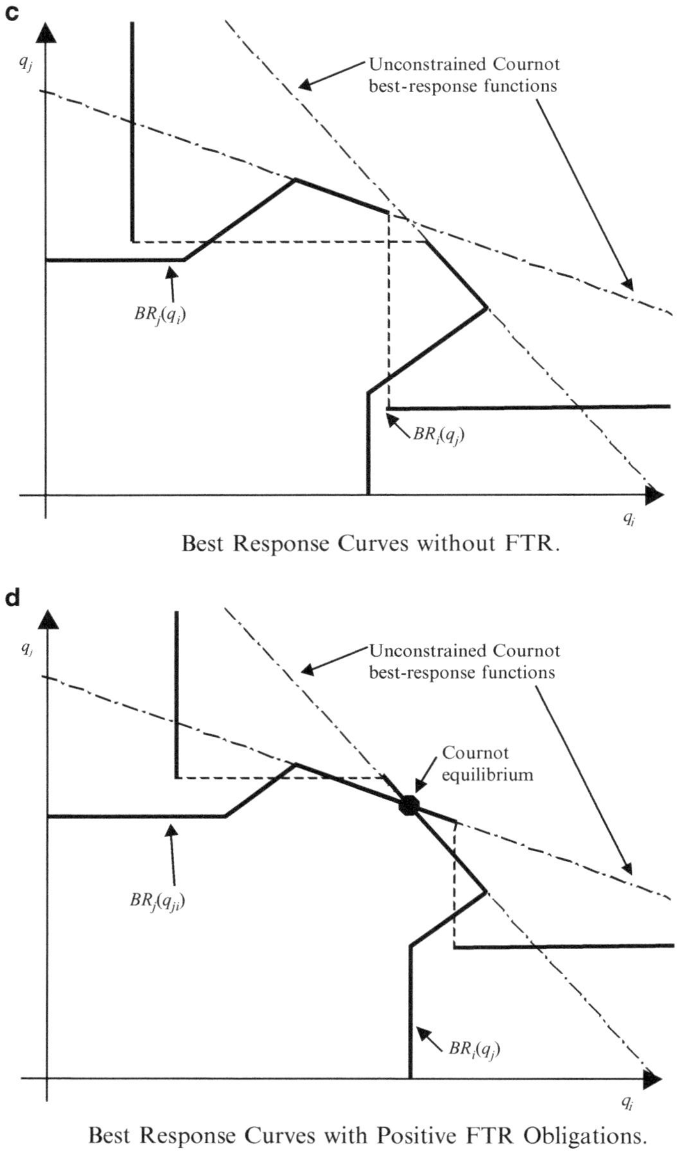


**Fig. 5.7** Illustration of the effects of positive FTR obligations on the Cournot equilibrium. (**a**) Best response curves without FTR. (**b**) Best response curves with positive FTR obligations. (**c**) Best response curves without FTR. (**d**) Best response curves with positive FTR obligations

On the other hand, Fig. 5.8 illustrates that negative FTR obligations may result in a passive/aggressive equilibrium while the unconstrained Cournot equilibrium is achieved without FTRs. Figure 5.8a illustrates that, due to the sufficient line capacity,

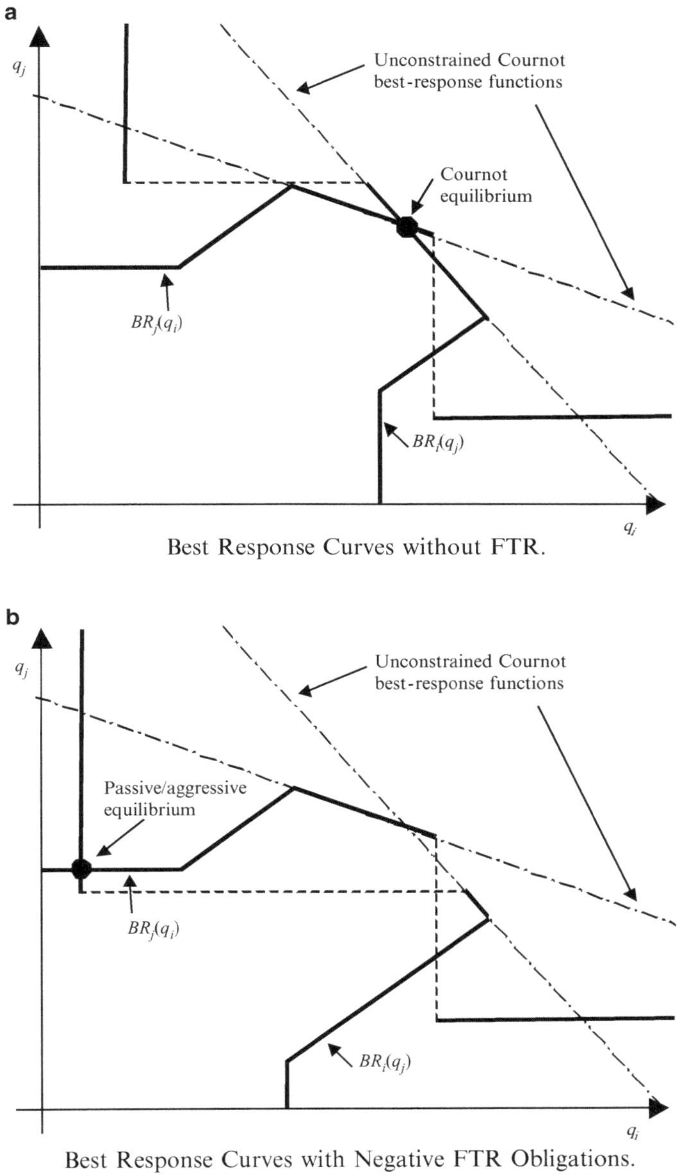


**Fig. 5.8** Illustration of the effects of negative FTR obligations on the Cournot equilibrium. (**a**) Best response curves without FTR. (**b**) Best response curves with negative FTR obligations

two best response curves intersect at the unconstrained Cournot equilibrium without FTRs. Here, we consider only the effect of firm $i$'s possession of FTR obligations. As shown in Fig. 5.8b, with $i$ possessing a negative FTR obligation, a passive/aggressive equilibrium is achieved.

## 5.5 Model Extensions

In this section, we comment on two model extensions: an asymmetric market and a competitive market.

### 5.5.1 Asymmetric Markets

Borenstein et al. (2000) showed that for the reference model, if markets are asymmetric enough, then even a very thin transmission line can provide a pure strategy equilibrium: a passive/aggressive equilibrium. Moreover, they showed that, with a sufficiently large line, the unconstrained Cournot equilibrium is the unique pure-strategy equilibrium and that this is the same as the case of symmetric markets.

With our other FTR models, under certain conditions, a passive/aggressive equilibrium is possible even in the case where, without ownership of FTRs, the unconstrained Cournot equilibrium is the unique pure-strategy equilibrium. This shows that FTRs may effectively increase asymmetry of markets that, otherwise, is not enough to yield a passive/aggressive equilibrium. However, by the same reasoning as for the reference model, with a sufficiently large line capacity, the unconstrained Cournot equilibrium will be the unique pure-strategy equilibrium even with FTRs.

Consider a case where, without FTRs, asymmetry of markets is small enough to achieve the unconstrained Cournot equilibrium. Figure 5.9a illustrates this case. Suppose that firm $i$ possesses an $\eta_i^{ji}$ FTR option. In this case, the resulting equilibrium may change from the unconstrained Cournot equilibrium to a passive/aggressive strategy equilibrium. This is illustrated in Fig. 5.9b. Figure 5.9b shows that by $i$ possessing an $\eta_i^{ji}$ FTR option, asymmetry of markets increases enough to result in a passive/aggressive equilibrium instead of the unconstrained Cournot equilibrium that is achieved without FTRs (Fig. 5.9a).

### 5.5.2 Competitive Market

In the work of Borenstein et al. (2000), the reference model was compared to a variant in which one of the markets is perfectly competitive. The result was that the effect of a transmission line on the reference model is greater than the effect on a model where one of the markets is competitive. This is mainly because the strategic interaction of two firms in the reference model leads to the Cournot duopoly quantity, while in the other model with a perfectly competitive market, the best response of a firm

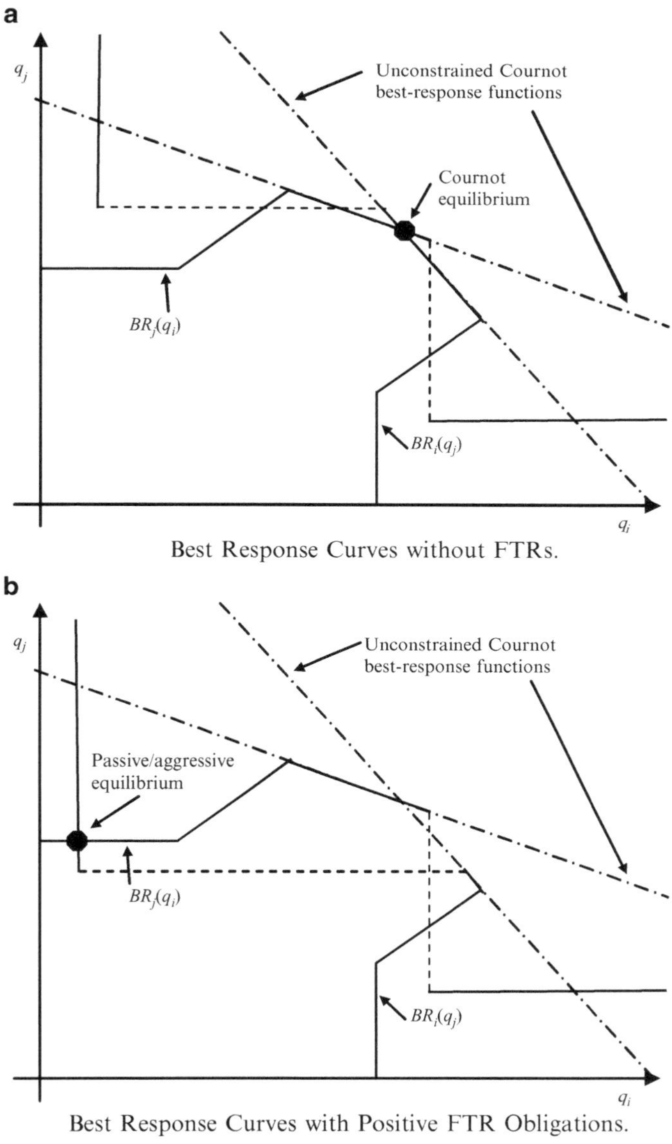


**Fig. 5.9** Illustration of the effects of FTR options on the Cournot equilibrium. (**a**) Best response curves without FTRs. (**b**) Best response curves with FTR option $\eta_i^{ji}$

confronting the competitive market will be the monopoly quantity, given imports from the competitive market equal to the line capacity.

Joskow and Tirole (2000) studied a similar model with FTRs. In their model, one market has a demand and a strategic supplier, while the other market has only competitive suppliers. Consequently, the direction of line flow is only in the direction to the market with demand and any strategic interaction among firms is not considered. They have shown using this setup that if only the strategic firm in the demand market holds FTRs, then these rights will enhance its market power.

Unlike Joskow and Tirole's model, in FTR models presented in this study, each market has both supply and demand and both directions of line flow must be considered. By assuming that one of the markets is competitive, there is no strategic interplay between firms. To correspond to Joskow and Tirole's model, we assume that firm $i$ is the only strategic firm and that firm $j$ is perfectly competitive. We consider the cases of FTR options and FTR obligations in Sects. 5.5.2.1 and 5.5.2.2.

Since firm $j$ is perfectly competitive, the following equation holds:

$$C'(q_j) = cq_j + b = p_j, \tag{5.16}$$

where $p_j$ is the price in market $j$. The price $p_j$ is determined by the supply quantities as follows:

$$p_j = \begin{cases} -\alpha q_j + \alpha K + \beta, & \text{if } q_i < q_j - 2K, \\ -\alpha \dfrac{q_i + q_j}{2} + \beta, & \text{if } q_j - 2K \le q_i \le q_j + 2K, \\ -\alpha q_j - \alpha K + \beta, & \text{if } q_i > q_j + 2K. \end{cases} \tag{5.17}$$

From (5.16) and (5.17), the quantity $q_j$ is represented by (5.18):

$$q_j = \begin{cases} \dfrac{1}{c+\alpha}(\beta + \alpha K - b), & \text{if } q_i < \dfrac{1}{c+\alpha}(\beta - b - (2c+\alpha)K), \\ -\dfrac{\alpha q_i}{2c+\alpha} + \dfrac{2}{2c+\alpha}(\beta - b), & \\ \qquad \text{if } \dfrac{1}{c+\alpha}(\beta - b - (2c+\alpha)K) \le q_i \le \dfrac{1}{c+\alpha}(\beta - b + (2c+\alpha)K), & \\ \dfrac{1}{c+\alpha}(\beta - \alpha K - b), & \text{if } q_i > \dfrac{1}{c+\alpha}(\beta - b + (2c+\alpha)K). \end{cases} \tag{5.18}$$

Consequently, $p_j$ can be rewritten as (5.19):

$$p_j = \begin{cases} \dfrac{\alpha b}{c+\alpha} + \dfrac{c}{c+\alpha}(\beta + \alpha K), & \text{if } q_i < \dfrac{1}{c+\alpha}(\beta - b - (2c+\alpha)K), \\ -\dfrac{\alpha c q_i}{2c+\alpha} + \dfrac{1}{2c+\alpha}(2c\beta + \alpha b), & \\ \qquad \text{if } \dfrac{1}{c+\alpha}(\beta - b - (2c+\alpha)K) \le q_i \le \dfrac{1}{c+\alpha}(\beta - b + (2c+\alpha)K), & \\ \dfrac{\alpha b}{c+\alpha} + \dfrac{c}{c+\alpha}(\beta - \alpha K), & \text{if } q_i > \dfrac{1}{c+\alpha}(\beta - b + (2c+\alpha)K). \end{cases} \tag{5.19}$$

Now, based on the above analytic results, each FTR model is analyzed in the following sections.

#### 5.5.2.1 FTR Option Model

We have:

**Lemma 4.** *Firm i's optimal output* $q_i^{\eta_i^{ij}}$ ($q_i^{ji}$) *with* $\eta_i^{ij}$ ($\eta_i^{ji}$) *is:*

$$q_i^{\eta_i^{ij}} = \frac{\beta - b + \alpha\left(1 + \eta_i^{ij}\right)K}{c + 2\alpha}, \tag{5.20}$$

$$q_i^{\eta_i^{ji}} = \frac{\beta - b - \alpha\left(1 + \eta_i^{ji}\right)K}{c + 2\alpha}. \tag{5.21}$$

*Proof.* See Appendix.

This shows that the larger $\eta_i^{ij}$, the larger $q_i^{\eta_i^{ij}}$, while the larger $\eta_i^{ji}$, the smaller $q_i^{\eta_i^{ji}}$. Joskow and Tirole's result (2000) corresponds only to (5.21) since they considered only the flow direction from *j* to *i*.

#### 5.5.2.2 FTR Obligation Model

We have:

**Lemma 5.** *Firm i's optimal output with an FTR obligation* $\gamma_i$ *will be either* $\frac{\beta - b + \alpha(1 + \gamma_i)K}{c + 2\alpha}$ or $\frac{\beta - b - \alpha(1 - \gamma_i)K}{c + 2\alpha}$.

*Proof.* See Appendix.

Note that $\gamma_i \in [-1, 1]$, where the sourcing direction is assumed positive. By investigating the analytic representation of the optimal output, we can easily see that by possessing larger $\gamma_i$, both the optimal aggressive and passive outputs will increase.

Although Joskow and Tirole's model (2000) also considers FTR obligations, their model cannot examine the whole characteristics of FTR obligations, i.e., the negative revenue from FTR obligations, since they limited the direction of line flow. As stated in 3.4.2.1, their model actually corresponds to the FTR option model with $\eta_i^{ji}$ corresponding to negative $\gamma_i$. They concluded that if the firm *i* "holds financial rights, these rights will enhance its market power", but this conclusion depends on the assumed directions of the flow and FTRs. The conclusion in this subsection is that, by possessing larger positive $\gamma_i$, both the optimal aggressive and passive outputs will increase. That is, larger FTR obligations will mitigate the right holding firm's market power in this case.

## 5.6 Summary and Conclusion

As stated in the work of Borenstein et al. (2000), the full benefits of competition can be achieved by connecting two markets with a sufficiently large capacity line so that each generator would compete over the merged market instead of over a residual market of its own. In this chapter, we have demonstrated how to analyze the impact of ownership of FTRs on competition, and showed that, by introducing FTRs in an appropriate manner, the physical capacity needed for the full benefits of competition can be reduced. It has also shown that, by introducing FTRs, we may reduce the required physical capacity of the transmission line that is necessary to achieve a pure strategy equilibrium, particularly for achieving the unconstrained Cournot equilibrium that gives the full benefits of competition of a merged market. We have provided separate results for FTR option models and for an FTR obligation model in this chapter. This enables the results to be applied to a market using a specific FTR model.

We also extended the FTR models by considering asymmetric markets and by assuming that one of the markets is perfectly competitive. Asymmetry of markets makes it possible for the ownership of FTRs to change market equilibrium from the unconstrained Cournot equilibrium to a passive/aggressive equilibrium. By constraining one market to be competitive, we could show a similar result to that in the work of Joskow and Tirole (2000). Moreover, other results from the same model were also obtained and some of them show that FTRs may reduce the firm's market power while Joskow and Tirole showed only the result of enhancing the firm's market power.

## Appendix

This appendix provides proofs of the Lemmas.

*Proof of Lemma 1.* First, we consider the direction of option share from market $i$ to $j$. Suppose that there is no congestion. In this case, prices are equated across markets so each market gets half of the total output of both firms. On the other hand, if there is congestion, then the prices of the markets are different. Two congested situations can be differentiated: one is to import $K$ MW with line congestion, and the other is to export $K$ MW with line congestion. Here, I notice that line congestion can occur only when the output difference of both firms is greater than $2K$ MW. More precisely, market $i$ imports with congestion when $q_i < q_j - 2K$, and exports with congestion when $q_i > q_j + 2K$. In this setting, the profit of firm $i$ is represented by the profit function $\pi_i$:

$$\pi_i = \begin{cases} P(q_i - K)q_i + \eta_i^{ij} K\big(P\big(q_j + K\big) - P(q_i - K)\big) - C(q_i), & \textit{if } q_i > q_j + 2K, \\ P(q_i + K)q_i - C(q_i), & \textit{if } q_i < q_j - 2K, \\ P\left(\dfrac{q_i + q_j}{2}\right)q_i - C(q_i), & \textit{if } q_j - 2K \le q_i \le q_j + 2K, \end{cases} \tag{5.22}$$

where $q_i$ is firm $i$'s output and $q_j$ is firm $j$'s output.

Using the explicit forms of demand and costs from (5.1) to (5.2), the profit function of firm $i$ is:

$$\pi_i = \begin{cases} q_i(-\alpha q_i + \alpha K + \beta) + \eta_i^{ij} K\big(\alpha\big(q_i - q_j\big) - 2\alpha K\big) - \dfrac{a}{2}q_i^2 - bq_i - c, & \textit{if } q_i > q_j + 2K, \\ q_i(-\alpha q_i - \alpha K + \beta) - \dfrac{a}{2}q_i^2 - bq_i - c, & \textit{if } q_i < q_j - 2K, \\ q_i\left(-\alpha\dfrac{q_i + q_j}{2} + \beta\right) - \dfrac{a}{2}q_i^2 - bq_i - c, & \textit{if } q_j - 2K \le q_i \le q_j + 2K. \end{cases} \tag{5.23}$$

From (5.23) and the definition, firm $i$'s optimal aggressive and passive outputs and Cournot best response output, which are denoted by $q_i^{uo_{ij}+}\left(K, \eta_i^{ij}\right)$, $q_i^{uo_{ij}-}(K)$, and $q_i^{obC}\left(q_j\right)$, respectively, are obtained by (5.24), (5.25), and (5.26):

$$q_i^{uo_{ij}+}\left(K, \eta_i^{ij}\right) = \arg\max_{q_i} \times \left[q_i(-\alpha q_i + \alpha K + \beta) + \eta_i^{ij} K\big(\alpha\big(q_i - q_j\big) - 2\alpha K\big) - \frac{a}{2}q_i^2 - bq_i - c\right], \tag{5.24}$$

$$q_i^{uo_{ij}-}(K) = \arg\max_{q_i} \left[q_i(-\alpha q_i - \alpha K + \beta) - \frac{a}{2}q_i^2 - bq_i - c\right], \tag{5.25}$$

$$q_i^{uo_{ij}C}\left(q_j\right) = \arg\max_{q_i} \left[q_i\left(-\alpha\frac{q_i + q_j}{2} + \beta\right) - \frac{a}{2}q_i^2 - bq_i - c\right]. \tag{5.26}$$

By solving (5.24), (5.25), and (5.26), the optimal aggressive and passive outputs and the Cournot best response output can be explicitly expressed as (5.27), (5.28), and (5.29):

$$q_i^{uo_{ij}+}\left(K, \eta_i^{ij}\right) = \frac{\beta + \left(1 + \eta_i^{ij}\right)\alpha K - b}{2\alpha + a}, \tag{5.27}$$

$$q_i^{uo_{ij}-}(K) = \frac{\beta - \alpha K - b}{2\alpha + a}, \tag{5.28}$$

$$q_i^{uo_{ij}C}(q_j) = -\frac{\alpha}{2(\alpha + a)} q_j + \frac{\beta - b}{\alpha + a}. \tag{5.29}$$

By comparing (5.27), (5.28), and (5.29) with (5.3), (5.4), and (5.5), we can easily observe that (5.6), (5.7), and (5.8) hold.

Similarly, for the case in which firm $i$ possesses an $\eta_i^{ji}$ FTR option in the other direction, the following results are obtained:

$$q_i^{uo_{ji}+}(K) = \frac{\beta + \alpha K - b}{2\alpha + a}, \tag{5.30}$$

$$q_i^{uo_{ji}-}\left(K, \eta_i^{ji}\right) = \frac{\beta - \left(1 + \eta_i^{ji}\right)\alpha K - b}{2\alpha + a}, \tag{5.31}$$

$$q_i^{uo_{ji}C}(q_j) = -\frac{\alpha}{2(\alpha + a)} q_j + \frac{\beta - b}{\alpha + a}, \tag{5.32}$$

Therefore, we also observe that (5.9), (5.10), and (5.11) hold.
**Q.E.D.**

*Proof of Lemma 3.* In the FTR obligation model, the profit of each firm $i$ is represented by the profit function $\pi_i$:

$$\pi_i = \begin{cases} P(q_i - K)q_i + \gamma_i K\left(P(q_j + K) - P(q_i - K)\right) - C(q_i), & \text{if } q_i > q_j + 2K, \\ P(q_i + K)q_i - \gamma_i K\left(P(q_i + K) - P(q_j - K)\right) - C(q_i), & \text{if } q_i < q_j - 2K, \\ P\left(\dfrac{q_i + q_j}{2}\right) q_i - C(q_i), & \text{if } q_j - 2K \le q_i \le q_j + 2K, \end{cases} \tag{5.33}$$

where $q_i$ is firm $i$'s output and $q_j$ is firm $j$'s output.

Using the explicit forms of demand and costs of (3.1) and (3.2), the profit function of firm $i$ is rewritten such that:

$$\pi_i = \begin{cases} q_i(-\alpha q_i + \alpha K + \beta) + \gamma_i K\left(\alpha(q_i - q_j) - 2\alpha K\right) - \dfrac{a}{2} q_i^2 - b q_i - c, & \text{if } q_i > q_j + 2K, \\ q_i(-\alpha q_i - \alpha K + \beta) - \gamma_i K\left(\alpha(q_j - q_i) - 2\alpha K\right) - \dfrac{a}{2} q_i^2 - b q_i - c, & \text{if } q_i < q_j - 2K, \\ q_i\left(-\alpha \dfrac{q_i + q_j}{2} + \beta\right) - \dfrac{a}{2} q_i^2 - b q_i - c, & \text{if } q_j - 2K \le q_i \le q_j + 2K. \end{cases} \tag{5.34}$$

From (5.34) and the definition, firm $i$'s optimal aggressive and passive outputs and Cournot best response output, which are denoted by $q_i^{ob+}(K,\gamma_i), q_i^{ob-}(K,\gamma_i)$, and $q_i^{obC}(q_j)$ respectively, are obtained by (5.35), (5.36), and (5.37).

$$q_i^{ob+}(K,\gamma_i) = \arg\max_{q_i}\left[q_i(-\alpha q_i + \alpha K + \beta) + \gamma_i K\left(\alpha\left(q_i - q_j\right) - 2\alpha K\right) - \frac{a}{2}q_i^2 - bq_i - c\right], \tag{5.35}$$

$$q_i^{ob-}(K,\gamma_i) = \arg\max_{q_i}\left[q_i(-\alpha q_i - \alpha K + \beta) - \gamma_i K\left(\alpha\left(q_j - q_i\right) - 2\alpha K\right) - \frac{a}{2}q_i^2 - bq_i - c\right], \tag{5.36}$$

$$q_i^{obC}(q_j) = \arg\max_{q_i}\left[q_i\left(-\alpha\frac{q_i + q_j}{2} + \beta\right) - \frac{a}{2}q_i^2 - bq_i - c\right]. \tag{5.37}$$

By solving (5.35), (5.36), and (5.37), the optimal aggressive and passive outputs and the Cournot best response output can be explicitly expressed as (5.16), (5.17), and (5.18).

**Q.E.D.**

*Proof of Lemma 4.* The profit of firm $i$, $\pi_i^{\eta_i^{ij}}$, with an FTR $\eta_i^{ij}$ is represented as:

$$\pi_i^{\eta_i^{ij}} = \begin{cases} -\left(\frac{c}{2}+\alpha\right)q_i^2 - (b+\alpha K-\beta)q_i - a, & \text{if } q_i < \frac{1}{c+\alpha}(\beta - b - (2c+\alpha)K), \\ -\left(\frac{c}{2}+\frac{\alpha c}{2c+\alpha}\right)q_i^2 - \left(b - \frac{1}{2c+\alpha}(2c\beta+\alpha b)\right)q_i - a, & \\ \quad \text{if } \frac{1}{c+\alpha}(\beta - b - (2c+\alpha)K) \le q_i \le \frac{1}{c+\alpha}(\beta - b + (2c+\alpha)K), & \\ -\left(\frac{c}{2}+\alpha\right)q_i^2 - \left(b - \alpha\left(1+\eta_i^{ij}\right)K - \beta\right)q_i - a - \frac{\alpha\eta_i^{ij}K}{c+\alpha}(\beta - b + (2c+\alpha)K), & \\ \quad \text{if } q_i > \frac{1}{c+\alpha}(\beta - b + (2c+\alpha)K). & \end{cases} \tag{5.38}$$

The profit of firm $i$, $\pi_i^{\eta_i^{ji}}$, with an FTR $\eta_i^{ji}$ is represented as:

$$\pi_i^{\eta_i^{ji}} = \begin{cases} -\left(\frac{c}{2}+\alpha\right)q_i^2 - \left(b+\alpha\left(1+\eta_i^{ji}\right)K-\beta\right)q_i - a - \frac{\alpha\eta_i^{ji}K}{c+\alpha}(\beta-b-(2c+\alpha)K), \\ \qquad if\ q_i < \frac{1}{c+\alpha}(\beta-b-(2c+\alpha)K), \\ -\left(\frac{c}{2}+\frac{\alpha c}{2c+\alpha}\right)q_i^2 - \left(b-\frac{1}{2c+\alpha}(2c\beta+\alpha b)\right)q_i - a, \\ \qquad if\ \frac{1}{c+\alpha}(\beta-b-(2c+\alpha)K) \le q_i \le \frac{1}{c+\alpha}(\beta-b+(2c+\alpha)K), \\ -\left(\frac{c}{2}+\alpha\right)q_i^2 - (b-\alpha K-\beta)q_i - a, \qquad if\ q_i > \frac{1}{c+\alpha}(\beta-b+(2c+\alpha)K). \end{cases} \tag{5.39}$$

Since there is no strategic response from market $j$, firm $i$ faces the above profit function to maximize. Each of $\pi_i^{\eta_i^{ij}}$ and $\pi_i^{\eta_i^{ji}}$ has three different regions with respect to $q_i$ and we need to compare the maximum profit in each region to identify firm $i$'s optimal output. Since we can observe that the possession of FTRs only affects the third row of (5.38) and the first row of (5.39), we need to consider only these two rows in order to assess the effect of FTR rights. So, suppose that the maximum profit is obtained by the third row of (5.38) or the first row of (5.39) with the FTR option $\eta_i^{ij}$ and $\eta_i^{ji}$, respectively. Then, firm $i$'s optimal output $q_i^{\eta_i^{ij}}$ ($q_i^{ji}$) with $\eta_i^{ij}$ ($\eta_i^{ji}$) will be (5.20) ((5.21)).

**Q.E.D.**

*Proof of Lemma 5.* We denote by $\pi_i^{\gamma_i}$ firm $i$'s profit with an FTR obligation $\gamma_i$. It is given by:

$$\pi_i^{\gamma_i} = \begin{cases} -\left(\frac{c}{2}+\alpha\right)q_i^2 - (b+\alpha(1-\gamma_i)K-\beta)q_i - a + \frac{\alpha\gamma_i K}{c+\alpha}(\beta-b-(2c+\alpha)K), \\ \qquad if\ q_i < \frac{1}{c+\alpha}(\beta-b-(2c+\alpha)K), \\ -\left(\frac{c}{2}+\frac{\alpha c}{2c+\alpha}\right)q_i^2 - \left(b-\frac{1}{2c+\alpha}(2c\beta+\alpha b)\right)q_i - a, \\ \qquad if\ \frac{1}{c+\alpha}(\beta-b-(2c+\alpha)K) \le q_i \le \frac{1}{c+\alpha}(\beta-b+(2c+\alpha)K), \\ -\left(\frac{c}{2}+\alpha\right)q_i^2 - (b-\alpha(1+\gamma_i)K-\beta)q_i - a - \frac{\alpha\gamma_i K}{c+\alpha}(\beta-b+(2c+\alpha)K), \\ \qquad if\ q_i > \frac{1}{c+\alpha}(\beta-b+(2c+\alpha)K). \end{cases} \tag{5.40}$$

To examine the effect of FTR obligations, we suppose that the maximum profit is obtained either by the first row or by the third row of (5.40). Then, firm $i$'s optimal output will be either $\frac{\beta - b + \alpha(1 + \gamma_i)K}{c + 2\alpha}$ or $\frac{\beta - b - \alpha(1 - \gamma_i)K}{c + 2\alpha}$.

**Q.E.D.**

# References

Borenstein S, Bushnell J, Stoft SE (2000) The competitive effects of transmission capacity in a deregulated electricity industry. Rand J Econ 31:294–325

California ISO (2011) Business practice manual for congestion revenue rights. Available: www.caiso.com

Chandley JD, Hogan WW (2002) Initial comments of John D. Chandley and William W. Hogan on the standard market design NOPR. 11 Nov 2002. Available: http://www.whogan.com

Cho I-K (2003) Competitive equilibrium in a radial network. Rand J Econ 34:438–460

ERCOT (2011) ERCOT nodal protocols. www.ercot.com

Federal Energy Regulatory Commission (2002) Notice of proposed rulemaking remedying undue discrimination through open access transmission service and standard market design. Docket No. RM01-12-000, 31 July 2002

Gilbert R, Neuhoff K, Newbery D (2004) Allocating transmission to mitigate market power in electricity networks. Rand J Econ 35:691–709

Joskow P, Tirole J (2000) Transmission rights and market power on electric power networks. Rand J Econ 31:450–487

Joung M (2008) Game-theoretic equilibrium analysis applications to deregulated electricity markets. Ph.D. thesis, The University of Texas at Austin

Léautier TO (2000) Regulation of an electric power transmission company. Energy J 21:61–92

Midwest ISO (2011) ARR and FTR overview. www.midwestiso.org

New England ISO (2011) Financial transmission rights (FTRs) frequently asked questions. www.iso-ne.com

New York ISO (2011) Market participants user's guide. www.nyiso.com

PJM (2011) FTR market frequently asked questions. www.pjm.com

Pritchard G, Philpott A (2005) On financial transmission rights and market power. Decis Supp Syst 40:507–515

Rosellón J (2003) Different approaches towards transmission expansion. Rev Net Econ 2(3): 238–269

Schweppe F, Caramanis M, Tabors R, Bohn R (1988) Spot pricing of electricity. Kluwer, Boston

Stoft SE (1999) Financial transmission rights meet Cournot: how TCC's curb market power. Energy J 20:1–23

Willems B (2002) Modeling Cournot competition in an electricity market with transmission constraints. Energy J 23:95–125

# Chapter 6
# A Merchant Mechanism for Electricity Transmission Expansion

**Tarjei Kristiansen and Juan Rosellón**

## 6.1 Introduction

The analysis of incentives for electricity transmission expansion is not easy. Beyond economies of scale and cost sub-additivity externalities in electricity transmission are mainly due to "loop flows" that come up from complex network interactions.[1] The effects of loop flows imply that transmission opportunity costs are a function of the marginal costs of energy at each location. Power costs and transmission costs depend on each other since they are simultaneously settled in electricity dispatch. Loop flows imply that certain transmission investments might have negative externalities on the capacity of other (perhaps distant) transmission links (see Bushnell and Stoft 1997). Moreover, the addition of new transmission capacity can sometimes paradoxically decrease the total capacity of the network (Hogan 2002a).

---

This paper was originally published as: Kristiansen T, Rosellón J (2006) A merchant mechanism for electricity transmission expansion. J Regul Econ 29(2):167–193. We are grateful to William Hogan for very useful suggestions and discussions. We also thank Ross Baldick, Ingo Vogelsang, and an anonymous referee for insightful comments. All remaining errors are our own. Part of this research was carried out while Kristiansen was a Fellow, and Rosellón a Fulbright Senior Fellow, at the Harvard Electricity Policy Group, Center of Business and Government, John F. Kennedy School of Government of Harvard University. Rosellón acknowledges support from the Repsol-YPF-Harvard Kennedy School Fellows Program and the Fundación México en Harvard, as well as from CRE and Conacyt.

[1] See Joskow and Tirole (2000), and Léautier (2001).

T. Kristiansen (✉)
KEMA Consulting GmbH, Kurt-Schumacher Strasse 8, Bonn D-53113, Germany
e-mail: tarjei_kristiansen_2003@alumni.ksg.harvard.edu

J. Rosellón
División de Economía, Centro de Investigación y Docencia Económicas (CIDE) and DIW Berlin, Carretera México-Toluca 3655, Lomas de Santa Fé, México D.F. 01210, Mexico
e-mail: juan.rosellon@cide.edu

J. Rosellón and T. Kristiansen (eds.), *Financial Transmission Rights*, Lecture Notes in Energy 7, DOI 10.1007/978-1-4471-4787-9_6,

The welfare effects of an increment in transmission capacity are analyzed by Léautier (2001). The welfare outcome of an expansion in the transmission grid depends on the weight in the welfare function of the generators' profits relative to the consumers' utility weight. Incumbent generators in load pockets are not in general the best agents to carry out transmission expansion projects. Even though an increase in transmission capacity might allow them to increase their revenues due to increased access to new markets and higher transmission charges, such gains are usually overcome by the loss of their local market power.

The literature on incentives for long-term expansion of the transmission network is scarce. The economic analysis of electricity markets has been reduced to short-run issues, and has typically assumed that transmission capacity is fixed (see Joskow and Tirole 2003). However, transmission capacity is random in nature, and it jointly depends on generation investment.

The way to solve transmission congestion in the short run is well known. In a power flow model, the price of transmission congestion is determined by the difference in nodal prices (see Hogan 1992, 2002b). Yet, there is no consensus with respect to the method to attract investment to finance the long-term expansion of the transmission network, so as to reconcile the dual opposite incentives to congest the network in the short run, and to expand it in the long run. Incentive structures proposed to promote transmission investment range from a "merchant" mechanism, based on long-term financial transmission right (LTFTR) auctions (as in Hogan 2002a), to regulatory mechanisms that charge the transmission firm the social cost of transmission congestion (see Léautier 2000; Vogelsang 2001; Joskow and Tirole 2002).

In practice, regulation has been used in England, Wales and Norway to promote transmission expansion, while a combination of planning and auctions of long-term transmission rights has been tried in the Northeast of the U.S. A mixture of regulatory mechanisms and merchant incentives is alternatively used in the Australian market.

In this paper we develop a merchant model to attract investment to *small-scale* electricity transmission projects based on LTFTR auctions. Locational prices give market players incentives to initiate transmission investments. FTRs provide transmission property rights, since they hedge the market player against future price differences. Our model further develops basic conditions under which FTRs and locational pricing provide incentives for long-term investment in the transmission network.

In meshed networks, a change in network capacity might imply negative externalities on transmission property rights. Then, in the process of allocation of incremental FTRs, the system operator must reserve certain unallocated FTRs so that the revenue adequacy of the transmission system is preserved. In order to deal with this issue, we develop a bi-level programming model for allocation of long-term FTRs and apply it to different network topologies.

The structure of the paper is as follows. In Sect. 6.2 we carry out an analytical review on the relevant literature on electricity transmission expansion. In Sect. 6.3 we develop our model. We first introduce FTRs and the feasibility rule, and then address the rationale for FTR allocation and efficient investments. We develop general optimality conditions as well. In Sect. 6.4, we carry out applications of our model to a radial line, and to a three-node network. Next, we describe the welfare implications in Sect. 6.5. In Sect. 6.6 we provide concluding comments.

## 6.2 Literature Review

There exist some hypotheses on structures for transmission investment: the market-power hypothesis, the incentive-regulation hypothesis, and the long-run financial-transmission-right hypothesis. The first approach seeks to derive optimal transmission expansion from the power-market structure of power generators, and takes into account the conjectures of each generator regarding other generators' marginal costs due to the expansion (Sheffrin and Wolak 2001; Wolak 2000, and The California ISO and London Economics International 2003). The generators' bidding behaviors are estimated before and after a transmission upgrade, and a real-option analysis is used to derive the net present value of transmission and generation projects together with the computation of their joint probability.

The model shows that there are few benefits of transmission expansion until added capacity surpasses a certain threshold that, in turn, is determined by the possibility of induced congestion by the strategic behavior of generators with market power. The generation market structure then determines when transmission expansion yield benefits. Additionally, many small upgrades of the transmission grid are preferable to large greenfield projects when cost uncertainty is added to the model.

The contribution of this method is that it models the existing interdependence of transmission investment and generation investment within a transportation model with no network loop flows. However, as pointed out by Hogan (2002b), the use of a transportation model in the electricity sector is inadequate since it does not deal with discontinuities in transmission capacity implied by the multidimensional character of a meshed network.

The second method for transmission expansion is a regulatory alternative that relies on a "Transco" that simultaneously runs system operation and owns the transmission network. The Transco is regulated through benchmark regulation or price regulation so as to provide it with incentives to invest in the development of the grid, while avoiding congestion. Léautier (2000), Grande and Wangensteen (2000), and Harvard Electricity Policy Group (2002) discuss mechanisms that compare the Transco performance with a measure of welfare loss due to its activities. Joskow and Tirole (2002) propose a surplus-based mechanism to reward the Transco according to the redispatch costs avoided by the expansion, so that the Transco faces the complete social cost of transmission congestion.

Another regulatory alternative is a two-part tariff cap proposed by Vogelsang (2001) that addresses the opposite incentives to congest the existing transmission grid in the short run, and to expand it in the long run. Incentives for investment in expansion of the network are achieved through the rebalancing of the fixed part and the variable part of the tariff. This method tries to deepen into the analysis of the cost and production functions for transmission services, which are not very well understood in the economics literature. Nonetheless, to achieve this goal Vogelsang needs to define an output (or throughput) for the Transco. As argued in the FTR literature (Bushnell and Stoft 1997; Hogan 2002a, b), this task is very difficult since the average physical flow through a meshed transmission network is not well defined.

The third approach is a "merchant" one based on LTFTR auctions by an independent system operator (ISO). This method deals with loop-flow externalities in that, to proceed with line expansions, the investor pays for the negative externalities it generates. To restore feasibility, the investor has to buy back sufficient transmission rights from those who hold them initially, or the ISO retains some unallocated transmission rights (*proxy awards*) during the LTFTR auction to protect unassigned rights while simultaneous feasibility of the system protects the rights of the existing FTR holders. This is the core of an LTFTR auction (see Hogan 2002a).

Joskow and Tirole (2003) criticize the LTFTR approach. They argue that the efficiency results of the *short-run* version of the FTR model rely on perfect-competition assumptions, which are not real for transmission networks. Moreover, defining an operational FTR auction is technically difficult[2] and, according to these authors, the FTR analysis is static (a contradiction with the dynamics of transmission investment). Joskow and Tirole analyze the implications of eliminating the perfect competition assumptions of the FTR model.

First, market power and vertical integration might impede the success of FTR auctions. Prices will not reflect the marginal cost of production in regions with transmission constraints. Generators in constrained regions will then withdraw capacity in order to increase their prices, and will overestimate the cost-saving gains from investments in transmission.[3]

Second, lumpiness in transmission investment makes the total value paid to investors through FTRs less than the social surplus created. The large and lumpy nature of major transmission upgrades requires long-term contracts before making the investment, or temporal property rights for the incremental investment.

Third, contingencies in electricity transmission impede the merchant approach to really solve the loop-flow problem. Moreover, existing transmission capacity and incremental capacity are stochastic. Even in a radial line, realized capacity could be less than expected capacity and the revenue-adequacy condition would not be met. Even more, the initial feasible FTR set can depend on random exogenous variables.

Fourth, an expansion in transmission capacity might negatively affect social welfare (as shown by Bushnell and Stoft 1997).

[2] No restructured electricity sector in the world has adopted a pure merchant approach towards transmission expansion. Australia has implemented a mixture of regulated and merchant approaches (see Littlechild 2003). Pope (2002), and Harvey (2002) propose LTFTR auctions for the New York ISO to provide a hedge against congestion costs. Gribik et al. (2002) propose an auction method based on the physical characteristics (capacity and admittance) of a transmission network.

[3] Generators can exert local power when the transmission network is congested. (See Bushnell 1999; Bushnell and Stoft 1997; Joskow and Tirole 2000; Oren 1997; Joskow and Schmalensee 1983; Chao and Peck 1997; Gilbert et al. 2002; Cardell et al. 1997; Borenstein et al. 1998; Wolfram 1998; Bushnell and Wolak 1999.)

Fifth, a moral hazard "in teams" problem arises due to the separation of transmission ownership and system operation in the FTR model. For instance, an outage can be claimed to be the consequence of poor maintenance (by the transmission owner) or of negligent dispatch (by the system operator).[4] Additionally, there is no perfect coordination of interdependent investments in generation and transmission, and stochastic changes in supply and demand conditions imply uncertain nodal prices. Likewise, there is no equal access to investment opportunities since only the incumbent can efficiently carry out deepening transmission investments.

Hogan (2003) responds to the above criticisms by arguing that LTFTRs only grant efficient outcomes under lack of market power, and non-lumpy marginal expansions of the transmission network. Furthermore, Hogan argues that regulation has an important role in fostering large and lumpy projects, and in mitigating market power abuses.

As argued by Pérez-Arriaga et al. (1995), revenues from nodal prices typically recover only 25 % of total costs. LTFTRs should then be complemented with a fix-price structure or, as in Rubio-Oderiz and Pérez-Arriaga (2000) a *complementary charge* that allows the recovery of fixed costs.[5] This fact is recognized by Hogan (1999) who states that complete reliance on market incentives for transmission investment is undesirable. Rather, Hogan (2003) claims that merchant and regulated transmission investments might be combined so that regulated transmission investment is limited to projects where investment is large relative to market size, and lumpy so that it only makes sense as a single project as opposed as to many incremental small projects.

Hogan also responds to contingency concerns.[6] On one hand, only those contingencies outside the control of the system operator could lead to revenue inadequacy of FTRs, but such cases are rare and do not represent the most important contingency conditions. On the other hand, most of remaining contingencies are foreseen in a security-constrained dispatch of a meshed network with loops and parallel paths. If one of "n" transmission facilities is lost, the remaining power flows would still be feasible in an "n − 1" contingency constrained dispatch.

[4] An example is the power outage of August 14, 2003, in the Northeast of the US, which affected six control areas (Ontario, Quebec, Midwest, PJM, New England, and New York) and more than 20 million consumers. A 9-s transmission grid technical and operational problem caused a cascade effect, which shut down 61,000 MW generation capacity. After the event there were several "finger pointings" among system operators of different areas, and transmission providers. The US-Canada System Outage Task Force identified in detail the causes of the outage in its final report of April, 2004. It shows that the main causes of the black out were deficiencies in corporate policies, lack of adherence to industry policies, and inadequate management of reactive power and voltage by First Energy (a firm that operates a control area in northern Ohio) and the Midwest Independent System Operator (MISO). See US-Canada Power System Outage Task Force (2004).

[5] In the US, transmission fixed costs are recovered through a regulated fixed charge, even in those systems that are based on nodal pricing and FTRs. This charge is usually regulated through cost of service.

[6] See Hogan (2002a, b, 2003).

Hogan (2003) also assumes that agency problems and information asymmetries are part of an institutional structure of the electricity industry where the ISO is separated from transmission ownership and where market players are decentralized. However, he claims that the main issue on transmission investment is the decision of the boundary between merchant and regulated transmission expansion projects. He argues that asymmetric information should not necessarily affect such a boundary.

Hogan (2002a) finally analyzes the implications of loop flows on transmission investment raised by Bushnell and Stoft (1997). He analytically provides some general axioms to properly define LTFTRs so as to deal with negative externalities implied by loop flows. We next present a model that develops the general analytical framework suggested by Hogan (2002a).

## 6.3 The Model

Assume an institutional structure where there are various established agents (generators, Gridcos, marketers, etc.) interested in the transmission grid expansion. Agents do not have market power in their respective market or, at least, there are in place effective market-power mitigation measures.[7] Also assume that transmission projects are incrementally small (relative to the total network) and non-lumpy so that the project does not imply a relatively large change in nodal-price differences. However, although projects are small, they might change or not the power transfer distribution factors (PTDFs) of the network.[8]

Under an initial condition of non-fully allocation of FTRs in the grid, the auctioning of incremental LTFTRs should satisfy the following basic criteria in order to deal with possible negative externalities associated with the expansion.

An LTFTR increment must keep being simultaneously feasible (*feasibility rule*).

1. An LTFTR increment remains simultaneously feasible given that certain currently unallocated rights (or *proxy awards*) are preserved.
2. Investors should maximize their objective function (*maximum value*).

[7] In fact, market power mitigation may be a major motive for transmission investment. A generator located outside a load pocket might want to access the high price region inside the pocket. Building a new line would mitigate market power if it creates new economic capacity (see Joskow and Tirole 2000).

[8] Examples of projects that do not change PTDFs include proper maintenance and upgrades (e.g. low sag wires), and the capacity expansion of a radial line. Such investments could be rewarded with flowgate rights in the incremental capacity without affecting the existing FTR holders (we assume however that only FTRs are issued). In our three-node example in Sect. 6.6.2, PTDFs change substantially. In certain cases, the change in PTDFs could not exist (see Appendix 3) or be small if, for example, a line is inserted in parallel with an already existing line (see Appendix 3). In a large-scale meshed network the change in PTDFs may not be as substantial as in a three-node network. However the auction problem is non-convex and nonlinear, and a global optimum might not be ensured. Only a local optimum might be found through methods such as sequential quadratic programming.

3. The LTFTR awarding process should apply both for decreases and increases in the grid capacity (*symmetry*).

The need for proxy awards arises whenever there is less than full allocation of the capacity of the existing grid. This occurs prominently during a transition to an electricity market when there is reluctance to fully allocate the existing grid for all future periods. Hence FTRs for the existing grid are short term (this period), but investors in grid expansion seek long term rights (next period). Full allocation of the existing grid seems necessary but not sufficient for defining and measuring incremental capacity. Hogan explains though that defining proxy awards is a difficult task. We next address this issue in a formal way in the context of an auction model designed to attract investment for transmission expansion.

## 6.4 The Power Flow Model and Proxy Awards

Consider the following economic dispatch model[9]:

$$\underset{Y,u\in U}{\text{Max}}\, B(d-g)$$

$$s.t. \tag{6.1}$$

$$Y = d - g, \tag{6.2}$$

$$L(Y,u) + \tau^T Y = 0 \tag{6.3}$$

$$K(Y,u) \leq 0 \tag{6.4}$$

where $d$ and $g$ are load and generation at the different locations. The variable $Y$ represents the real power bus net loads, including the swing bus $S(Y^T = (Y_s, \bar{Y}^T))$. $B(d-g)$ is the net benefit function,[10] and $\tau$ is a unity column vector, $\tau^T = (1, 1, \ldots, 1)$ All other parameters are represented in the control variable $u$. The objective equation (6.1) includes the maximization of benefit to loads and the minimization of generation costs. Equation (6.2) denotes the net load as the difference between load and generation. Equation (6.3) is a loss balance constraint where $L(Y,u)$ is a vector which denotes the losses in the network. In (6.4) $K(Y,u)$, is a vector of power flows in the lines, which are subject to transmission capacity limits. The corresponding multipliers or shadow

[9] Hogan (2002b) shows that the economic dispatch model can be extended to a market equilibrium model where the ISO produces transmission services, power dispatch, and spot-market coordination, while consumers have a concave utility function that depends on net loads, and on the level of consumption of other goods.

[10] Function $B$ is typically a measure of welfare, such as the difference between consumer surplus and generation costs (see Hogan 2002b).

prices for the constraints are $(P,\ \lambda_{ref},\ \lambda_{tran})$ for net loads, reference bus energy (or loss balance) and transmission constraints, respectively.[11]

The locational prices $P$ are the marginal generation cost or the marginal benefit of demand, which in turn equals the reference price of energy plus the marginal cost of losses and congestion. With the optimal solution $(d^*,\ g^*,\ Y^*,\ u^*)$ and the associated shadow prices, we have the vector of locational prices as:

$$P^T = \nabla C(g^*) = \nabla B(d^*) = \lambda_{ref}\tau^T + \lambda_{ref}\nabla L_Y(Y^*, u^*) + \lambda_{tran}^T \nabla K_Y(Y^*, u^*) \quad (6.5)$$

If losses[12] are ignored, only the energy price at the reference bus and the marginal cost of congestion contribute to set the locational price.

FTR obligations[13] hedge market players against differences in locational prices caused by transmission congestion.[14] FTRs are provided by an ISO, and are assumed to redistribute the congestion rents. The pay-off from these rights is given by:

$$FTR = (P_j - P_i)Q_{ij} \quad (6.6)$$

where $P_j$ is the price at location $j$, $P_i$ is the price at location $i$, and $Q_{ij}$ is the directed quantity injected at point $i$ and withdrawn at point $j$ specified in the FTR. The FTR payoffs can take negative, positive or zero values.

A set of FTRs is said to be simultaneously feasible if the associated set of net loads is simultaneously feasible, that is if the net loads satisfy the loss balance and transmission capacity constraints as well as the power flow equations given by:

$$\begin{aligned} Y &= \sum_k t_k^f \\ L(Y, u) &+ \tau^t Y = 0, \\ K(Y, u) &\leq 0 \end{aligned} \quad (6.7)$$

where $\sum_k t_k^f$ is the sum over the set of point-to-point obligations.[15]

[11] When security constraints are taken into account (n − 1 criterion) this is a large-scale problem, and it prices anticipated contingencies through the security-constrained economic dispatch. In operations the n − 1 criterion can be relaxed on radial paths, however, doing the same in the FTR auction of large-scale meshed networks may result in revenue inadequacy. We do not use the n − 1 criterion in our paper.

[12] In the PJM (Pennsylvania, New Jersey and Maryland) market design, the locational prices are defined without respect to losses (DC-load flow model), while in New York the locational prices are calculated based on an AC-network with marginal losses.

[13] FTRs could be options with a payoff equal to max $((P_j - P_i)\ Q_{ij}, 0)$.

[14] See Hogan (1992).

[15] The set of point-to-point obligations can be decomposed into a set of balanced and unbalanced (injection or withdrawal of energy) obligations (see Hogan 2002b).

If the set of FTRs is simultaneously feasible and the system constraints are convex,[16] then the FTRs satisfy the *revenue adequacy* condition in the sense that equilibrium payments collected by the ISO through economic dispatch will be greater than or equal to payments required under the FTR forward obligations.[17]

Assume now investments in new transmission capacity. The associated set of new FTRs for transmission expansion has to satisfy the simultaneous feasibility rule too. That is, the new and old FTRs have to be simultaneously feasible after the system expansion. Assume that $T$ is the current partial allocation of long-term FTRs, then by assumption it is feasible $(K(T, u) \leq 0)$. Suppose there is to be a total possible incremental award, and that a fraction of the possible awards is reserved as proxy awards for the existing grid with the remainder provided to the incremental investor as representing the proportion that could only be awarded as a result of the investment. Let $a$ be the scalar amount of incremental FTR awards, and $\hat{t}$ the scalar amount of proxy awards. Furthermore let $\delta$ be directional vector[18] such that $a\delta$ is the MW amount of incremental FTR awards, and $\hat{t}\delta$ is the MW amount of proxy awards between different locations. Any incremental FTR award $a\delta$ should comply with feasibility rule in the expanded grid. Hence we must have $K^{+}(T + a\delta, u) \leq 0$, where $K^{+}$ corresponds to capacity of the expanded grid.

When certain currently unallocated rights (proxy awards) $\hat{t}\delta$ in the existing grid must be preserved, combined with existing rights they sum up to $T + \hat{t}\delta$.[19] Then $K^{+}$ should also satisfy simultaneous feasibility so that $K(T + \hat{t}\delta, u) \leq 0$, $K^{+}(T + a\delta, u) \leq 0$, and $K^{+}(T + \hat{t}\delta + a\delta, u) \leq 0$ for incremental awards $a\delta$.

A question then arises regarding the way to best define proxy awards. One possibility is to define them as the "best use" of the current network along the same direction as the incremental awards.[20] This includes both positive and negative incremental FTR awards. The best use in a three-node network may be thought of as a single incremental FTR in one direction or a combination of incremental

---

[16] This has been demonstrated for lossless networks by Hogan (1992), extended to quadratic losses by Bushnell and Stoft (1996), and further generalized to smooth nonlinear constraints by Hogan (2000). As shown by Philpott and Pritchard (2004) negative locational prices may cause revenue inadequacy. Moreover, in the general case of an AC or DC formulation to ensure revenue adequacy the transmission constraints must satisfy optimality conditions (in particular, if such constraints are convex they satisfy optimality). See O'Neill et al. (2002), and Philpott and Pritchard (2004).

[17] Revenue adequacy is the financial counterpart of the physical concept of availability of transmission capacity (see Hogan 2002a).

[18] Each element in the directional vector represents an FTR between two locations and the directional vector may have many elements representing combinations of FTRs.

[19] Proxy awards are then currently unallocated FTRs in the pre-existing network that basically facilitate the allocation of incremental FTRs and help to preserve revenue adequacy by reserving capacity for hedges in the expanded network.

[20] Another possibility would be to define every possible use of the current grid as a proxy award. However, this would imply that any investment beyond a radial line would be precluded, and that incremental award of FTRs might require adding capacity to every link on every path of a meshed network. The idea of defining proxy awards along the same direction as incremental awards originates from a proposal developed for the New Zealand electricity market by Transpower.

FTRs defined by the directional vector $\delta$, depending on the investor preference. Hogan (2002a) suggests two ways of defining "best use":

$$\begin{aligned}&\text{Preset proxy preferences } (p)\\ &\hat{y} = T + \hat{t}\delta,\\ &\hat{t} \in \arg\max_{t} \{\hat{t}p\delta | K(T + t\delta) \leq 0\}\end{aligned} \tag{6.8}$$

or,

$$\begin{aligned}&\text{Investor preferences } (\beta(\mathrm{a}\delta))\\ &\hat{y} = T + \hat{t}\delta,\\ &\hat{t} \in \arg \min_{\substack{t\\ K(T+t\delta)\leq 0}} \left\{ \max_{a\geq 0} \{\beta(a\delta) | K^{+}(T + t\delta + a\delta) \leq 0\} \right\}\end{aligned}$$

In the preset proxy formulation the objective is to maximize the value (defined by prices $p$) of the proxy awards given the pre-existing FTRs, and the power flow constraints in the pre-expansion network. In the investor preference formulation the objective is to maximize the investor's value (defined by the bid functions for different directions, $\beta(a\delta)$) of incremental FTR awards given the proxy and pre-existing FTRs and the power flow constraints in the expanded network, while simultaneously calculating the minimum proxy scalar amount that satisfies the power flow constraints in the pre-expansion network.

We will use as a proxy protocol the first definition. We next analyze the way to use this protocol to carry out an allocation of LTFTRs that stimulates investment in transmission.

## 6.5 The Auction Model

Assume the preset proxy rule is used to derive prices that maximize the investor preference $\beta(a\delta)$ for an award of $a$ MWs of FTRs in direction $\delta$. We then have the following auction maximization problem:

$$\begin{aligned}&\max_{a,\hat{t},\delta} \quad \beta(a\delta)\\ &s.t.\\ &K^{+}(T + a\delta) \leq 0,\\ &K^{+}(T + \hat{t}\delta + a\delta) \leq 0,\\ &\hat{t} \in \arg\max_{t} \{tp\delta | K(T + t\delta) \leq 0\},\\ &\|\delta\| = 1,\\ &a \geq 0.\end{aligned} \tag{6.9}$$

In this model, the investor's preference is maximized subject to the simultaneous feasibility conditions, and the best use protocol. We add a constraint on the (two-) norm[21] of the directional vector to preclude the trivial case $\delta = 0$. We want to explore if such an auction model approach can produce acceptable proxy and incremental awards. We next analyze this issue within a framework that ignores losses, and utilizes a DC-load flow approximation.

The auction model is a nonlinear optimization problem of "bi-level" nature.[22] There are two optimization stages. Maximization is non-myopic since the result of the lower problem (first stage) depends on the direction chosen in the upper problem (second stage).[23] Bi-level problems may be solved by first transforming the lower problem (i.e. the allocation of proxy awards) into to a set of Kuhn-Tucker equations that are subsequently substituted in the upper problem (i.e. the maximization of the investors' preference). The model can then be understood as a Stackelberg problem although it is not intending to optimize the same type of objective function at each stage.[24]

The Lagrangian ($L$) for the lower problem is:

$$L(\hat{t}, \delta, \lambda) = \hat{t}p\delta - \lambda^T(K(T + \hat{t}\delta))$$

where $\lambda^T$ is the Lagrange multiplier vector associated with transmission capacity on the respective transmission lines before the expansion. It is the Lagrange multiplier of the simultaneous feasibility restriction for proxy awards. The Kuhn-Tucker conditions are:

$$\frac{\partial L(\hat{t}, \delta, \lambda)}{\partial \hat{t}} = 0, \quad \frac{\partial L(\hat{t}, \delta, \lambda)}{\partial \lambda} \geq 0,$$
$$\lambda^T \frac{\partial L(\hat{t}, \delta, \lambda)}{\partial \lambda} = 0, \quad \lambda \geq 0$$

The transformed problem is then written as:

$$\begin{aligned}
&\underset{a,\hat{t},\delta,\lambda}{\text{Max}} \quad \beta(a\delta) \\
&s.t. \\
&K^+(T + a\delta) \leq 0, \qquad (\omega) \\
&K^+(T + \hat{t}\delta + a\delta) \leq 0, \qquad (\gamma)
\end{aligned}$$

[21] We use "two-norm" to guarantee differentiability.

[22] See Shimizu et al. (1997).

[23] The model could also be interpreted as having multiple periods. Although we do not explicitly include in our model a discount factor, we assume that it is included in the investor's preference parameter $b$.

[24] Other examples in the economics literature where an upper level maximization takes the optimality conditions of another problem as constraints are given in Mirrlees (1971), Brito and Oakland (1977), and Rosellón (2000).

$$\frac{\partial L(\hat{t},\delta,\lambda)}{\partial \hat{t}} = 0, \qquad (\theta) \tag{6.10}$$

$$\lambda^T \frac{\partial L(\hat{t},\delta,\lambda)}{\partial \lambda} = 0, \qquad (\zeta) \tag{6.11}$$

$$\frac{\partial L(\hat{t},\delta,\lambda)}{\partial \lambda} \geq 0, \qquad (\varepsilon)$$

$$\begin{aligned} \|\delta\| &= 1, && (\varphi) \\ a &\geq 0, && (\kappa) \\ \lambda &\geq 0 && (\pi) \end{aligned}$$

where $\omega, \gamma, \theta, \zeta, \varepsilon, \varphi, \kappa$ and $\pi$ are Lagrange multipliers associated with each constraint. More specifically, $\omega$ is the shadow price of the simultaneous feasibility restriction for existing and incremental FTRs; $\gamma$ is the shadow price of the simultaneous feasibility restriction for existing FTRs, proxy awards and incremental FTRs; $\theta$, $\zeta$, $\varepsilon$ are the shadow prices of the restriction on optimal proxy FTRs; $\varphi$, $\kappa$ are the shadow prices of the non-negativity constraints for $a$ and $\lambda$, respectively; and $\pi$ is the shadow price of the unit restriction on $\delta$.

The Lagrangian of the auction problem is:

$$\begin{aligned} L(a,\hat{t},\delta,\lambda,\Omega) =& \beta(a\delta) - \omega^T(K^+(T+a\delta)) \\ & - \gamma^T(K^+(T+\hat{t}\delta+a\delta)) - \theta^T \frac{\partial L(\hat{t},\delta,\lambda)}{\partial \hat{t}} \\ & - \zeta^T\left(\lambda^T \frac{\partial L(\hat{t},\delta,\lambda)}{\partial \lambda}\right) + \varepsilon^T \frac{\partial L(\hat{t},\delta,\lambda)}{\partial \lambda} \\ & + \varphi^T(1-\|\delta\|) + \kappa^T a + \pi^T \lambda \end{aligned} \tag{6.12}$$

where $\Omega = (\omega, \gamma, \theta, \zeta, \varepsilon, \varphi, \kappa, \pi)$ denotes the vector of Lagrange multipliers. Kuhn-Tucker conditions for the upper problem are:

$$\begin{aligned} \frac{\partial L(a,\hat{t},\delta,\lambda,\Omega)}{\partial a} =& \frac{\partial \beta(a\delta)}{\partial a} - \left[\frac{\partial K^+(T+a\delta)}{\partial a}\right]^T \omega \\ & - \left[\frac{\partial K^+(T+\hat{t}\delta+a\delta)}{\partial a}\right]^T \gamma + \kappa = 0 \end{aligned} \tag{6.13}$$

$$\begin{aligned} \frac{\partial L(a,\hat{t},\delta,\lambda,\Omega)}{\partial \hat{t}} =& - \left[\frac{\partial K^+(T+\hat{t}\delta+a\delta)}{\partial \hat{t}}\right]^T \gamma \\ & - \left[\frac{\partial^2 L(\hat{t},\delta,\lambda)}{\partial \hat{t}\partial \lambda}\right]^T \lambda\zeta + \left[\frac{\partial L^2(\hat{t},\delta,\lambda)}{\partial \hat{t}\partial \lambda}\right]^T \varepsilon = 0 \end{aligned} \tag{6.14}$$

$$\frac{\partial L(a,\hat{t},\delta,\lambda,\Omega)}{\partial \delta} = \frac{\partial \beta(a\delta)}{\partial \delta} - \left[\frac{\partial K^{+}(T+a\delta)}{\partial \delta}\right]^{T} \omega - \left[\frac{\partial K^{+}(T+\hat{t}\delta+a\delta)}{\partial \delta}\right]^{T} \gamma - \left[\frac{\partial^{2} L(\hat{t},\delta,\lambda)}{\partial \delta \partial \hat{t}}\right]^{T} \theta - \left[\frac{\partial^{2} L(\hat{t},\delta,\lambda)}{\partial \delta \partial \lambda}\right]^{T} \lambda\zeta + \left[\frac{\partial^{2} L(\hat{t},\delta,\lambda)}{\partial \delta \partial \lambda}\right]^{T} \varepsilon - \left[\frac{\partial \|\delta\|}{\partial \delta}\right]^{T} \varphi = 0 \quad (6.15)$$

$$\frac{\partial L(a,\hat{t},\delta,\lambda,\Omega)}{\partial \lambda} = -\left[\frac{\partial^{2} L(\hat{t},\delta,\lambda)}{\partial \lambda \partial \hat{t}}\right]^{T} \theta - \left[\frac{\partial L(\hat{t},\delta,\lambda)}{\partial \lambda}\right]^{T} \zeta + \pi = 0, \quad (6.16)$$

$$\frac{\partial L(a,\hat{t},\delta,\lambda,\Omega)}{\partial \omega} = -K^{+}(T+a\delta) \geq 0, \quad (6.17)$$

$$\frac{\partial L(a,\hat{t},\delta,\lambda,\Omega)}{\delta \gamma} = -K^{+}(T+\hat{t}\delta+a\delta) \geq 0 \quad (6.18)$$

$$\frac{\partial L(a,\hat{t},\delta,\lambda,\Omega)}{\partial \theta} = -\frac{\partial L(\hat{t},\delta,\lambda)}{\partial \hat{t}} = 0, \quad (6.19)$$

$$\frac{\partial L(a,\hat{t},\delta,\lambda,\Omega)}{\partial \zeta} = -\lambda^{T} \frac{\partial L(\hat{t},\delta,\lambda)}{\partial \lambda} = 0, \quad (6.20)$$

$$\frac{\partial L(a,\hat{t},\delta,\lambda,\Omega)}{\partial \varepsilon} = \frac{\partial L(\hat{t},\delta,\lambda)}{\partial \lambda} \geq 0, \quad (6.21)$$

$$\frac{\partial L(a,\hat{t},\delta,\lambda,\Omega)}{\partial \varphi} = 1 - \|\delta\| = 0, \quad (6.22)$$

$$\frac{\partial L(a,\hat{t},\delta,\lambda,\Omega)}{\partial \kappa} = a \geq 0, \quad (6.23)$$

$$\frac{\partial L(a,\hat{t},\delta,\lambda,\Omega)}{\partial \pi} = \lambda \geq 0, \quad (6.24)$$

$$\omega^{T} \frac{\partial L(a,\hat{t},\delta,\lambda,\Omega)}{\partial \omega} = 0, \;\; \omega \geq 0, \quad (6.25)$$

$$\gamma^{T} \frac{\partial L(a,\hat{t},\delta,\lambda,\Omega)}{\partial \gamma} = 0, \;\; \gamma \geq 0, \quad (6.26)$$

$$\varepsilon^{T} \frac{\partial L(a,\hat{t},\delta,\lambda,\Omega)}{\partial \varepsilon} = 0, \;\; \varepsilon \geq 0, \quad (6.27)$$

$$\kappa^T a = 0, \ \kappa \geq 0, \tag{6.28}$$

$$\pi^T \lambda = 0, \ \pi \geq 0 \tag{6.29}$$

The constraint $\dfrac{\partial L(\hat{t}, \delta, \lambda)}{\partial \lambda} = 0$ is redundant when the preset proxy preference ($p$) is non-zero, since it is a sub-gradient of the constraint $\lambda^T \dfrac{\partial L(\hat{t}, \delta, \lambda)}{\partial \lambda} = 0$, and $\varepsilon$ is therefore zero when $p$ is non-zero. We show in a later example that $\theta$ and $\varphi$ are zero because the associated constraints are redundant. The binding constraint in the lower level problem is $\lambda^T \dfrac{\partial L(\hat{t}, \delta, \lambda)}{\partial \lambda} = 0$, since some transmission constraints are fully utilized by proxy awards.

This is a nonlinear and non-convex problem, and its solution depends on the investor-preference parameters, the current partial allocation ($T$), and the topology of the network prior to and after the expansion.[25] A general solution method utilizing Kuhn-Tucker conditions would be through checking which of the constraints are binding.[26] One way to identify the active inequality constraints is the active set method.[27] In this paper we solve the problem in detail for different network topologies, including a radial line and a three-node network.

## 6.6 Simulations

### 6.6.1 Radial Line

Let us first analyze a radial transmission line that is expanded as in Fig. 6.1.

The corresponding optimization problem is:

$$\begin{aligned}
&\underset{a,\hat{t},\delta}{\text{Max}} \quad b_{12} a \delta_{12} \\
&s.t. \\
&T_{12} + a\delta_{12} \leq C_{12}^{+}
\end{aligned}$$

[25] According to Shimizu et al. (1997), the necessary optimality conditions for this problem are satisfied. The objective function and the constraints are differentiable functions in the region bounded by the constraints. A local optimal solution and Kuhn-Tucker vectors then exist.

[26] There are other methods available such as transformation methods (penalty and multiplier), and non-transformation methods (feasible and infeasible). See Shimizu et al. (1997).

[27] This method considers a tentative list of constraints that are assumed to be binding. This is a working list, and consists of the indices of binding constraints at the current iteration. Because this list may not be the solution list, the list is modified either by adding another constraint to the list or by removing one from the list. Geometrically, the active set method tends to step around the boundary defined by the inequality constraints. (See Nash and Sofer 1988).

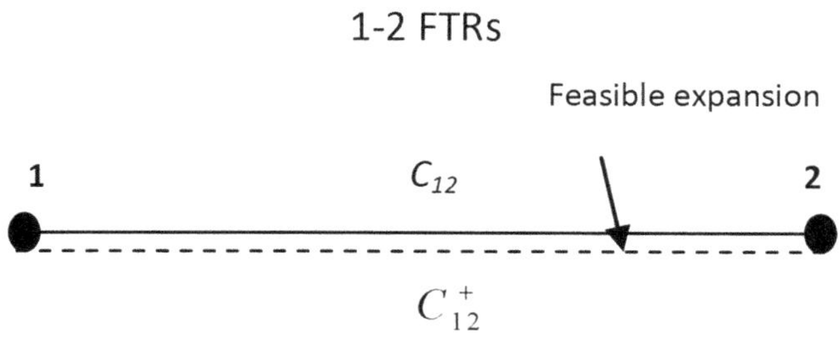


**Fig. 6.1** An expanded line and its feasible expansion

$$
\begin{aligned}
&T_{12} + \hat{t}\delta_{12} + a\delta_{12} \leq C^+_{12} \\
&\hat{t}(\delta_{12}) \in \arg\max_{t} \{tp_{12}\delta_{12} | T_{12} + t\delta_{12} \leq C_{12}\} \\
&\|\delta_{12}\| = 1, \\
&a \geq 0
\end{aligned}
\tag{6.30}
$$

where $C_{12}$ is the transmission capacity of the network before the expansion, $C^+_{12}$ is the transmission capacity of the network after the expansion, and $b_{12}$ is the investor preference. The first order conditions of the lower maximization problem can then be added as constraints to the upper problem:

$$
\begin{aligned}
&\underset{a,\hat{t},\delta_{12},\lambda}{\text{Max}} \quad b_{12}a\delta_{12} \\
&s.t. \\
&T_{12} + a\delta_{12} \leq C^+_{12} \\
&T_{12} + \hat{t}\delta_{12} + a\delta_{12} \leq C^+_{12} \\
&p_{12}\delta_{12} - \lambda\delta_{12} = 0 \\
&\lambda(C_{12} - T_{12} - \hat{t}\delta_{12}) = 0
\end{aligned}
\tag{6.31}
$$

$$
\begin{aligned}
&T_{12} + \hat{t}\delta_{12} \leq C_{12} \\
&\delta^2_{12} = 1 \\
&a, \lambda \geq 0
\end{aligned}
$$

Since the grid is being expanded, the constraint on simultaneous feasibility of incremental FTRs $T_{12} + a\delta_{12} \leq C^+_{12}$ is non-binding. The solution to this problem provides the values for the decision variables, and shadow prices.[28] First, $\delta_{12} = 1$, because the network is being expanded. Additionally $\gamma = b_{12}$ which implies that the higher the value of the investor-preference parameter $b_{12}$ the more the investor values post-expansion transmission capacity (its marginal valuation of transmission capacity increases with the bid value).

[28] The mathematical derivation of these values is presented in Appendix 1.

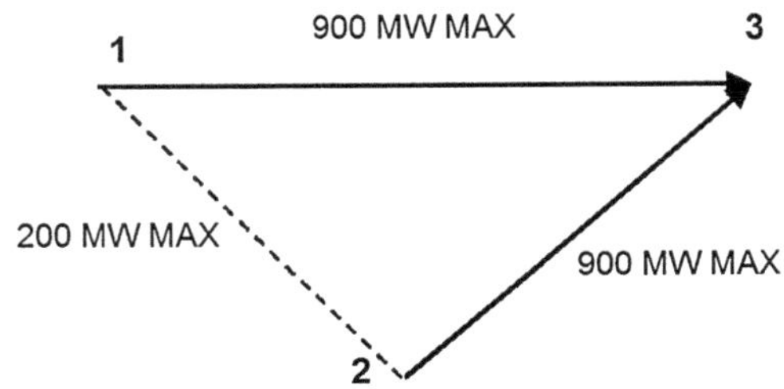


**Fig. 6.2** Three-node network with expansion of line 1–2

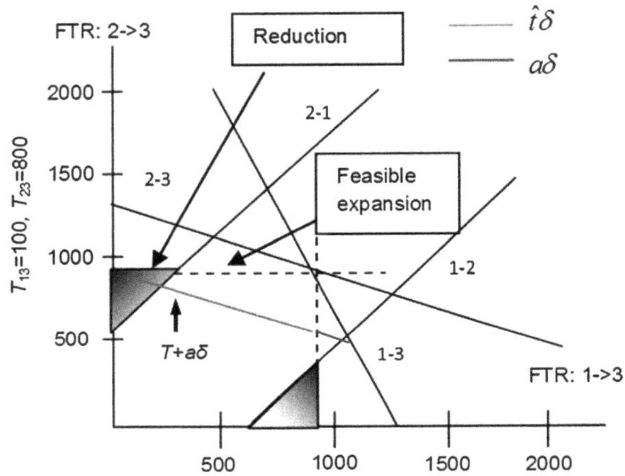


**Fig. 6.3** Feasible expansion of FTRs

Similarly, we get $\lambda = p_{12}$ which implies that the higher the value of the preset proxy preference parameter $p_{12}$ the higher marginal valuation of pre-expansion transmission capacity. Other results are $\theta = 0$, $\zeta = \gamma/p_{12} = b_{12}/p_{12}$ and $\varepsilon = 0$. This was expected since only one restriction for the lower problem is binding because the two other are redundant. The value of the binding Lagrange multiplier equals the ratio between the investor's bid value and the preset proxy parameter.

It also follows that $\varphi = 0$ which is to be expected because the directional vector $\delta$ is non-zero. Furthermore, $\hat{t} = C_{12} - T_{12}$, which means that for given existing rights the higher the current capacity the larger the need for reserving some proxy FTRs for possible negative externalities generated by the expansion. Proxy awards are auctioned as a hedge against externalities generated by the expanded network.

We finally get $a = C_{12}^{+} - T_{12} - \hat{t} = C_{12}^{+} - C_{12}$, which shows that the optimal amount of additional MWs of FTRs in direction $\delta$ directly depends on the amount of capacity expansion. Transmission capacity is in fact fully utilized by proxy awards (in the pre-expansion network), and by incremental FTRs (in the expanded network). Likewise, the investor receives a reward equal to the MW amount of new transmission capacity that it creates.

### 6.6.2 Three-Node Network with Two Links

We now consider a three-node network example from Bushnell and Stoft (1997) where there is an expansion of line 1–2. The network is illustrated in Fig. 6.2 and the feasible expansion in Fig. 6.3.

The network expansion problem for identical links and FTRs between buses 1–3 and 2–3 is formulated as:

$$\max_{a,\hat{t},\delta} \quad a(b_{13}\delta_{13} + b_{23}\delta_{23})$$

$$s.t.$$

$$\begin{aligned}
&\frac{2}{3}(T_{13} + a\delta_{13}) + \frac{1}{3}(T_{23} + a\delta_{23}) \leq C_{13}\\
&\frac{2}{3}(T_{13} + \hat{t}\delta_{13} + a\delta_{13}) + \frac{1}{3}(T_{23} + \hat{t}\delta_{23} + a\delta_{23}) \leq C_{13}\\
&\frac{1}{3}(T_{13} + a\delta_{13}) + \frac{2}{3}(T_{23} + a\delta_{23}) \leq C_{23}\\
&\frac{1}{3}(T_{13} + \hat{t}\delta_{13} + a\delta_{13}) + \frac{2}{3}(T_{23} + \hat{t}\delta_{23} + a\delta_{23}) \leq C_{23}
\end{aligned} \tag{6.32}$$

$$\begin{aligned}
&\frac{1}{3}(T_{13} + a\delta_{13}) - \frac{1}{3}(T_{23} + a\delta_{23}) \leq C_{12}\\
&\frac{1}{3}(T_{13} + \hat{t}\delta_{13} + a\delta_{13}) - \frac{1}{3}(T_{23} + \hat{t}\delta_{23} + a\delta_{23}) \leq C_{12}
\end{aligned} \tag{6.33}$$

$$\begin{aligned}
&-\frac{1}{3}(T_{13} + a\delta_{13}) + \frac{1}{3}(T_{23} + a\delta_{23}) \leq C_{21}\\
&-\frac{1}{3}(T_{13} + \hat{t}\delta_{13} + a\delta_{13}) + \frac{1}{3}(T_{23} + \hat{t}\delta_{23} + a\delta_{23}) \leq C_{21}
\end{aligned}$$

$$\begin{aligned}
&\hat{t}(\delta) \in \arg\max_{t} \{t(p_{13}\delta_{13} + p_{23}\delta_{23})\}\\
&(T_{13} + t\delta_{13}) \leq C_{13}\\
&(T_{23} + t\delta_{23}) \leq C_{23}\\
&\|\delta\| = 1\\
&a \geq 0
\end{aligned}$$

Appendix 2 presents the calculations to obtain the power transfer distribution factors (PTDFs) for the post expansion network. In Fig. 6.3 the pre-existing FTRs in the direction 2–3 do not use the full capacity of the pre-expansion network and become infeasible after inserting line 1–2. The preference is for FTRs in the direction 1–3 for transmission expansion. As seen from Fig. 6.3 the maximum amount of proxy and incremental FTRs in the direction 1–3 that can be obtained is 1,100, and corresponds to the point where the 1–3 and 1–2 transmission capacity constraints intersect.

In solving this problem, we get[29]:

$$\delta_{13} = \frac{(1/3\gamma_1 - 1/3\gamma_2)}{\left((2/3\gamma_1 + 1/3\gamma_2 - \zeta\lambda)^2 + (1/3\gamma_1 - 1/3\gamma_2)^2\right)^{1/2}}$$

$$\delta_{23} = \frac{-(2/3\gamma_1 + 1/3\gamma_2 - \zeta\lambda)}{\left((2/3\gamma_1 + 1/3\gamma_2 - \zeta\lambda)^2 + (1/3\gamma_1 - 1/3\gamma_2)^2\right)^{1/2}}$$

$$a = \frac{C_{12}}{\delta_{13}}$$

$$\hat{t} = \frac{(C_{13} - T_{13})}{\delta_{13}}$$

$$\gamma_1 = \frac{(b_{13} + Bb_{23} + \gamma_2(B/3 - 1/3))}{(2/3 + B/3)}$$

$$\gamma_2 = \frac{1}{(1 - B - AB + A)}[b_{13}(1 + 3A - B - 2A - AB) + b_{23}(B + 3AB - B^2 - 2A - AB)]$$

$$\zeta\lambda = (1 + A)\gamma_1 - A(b_{13} + b_{23})$$

with

$$A = \frac{C_{12}}{(C_{13} - T_{13})}$$

$$B = \frac{1}{(1 + A)} \frac{(C_{13} - 2C_{12} - T_{23})}{(C_{13} - T_{13})}$$

where $\gamma_1$ and $\gamma_2$ are the Lagrange multipliers associated with transmission capacity on the lines 1–3 and 1–2, respectively, in the expanded network, and $\zeta$ is the multiplier associated with the Kuhn-Tucker condition regarding transmission capacity in the pre-expansion network for the line 1–3. This line has the Lagrange multiplier $\lambda$ associated with it before expansion. So as to characterize the solution to our model, we now calculate the Lagrange multipliers and decision variables for particular parameter values. In particular, we find the solution for the allocation presented in Fig. 6.3. We assume the following bid values, preset proxy preferences and pre-existing amount of FTRs:

$$b_{13} = 40, \quad b_{23} = 10,$$
$$p_{13} = 60, \quad p_{23} = 10,$$
$$T_{13} = 100, \quad T_{23} = 800$$

[29] The detailed mathematical derivation of solutions to program (6.32) is presented in Appendix 1.

From these parameters we find that the marginal value of transmission capacity on line 1–3 and line 1–2 are $\gamma_1 = 39.6$ and $\gamma_2 = 33.6$, respectively. Thus the investor values transmission capacity on line 1–3 more than on line 1–2. We find that the product of the Kuhn-Tucker multiplier and the transmission capacity multiplier for the line 1–3 is $\zeta\lambda = 37$.

Likewise, the values of the decision variables are calculated as:

$$\delta_{13} = 0.958, \quad \delta_{23} = -0.287,$$
$$a = 208, \quad \hat{t} = 835$$

The MW amount of awarded proxy FTRs in the direction 1–3 is $\hat{t}\delta_{13} = 800$, and the amount of awarded incremental FTRs is $a\delta_{13} = 200$. The amount of incremental 1–3 FTRs corresponds to the new transmission capacity on line 1–2 that the investor has created. There is also an allocation of proxy FTRs such that the full capacity of line 1–3 is utilized. Similarly the proxy awards in direction 2–3 is $\hat{t}\delta_{23} = -240$, and the amount of awarded incremental FTRs is $a\delta_{23} = -60$. The amount of incremental 2–3 FTRs is minimized and corresponds to 20 % of the reduction (300) in pre-existing FTRs. The incremental 2–3 awards are mitigating FTRs, and are necessary to restore feasibility. The investor is then responsible for additional counterflows so that it pays back for the negative externalities it creates. The solution is indicated by the black arrow in Fig. 6.3 and consists of both pre-existing and incremental FTR awards amounting to $T_{13} + a\delta_{13} = 300$ and $T_{23} + a\delta_{23} = 740$. The allocation of incremental 2–3 FTRs is minimized because the model takes into account that one line is expanded, and some of the pre-existing FTRs become infeasible after the expansion.

This illustrates that the amount of incremental FTRs in the preference direction must be greater than zero such that feasibility is restored. Both the proxy and incremental FTRs exhaust transmission capacity in the pre-expansion and expanded grid, respectively. The proxy FTRs help allocating incremental FTRs by preserving capacity in the pre-expansion network, which results in an allocation of incremental FTRs amounting to the new transmission capacity created in 1–2 direction.[30] The proxy awards are transmission congestion hedges that can be auctioned to electricity market players in the expanded network.[31]

In the example provided by Bushnell and Stoft (1997), the investor with pre-existing FTRs chooses the most profitable incremental FTR based on optimizing its

[30] Note that this result will depend on the network interactions. In some cases the amount of incremental FTRs in the preference direction will differ from the new capacity created on a specific line. However, it will always amount to the new capacity created as defined by the scalar amount of incremental FTRs times the directional vector.

[31] Whenever there is an institutional restriction to issue LTFTRs there will be an additional (expected congestion) constraint to the model. A proxy for the shadow price of such a constraint would be reflected by the preferences of the investor that carries out the expansion project (assuming risk neutrality and a price taking behavior). The proxy award model takes the "linear" incremental and proxy FTR trajectories to the after-expansion equilibrium point in the ex-post FTR feasible set to ensure the minimum shadow value of the constraint.

final benefit. The investor is then awarded a mitigating incremental 1–2 FTR with associated power flows corresponding to the difference between the ex-ante and ex-post optimal dispatches. The pre-existing FTRs correspond to the actual dispatch of the system and become infeasible after expanding line 1–2, and therefore a mitigating 1–2 FTR[32] is allocated so that feasibility is exactly restored (that is, the investor "pays back" for the negative externalities to other agents). There is no allocation of proxy awards because the pre-expansion network is fully allocated by FTRs before the expansion. The amount of incremental FTRs is minimized because they represent a negative value to the investor and decrease its revenues from the pre-existing FTRs.

## 6.7 The Auction Model and Welfare

Bushnell and Stoft (1997) demonstrate that the increase in social welfare will be at least as large as the ex-post value of new contracts, when the FTRs initially match dispatch in the aggregate and new FTRs are allocated according to the feasibility rule. In particular, if social welfare is decreased by transmission expansion, the investor will have to take FTRs with a negative value (If social welfare is increased there will be free riding). With only the aggregate match of FTRs and dispatch, some agents might still benefit from investments that reduce social welfare, whenever their own commercial interests improve to an extent that more than offsets the negative value of the new FTRs. Further, Bushnell and Stoft show that incentives for expansion that reduce social welfare would be removed if FTRs for each agent as a perfect hedge and match their individual net loads. In such a case, FTRs allocated under the feasibility rule ensure that no one will benefit from an expansion that reduces welfare.

Although apparently similar, our mechanism and its implications on welfare are different from those in the Bushnell and Stoft (1997) model. Bushnell and Stoft analyze the welfare implications of transmission expansion given matching of dispatch both in the aggregate and individually. In our model, we assume unallocated FTRs both before and after the expansion, so that there is no match in dispatch.[33]

However, the proxy award mechanism developed in this paper implies nonnegative effects on welfare in the sense that future investments in the grid cannot reduce the welfare of aggregate use for *FTR holders*. The reason is that simultaneous feasibility is guaranteed before and after the enhancement project so that revenue adequacy is also guaranteed after expansion. Only those non-hedged agents in the spot market might be exposed to rent transfers. For feasible long term transactions identified ex ante, the FTRs provide perfect congestion hedges for the existing grid or for any future grid that develops under the feasibility rule. However, FTRs cannot provide perfect hedges

[32] The incremental 1–2 FTR can be decomposed into a 1–3 FTR and a 3–2 FTR.

[33] Additionally, Bushnell and Stoft explicitly define loads, nodal prices, and generation costs so that the effects on welfare are measured as the change in net generation costs. In contrast, we do not define a net benefit function of the users of the grid in terms of prices, generation costs or income from loads. Alternatively, our model maximizes the investors' objective function in terms of incremental FTRs.

ex post for all possible transactions. A similar property carries over to any welfare analysis under FTRs.

More formally, suppose we have a social welfare function $B$ for dispatch in a single period. Also assume that there is no uncertainty, that all functions are known, and that agents are price takers in the electricity and FTR markets. A simple welfare model associated with transmission expansion $\Delta$ is[34]:

$$\begin{aligned} &\underset{\Delta}{Max}\ B(Y^* + \Delta) \\ &\text{s.t.} \\ &K^+(Y^* + \Delta) \le 0 \end{aligned} \tag{6.34}$$

where

$$Y^* \in \arg\max\{B(Y)|K(Y) \le 0\}$$

Then $Y^*$ is the dispatch that maximizes social welfare without the expansion. Let $\Delta^+$ be the dispatch that would be provided as an increment due to transmission expansion. $\Delta^+$ solves program (6.34). Note that if $P^+ = \nabla B(Y^* + \Delta^+)$, then under reasonable regularity conditions $\Delta^+$ is also a solution to:

$$\begin{aligned} &\underset{\Delta}{Max}\ P^+\Delta \\ &\text{s.t.} \\ &K^+(Y^* + \Delta) \le 0 \end{aligned} \tag{6.35}$$

Formulation (6.35) is interpreted as the maximization of congestion rents for the incremental allocation $\Delta$.

In the context of Bushnell and Stoft assumptions,[35] suppose no w that the current allocation of FTRs $T$ satisfies $T = Y^*$, then (6.35) would award the maximum value of incremental FTRs. In this case, we need not know the full benefit function. We could rely on the expander to estimate $P^+$, and provide this preference ranking function as part of the bid. Then solving (6.35) would give the maximum value incremental award for expansion $K^+$, and this award would preserve the welfare maximizing property of the FTRs for the expanded grid.[36]

[34] We are grateful to William Hogan for the insights in the formulation of the following model.

[35] See Bushnell and Stoft (1997, pp. 100–106).

[36] This is however a particular type of welfare maximization since, as opposed to Bushnell and Stoft, costs of expansion are not addressed.

Now suppose that (for some reason) $T \neq Y^*$. To preserve simultaneous feasibility the constraint $K^+(T + \Delta) \leq 0$ should be imposed. A natural (second best) rule might be:

$$\underset{\Delta}{Max}\ P^+\Delta$$
$$s.t. \tag{6.36}$$
$$K^+(Y^* + \Delta) \leq 0$$
$$K^+(T + \Delta) \leq 0$$
$$Y^* \in \arg\max\{B(Y)|K(Y) \leq 0\}$$

Hence, the existing users of the grid could continue to do as before the expansion, and the expander receives the incremental values arising from the expansion. Then the example in Hogan (2002a, p. 12; see also Appendix 3) illustrates the case of a beneficial expansion where the only solution to (6.36) is $\Delta = 0$ so that the expansion project does not occur. In fact, $K^+(T + \Delta) \leq 0$ cannot be relaxed without violating the critical property of simultaneous feasibility. We illustrate the argument in the following examples.

Consider the example in Hogan (2002a, p. 12) illustrated in Fig. 6.4. Here the dispatch in the pre-existing network does not match the current allocation of FTRs. The limiting constraints for the dispatch are the 1–3 and 2–3 constraints. Likewise, the limiting constraints for the current allocation of FTRs are the 1–2 and 1–3 constraints. Assume that the incremental dispatch in the 1–3 and 2–3 directions may be caused by the increased capacity of line 1–3. The relevant constraints are $K^+(Y^* + \Delta) \leq 0$ for the current dispatch and $K^+(T + \Delta) \leq 0$ for the current allocation of FTRs. The corresponding objective is $Max\left(P_{13}^+\Delta_{13} + P_{23}^+\Delta_{23}\right)$. Then the following constraints would apply for the specific network topology:

$$\frac{2}{3}(T_{13} + \Delta_{13}) + \frac{1}{3}(T_{23} + \Delta_{23}) \leq C_{13}^+$$

$$\frac{2}{3}(Y_{13} + \Delta_{13}) + \frac{1}{3}(Y_{23} + \Delta_{23}) \leq C_{13}^+$$

$$\frac{1}{3}(T_{13} + \Delta_{13}) + \frac{2}{3}(T_{23} + \Delta_{23}) \leq C_{23}$$

$$\frac{1}{3}(Y_{13} + \Delta_{13}) + \frac{2}{3}(Y_{23} + \Delta_{23}) \leq C_{23}$$

$$\frac{1}{3}(T_{13} + \Delta_{13}) - \frac{1}{3}(T_{23} + \Delta_{23}) \leq C_{12}$$

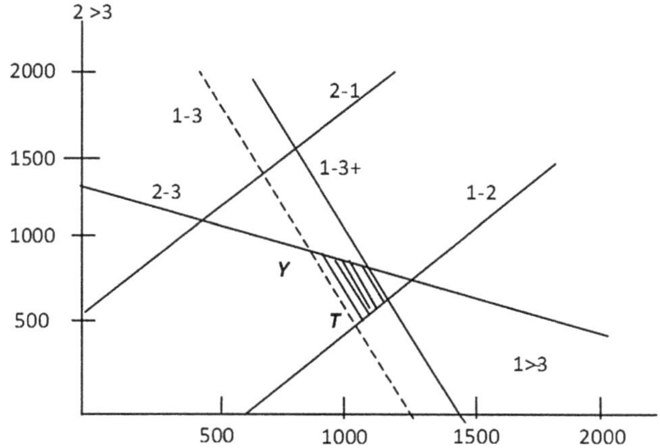


**Fig. 6.4** Dispatch $Y$ does not match the current allocation of FTRs

$$\frac{1}{3}(Y_{13} + \Delta_{13}) - \frac{1}{3}(Y_{23} + \Delta_{23}) \le C_{12}$$

$$-\frac{1}{3}(T_{13} + \Delta_{13}) + \frac{1}{3}(T_{23} + \Delta_{23}) \le C_{21}$$

$$-\frac{1}{3}(Y_{13} + \Delta_{13}) + \frac{1}{3}(Y_{23} + \Delta_{23}) \le C_{21}$$

First assume that $T_{13} = 1100$, $T_{23} = 500$ and $Y_{13} = 900$, $Y_{23} = 900$ and that the incremental benefit of expansion is greater in the 1–3 direction than in the 2–3 direction. Also assume that $C_{13}^{+} = 1000$. We notice that the mismatch between the dispatch and existing FTRs is $Y_{13} - T_{13} = -200$ and $Y_{23} - T_{23} = 400$. Furthermore, the marginal expansion occurs from the current dispatch to where the 1–3+ and 2–3 transmission constraints intersect. This amounts to the incremental dispatch $\Delta_{13} = 200$ and $\Delta_{23} = -100$. If the above numbers are substituted in the constraints we find that the transmission capacity constraint for line 1–2 and existing FTRs are violated because $\frac{1}{3}(T_{13} + \Delta_{13}) - \frac{1}{3}(T_{23} + \Delta_{23}) = \frac{1}{3}(1100 + 200) - \frac{1}{3}(500 - 100) = 300 > C_{12} = 200$. Hence the expansion does not occur. Conversely, assume that the location of the current dispatch and existing FTRs are interchanged so the mismatch between the dispatch and existing FTRs is $Y_{13} - T_{13} = 200$ and $Y_{23} - T_{23} = -400$ and assume that the marginal benefit of the expansion is greater in the 2–3 direction than in the 1–3 direction. Then the incremental dispatch would be $\Delta_{13} = 0$ and $\Delta_{23} = 300$. In this case the 2–3 transmission capacity constraint would be violated for the existing FTRs $\frac{1}{3}(T_{13} + \Delta_{13}) + \frac{2}{3}(T_{23} + \Delta_{23}) = \frac{1}{3}900 + \frac{2}{3}(900 + 300) = 1100 > C_{23} = 900$.

Similar problems would arise with rules such as preserving proxy awards to allow for any possible dispatch on the existing grid, where the only expansions incented would be those that added to every constraint in the system, virtually foreclosing the possibility of investment under this rule.

Given the complicated externalities of electric grid, a first best system based on decentralized property rights is not known. Traditionally, all investment decisions relied on central decisions by regulators under certification of need. This often produced regulatory gridlock precisely because of the grid externalities considered here (not to mention the siting and environmental issues). The FTR feasibility rule always preserves the property that the incidence of any welfare reductions falls to those whose transaction were not selected ex ante to be hedged by FTRs. In dealing with the aggregate welfare effects, the second best motivation is shown in (6.35) (without going all the way to (6.36)). In the absence of the known welfare function or the possibility of allocating all the existing grid, the total award is divided between proxy awards and incremental awards for the investor. The proportional part of the resulting total award that could be achieved with the existing grid is preserved as a proxy award. The remainder is assigned to the investor. Subject to this rule, the total award is chosen to maximize the market value of the incremental award to the investor. Presumably this would reinforce the incentive for the investor to provide an accurate estimate of the market value. Given the prices, the special case of FTRs matching dispatch or $T = Y^*$ (considered by Bushnell and Stoft 1996, and Bushnell and Stoft 1997) is consistent with this rule, and the welfare maximizing results apply. In the case where there is not a full allocation of the existing grid, the likely result is that there would be more scope for welfare reducing investments. The need for regulatory oversight would not be eliminated, but the intent is that the scope of the regulatory issues would be reduced. Since proxy award mechanisms are in use and more are under development, further investigation of the private incentives, welfare effects and regulatory implications would be of value.

## 6.8 Concluding Remarks

We proposed a merchant mechanism to expand electricity transmission. Proxy awards (or reserved FTRs) are a fundamental part of this mechanism. We defined them according to the best use of the current network along the same direction of the incremental expansion. The incremental FTR awards are allocated according to the investor preferences, and depend on the initial partial allocation of FTRs and network topology before and after expansion.

Our examples showed that the internalization of possible negative externalities caused by potential expansion is possible according to the rule proposed by Hogan (2002a): allocation of FTRs before (proxy FTRs) and after (incremental FTRs) the expansion is in the same direction and according to the feasibility rule. Under these circumstances, the investor will have the proper incentives to invest in transmission expansion in its preference direction given by its bid parameters. Likewise the larger the existing current capacity the greater the number of FTRs that must be reserved in order to deal with potential negative externalities depending on post network topology.

Our mechanism of long term FTRs is basically a way to hedge market players from long-run nodal price fluctuations by providing them with the necessary property transmission rights. The main purpose of the four basic criteria that support our model (*feasibility rule*, *proxy awards*, *maximum value* and *symmetry*) were to define property rights for increased transmission investment according to the preset proxy rule. However, the general implications on welfare, and incentives for gaming are still an open research question.

Although our model is specifically designed to deal with loop flows, and the security-constrained version of our model can take care of contingency concerns, our proposed mechanism is to be applied to small line increments in meshed transmission networks. LTFTRs are efficient under non-lumpy marginal expansions of the transmission network, and lack of market power. Regulation has then an important complementary role in fostering large and lumpy projects where investment is large relative to market size, and in mitigating market power. Since revenues from nodal prices only recover a small part of total costs, LTFTRs must be complemented with a regulated framework that allows the recovery of fixed costs. The challenge is to effectively combine merchant and regulated transmission investments or, as Hogan (2003) puts it, to establish a rule in practice for drawing a line between merchant and regulated investment.

# Appendix 1

## *Solution to Program (6.30)*

The Lagrangian of the problem is:

$$\begin{aligned} L(a,\hat{t},\delta_{12},\lambda,\Omega) &= b_{12}a\delta_{12} + \gamma\left(C_{12}^{+} - T_{12} - (a+\hat{t})\delta_{12}\right) \\ &- \theta(p_{12}\delta_{12} - \lambda\delta_{12}) - \zeta(\lambda(C_{12} - T_{12} - \hat{t}\delta_{12})) \\ &+ \varepsilon(C_{12} - T_{12} - \hat{t}\delta_{12}) + \varphi\left(1 - \delta_{12}^{2}\right) + \kappa a + \pi\lambda \end{aligned} \tag{6.37}$$

where $\gamma, \theta, \zeta, \varepsilon, \varphi, \kappa,$ and $\pi$ are the multipliers associated with the respective constraints.

At optimality the Kuhn-Tucker conditions are:

$$\frac{\partial L(a,\hat{t},\delta_{12},\lambda,\Omega)}{\partial a} = b_{12}\delta_{12} - \gamma\delta_{12} = 0, \tag{6.38}$$

$$\begin{aligned} \frac{\partial L(a,\hat{t},\delta_{12},\lambda,\Omega)}{\partial \delta_{12}} &= ab_{12} - (\hat{t}+a)\gamma - (p_{12} - \lambda)\theta \\ &+ \lambda\zeta\hat{t} - \varepsilon\hat{t} - 2\delta_{12}\varphi = 0, \end{aligned} \tag{6.39}$$

$$\frac{\partial L(a,\hat{t},\delta_{12},\lambda,\Omega)}{\partial \hat{t}} = -\gamma\delta_{12} + \lambda\zeta\delta_{12} - \varepsilon\delta_{12} = 0, \tag{6.40}$$

$$\frac{\partial L(a, \hat{t}, \delta_{12}, \lambda, \Omega)}{\partial \lambda} = \delta_{12}\theta - (C_{12} - T_{12} - \hat{t}\delta_{12})\zeta = 0, \tag{6.41}$$

$$\frac{\partial L(a, \hat{t}, \delta_{12}, \lambda, \Omega)}{\partial \gamma} = \left(C_{12}^{+} - T_{12} - (a + \hat{t})\delta_{12}\right) = 0, \tag{6.42}$$

$$\frac{\partial L(a, \hat{t}, \delta_{12}, \lambda, \Omega)}{\partial \theta} = -(p_{12}\delta_{12} - \lambda\delta_{12}) = 0, \tag{6.43}$$

$$\frac{\partial L(a, \hat{t}, \delta_{12}, \lambda, \Omega)}{\partial \zeta} = -\lambda(C_{12} - T_{12} - \hat{t}\delta_{12}) = 0, \tag{6.44}$$

$$\frac{\partial L(a, \hat{t}, \delta_{12}, \lambda, \Omega)}{\partial \varepsilon} = (C_{12} - T_{12} - \hat{t}\delta_{12}) = 0, \tag{6.45}$$

$$\frac{\partial L(a, \hat{t}, \delta_{12}, \lambda, \Omega)}{\partial \varphi} = \left(1 - \delta_{12}^{2}\right) = 0, \tag{6.46}$$

$$\frac{\partial L(a, \hat{t}, \delta_{12}, \lambda, \Omega)}{\partial \kappa} = a > 0, \quad \kappa = 0, \tag{6.47}$$

$$\frac{\partial L(a, \hat{t}, \delta_{12}, \lambda, \Omega)}{\partial \pi} = \lambda > 0, \quad \pi = 0, \tag{6.48}$$

$$\gamma, \varepsilon \geq 0 \tag{6.49}$$

Equation (6.46) gives $\delta_{12} = 1$. Equation (6.38) gives $\gamma = b_{12}$. Equation (6.43) gives $\gamma = b_{12}$, (6.40) $\zeta = \gamma/p_{12} = b_{12}/p_{12}$ ($\varepsilon$ is zero because the constraint is redundant), and (6.41) $\theta = 0$. From this it follows (6.39) that $\varphi = 0$ Furthermore (6.44) gives $\hat{t} = C_{12} - T_{12}$. Equation (6.42) implies that $a = C_{12}^{+} - T_{12} - \hat{t} = C_{12}^{+} - C_{12}$.

## Solution to Program (6.32)

The Lagrangian of the problem is:

$$L(a, \hat{t}, \delta, \lambda, \Omega) =$$

$$a(b_{13}\delta_{13} + b_{23}\delta_{23}) + \gamma_1\left(C_{13} - \frac{2}{3}(T_{13} + (\hat{t} + a)\delta_{13}\right) - \frac{1}{3}(T_{23} + (\hat{t} + a)\delta_{23}))$$
$$+ \gamma_2\left(C_{12} - \frac{1}{3}(T_{13} + (\hat{t} + a)\delta_{13}\right) + \frac{1}{3}(T_{23} + (\hat{t} + a)\delta_{23}))$$
$$- \zeta(\lambda(C_{13} - (T_{13} + \hat{t}\delta_{13})) \tag{6.50}$$

$$+\varepsilon(C_{13}-(T_{13}+\hat{t}\delta_{13}))$$
$$+\varphi\left(1-\delta_{13}^{2}-\delta_{23}^{2}\right)+\kappa a+\pi\lambda$$

where $\gamma_1$ and $\gamma_2$ are the Lagrange multipliers associated with transmission capacity on the lines 1–3 and 1–2 in the expanded network, respectively. $\zeta$ is the multiplier associated with the Kuhn-Tucker condition of transmission capacity in the pre-expansion network for line 1–3. This line has the Lagrange multipliers $\lambda$ associated with it before expansion. $\varepsilon$ is the investor's marginal value of transmission capacity in the pre-expansion network when allocating incremental FTRs. The normalization condition has the multiplier $\varphi$ and the non-negativity conditions have the associated multipliers $\kappa$ and $\pi$. The first order conditions are:

$$\frac{\partial L(a,\hat{t},\delta,\lambda,\Omega)}{\partial a}=(b_{13}\delta_{13}+b_{23}\delta_{23})-\left(\frac{2}{3}\delta_{13}+\frac{1}{3}\delta_{23}\right)\gamma_1 -\left(\frac{1}{3}\delta_{13}-\frac{1}{3}\delta_{23}\right)\gamma_2=0, \tag{6.51}$$

$$\frac{\partial L(a,\hat{t},\delta,\lambda,\Omega)}{\partial \delta_{13}}=ab_{13}-\frac{2}{3}(\hat{t}+a)\gamma_1-\frac{1}{3}(\hat{t}+a)\gamma_2 +\zeta\lambda\hat{t}-\varepsilon\hat{t}-2\varphi\delta_{13}=0, \tag{6.52}$$

$$\frac{\partial L(a,\hat{t},\delta,\lambda,\Omega)}{\partial \delta_{23}}=ab_{23}-\frac{1}{3}(\hat{t}+a)\gamma_1+\frac{1}{3}(\hat{t}+a)\gamma_2 -2\varphi\delta_{23}=0, \tag{6.53}$$

$$\frac{\partial L(a,\hat{t},\delta,\lambda,\Omega)}{\partial \hat{t}}=-\left(\frac{2}{3}\delta_{13}+\frac{1}{3}\delta_{23}\right)\gamma_1-\left(\frac{1}{3}\delta_{13}-\frac{1}{3}\delta_{23}\right)\gamma_2 +\delta_{13}\zeta\lambda-\delta_{13}\varepsilon=0, \tag{6.54}$$

$$\frac{\partial L(a,\hat{t},\delta,\lambda,\Omega)}{\partial \lambda}=-\zeta(C_{13}-T_{13}-\hat{t}\delta_{13})=0, \tag{6.55}$$

$$\frac{\partial L(a,\hat{t},\delta,\lambda,\Omega)}{\partial \gamma_1}=C_{13}-\frac{2}{3}(T_{13}+(\hat{t}+a)\delta_{13}) -\frac{1}{3}(T_{23}+(\hat{t}+a)\delta_{23})=0, \tag{6.56}$$

$$\frac{\partial L(a,\hat{t},\delta,\lambda,\Omega)}{\partial \gamma_2}=C_{12}-\frac{1}{3}(T_{13}+(\hat{t}+a)\delta_{13}) +\frac{1}{3}(T_{23}+(\hat{t}+a)\delta_{23})=0, \tag{6.57}$$

$$\frac{\partial L(a,\hat{t},\delta,\lambda,\Omega)}{\partial \zeta} = -\lambda(C_{13} - (T_{13} + \hat{t}\delta_{13})) = 0, \tag{6.58}$$

$$\frac{\partial L(a,\hat{t},\delta,\lambda,\Omega)}{\partial \varepsilon} = (C_{13} - T_{13} - \hat{t}\delta_{13}) = 0, \tag{6.59}$$

$$\frac{\partial L(a,\hat{t},\delta,\lambda,\Omega)}{\partial \varphi} = 1 - \delta_{13}^2 - \delta_{23}^2 = 0, \tag{6.60}$$

$$\frac{\partial L(a,\hat{t},\delta,\lambda,\Omega)}{\partial \kappa} = a > 0, \quad \kappa = 0, \tag{6.61}$$

$$\frac{\partial L(a,\hat{t},\delta,\lambda,\Omega)}{\partial \pi} = \lambda > 0, \quad \pi = 0, \tag{6.62}$$

The solution for the first order conditions is given by:

$$\delta_{13} = \frac{(1/3\gamma_1 - 1/3\gamma_2)}{\left((2/3\gamma_1 + 1/3\gamma_2 - \zeta\lambda)^2 + (1/3\gamma_1 - 1/3\gamma_2)^2\right)^{1/2}}$$
$$\delta_{23} = \frac{-(2/3\gamma_1 + 1/3\gamma_2 - \zeta\lambda)}{\left((2/3\gamma_1 + 1/3\gamma_2 - \zeta\lambda)^2 + (1/3\gamma_1 - 1/3\gamma_2)^2\right)^{1/2}}$$

$$a = \frac{C_{12}}{\delta_{13}}$$
$$\hat{t} = \frac{(C_{13} - T_{13})}{\delta_{13}}$$

$$\gamma_1 = \frac{(b_{13} + Bb_{23} + \gamma_2(B/3 - 1/3))}{(2/3 + B/3)}$$

$$\gamma_2 = \frac{1}{(1 - B - AB + A)}[b_{13}(1 + 3A - B - 2A - AB) + b_{23}(B + 3AB - B^2 - 2A - AB)]$$

$$\zeta\lambda = (1 + A)\gamma_1 - A(b_{13} + b_{23})$$

with

$$A = \frac{C_{12}}{(C_{13} - T_{13})}$$
$$B = \frac{1}{(1 + A)}\frac{(C_{13} - 2C_{12} - T_{23})}{(C_{13} - T_{13})}$$

## Appendix 2

This appendix derives the power transfer distribution factors (PTDFs) for the three-node network with two parallel lines, and where all lines have identical reactance. The net injection (or net generation) of power at each bus is denoted $P_i$. We have the following relationship between the net injection, the power flows $P_{ij}$ and phase angles $\theta_i$ (Wood and Wollenberg 1996):

$$P_i = \sum_j P_{ij} = \sum_j \frac{1}{x_{ij}}(\theta_i - \theta_j)$$

where $x_{ij}$ is the line inductive reactance in per unit.

We can write the power flow equations as:

$$\begin{bmatrix} P_1 \\ P_2 \\ P_3 \end{bmatrix} = \begin{bmatrix} 2 & -1 & -1 \\ -1 & 2 & -1 \\ -1 & -1 & 2 \end{bmatrix} \begin{bmatrix} \theta_1 \\ \theta_2 \\ \theta_3 \end{bmatrix}$$

The matrix is called the susceptance matrix. The matrix is singular, but by declaring one of the buses to have a phase angle of zero and eliminating its row and column from the matrix, the reactance matrix can be obtained by inversion. The resulting equation then gives the bus angles as a function of the bus injection:

$$\begin{bmatrix} \theta_2 \\ \theta_3 \end{bmatrix} = \begin{bmatrix} 2/3 & 1/3 \\ 1/3 & 2/3 \end{bmatrix} \begin{bmatrix} P_2 \\ P_3 \end{bmatrix}$$

The *PTDF* is the fraction of the amount of a transaction from one node to another node that flows over a given line. $PTDF_{ij,mn}$ is the fraction of a transaction from node $m$ to node $n$ that flows over a transmission line connecting node $i$ and node $j$. The equation for the *PTDF* is:

$$PTDF_{ij,mn} = \frac{x_{im} - x_{jm} - x_{in} + x_{jn}}{x_{ij}}$$

where $x_{ij}$ is the reactance of the transmission line connecting node $i$ and node $j$ and $x_{im}$ is the entry in the *ith* row and the *mth* column of the bus reactance matrix. Utilizing the formula for the specific example network gives:

$$PTDF_{12,13} = 1/3, \ PTDF_{13,13} = 2/3, \ PTDF_{23,13} = 1/3,$$
$$PTDF_{12,23} = -1/3, \ PTDF_{13,23} = 1/3, \ PTDF_{23,23} = 2/3$$
$$PTDF_{21,13} = -1/3, \ PTDF_{21,23} = 1/3$$

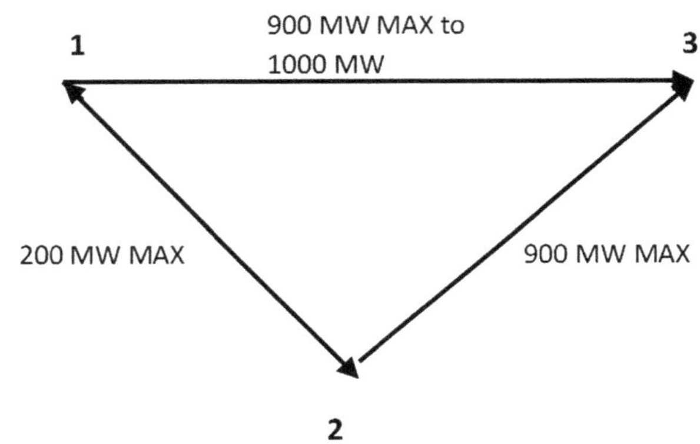


**Fig. 6.5** Three-node network with expansion in one line

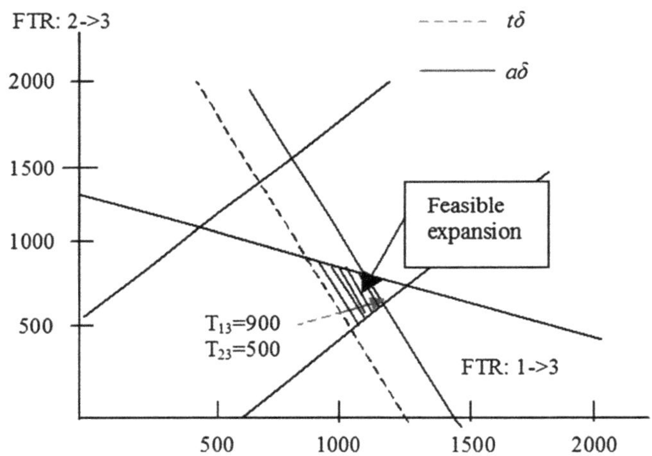


**Fig. 6.6** Feasible expansion FTR set

# Appendix 3

## *Transmission Investment That Does Not Change PTDFs*

An example on an investment that does not change the PTDFs of the network is shown in Fig. 6.5 where there is an expansion of line 1–3 from 900 to 1,000 MW transmission capacity. The associated feasible expansion FTR set is shown in Fig. 6.6. We observe that whatever feasible FTRs that existed before expansion, none of these will become infeasible after the expansion.

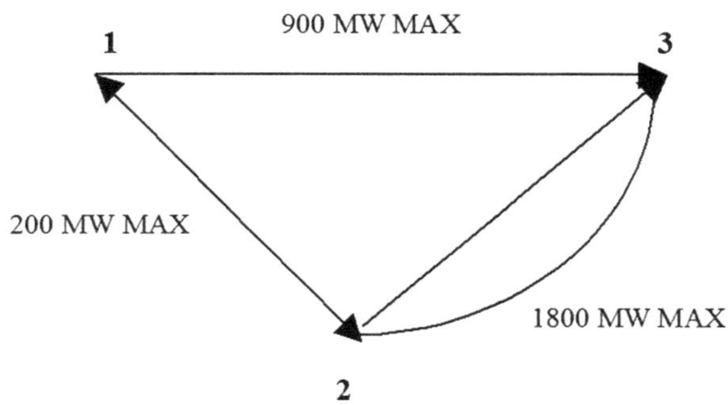


**Fig. 6.7** Three-node network where a line is inserted in parallel with an existing line

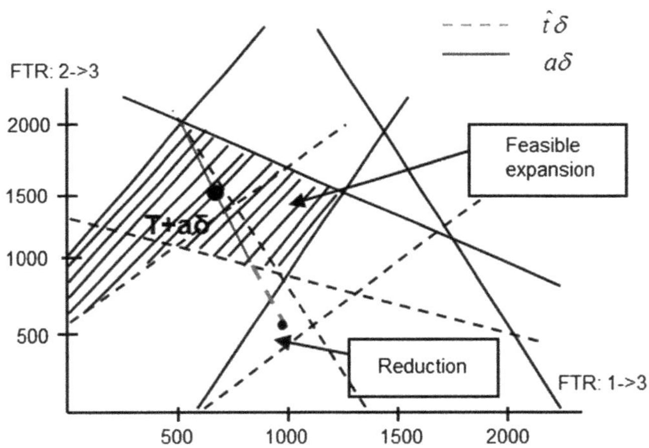


**Fig. 6.8** Feasible expansion FTR set

## *Transmission Investment That Does Change PTDFs*

Figure 6.7 shows a three-node network where a line is inserted in parallel with an existing line between the nodes 2 and 3. Inserting a parallel line with identical reactance as the existing line halves the total reactance between nodes 2 and 3. As a result the PTDFs of the expanded network change.

$$PTDF_{12,13} = 1/3 \text{ and } PTDF_{13,13} = 2/3$$

change to

$$PTDF_{12,13} = 0.4 \text{ and } PTDF_{13,13} = 0.6.$$

Furthermore, the inserted line has identical transmission capacity to the existing one so that the total transmission capacity is doubled between the buses 2 and 3. However, the simultaneous interaction of the reactances and transmission capacities changes the feasible expansion FTR set as illustrated in Fig. 6.8. Then some of the pre-existing FTRs may become infeasible.

## References

Borenstein S, Bushnell J, Stoft S (1998) The competitive effects of transmission capacity in a deregulated electricity industry. POWER working paper PWP-040R, University of California Energy Institute. http://www.ucei.berkely.edu/ucei

Brito DL, Oakland WH (1977) Some properties of the optimal income tax. Int Econ Rev 18: 407–423

Bushnell J (1999) Transmission rights and market power. Electr J 12:77–85

Bushnell JB, Stoft S (1996) Electric grid investment under a contract network regime. J Regul Econ 10:61–79

Bushnell JB, Stoft S (1997) Improving private incentives for electric grid investment. Res Energy Econ 19:85–108

Bushnell JB, Wolak F (1999) Regulation and the leverage of local market power in the California electricity market. POWER working paper PWP-070R, University of California Energy Institute. http://www.ucei.berkely.edu/ucei

Cardell C, Hitt C, Hogan W (1997) Market power and strategic interaction in electricity networks. Res Energy Econ 19:109–137

Chao H-P, Peck S (1997) An institutional design for an electricity contract market with central dispatch. Energy J 18(1):85–110

Gilbert R, Neuhoff K, Newbery D (2002) Mediating market power in electricity networks. Mimeo, Cambridge

Grande OS, Wangensteen I (2000) Alternative models for congestion management and pricing impact on network planning and physical operation. CIGRE, Paris

Gribik PR, Graves JS, Shirmohammadi D, Kritikson G (2002) Long term rights for transmission expansion. Mimeo

Harvard Electricity Policy Group (2002) Transmission expansion: market based and regulated approaches. Rapporteur's summaries of HEPG twenty-seventh plenary sessions, Session 2, January 24–25, Cambridge MA

Harvey SM (2002) TCC expansion awards for controllable devices: initial discussion. Mimeo, Cambridge MA

Hogan W (1992) Contract networks for electric power transmission. J Regul Econ 4:211–242

Hogan W (1999) Market-based transmission investments and competitive electricity markets. Mimeo, JFK School of Government, Harvard Electricity Policy Group, Harvard University. http://www.ksg.harvard.edu/people/whogan

Hogan W (2000) Flowgate rights and wrongs. Mimeo, JFK School of Government, Harvard Electricity Policy Group, Harvard University. http://www.ksg.harvard.edu/people/whogan

Hogan W (2002a) Financial transmission right incentives: applications beyond hedging. Presentation to HEPG twenty-eight plenary sessions, May 31. http://www.ksg.harvard.edu/people/whogan

Hogan W (2002b) Financial transmission right formulations. Mimeo, JFK School of Government, Harvard Electricity Policy Group, Harvard University. http://www.ksg.harvard.edu/people/whogan

Hogan W (2003) Transmission market design. Mimeo, JFK School of Government, Harvard Electricity Policy Group, Harvard University. http://www.ksg.harvard.edu/people/whogan

Joskow P, Schmalensee S (1983) Markets for power: an analysis of electric utility deregulation. MIT Press, Cambridge, MA

Joskow P, Tirole J (2000) Transmission rights and market power on electric power networks. RAND J Econ 31(3):450–487
Joskow P, Tirole J (2002) Transmission investment: alternative institutional frameworks. Mimeo, IDEI (Industrial Economic Institute), Toulouse, France
Joskow P, Tirole J (2003) Merchant transmission investment. Mimeo, MIT and IDEI
Léautier T-O (2000) Regulation of an electric power transmission company. Energy J 21(4):61–92
Léautier T-O (2001) Transmission constraints and imperfect markets for power. J Regul Econ 19(1):27–54
Littlechild S (2003) Transmission regulation, merchant investment, and the experience of SNI and Murraylink in the Australian National Electricity Market. Mimeo, Cambridge
Mirrlees JA (1971) An explanation in the theory of optimum income taxation. Rev Econ Stud 38: 175–208
Nash SG, Sofer S (1988) Linear and nonlinear programming. Wiley, New York
O'Neill RP, Helman U, Hobbs BF, Stewart WR, Rothkopf MH (2002) A joint energy and transmission rights auction: proposal and properties. IEEE Trans Power Syst 17(4):1058–1067
Pérez-Arriaga IJ, Rubio FJ, Puerta Gutiérrez JF et al (1995) Marginal pricing of transmission services: an analysis of cost recovery. IEEE Trans Power Syst 10(1):546–553
Philpott A, Pritchard G (2004) Financial transmission rights in convex pool markets. Operat Res Lett 32(2):109–113
Pope S (2002) TCC awards for transmission expansions. Mimeo, Cambridge MA
Rosellón J (2000) The economics of rules of origin. J Int Trade Econ Dev 9(4):397
Rubio-Oderiz J, Pérez-Arriaga IJ (2000) Marginal pricing of transmission services: a comparative analysis of network cost allocation methods. IEEE Trans Power Syst 15:448–454
Sheffrin A, Wolak FA (2001) Methodology for analyzing transmission upgrades: two alternative proposals. Mimeo, Cambridge, MA
Shimizu K, Ishizuka Y, Bard JF (1997) Nondifferentiable and two-level mathematical programming. Kluwer, Norwell
The California ISO and London Economics International LLC (2003) A proposed methodology for evaluating the economic benefits of transmission expansions in a restructured wholesale electricity market. http://www.caiso.com/docs/2003/03/25/2003032514285219307.pdf. Accessed June 2003
US-Canada Power System Outage Task Force (2004) Final report on the August 14, 2003 blackout in the United States and Canada: causes and recommendations. Available at: https://reports.energy.gov/BlackoutFinal-Web.pdf. Accessed Dec 2004
Vogelsang I (2001) Price regulation for independent transmission companies. J Regul Econ 20(2):141
Wolak FA (2000) An empirical model of the impact of hedge contract on bidding behavior in a competitive electricity market. Int Econ J 14:1–40
Wolfram C (1998) Strategic bidding in a multi-unit auction: an empirical analysis of bids to supply electricity in England and Wales. RAND J Econ 29:703–725
Wood AJ, Wollenberg BF (1996) Power generation, operation and control. Wiley, New York

# Chapter 7
# Mechanisms for the Optimal Expansion of Electricity Transmission Networks

**Juan Rosellón**

## 7.1 Introduction

Electricity transmission grid expansion and pricing have received increasing attention in recent years.[1] Transmission networks provide the fundamental support upon which competitive electricity markets depend. Congestion of transmission networks might increase market power in certain regions, put entry barriers to potential competitors in the generation business, and in general reduce the span of competitive effects. A well functioning transmission network is a critical component of wholesale and retail markets for electricity.

The formal analysis of adequate incentives for network expansion in the electricity industry is complicated due to externalities generated by the physical characteristics of electricity itself as well as due to cost sub-additivity and economies-of-scale features

This paper was previously published as: Rosellón J (2009) Mechanisms for the optimal expansion of electricity transmission networks. In: Evans J, Hunt LC (eds) International handbook on the economics of energy. Edward Elgar Publishing, UK.

Support from Conacyt (p. 60334) and from Pieran-Colegio-de-México is gratefully acknowledged.

[1] Problems related with coordination and capacity to the transmission network partly caused power outages in the northeast of the US during 2003, which affected more than 20 million consumers and six control areas (Ontario, Quebec, Midwest, PJM, New England, and New York), and shut down 61,000 MW of generation capacity. Similar recent events in other parts of the world such as UK, Italy, Sweden, Brazil, Argentina, Chile New Zealand, and Germany (incidence of E.ON Netz that blacked out large chunks of Europe in 2006) also awakened the interest in the factors that ensure reliability of transmission grids.

J. Rosellón (✉)
División de Economía, Centro de Investigación y Docencia Económicas (CIDE), Carretera México-Toluca 3655, Mexico, D.F 01210, Mexico

German Institute for Economic Research (DIW Berlin), Mohrenstrasse 58, Berlin 10117, Germany
e-mail: juan.rosellon@cide.edu; jrosellon@diw.de

J. Rosellón and T. Kristiansen (eds.), *Financial Transmission Rights*, Lecture Notes in Energy 7, DOI 10.1007/978-1-4471-4787-9_7,

of the grid.[2] Externalities in electricity transmission are mainly due to "loop flows",[3] which arise from interactions in the transmission network.[4] The effects of loop flows imply that transmission opportunity costs and pricing critically depend on the marginal costs of power at every location. Energy costs and transmission costs are not independent since they are determined simultaneously in the electricity dispatch and the spot market. Then, certain transmission investments in a particular link might have negative externalities on the capacity of other transmission links.

The analysis of incentives for transmission investment is further complicated since equilibria in forward electricity transmission markets has to be coordinated with equilibria in other markets such as the energy spot market, the forward energy market, and the generation capacity-reserves market.[5] Likewise, electricity pricing is a complex issue since electricity is not storable, and because it has to simultaneously guide long-term investment decisions by transmission companies as well as to ration demands in the short run due to congestion. Furthermore, the effects of an increase in transmission capacity are uncertain. For instance, the net welfare outcome of an expansion in the transmission grid depends on the weight in the welfare preferences of the generators' profits relative to the consumers' weight.[6] Generation revenues gains, due to improved access to increased transmission charges and new markets, might be overcome by the loss of local market power.

The institutional structure of the system operator, and its relationship with the transmission network, are also key components that define the alternatives that might attract new investment to the grid. There exist three possible structures for a system operator.[7] The first structure is an independent system operator (ISO) – different from the company that owns the transmission grid – that is *decentralized* and intrudes to the least possible extent in the markets. The second is a *centralized* ISO that controls and coordinates the markets. The third is an integrated company, the transmission company (*Transco*), which combines ownership of the transmission network with system operation.[8]

---

[2] Vogelsang (2006).

[3] Loop flow is the characteristic of electricity that takes it through all available routes (path of least resistance) to get from one point to another. For instance, if a second line becomes available that is identical to a first line, the electricity that had been flowing over the first line will "divide" itself so that half of it will remain flowing through the first line and the other half will flow over the second (see Brennan et al. 1996).

[4] Joskow and Tirole (2000), and Léautier (2001).

[5] Wilson (2002).

[6] Léautier (2001),

[7] Wilson (2002).

[8] In practice, the ISO model has been used in Argentina, and Australia. System operation is carried out by the ISO and transmission ownership is carried out by another independent company, the Gridco. ISOs also exist in California, New England, New York, Pennsylvania-New Jersey-Maryland (PJM), and the Texas. ISO practical experiences and proposals have been centralized. The Transo model has been typically used in practice in the UK, Spain and the Scandinavian countries.

The economic analysis of electricity markets has typically concentrated on short-term issues such as short-run congestion management, and nodal pricing. However, investment in transmission capacity is long-run in nature as well as stochastic. In the short run, the difference of electricity prices between two nodes in a power flow model defines the price of congestion.[9] Nevertheless, an "optimal" way to attract investment for the long-term expansion of the transmission network is still an open question both formally, and in practice.[10]

There are two main disparate (*non-Bayesian*) analytical approaches to transmission investment: [11] one employs the theory based on long-run financial rights (LTFTR) to transmission (merchant approach), while the other is based on the incentive-regulation hypothesis (performance-based-regulation (PBR) approach).[12] Practical approaches to transmission expansion have then to a large extent been designed according to particular criteria as opposed to being based on general economic theory, or on the more specific regulatory economics literature. In this paper, I review in Sects. 7.2 and 7.3 recently developed approaches for PBR and merchant mechanisms. Likewise, in Sect. 7.4 I provide insights so as to build a comprehensive approach that combines both mechanisms in a setting of price-taking electricity generators and loads.

## 7.2 The Incentive-Regulation Approach to Transmission Expansion

The PBR approach to transmission expansion relies on incentive-compatible regulatory mechanisms for a Transco. Such mechanisms provide the firm with incentives to make efficient investment decisions as well as to earn enough

---

[9] Hogan (2002b).

[10] Vogelsang (2006).

[11] A third alternative method for transmission expansion seeks to derive optimal transmission expansion from the power-market structure of electricity generation, and considers conjectures made by each generator on other generators' marginal costs due to the expansion (Wolak 2000). This method uses a real-option analysis to derive the net present value of both transmission and generation projects through the calculation of their joint probability. Transmission expansion only yields benefits until it is large enough compared to a given generation market structure. Likewise, many small upgrades are preferable to large greenfield project.

[12] Vogelsang (2006) makes a division between *Bayesian* and *non-Bayesian* mechanism for transmission expansion. The Bayesian approach derives from the merger of the principal agent theory and the optimal pricing approach, and implies a theoretical framework supported by the *Revelation Principle* but that does not in general translate into rules that regulators can apply directly. According to the canonical model of regulation, under asymmetric information the need for prices to provide incentives arises when transfers from the regulator are not possible (Laffont 1994). Non-Bayesian mechanisms arise from more practical reasons so as to improve inadequacies associated to rate-of-return regulation. Then PBR regulation, including price-caps and yardsticks, were developed as non-Bayesian instruments to promote cost-minimization. However, the application of PBR to network industries has been scarce, mainly due the lumpy and long-term nature of networks, such as the electricity grid.

revenues to recover capital and operating costs.[13] In the international practice, PBR schemes for transmission expansion have been basically applied in England, Wales and Norway to guide the expansion of the transmission network. In the case of the two first countries, transmission pricing has been typically separated from energy pricing. A regulatory mechanism based on an "out-turn" has been used there. The out-turn is the difference between the price actually paid to generators and the price that would have been paid absent congestion. An "uplift management rule" is then applied to the Transco responsible for the full cost of the out-turn, plus any transmission losses. In Australia, a combination of regulatory mechanisms and merchant incentives has been implemented.[14] Argentina has also relied on a combined regulatory-merchant approach under an ISO regime with nodal pricing.[15]

The formal analyses of PBR mechanisms for transmission expansion basically rely on comparing a Transco's performance with a measure of welfare.[16] The Transco is penalized for increasing congestion costs in the network, and is responsible for the costs of congestion it creates and the needed investment to relieve it. For instance, Joskow and Tirole (2002) propose a simple surplus-based mechanism to provide the Transco with enough incentives to expand the transmission network. The idea is to reward the Transco according to the redispatch costs avoided by the expansion, so that the Transco faces the entire social cost of congestion. Such a mechanism would presumably eliminate the problems associated with lumpiness and loop flows, but it could still be subject to manipulation of bids in the energy market by a Transco that is vertically integrated with generation. Even with no vertical integration, generators might invest no more than what is needed to match existing transmission capacity.

An alternative PBR approach is to explicitly study the nature of transmission cost and demand functions (Vogelsang 2001). The monopolistic nature of a for-profit Transco that owns the complete transmission network is isolated. This scheme might also be applied in a combined institutional structure where a (centralized) ISO takes care of the short-run market, and an independent transmission company handles investment issues. Regulation of transmission must then solve the duality of incentives for the transmission firm both in the short run (congestion), and in the long run (investment in network expansion). Conditions for optimal capacity expansion have been studied by the peak-load pricing literature: the per-unit marginal cost of new capacity must be equal to the expected congestion cost of not adding an additional unit of capacity.[17] However, there is still the question on how price regulation can provide incentives to reach such a stage.

---

[13] Léautier (2000), Grande and Wangensteen (2000), Vogelsang (2001), and Joskow and Tirole (2005).

[14] Littlechild (2003).

[15] Littlechild and Skerk (2004a, b).

[16] Gans and King (1999), Léautier (2000), Grande and Wangensteen (2000), Joskow and Tirole (2002).

[17] Crew et al. (1995).

Price-cap mechanisms deal with regulation of "price level" and regulation of "price structure."[18] Price level regulation refers to the long-run distribution of rents and risks between consumers and the regulated firm. Applied alternatives for level regulation typically include cost-of-service, price-cap, and yardstick regulations. Price structure regulation refers to the short-run allocation of benefits and costs among distinct types of consumers. Alternatives for regulation of price structure include price bands, flexible price structures as well as fixed or non-fixed weight regulation.[19] As in other network industries, electricity transmission price-level regulation is applied together with inflation (*RPI*) and efficiency factors (*X*), and a cost-of-service check every 5 years.

Price structure regulation is used by Vogelsang (2001) to solve transmission congestion, in the short run, as well as capital costs and investment issues in the long run. In a two-part tariff regulatory model with a variable (or usage) charge, and a fixed (or capacity) charge, the variable charge is mainly based on nodal prices and relieves congestion. Recuperation of long-term capital costs is achieved through the fixed charge that can be interpreted as the price for the right to use the transmission network. The fixed charge can also provide incentives for productive efficiency and, if it does not affect the number of transmission consumers, allocative efficiency – i.e., convergence to the Ramsey price structure – can be intertemporally achieved.[20] The basic model proposed in Vogelsang (2001) is:

$$\max\ \pi^t = p^t q^t + F^t N - C(q^t, K^t)$$

subject to

$$F^t \leq F^{t-1} + (p^{t-1} - p^t) q^w / N$$

$$q^t \leq K^t$$

where:

$F_t$ = fixed fee in period t.
$p_t$ = variable fee in period t.
$q_t$ = real oriented energy flow in period t (in kWh).
$K_t$ = available transmission capacity in period t.
$w$ = type of weight.
$N$ = number of consumers

The transmission cost function $c(q,K)$ reflects the sunk cost nature of transmission investment and has the following form:

---

[18] Brown et al. (1991).

[19] Vogelsang (1999).

[20] Baldick et al. (2007), provide practical guidelines for allocation among consumers of the costs of transmission expansion.

$$C(q^t, K^{t-1}) = C(q^{t-1}, K^{t-1}),\ \forall\ q^t, q^{t-1} \leq K^{t-1}$$

$$C(q^t, K^t) = C(q^t, K^{t-1}) + f(K^{t-1}, I^t),\ for\ q^t > K^{t-1}$$

where investment $I_t$ is such that

$$I^t = K^t - K^{t-1}$$

Assuming that constraints are binding, and that $\mu_t$ is the Lagrange multiplier of the capacity constraint, the first order condition with respect to $p_t$ is:

$$\left(\frac{\partial q^t}{\partial p^t}\right)\left(p^t + \mu^t - \frac{\partial C}{\partial q^t}\right) = q^w - q^t$$

For the optimal level of investment, $q^* = K^*$ it is true that $\mu_t = 0$, so that the first order condition yields the (equilibrium) Ramsey rule:

$$\left(p^t - \frac{\partial C}{\partial q^t}\right) = -\left(\frac{q^w}{q^t - 1}\right)\Big/\varepsilon$$

where $\varepsilon$ is the price elasticity of demand.

The proper incentives for efficient investment in the expansion of the network in the Vogelsang model are reached by the rebalancing of fixed and variable charges. Likewise, incentives for investment crucially depend on the type of weights used. For instance, a Laspeyres index uses the quantity of the previous period as weight for the price so that the Transco will intertemporally invest until its transmission tariffs converge to Ramsey prices. However, this will not occur automatically since the firm faces a tension between short-run gains from congestion, and increases in capacity investment. These results are true only if it is assumed that cost and demand functions are stable, and that the Transco does not use strategic conduct in setting its prices.[21] In the case of changing cost and demand functions, or non myopic profit maximization, convergence to Ramsey prices under the Laspeyres index cannot be guaranteed.[22] Thus, when there is congestion in capacity the Transco will expand the network because its profits increase with network expansion when congestion variable charges are marginally larger than the marginal costs of expanding capacity. On the contrary, in times of excess capacity, the variable charge of the two-part tariff will be reduced causing an increase in consumption. The fixed charge, in turn, increases so that total income augments despite the decrease in the variable charge. As a consequence, the Transco ceases to invest in capacity expansion, and net profits expand since costs do not increase.

---

[21] See Vogelsang 1999, pp. 28–31.

[22] See Ramírez and Rosellón (2002).

The pure price-cap approach in Vogelsang (2001) however relies on simplifying assumptions that are rarely met in practice. Transmission demand functions are assumed differentiable and downward sloping, while transmission marginal costs curves are supposed to cut demand only once. These assumptions are generally invalid since, under loop flows, an expansion in a certain transmission link can result in decreases of other network links leading to discontinuities in the marginal-cost function.[23] Likewise, transmission activity is considered as a physical output (or throughput) process as opposed to a transmission output defined in terms of point-to-point transactions.[24] This task is impossible since the physical flow through a meshed transmission network cannot be traced (Bushnell and Stoft 1997; Hogan 2002b, c).[25]

One of the main problems of PBR mechanisms is their inconsistency with timing issues of transmission networks. Vogelsang (2006) then proposes a framework based on the distinction of utra-short periods, short periods and long periods. The ultra short-period is motivated by real-time pricing of point-to-point transmission services, and there are no possibilities within this period for costs reductions. So, the main allocative-efficiency problem is price rationing of congested inputs. The short-period coincides with the application of (RPI-X) factors, and is also the period for the calculation of fixed fees. The long period is given by the regulatory lag of the PBR mechanisms; that is, the time between (cost-of-service) tariff revisions (of typically 5 years). The long period crucially depends on the regulatory commitment so as to avoid ratcheting.

In the Vogelsang (2001) mechanism, investment in the grid occurs at the beginning of each period while fixed fees are calculated at the end of the period. Therefore, this mechanism implicitly lumps together the short period and the long period, and assumes that investments do not occur beyond such period. The Vogelsang (2006) mechanism on the contrary combines the ultra-short, short and long periods and allows for the possibility of no investment for several short periods or even for times beyond a long period. This mechanism then depends on previous price performance of the mechanism in the past as well as on the long-run certainty provided by revisions based on rate-of-return regulation.

---

[23] Hogan (2002a)

[24] See Hogan et al. (2010) for a redefinition of transmission outputs in terms of point-to-point FTRs.

[25] An application of the Vogelsang (2001) PBR model is carried out in Rosellón (2007) for the electricity transmission system in Mexico, under stable demand growth for electricity. Three scenarios are studied: (a) a single Transco providing transmission services nationally and that applies postage-stamp tariffs; (b) several regional companies that separately operate in each of the areas of the national transmission system, and that charge different prices; and (c) a single Transco owns all the regional systems in the nation but that charges different prices in each region. Achieved capacity and network increases are highest under the first scenario, while higher profits are implied by the second approach. These results are found to critically depend on two basic effects; namely, the "economies-of-scale effect" and the "discriminatory effect". The economies-of-scale effect produces greater capacity and network expansion whereas the discriminatory effect increases profit.

The combined approach for all types of periods in Vogelsang (2006) relies on a combination of Vogelsang (2001) and the incremental surplus subsidy scheme (ISS).[26] According to the ISS, the firm receives a subsidy in each period equal to the difference between the last period´s profit and the current-period´s consumer-surplus increase. In Vogelsang (2006), the subsidy is financed through the fixed fee of a two-part tariff and consumer surplus is calculated with a verifiable approximation. The Vogelsang (2001) price-cap constraint is then used in Vogelsang (2006) for pricing in the ultra-short and short periods, together with an (RPI-X) adjustment for short periods and a profit adjustment at the end of long periods. Prices would then be average revenues from ultra-short periods. The (RPI-X) adjustments would affect only the fixed fees, and partially counteract any consumer-surplus increases handed to the Transco.

## 7.3 Merchant Transmission Investment

The merchant approach to transmission expansion is based on auctions of financial transmission rights (FTRs) that seek to attract voluntary participation by potential investors. Incremental FTRs provide market-based transmission pricing that attracts transmission investment since it implicitly defines property rights. FTR auctions are carried out within a bid-based security-constrained economic dispatch with nodal pricing (which includes a short-run spot market for energy and ancillary services) of an ISO. The ISO runs a power flow model that provides nodal prices derived from shadow prices of the model's constraints.[27] FTRs are subsequently derived from nodal price differences. Due to the long-run nature of electricity transmission, the ISO allocates long-term (LT) FTRs through an auction so as to protect the holders from future unexpected changes in congestion costs. Therefore, LTFTR auctions work in parallel with LT generation contracts.[28] The long-run concept is important for transmission expansion projects for investors. They usually have a useful life of approximately 30 years, so that auctions allocate FTRs with durations of several years. An FTR can in practice materialize in an obligation, a flowgate right or an option. "Point-to-point" (PTP) forward obligations are in practice the most feasible

[26] Sappington and Sibley (1988), and Gans and King (1999).

[27] The typical power-flow model framework is that of a centralized ISO that maximizes social welfare subject to transmission-loss and flow-feasibility constraints in a spot market. In practice, this model has been applied in Argentina, Australia, and several regions in the United States (Pennsylvania-Maryland-New Jersey (PJM), New York, Texas, California). The economic dispatch model can also be understood within a static competitive equilibrium model. The producing entity is an ISO that provides transmission services, receives and delivers power, and coordinates the spot market. Meanwhile, consumers inject power into the grid at some nodes and remove power out at other points. See Hogan (2002b).

[28] FTRs give their holders a share of the congestion surpluses collected by the ISO under a binding constraint. The quantity of FTRs is normally fixed ex ante and allocated to holders. This reflects the capacity of the network. The difference between allocated FTRs and actual transmission capacity provides congestion revenues for the ISO. FTRs are defined in terms of the difference in nodal prices. See Joskow and Tirole (2002).

instruments, while PTP options and flowgate rights are of limited applicability.[29] PTP-FTR obligations can be either "balanced" or "unbalanced". Through a balanced PTP-FTR a perfect hedge is achieved, while an unbalanced PTP-FTR obligation is a forward sale of energy.

An example of an FTR auction is the New York ISO's allocation of transmission congestion contracts as a hedge for congestion costs, both in the short run and long run.[30] Incremental FTRs are allocated to parties that pay for the expansion only if the new FTRs are made possible by the expansion. FTR awards are mainly derived from investors' choices but the ISO might also identify some needed incremental FTRs. When investors choose new FTRs for transmission expansion, simultaneous feasibility[31] of both the already existing FTRs and the new FTRs must be satisfied, because both flows and amount of transferred power among nodes is modified by the expansion. The ISO also temporarily reserves some feasible FTRs prior to the expansion project. Auctions are carried out both for short-term FTRs (6 months) and LTFTRs (20 years). LTFTRs are allocated before short-term FTRs through auctioned and unauctioned mechanisms. The unauctioned mechanism simply reserves capacity for sales in later auctions, while in an LTFTR auction investors reveal their preferences for expansion FTRs by assigning to each one a certain positive weight. Investors' preferences are maximized preserving simultaneous feasibility together with all pre-expansion FTRs. Losses are included in the dispatch and only balanced PTP-FTRs are defined to provide payments for congestion costs but not for losses. [32]

A mixture of planning and auctions of long-term transmission rights has also been applied in PJM. The centralized PJM-ISO applies an LTFTR approach within a DC (Direct Current)-load dispatch model where locational prices differ according to congestion. PTP-FTRs are thus defined for congestion-cost payments. Revenues from FTRs are returned to owners of the transmission capacity in order to defray capital, operation and maintenance costs. Secondary FTR markets have also developed in several regions of the Northeast of the USA. FTR secondary markets are generally imbedded in the ISO's dispatch process so that their revenue adequacy is met.[33] Whenever there is need for an FTR between any two nodes, it is usually possible to derive it from nodal-price differences. Likewise, FTRs can be traded within various time frameworks (such as weeks, months and years). Nonetheless, no restructured electricity sector in the world has adopted a pure merchant approach to transmission expansion. The auction-planning combination has also being considered in

[29] Flowgate rights are defined in terms of the constraints implied from limits in the selling of capacity (Hogan 2000).

[30] Pope (2002).

[31] A set of FTRs is simultaneously feasible if the associated set of net loads satisfies the energy balance and transmission capacity constraints, as well as the power flow equations.

[32] Other LTFTR allocation practical mechanisms are provided by Harvey (2002), and Gribik et al. (2002).

[33] Revenue adequacy is the financial counterpart of the physical concept of availability of transmission capacity. FTRs meet the revenue-adequacy condition when they are also simultaneously feasible (Hogan 1992).

New Zealand and Central America, while in Australia a combined merchant-regulatory approach has been attempted.[34]

The formal analyses of FTR auctions can be subdivided into long-term and short-term models. The short-run FTR models provide efficiency results only under strong assumptions of perfect competition, such as: absence of market power and sunk costs, an ISO without an internal preference on effective transmission capacity, complete future markets, lack of uncertainty over congestion rents, nodal prices that internalize network externalities and that reflect consumers' willingness to pay, as well as non-increasing returns to scale.[35] The lifting of these assumptions would imply inefficient results on the use of FTRs. For instance, whenever market power exists, prices will not reflect the marginal cost of production. Generators in constrained regions will withdraw capacity (increasing generation prices), which would overestimate the cost-saving gains from investments in transmission. Likewise, market power in the FTR market by a generator provides an incentive to curtail output so as to make FTRs more valuable.[36]

Additionally, increasing returns and lumpiness in transmission investment imply that social surplus created by transmission investments will be greater than the value paid to investors through FTRs. This is why investors in transmission expansion projects would prefer LT contracts, and exclusive property rights (at least temporarily) in the use of increased capacity. To this, it must be added that existing transmission and incremental capacities cannot be well defined since they are of stochastic nature. Even in a radial line, realized capacity could be less than expected capacity so that the revenue-adequacy condition is not met. Stochastic changes in supply and demand conditions imply uncertain nodal prices as well.

More importantly, for meshed networks with loop flows an addition in transmission capacity in a link of the network might result in an actual reduction of capacity of other links. This, combined with asymmetry of information among the different agents in the electricity industry (generators, ISO, and transmission owners), might result in negative social value.[37]

All these insights are deemed as relevant by the long-term FTR model. LTFTR auctions grant efficient outcomes under lack of market power and non-lumpy marginal expansions of the transmission network.[38] Regulation thus has an important role in large and lumpy projects in order to mitigate market power and let LTFTR auctions efficiently attract investors. In particular, market-power alleviation in the FTR market could be fostered by keeping transmission-owner buyers and sellers of LTFTRs under strict enforcement of open access to their grid facilities.

[34] Littlechild (2003).

[35] Joskow and Tirole (2005).

[36] Joskow and Tirole (2000), Léautier (2001), and Gilbert et al. (2002).

[37] Hogan (2002a), and Kristiansen and Rosellón (2006).

[38] Hogan (2003).

Additionally, contingency and stochastic concerns are mainly taken care by a security-constrained dispatch of a meshed network with loops and parallel paths.[39] Likewise, agency problems and information asymmetries are indeed part of an institutional structure of an electricity industry where the ISO is separated from transmission ownership, and where market players are decentralized. However, the boundary between merchant and regulated expansion projects can hardly be affected by asymmetry of information. The need for regulation is therefore acknowledged in LTFTR auction mechanisms, and complete reliance on market incentives for transmission investment is thus undesirable. Rather, merchant and regulated transmission LTFTR mechanisms could be combined so that regulated transmission is used for projects that are lumpy (where only a single project makes sense as opposed to many small projects), and large (relative to the market size). [40]

The implications of loop flows on transmission investment have also received detailed consideration by the LTFTR literature. A first seminal idea is to require the agent making an expansion to "pay back" for the possible loss of property rights of other agents.[41] A new transmission link creates in turn a new feasible set that requires a redispatch of the net loads at each node. Loads and associated FTRs that were not previously feasible (pre-investment) become feasible (post-investment), while other pairs of loads (and associated FTRs) that were feasible might become infeasible. In this process, the expansion link might reduce social welfare when it is a binding constraint on low-cost generation schedules. Thus, to restore feasibility, the investor in the new link must buy back sufficient rights from initial holders.

Further, LTFTR auctions designed for small-scale networks subject to relatively marginal expansions, might rely on several axioms in order to solve the loop-flow dilemma.[42] The LTFTR auction should maximize the investors' objective function, both for decreases and increases in grid capacity. More importantly, under an initial condition of incomplete allocation of FTRs the transmission energy balance and capacity constraints, as well as the power flow equations, must be satisfied for the existing and incremental FTRs. Simultaneous feasibility should also prevail given that certain currently unallocated rights – or *proxy awards* – are preserved. Under these assumptions, and when FTRs are simply defined as point-to-point obligations, LTFTR will not reduce social welfare of the *hedged* agents.

Additionally, proxy awards are to be defined according to the *best* use of the current grid along the same direction that the incremental FTRs were awarded. "Best" is defined in terms of preset proxy references so that proxy awards maximize the value of such references. Given a proxy rule, the auction is carried out in order to maximize the investor's preferences to award the needed FTRs in the direction of the expansion, subject to the simultaneously feasibility conditions and the "best" rule. Kristiansen and

---

[39] Hogan (2002b).

[40] Hogan (1999, 2000).

[41] Bushnell and Stoft (1997).

[42] Hogan (2002a), and Kristiansen and Rosellón (2006).

Rosellón (2006) develop a bi-level programming model for allocation of long-term FTRs according to the best rule, and apply it to different network topologies.

When the preset proxy rule is used, Kristiansen and Rosellón (2006) derive prices that maximize the investor preference $\beta(a\delta)$ for an award of $a$ MWs of FTRs in direction $\delta$:

$$\begin{aligned}
&\underset{a,\hat{t},\delta}{\text{Max}} \quad \beta(a\delta) \\
&s.t. \\
&K^{+}(T + a\delta) \leq 0, \\
&K^{+}(T + \hat{t}\delta + a\delta) \leq 0, \\
&\hat{t} \in \arg\max_{t} \{tp\delta | K(T + t\delta) \leq 0\}, \\
&\|\delta\| = 1, \\
&a \geq 0.
\end{aligned}$$

where $K^{+}(T + a\delta) \leq 0$ and $K^{+}(T + \hat{t}\delta + a\delta) \leq \; 0 \; \delta$ are the feasibility constraints for "existing plus incremental FTRs ($T + a\delta$)" and "existing plus proxy plus incremental FTRs ($T + \hat{t}\delta + a\delta$)", respectively. This is a nonlinear and non-convex problem, and its solution depends on the parameter values, the current partial allocation ($T$), and the topology of the network prior to and after the expansion.[43] Simultaneous technical feasibility is shown to crucially depend on the investor-preference and the proxy-preference parameters. Likewise, the larger the current capacity the greater the need to reserve some FTRs for possible negative externalities generated by the expansion changes.

However, as previously argued, the described LTFTR mechanism implies that future investments in the grid cannot decrease the welfare of FTR *holders* only. FTRs cannot provide perfect hedges ex post for all possible transactions. The FTR feasibility rule always preserves the property that the incidence of any welfare reductions falls to those whose transaction were not selected ex ante to be hedged by FTRs. The special case of FTRs matching dispatch is consistent with welfare maximization, but in the case where there is not a full allocation of the existing grid, the likely result is that there would be more scope for welfare reducing investments. The need for regulatory oversight would then not be eliminated with FTR auctions, but the intent is that the scope of the regulatory intervention would typically be reduced.

In an applied European transmission-market context, Brunekreeft et al. (2005) argue that unregulated merchant investment should also be complemented with a light-handed regulatory approach so as to increase welfare. In the welfare-competition trade-off, welfare should be more relevant so that third-party-access and must-offer

[43] A general solution method utilizing Kuhn-Tucker conditions would check which of the constraints are binding. One way to identify the binding inequality constraints is the active set method. Kristiansen and Rosellón (2006) solve the problem in detail with simulations for different network topologies, including a radial line and three-node networks.

provisions are not necessary in the European Union regulations that promote unregulated merchant investments in electricity transmission (see also Brunekreeft and Newbery 2006). Likewise, cross-border transmission issues are much relevant in the European case. Market coupling mechanisms with voluntary participation are necessary due to the politically infeasibility of implementing a location-marginal-pricing mechanisms. Kristiansen and Rosellón (2010) carry out an application of the merchant FTR model for transmission expansion to the trilateral market coupling (TLC) border arrangements in Europe (such as the TLC among Netherlands, Belgium and France). The potential introduction of FTRs to the TLC is part of a wider interest in Europe for hedging products for cross-border trade, and congestion management by several regulatory bodies at the European continental level as well as at the national levels (e.g., Spain, France, and Italy). The model of an ISO that reserves some proxy FTR awards and resolves the negative externalities derived from transmission expansion is simulated for the interconnector between France and Belgium. Such a project is shown to be feasible under the proposed FTR auction system. Other likely projects – such as an interconnector that invests in parallel to an existing line, or a third interconnector that links to the TLC arrangement thus forming a three-node network (such as an undersea cable from France to the Netherlands, or the links with Nord Pool or Germany) – are possible. These examples show that FTR-supported expansion projects in Europe could be technically and financially feasible. However, the actual employment of FTRs in TLC arrangements would also require clear definitions of the roles of system operators and power exchanges, daily settlements in implicit auctions between power exchanges, as well as the identification and provision of appropriate risk-sharing and regulatory incentives.[44]

## 7.4 The Combined Merchant-Regulatory Approach

As seen in the previous sections, there is not yet in theory or in practice a single system that guarantees an optimal long-term expansion of all types of electricity transmission networks. This is especially true for non-Bayesian mechanisms, which are usually designed for allocative and productive efficiency improvements in the short run. However, the distinct study efforts suggest a second-best standard that combines the above seen merchant and PBR transmission models so as to reconcile the dual short-run incentives to congest the grid, and the long-run incentives to invest in expanding the network.[45] The merchant mechanisms are easiest to understand for incrementally

[44] See also Brunekreeft et al. (2005).

[45] This would be an alternative approach to the previously seen model in Vogelsang (2006). A main difference would be that the combined merchant-regulatory approach mainly focuses in generalizing the price-cap constraints for electricity transmission (as in Vogelsang 2006) within a power flow model. Likewise, this combined model aims to redefine the output of transmission in terms of PTP transactions (or incremental FTRs) as well as to seriously tackle the "heroic" assumption of smooth well-behaved transmission cost functions of the models in Vogelsang 2001, 2006, and Tanaka 2007.

small expansions in meshed networks under an ISO environment. The price-cap method seeks to regulate a monopoly Transco. Thus, "small" transmission expansion projects might rely on the merchant approach while "large and lumpy" projects could be developed through PBR incentive regulation, that combines price-level and price structure regulation so as to reconcile short-run and long-run incentives.[46] More specifically, LTFTR auctions could be used within regions with meshed transmission networks regions of the country for marginal expansions,[47] while price-cap methodology – that also takes care of the loop flow issue – could be applied to develop the large lumpy links among such regions. In this section, I analyze the basic elements needed to construct a coherent framework for the latter issue.

As previously discussed, the basic PBR model on a regulatory approach to transmission expansion postulates cost and demand functions with fairly general smooth properties, and then adapts some known regulatory adjustment processes to the electricity transmission problem. A concern with this approach is that the properties of transmission cost and demand functions are scarcely known and suspected to differ from usual functional forms. The assumed well-behaved cost and demand properties may actually not hold for a transmission firm. Loop-flows imply that certain investments in transmission upgrades might cause negative network effects on other transmission links, so that capacity is multidimensional. Thus, the transmission capacity cost function can be discontinuous.

There have been some recent developments that tackle the issue of defining a price-cap model for transmission expansion within a power flow-model, so as to define a system that is coherent under loop flows. One such attempt, Tanaka (2007), derives optimal transmission capacity from the effects of capacity expansion on flows and welfare. A welfare function is maximized with respect to capacity subject to the Transco's budget constraint, which is further defined as the difference between capacity cost and congestion rents. Various incentive mechanisms are then analyzed since the Transco alone would prefer to maximize the difference between congestion rents and costs, rather than social welfare. A Laspeyres-type price-cap on nodal prices is shown to converge to optimal transmission capacity over time under its budget constraint. A second mechanism is a two-part tariff cap also based on Laspeyres weights. Finally, another mechanism based on an incremental surplus subsidy, where the regulator observes the actual cost but not the complete cost function, is analyzed. These two last mechanisms are also shown to achieve optimal transmission capacity over time but without a budget constraint. However, Tanaka (2007) still relies on the big assumption of a well-behaved capacity cost functions for electricity transmission.

Another recent model, Hogan et al. (2010) (HRV), combines the merchant and regulatory approaches in an environment of price-taking generators and loads.

[46] Of course, this includes (RPI-X) adjustments together with cost-of-service tariff reviews at the end of each regulatory lag.

[47] The Kristiansen and Rosellón (2006) model is an example of a concrete merchant mechanism designed for small line increments in meshed transmission networks.

A crucial aspect is to redefine the transmission output in terms of incremental LTFTRs in order to apply the basic price-cap mechanism in Vogelsang (2001) to large and lumpy meshed networks, and within a power flow model. Very importantly, the HRV model does not make any previous assumption on the behavior of cost and demand transmission functions. In this model, the Transco intertemporally maximizes profits subject to a cap on its two-part tariff, so that choice variables are the fixed and the variable fees. The fixed part of the tariff plays the role of a complementary charge that recovers fixed costs, while the variable part is the price of the FTR output, and is then based on nodal prices.

In the HRV model there is a sequence of auctions at each period $t$ where participants buy and sell LTFTRs. LTFTRs are assumed to be point-to-point balanced financial transmission right obligations. The Transco maximizes expected profits at each auction subject to simultaneous feasibility constraints, and a two-part-tariff cap constraint. The transmission outputs are the incremental LTFTRs between consecutive periods. The model first defines the least cost solution for the network configuration that meets a given demand. Over the domain where $\iota' q = 0$ (i.e., no losses):

$$c^*\left(q,\, K^{t-1},\, H^{t-1}\right) = \underset{K^t \in \mathrm{K}, H^t \in \mathrm{H}}{Min} \left\{c\left(K^t,\, K^{t-1},\, H^t,\, H^{t-1}\right) \middle| H^t\, q \leq K^t\right\}.$$

where:

$q^t$ = the net injections in period $t$ (FTRs are derived from: $\sum_j \tau_j^t = q^t;\ \tau_j^t = \begin{bmatrix} -x \\ 0 \\ 0 \\ . \\ . \\ +x \\ 0 \end{bmatrix}$)

$K^t$ = available transmission capacity in period $t$
$H^t$ = transfer admittance matrix at period $t$

$\iota^t$ = a vector of ones

$c(K^t, K^{t-1}, H^t, H^{t-1})$ is the cost of going from one configuration to the next. For a DC load approximation model, the Transco's profit maximization problem is then given by:

$$\underset{t^t, F^t}{Max}\, \pi^t = \tau^t\left(q(\tau^t) - q^{t-1}\right) + F^t N^t - c(K^t, K^{t-1}, H^t, H^{t-1})$$

subject to

$$\tau^t Q^w + F^t N^t \leq \tau^{t-1} Q^w + F^{t-1} N^t$$

where:

$\tau^{t}$ = vector of transmission prices between locations in period $t$
$F^{t}$ = fixed fee in period $t$
$N^{t}$ = number of consumers in period $t$

$$Q^{w} = (q^{t} - q^{t-1})^{w}$$

$w$ = type of weight.

The price cap index is defined on two-part tariffs: a variable fee $\tau^{t}$ and a fixed fee $F$ where the output is incremental LTFTRs. The weighted number of consumers $N^{t}$ is assumed to be determined exogenously. When the demand and optimized cost functions are differentiable the first order optimality conditions yield:

$$\nabla q(\tau - \nabla_{q}c^{*}) = Q^{w} - (q(\tau) - q^{t-1})$$

The results of this model then show convergence to marginal-cost pricing (and to Ramsey pricing) under idealized weights, while under Laspeyres weights there is evidence of such a convergence under more restrictive conditions.[48] Likewise, transmission cost functions are shown to have typical economic properties under a variety of circumstances. This holds, in particular, if the topology of all nodes and links is given and only the capacity of lines can be changed, which implies that unusually behaved cost functions require modification of the network topology.

The HRV mechanism is further tested for different network topologies in Rosellón and Weigt (2011). Firstly, the behavior of cost functions (in terms of FTRs) for distinct network topologies is studied. Secondly, the HRV regulatory model is incorporated in a MPEC (mathematical program with equilibrium constraints) problem and tested for three-node networks. Finally, the HRV mechanism is applied to Northwestern Europe. The results of the cost function analysis in Rosellón and Weigt (2011) show how, due to loop flows, rather simplistic extension functions can lead to mathematical problematic global cost function behavior. Furthermore, the linkage between capacity extension and line reactances, and thus the flow patterns, leads to complex results that are highly sensitive to the underlying grid structure. However, for modeling purposes the logarithmic cost form leads to nonlinearities with non-smooth behavior thus making it demanding with respect to calculation effort and solver capability. Quadratic cost

[48] Under Laspeyres weights – and assuming that cross-derivatives have the same sign – if goods are complements and if prices are initially above to marginal costs, prices will intertemporally converge to marginal costs. When goods are substitutes, this effect is only obtained if the cross effects are smaller than the direct effects. If prices are below marginal costs the opposite results are obtained.

functions show a generally continuous behavior that makes them suitable for modeling purposes. In an overall analysis, the piecewise linear nature of the resulting global costs functions makes the derivation of global optima feasible. Hence, the testing of HRV regulatory model as an MPEC problem in Rosellón and Weigt (2011) results in a Transco expanding the network so that prices develop in the direction of marginal costs. These results are confirmed when the MPEC approach is tested using a simplified grid of Northwestern Europe with a realistic generation structure. The nodal prices that were subject to a high level of congestion converge to a common marginal price level.

These results show then that the HRV mechanism has the potential to foster investment into congested networks in an overall desirable direction, satisfying the simultaneous-feasibility and revenue-adequacy constraints. However, further analysis is needed to estimate impacts of externalities such as the generation implications on the Transco's behavior. Furthermore, the extension functions and restrictions have to be adjusted for a better representation of real world conditions, particularly with regards to the lumpiness of investments as well as property-right issues, and existing long term transaction contracts.

## 7.5 Concluding Remarks

Network expansions are relevant in many parts of the world such as in the European electricity market. Due to the liberalization processes initiated in the late 1990s, former national electricity networks with only limited cross border capacities should now build the infrastructure for emerging wide energy markets. However, in Europe 10 years after the first liberalization efforts the extended network is still segmented into several regional and national sub networks with little expansion incentives between countries. Other regions in the world face similar problems too. Deeper understanding of the factors that determine a reliable framework for the investment in transmission networks is therefore of utmost importance.

In this chapter, I addressed the developments in the literature regarding merchant and PBR non-Bayesian mechanisms, as well as their combination, for non-vertically integrated firms. A combined merchant-regulatory mechanism to expand electricity transmission was analyzed. The merchant mechanism in Kristiansen and Rosellón (2006) for marginal increments in small links of severely meshed networks is such that internalization of possible negative externalities caused by potential expansion is possible according to the proxy rule: allocation of FTRs before (proxy FTRs) and after (incremental FTRs) the expansion is in the same direction and according to the feasibility rule. For large and lumpy networks, the HRV mechanism redefines transmission output in terms of incremental LTFTRs in order to solve the loop-flow issue. Constructing the output measure and property rights model in terms of FTRs provides the regulatory model with a connection to the merchant investment theory, and adapts the known regulatory adjustment processes in the network economics literature to the electricity transmission problem.

Of course much future research effort would be of value. Although some progress have been made,[49] the HRV model needs to characterize in detail piecewise cost functions when changes in topology are incorporated, as well as to address global rather than local optimality properties of incentives. Likewise, since proxy award mechanisms are in use and more are under development, further analytical investigation of the private incentives, welfare effects and regulatory implications would very useful. Finally, formal research on the relationship between FTR auctions and social welfare is needed. Such analysis would require a new model that from its origin provides an FTR mechanism that simultaneously addresses the expansion problem, and that maximizes social preferences as well.

## References

Baldick R, Brown A, Bushnell J, Tierney S, Winter T (2007) A national perspective on allocating the costs of new transmission investment: practice and principles, white paper prepared by The Blue Ribbon Panel on cost allocation for WIRES (Working Group for Investment in Reliable and Economic electric Systems), September. Available from http://www.ksg.harvard.edu/hepg/Papers/Rapp_5-07_v4.pdf. Accessed Jan 2007

Brennan TJ, Palmer KL, Kopp RJ, Krupnick AJ, Stagliano V, Burtraw D (1996) A shock to the system: restructuring America's electricity industry. Resources for the Future, Washington, DC. ISBN 0915707802

Brown L, Einhorn MA, Vogelsang I (1991) Toward improved and practical incentive regulation. J Regul Econ 3:313–338

Brunekreeft G, Newbery D (2006) Should merchant transmission investment be subject to a must-offer provision. J Regul Econ 30:233–260

Brunekreeft G, Neuhoff K, Newbery D (2005) Electricity transmission: an overview of the current debate. Utilit Policy 13(2):73–93

Bushnell JB, Stoft SE (1997) Improving private incentives for electric grid investment. Res Energy Econ 19:85–108

Crew MA, Fernando CS, Kleindorfer PR (1995) The theory of peak-load pricing: a survey. J Regul Econ 8:215–248

Gans JS, King SP (1999) Options for electricity transmission regulation in Australia. Working paper, September 10, University of Melbourne

Gilbert R, Neuhoff K, Newbery D (2002) Mediating market power in electricity networks. Center for Competition Policy working paper no. CPC02-32, UC Berkeley

Grande OS, Wangensteen I (2000) Alternative models for congestion management and pricing. Impact on network planning and physical operation. CIGRE (International Council on Large Electric Systems), 37-203_2000, Paris, August/September

Gribik PR, Graves JS, Shirmohammadi D, Kritikson G (2002) Long term rights for transmission expansion. P.A. Consulting Group, Mimeo

Harvey SM (2002) TCC expansion awards for controllable devices: initial discussion. LECG (Law and Economics Consulting Group), Mimeo

Hogan W (1992) Contract networks for electric power transmission. J Regul Econ 4:211–242

Hogan W (1999) Restructuring the electricity market: coordination for competition. Mimeo, JFK School of Government, Harvard University, mimeo, Harvard Electricity Policy Group Harvard University. http://www.ksg.harvard.edu/people/whogan. Accessed Aug 2007

[49] As in Rosellón and Weigt (2011).

Hogan W (2000) Flowgate rights and wrongs. Mimeo, JFK School of Government, Harvard Electricity Policy Group Harvard University. http://www.ksg.harvard.edu/people/whogan. Accessed Aug 2007

Hogan W (2002a) Financial transmission right incentives: applications beyond hedging. Presentation to HEPG twenty-eight plenary sessions, May 31. http://www.ksg.harvard.edu/people/whogan. Accessed Aug 2007

Hogan W (2002b) Financial transmission right formulations. Mimeo, JFK School of Government, Harvard Electricity Policy Group Harvard University. http://www.ksg.harvard.edu/people/whogan. Accessed Aug 2007

Hogan W (2002c) Electricity market restructuring: reform of reforms. J Regul Econ 21:103–132

Hogan W (2003) Transmission market design. Mimeo, JFK School of Government, Harvard Electricity Policy Group, Harvard University. http://www.ksg.harvard.edu/people/whogan. Accessed Aug 2007

Hogan W, Rosellón J, Vogelsang I (2010) Toward a combined merchant-regulatory mechanism for electricity transmission expansion. J Regul Econ 38(2):113–143

Joskow P, Tirole J (2000) Transmission rights and market power on electric power networks. RAND J Econ 31:450–487

Joskow P, Tirole J (2002) Transmission investment: alternative institutional frameworks. IDEI (Industrial Economic Institute), Tolouse. http://idei.fr/doc/conf/wme/tirole.pdf. Accessed Aug 2007, Mimeo

Joskow P, Tirole J (2005) Merchant transmission investment. J Ind Econ 53(2):233–264

Kristiansen T, Rosellón J (2006) A merchant mechanism for electricity transmission expansion. J Regul Econ 29(2):167–193

Kristiansen T, Rosellón J (2010) Merchant mechanism electricity transmission expansion: a European case study. Energy 35(10):4107–4115

Laffont J-J (1994) The new economics of regulation ten years after. Econometrica 62(3):507–537

Léautier T-O (2000) Regulation of an electric power transmission company. Energy J 21:61–92

Léautier T-O (2001) Transmission constraints and imperfect markets for power. J Regul Econ 19:27–54

Littlechild S (2003) Transmission regulation, merchant investment, and the experience of SNI and Murraylink in the Australian National Electricity Market. Electricity Policy Research Group, University of Cambridge. http://www.eprg.group.cam.ac.uk/wp-content/uploads/2008/11/littlechildtransmission.pdf. Accessed Aug 2007, Mimeo

Littlechild S, Skerk CJ (2004a) Regulation of transmission expansion in Argentina. Part I: State ownership, reform and the fourth line. Cambridge working papers in economics CWPE 0464, Mimeo

Littlechild S, Skerk CJ (2004b) Regulation of transmission expansion in Argentina. Part II: Developments since the fourth line. Cambridge working papers in economics CWPE 0465, Mimeo

Pope S (2002) TCC awards for transmission expansions. Mimeo, LECG LECG: Law and Economics Consulting Group, 20 Mar 2002. Available at www.nyiso.com. Accessed Aug 2007

Ramírez JC, Rosellón J (2002) Pricing natural gas distribution in Mexico. Energy Econ 24(3):231–248

Rosellón J (2007) A regulatory mechanism for electricity transmission in Mexico. Energy Policy 35(5):3003–3014

Rosellón J, Weigt H (2011) A combined merchant-regulatory mechanism for electricity transmission expansion in Europe. Energy J 32(1):119–148

Sappington D, Sibley D (1988) Regulating without COST information: the incremental surplus subsidy scheme. Int Econ Rev 29:297–306

Tanaka M (2007) Extended price cap mechanism for efficient transmission rxpansion under nodal pricing. Netw Spatial Econ 7:257–275

Vogelsang I (1999) Optimal price regulation for natural and legal monopolies. Economía Mexicana Nueva Época 8(1):5–43

Vogelsang I (2001) Price regulation for independent transmission companies. J Regul Econ 20 (2):141–165
Vogelsang I (2006) Electricity transmission pricing and performance-based regulation. Energy J 27(4):97–126
Wilson R (2002) Architecture of power markets. Econometrica 70(4):1299–1340
Wolak FA (2000) An empirical model of the impact of hedge contract on bidding behavior in a competitive electricity market. Int Econ J (Summer):1–40

# Chapter 8
# Long Term Financial Transportation Rights: An Experiment

**Bastian Henze, Charles N. Noussair, and Bert Willems**

## 8.1 Introduction

One challenge facing operators of network infrastructure, such as gas pipelines and electricity grids, is that large new investments in capacity must be undertaken as overall demand increases. In the European Union alone, roughly 200 Billion Euro must be invested in the energy transport networks (gas and electricity) by 2020 (MEMO/10/582). However, there is a considerable risk that an operator's estimates of future demand might prove too optimistic, irreversible investments would be undertaken, and some of the capacity would sit idle or underused. On the other hand, failing to expand capacity sufficiently would result in lost profits and lower welfare than under optimal capacity provision.

One possible method for encouraging better infrastructure investment decisions is to attempt to reveal private information that users have about future demand. One way to do this is to organize a market for forward contracts. In addition to its information revelation function, forward contracting has other attractive properties. It can reduce risk for the network operator, because it makes his income less dependent on spot market prices, which might be quite volatile. The contracts can also reduce the risk for network users, who can use forward contracts to hedge against spot market price and quantity fluctuations. The forward market might also

B. Henze • B. Willems (✉)
Tilec & CentER, Department of Economics, Tilburg University, P.O. Box 90153, 5000 LE Tilburg, The Netherlands
e-mail: bastian.henze@googlemail.com; b.r.r.willems@tilburguniversity.edu

C.N. Noussair
CentER, Department of Economics, Tilburg University, P.O. Box 90153, 5000 LE Tilburg, The Netherlands
e-mail: c.n.noussair@tilburguniversity.edu

J. Rosellón and T. Kristiansen (eds.), *Financial Transmission Rights*, Lecture Notes in Energy 7, DOI 10.1007/978-1-4471-4787-9_8,

make the spot market more efficient as traders exploit arbitrage opportunities between the spot and forward markets.[1]

A *Long-Term Financial Transmission (Transportation) Right* or LTFTR (Hogan 1992; Bushnell and Stoft 1996; Hogan et al. 2010) is a type of forward contract that has been proposed specifically for energy markets. The holder of an LTFTR obtains a payment equal to the spot price of the commodity, in this case access to the network. The payment is received regardless of whether or not the owner of the LTFTR obtains units on the spot market. Bushnell and Stoft (1996, 1997) have proposed the use of such financial transmission rights (FTRs) in the American electricity sector, and Kristiansen and Rosellón (2006) have done so for the electricity sector. In both cases, the authors argue that investment decisions would improve. Joskow and Tirole (2000) show that a financial transmission right, is strategically equivalent to a physical transmission right with a use-it-or-lose-it condition. This is a requirement that a user purchasing capacity must pay for it regardless of whether or not he uses it, with the unused capacity resold to other users.

In this chapter we describe a laboratory experiment that considers the behavior of LTFTR. The structure of the experimental environment is informed by European policy issues. The parameters are chosen based on conditions characteristic of the European gas and electricity markets. The LTFTR are allocated with a uniform price sealed bid auction with lowest-accepted-bid pricing. Although this auction type is not incentive compatible (see for example Draaisma and Noussair 1997), experiments show that it typically allocates the items sold to the demanders with the greatest valuations (Alsemgeest et al. 1998). The auction also has the advantage that it yields a revealed demand curve, rather than only market prices, and thus can be especially informative to the network operator about future demand.

Gas pipelines and electricity grids are typically natural monopolies which are regulated in some manner to limit market power. In the European Union, incentive regulation in the form of revenue and price caps is increasingly common in energy transmission and distribution (Cambini and Rondi 2010). Therefore, in our experiment, we impose a price cap on the network operator. Thus, we study the implementation of an LTFTR within a framework of incentive regulation. While cap regulation without LTFTR encourages investment that reduces marginal cost, it does not create additional incentive on its own to expand capacity. Indeed, under some conditions it can lower the expected profits from capacity expansion (Vogelsang 2010). Thus, there is scope for a system of LTFTR to improve capacity investment decisions.

For the demand and cost functions chosen for our experiment, the LTFTR auctions are expected to often yield market prices that exceed the price cap. By requiring bidders to pay the market price, we preserve the relationship between bidders' payoffs and bidding strategies that would exist without the price cap. Thus, the array of bids can remain informative about underlying demand. The difference between the price that the buyers pay and the price cap is kept by the experimenter, and can be thought of as government revenue.

[1] Other benefits of forward contracting appear in oligopoly settings (see for example Allaz and Vila 1993 and Holmberg 2011).

In Sect. 8.2, we provide a brief introduction to the methodology of economic laboratory experiments. In Sect. 8.3, we review experimental studies on electricity markets and forward contracts. We detail the design of our experiment in Sect. 8.4; our results are presented in Sect. 8.5. Section 8.6 concludes.

## 8.2 The Experimental Approach

An experimental economy is one that the researcher creates for the purpose of answering one or more specific research questions. The traditional approach, laboratory experimentation, typically involves human participants interacting in a paradigm that reproduces the features the researcher deems essential in capturing the economics of the setting of interest. The setting could be that of a theoretical model, an extension or modification of a previously-studied experimental environment, or an industry or market of interest outside the laboratory. Experiments may be conducted with the purpose of testing theories, generating new theories, comparison of different institutional arrangements, or evaluation of policy proposals, among other objectives (see Smith 1994).

In our opinion, experiments are appropriately viewed as complementary to theoretical and other empirical approaches in economics. They allow the researcher to control the environment, observe key variables with certainty, and eliminate measurement error. The experiment can be replicated by other researchers. Some environments that are intractable to theoretical analysis can be studied. New policy proposals that have never been implemented, and thus for which there are no data, can be evaluated before they are taken the field. Their performance can be compared to theoretical benchmarks, to the status quo, or to alternative proposals, holding all else equal. The experimenter can change the environment or institutional structure exogenously to establish causal relationships. This approach parallels the use of agent-based modeling, which can achieve the same ends with behavioral assumptions that the researcher specifies exogenously. Experiments with human subjects allow the behavior to be generated endogenously from the interaction of human agents with the environment and institutions, in which they have been placed.

Experiments considering policy questions typically differ in a number of ways from the field environments they are meant to represent. These differences include the size of monetary stakes, the scale of the economy, some aspects of the economic environment, and in the fact that the experiment is conducted in a laboratory. The stakes in the experiment are on the order of US$15 an hour, which is thought to be sufficient to motivate participating individuals to attempt to attain high values of the objective functions they have been assigned. The participants are typically university students. This subject pool is used because it is a large pool that is accessible to researchers at relatively low cost, but also because it facilitates replication, which can be done at other universities. Some aspects of the field environment may be difficult to create or not essential to the *economics* of the task at hand. In such cases the experimenter resorts to similar simplifications as theorists do, for example modeling a firm as if it were one profit-maximizing individual, using a partial equilibrium framework, or introducing permanent commitment to a current policy. Finally, the experiments are typically conducted in an experimental laboratory where the researcher can exercise control over participant communication and answer the questions that subjects have.[2] In our view, it falls on those skeptical of experimental methods to argue how any of these differences would change the conclusion of a specific study.

## 8.3 Previous Experiments on Electricity Markets and Forward Contracting

A complete overview of all relevant experimental studies on electricity markets and forward contracting is beyond the scope of this chapter, and we review only the most closely related studies. For a more complete overview of experiments on electricity markets, see Kiesling (2005) or Staropoli and Jullien (2006). For a more general review of policy-oriented experimental work, see Normann and Riccuti (2009).

In our experiment, which is described in Sect. 8.4, we use uniform-price sealed-bid auctions to allocate both LTFTR and capacity on the spot market. Some other experimental studies have also studied environments modeled on electricity markets, and used auctions on either the demand or the supply side. Rassenti et al. (2003a) find that demand side bidding on a spot market successfully counteracts the exploitation of market power on the part of a network operator. Rassenti et al. (2003b) find that a uniform price auction leads to more efficient allocations than a discriminatory auction in an experimental electricity market. Noussair and Porter (1992) compare uniform-price sealed-bid auctions with highest-rejected-bid pricing and English clock auctions in an experimental electricity market, in which supply shortfalls may occur and rationing must take place ex-post. They find that the uniform price auction generates more efficient allocations than the English clock.

Other studies have considered the behavior of auctions where generators make offers to sell electricity to a power grid (see for example Denton et al. 2001; Abbink et al. 2003; and Vossler et al. 2009). This work suggests that our choice of auction would not generate undue inefficiency compared to alternative auction forms.

Two previous experimental studies focus specifically on investment in supply capacity in energy markets. They differ from our work in that they consider the generation and supply of energy rather than transport. They also differ in that they both consider an oligopolistic industry while our interest here is in a monopoly network operator. Kiesling and Wilson (2007) study an automated mitigation procedure (AMP), an alternative to a price cap, as a mechanism to control price

[2] There is an ongoing debate about whether the scrutiny placed on subjects in the laboratory affects the observed level of socially-oriented behavior in non-market interactions (see Levitt and List 2007). We are not aware of an argument in a similar vein that has been made with regard to market experiments.

spikes in wholesale power markets. They find that AMP does not decrease capacity investment compared to a benchmark of unregulated prices. Williamson et al. (2006) also consider investment in additional capacity in an experiment designed to study wholesale electricity markets. The market is an unregulated oligopoly. They observe investment close to the Cournot-Nash equilibrium level on average, but with a bias in the mix between marginal and baseload capacities.

There is also an experimental literature that investigates forward contracting. Krogmeier et al. (1997) and Phillips et al. (2001) compare markets, in which contracts are concluded before production. These can be thought of as forward markets. Le Coq and Orzen (2006) study an environment with an explicit forward market structure. These studies all report that a forward market is characterized by lower prices, greater quantity traded, and greater efficiency than is the case when contracting occurs on a spot market. Brandts et al. (2008) consider, in an experimental setting designed to model an electricity market, the effect of adding a forward wholesale market for electricity. They study a situation with imperfect competition between sellers and no demand uncertainty, so that forward contracting has no risk hedging function. They report that the addition of a forward market lowers prices and increases production, for both quantity and supply function competition.

We are aware of only one previous experiment that considers financial transmission rights. Kench (2004) compares a system of financial rights to one of physical rights. In his environment, network users obtain a random initial allocation of rights. They can then trade the rights with other users. This differs from our experiment, where users purchase rights from the network operator. He finds that physical rights provide more accurate market signals than financial rights. Physical transmission rights are allocated more efficiently than the financial transmission rights. Network users pay a penalty if physical transmission capacity remains unused. Network users compete more aggressively to obtain physical rights, because they would be unable to transport energy without them. In a setting with financial rights, generators are less active in the transmission rights market, because they also have the option to wait, and to trade energy in the spot market. Furthermore, network users that have not bought financial transmission rights and are therefore unhedged, bid strategically in the spot market and reduce efficiency.

## 8.4 The Experiment

Four sessions were conducted at the CentERlab experimental facility, at Tilburg University, Tilburg, The Netherlands. The experiment was computerized, and employed the Ztree platform (Fischbacher 2007). All participants were undergraduate students at Tilburg University, most of whom were majoring in economics or business.

Subjects participated in three independent sequences of periods. One participant was in the role of network operator and four were in the role of network user in each session. Each individual remained in the same role for the entire session. Initially, there was a twelve period training sequence, which did not count toward participants' earnings, followed by two 30 period sequences which did count.

The data from the last 30-period sequence, during which the subjects were the most experienced, is presented here. The experimental sessions lasted from 3 to 3.5 h. A quiz on the instructions was used to select participants. Out of eight subjects recruited for each session, only five were allowed to participate in the remainder of the experiment. The top performer on the quiz was assigned to the role of the network operator. The next four highest scorers were the network users. Those with the three lowest scores were asked to leave the experiment. The instructions and the quiz took on average between 60 and 75 min to complete. Earnings averaged 30.94 Euro for participants in the role of network operators and 24.31 Euro for those in the role of users.

The regulator was automated, and kept a fixed price cap policy in place throughout the session. Regulator revenue was not rebated to participants and is thus assumed to be spent outside the sector.

Market demand for access to the network in each period $t$ is a discrete approximation of a function of the form $D_t = a - \frac{2b}{g_t}q_t$, where $a$ and $b$ are constants, $q_t$ is the quantity of access supplied, and $g_t$ is a growth parameter. Individual demand is privately known to users.

Access to the network is supplied by one monopolistic operator, who can sell a quantity up to its current capacity. The installation of additional network capacity is in itself costless for the operator. However, each unit of capacity carries a maintenance cost of $c$ in each period, regardless of whether or not it was actually sold. This cost $c$ can be interpreted as the leasing cost or rental price of network capacity. Network capacity cannot be reduced at any time. There is no depreciation or scrap value for capacity. $K_t$ denotes the total capacity of the network in period $t$, and $K_0$ is the initial network capacity. We require that $K_t \geq K_{t-1}$ for all $t$.

The 30 periods are divided into five blocks of six periods each (Table 8.2). At the beginning of the first period of each block, each of the four users observes her individual demand for each period in the current block. The users then decide whether or not to increase the valuations from their initial level for their first two units. They can do so by either a relatively small fixed value $\kappa_{LOW}$, a relatively large fixed value $\kappa_{HIGH}$, or 0. This decision remains in effect for every period of the current block. To increase their valuations, users incur per-period costs of $\gamma_{LOW}$, $\gamma_{HIGH}$, or 0, respectively.

This opportunity to increase one's valuations is meant to represent the take-or-pay contracts that are common in Europe. These are long-term contracts that a network user, typically a gas company, makes with upstream suppliers. These contracts can be very profitable, but are also costly to break, which occurs if delivery of the contracted quantity does not occur. These contracts lead to greater user valuations for the corresponding units of network capacity. The cost to users of increasing these valuations represents the various costs of concluding such a contract and the penalty that one incurs if the contracted quantity is not exchanged.

At the beginning of the first and fourth periods of each block, the network operator decides whether or not to increase the capacity of the network (Table 8.2). $\Delta K^{MAX}$ denotes the maximum amount of additional capacity which can be installed

**Table 8.1** Parameters of the experiment

| Parameter | Value | Description |
|---|---|---|
| $a$ | 80 | Intercept parameter in aggregate demand function |
| $b$ | 5 | Slope parameter in aggregate demand function |
| $c$ | 10 | Per period cost for one unit of network capacity |
| $p^{cap} = f^{cap}$ | 15 | Price cap |
| $\kappa_{Low}$ | 20 | Low optional increase of the highest two valuations of a network user |
| $\kappa_{High}$ | 40 | High optional increase of the highest two valuations of a network user |
| $\gamma_{low}$ | 10 | Per period cost for raising highest two valuations by $\kappa_{Low}$ |
| $\gamma_{High}$ | 20 | Per period cost for raising highest two valuations by $\kappa_{High}$ |
| $K_0$ | 4 | Initial network capacity |
| $\Delta K^{Max}$ | 5 | Maximum possible investment at each investment opportunity |

**Table 8.2** Timing of activity

| | Period within six period block | | | | | |
|---|---|---|---|---|---|---|
| | 1 | 2 | 3 | 4 | 5 | 6 |
| Users learn their individual valuations for periods 1–6 | X | | | | | |
| Network users bid for LTFTR capacity | X | | | | | |
| Operator builds additional network capacity, sells all capacity in LTFTR | X | | | X | | |
| Users decide whether to raise their highest two valuations at a cost | X | | | | | |
| Spot auction | X | X | X | X | X | X |

at one time. Users are informed of changes in network capacity before they submit their bids in the spot auction.

The parameters of the experimental environment are listed in Table 8.1. They are chosen to ensure that key variables and ratios take on similar values in the experiment as in the field. See Henze et al. (2012) for details.

During each period, units of access to capacity are sold in a multi-unit uniform-price sealed-bid auction with lowest-accepted-bid pricing (Table 8.2). Users can submit one bid for each one of their valuations. The network operator then chooses the quantity of access to offer. If the operator offers $q$ units, the $q$ highest spot auction bids are accepted. Accepted bidders pay a per-unit price equal to the $q$-th highest bid. This is the market price $p$ for the current period.

An auction for LTFTR is also conducted in the first period of each six-period block. This auction takes place after the network users have been informed about their valuations, but prior to their decision about whether to increase their valuations, as well as prior to the network operator's decision to install additional capacity. The LTFTR are forward contracts which pay the network user who obtains them the spot price of one unit of network access in each period of the current block. The payment is made whether or not the LTFTR holder obtains units in the spot market.

The forward auction for LTFTR is also organized as a uniform price sealed-bid auction with lowest-accepted-bid pricing. All network users pay the same per-unit,

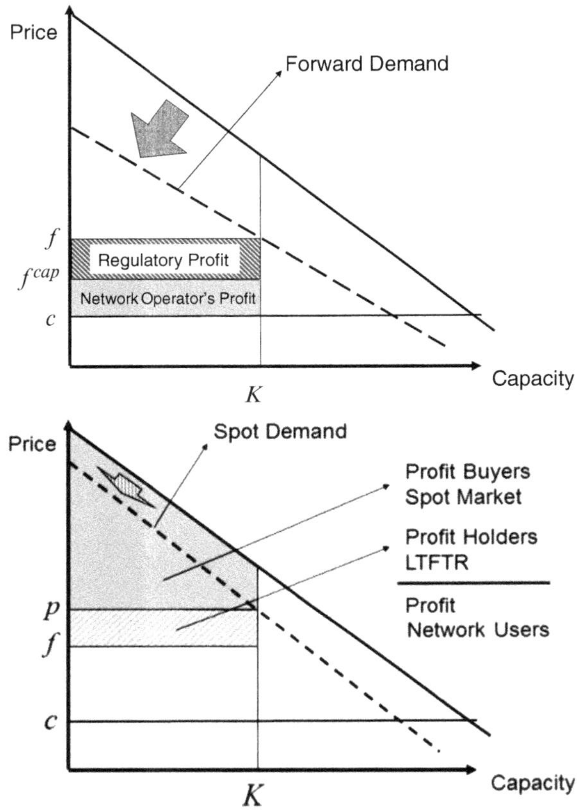


**Fig. 8.1** Forward market (*Upper panel*) and spot market (*Lower panel*)

per-period market price $f$ and there is a price-cap of $f^{cap}$ ECU (Experimental Currency Units, the unit of account in the experiment) in place. If the market price exceeds the cap, the operator receives $f^{cap}$ ECU per unit of the product while network users pay the market price $f$. If the market price exceeds the cap, the difference is kept by the regulator and considered as government revenue.

The network operator must offer to sell every unit of current capacity in both the forward and the spot auctions. All of the revenue from the spot auction is transferred to individuals holding forward contracts. Thus, its profit is determined exclusively in the forward market. The spot market is in essence a secondary market.

The allocation of surplus to users, the operator and the government is illustrated in Fig. 8.1. In the left panel of the figure, the revealed demand in the forward LTFTR auction is shown as the dashed line labeled "Forward Demand". The profit to the network operator is given by $(f^{cap} - c) \cdot K$, and regulator payoff by $(f - f^{cap}) \cdot K$. The panel on the right presents the spot market. The revealed demand in the spot market is illustrated with the dashed line labeled "Spot Demand". The surplus of the buyers in the spot market is given by the darker area. The auction revenue, indicated as $p \cdot K$ is transferred from buyers in the spot market to holders of LTFTRs, and thus

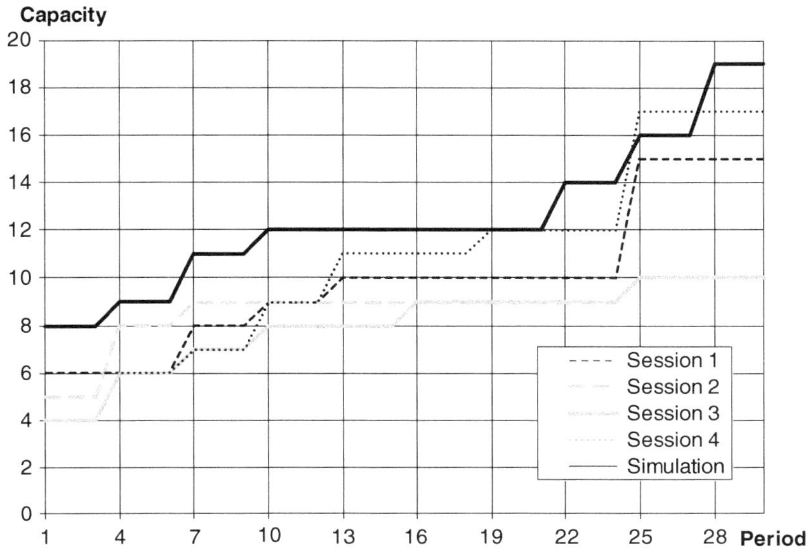


**Fig. 8.2** Time path of capacity in each session and benchmark simulation

$(p - f) \cdot K$ indicates the profit accruing to holders of LTFTR. The sum of both areas is the total profit of network users.

## 8.5 Results

Figure 8.2 illustrates the time path of capacity in each of the four sessions. The fifth time series in the figure, entitled simulation, is the capacity trajectory generated by a simulated profit maximizing network operator. This simulated agent is assumed to have perfect foreknowledge of future demand, but is subject to the price cap of 15 ECU that was present in the experiment. The simulation assumes that all four network users committed to the largest possible demand increase in each period.

The figure shows that at the outset, capacity averages 5.25 compared to the simulated level of 8, the difference likely due to the fact that the forward-looking simulation anticipates future demand growth. The ratio of actual to benchmark capacity improves over time from an average of 0.656 in period 1 to 0.833 in period 20 before decreasing again to 0.764 in period 30. There is a clear separation between sessions 1 and 4, which achieve close to the optimal capacity trajectory and sessions 2 and 3, which are characterized by low investment. The forward-looking network operator under the simulation anticipates a decrease in demand that occurs from periods 14 and 18 and slows her capacity expansion, while the human operators slow their investment somewhat later. When demand picks up again in the late periods, the network operators in sessions 1 and 4 expand capacity as a result, while the other two fail to do so sufficiently.

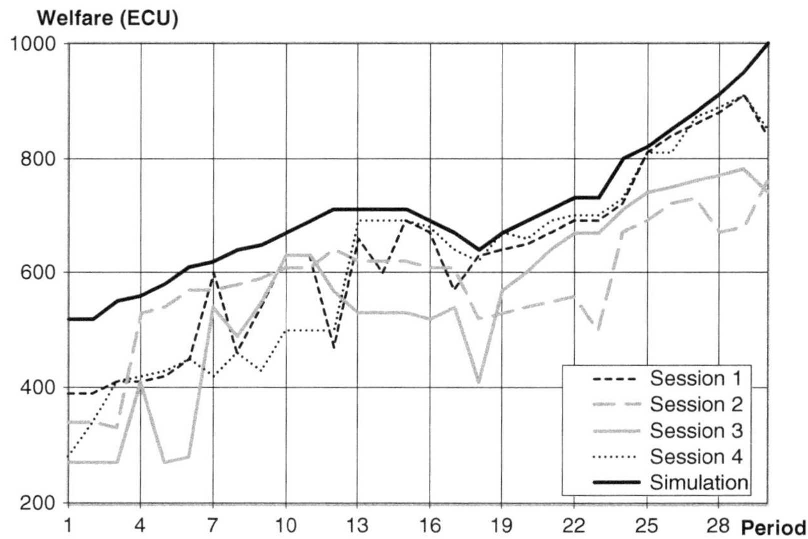


**Fig. 8.3** Time path of welfare in each session and benchmark simulation

Figure 8.3 illustrates the total welfare realized in the market, given as the total of consumer surplus, producer surplus, and regulator revenue, in each of the four sessions. The figure also includes the welfare along the simulated optimal trajectory. The figure shows that sessions 1 and 4 attain welfare levels close to the optimal level after a few periods, while sessions 2 and 3 consistently attain levels considerably lower than the benchmark level. The greater capacity in sessions 1 and 4 relative to the other two sessions is closely associated with greater welfare.

The distribution of surplus among agents follows a similar pattern in the experimental data as in the simulation. In the simulation, the payoff to the network operator averages 1,710 ECU (8.1 % of total surplus), that of the four users together averages 16,460 (each user receiving 19.5 % of total surplus), and the revenue to the regulator averages 2,910 (13.8 % of total). In the actual experiment, the shares were 7.5 %, 17.6 %, and 22.1 %, for the network operator, average individual user, and regulatory authority. The greater than predicted prices in the experiment result in more revenue to the regulator, since the price the operator receives is capped.

As a measure of welfare, we consider the efficiency of the outcome. The efficiency is the fraction of the maximum possible surplus that is actually realized in the experiment, and is a standard measure of welfare in experimental economics. We define Total Efficiency $\eta_t^{Tot}$ in period $t$ as the total welfare realized in each period, $W_t$, the sum of consumer surplus, network operator's profit and regulatory revenue, divided by the maximum possible total welfare for the same period given the demand and cost profile in the experiment, $W_t^*$, and thus $\eta_t^{Tot} = W_t/W_t^*$.

$W_t^*$ is calculated by simulating the decisions of a benevolent social planner under the assumption that there is no price cap in place. The simulated benchmark for a profit maximizing network operator, with perfect foresight but subject to the price cap we set in the experiment, generates efficiency of 99.5 %. This indicates that the price cap is set nearly optimally, and represents close to a best case scenario of incentive regulation. Total efficiency equals 82.3 % and 78.5 % in sessions 2 and 3, while it is 88.8 % and 87.0 % in sessions 1 and 4.

We distinguish between two components within the efficiency measure: (1) allocative and (2) dynamic efficiency. Allocative efficiency is a measure of the ability of the market to award the current existing capacity to the demanders with the greatest valuations. If all units are not sold to the highest-valued users, there is allocative inefficiency. The allocative efficiency $\eta_t^a$ in period $t$ is defined as $\eta_t^a = W_t/W_t^o(K_t)$. $W_t^o(K_t)$ is the welfare level resulting from allocating the current capacity $K_t$ to the users with the highest valuations.

The other component, dynamic efficiency, is a measure of the optimality of the time profile of capacity investment. Dynamic efficiency is the fraction of the globally optimal welfare level that could be attained with an efficient allocation of the actual current capacity.[3] Global optimality requires both optimal allocation of existing capacity and a socially optimal investment time trajectory. Dynamic efficiency is defined as $\eta_t^d = W_t^o(K_t)/W_t^*$.

In our data, the average allocative efficiency is 94.6 % and dynamic efficiency is 88.0 %. Figure 8.4 gives the time path of dynamic and static efficiency in each session, and reports as well the average level over all periods. Most of the efficiency loss in sessions 2 and 3 is because investment is low, and this is reflected in low dynamic efficiency. The forward auction, in conjunction with the spot market which serves the function of a secondary market, performs well in allocating the existing capacity to the demanders with the highest valuations. However, the forward auction in some sessions does not induce the appropriate level of investment.

Figures 8.5 and 8.6 show the time series of spot and forward prices in the four sessions, respectively, as well as under the benchmark simulation. There are a variety of patterns evident in the figures. In the simulation, the spot price begins at 50, falls to 10 in period 4, and remains between 10 and 20 for the rest of the 30-period horizon. In general, observed prices are considerably greater than along this benchmark trajectory. In the early periods, spot prices in sessions 1 and 3 are much greater than the benchmark level early on, reaching more than 3.5 times that level by period 11. In session 3, the spot price remains high for the rest of the session. Session 2 tracks the baseline level closely until period 18 and then exhibits an increase to a higher level, which is then sustained. In session 4, prices conform closely to the benchmark level.

Forward prices exhibit similar heterogeneity across sessions, both with regard to their differences from the benchmark prediction and differences from concurrent spot prices. In session 3, forward prices are much higher than the simulated level,

[3] Dynamic efficiency in a particular period can exceed 100 % if a network operator overinvests compared to the social planner, who maximizes the sum of welfare over all 30 periods.

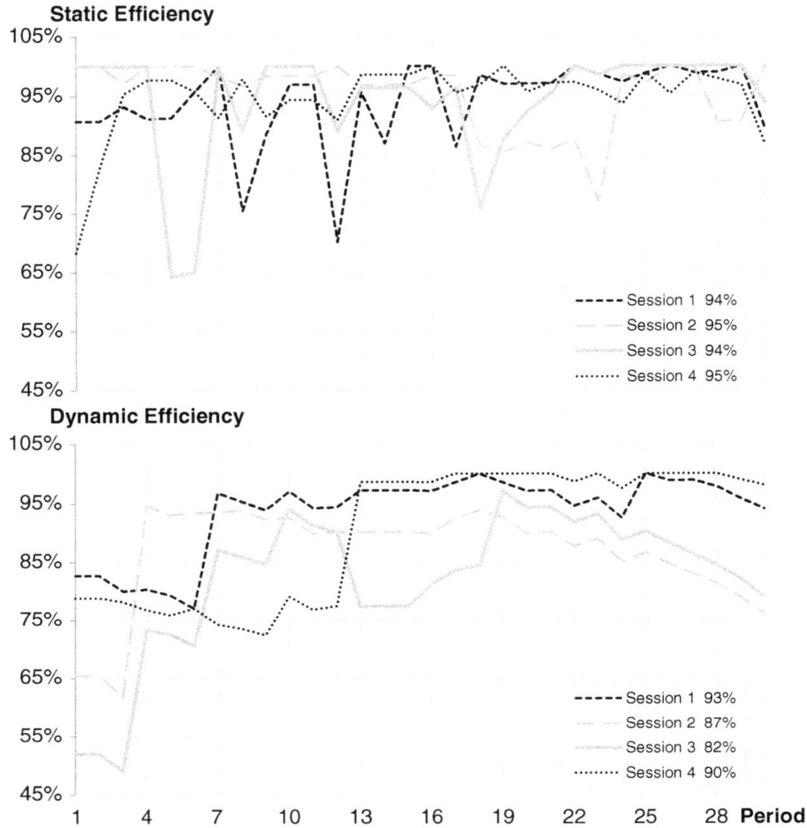


**Fig. 8.4** Time path of static and dynamic efficiency in each session

though they are also modestly lower than spot prices. In session 1, forward prices are much lower than spot prices, though still somewhat greater than the benchmark level. In session 4, forward prices are modestly lower than spot prices, but track each other fairly well, converging to close to the benchmark. In session 2, forward prices are much lower than spot rises, though they are close to the benchmark scenario.

Table 8.3 gives the value of (a) $f_t - s_t$, (b) $|f_t - s_t|$, and (c) $\sigma(f_t - s_t)$, averaged over the 30 periods of each session. These are the average difference between forward and spot prices, as well as the average absolute value and standard deviation of the difference. Also included in the table are the number of times that forward prices change over the 30 periods (9 is the maximum possible number of instances), and the average absolute value of such changes. The table reveals some interesting patterns.

The first is that the difference is negative in all four sessions, indicating that forward prices are typically lower than eventual spot prices. This might be due to the use of an auction to allocate the LTFTR. A one-sided auction allows for

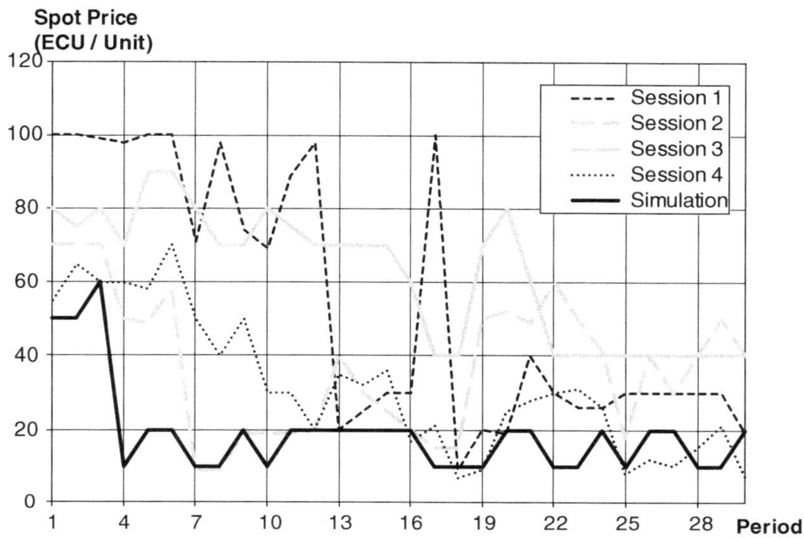


**Fig. 8.5** Time path of spot prices in each session and benchmark simulation

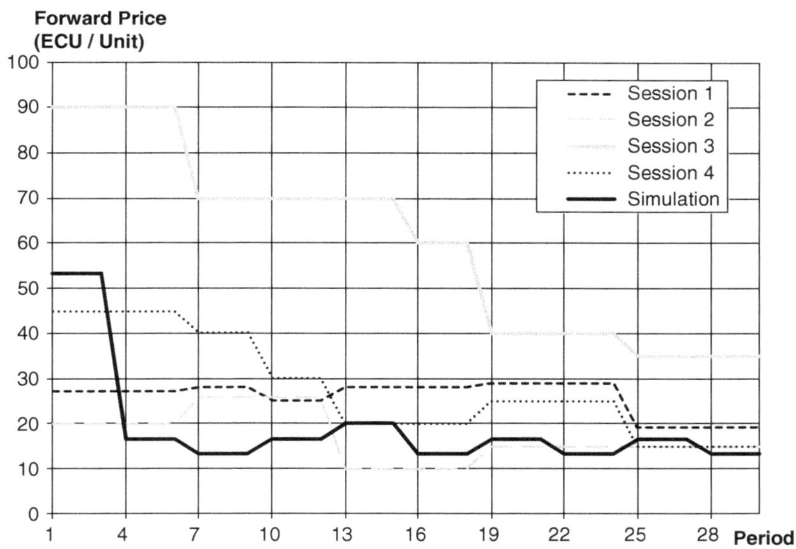


**Fig. 8.6** Time path of forward prices in each session and benchmark simulation

strategic underbidding, which could lower prices, while the passive seller does not behave strategically to offset this effect. The resulting low prices, if not properly interpreted by the network operator, might create an impression on the part of the operator that future demand is lower than it really is, and dampen investment.

Table 8.3 Differences between spot and forward prices and changes in forward prices

| | Session 1 | Session 2 | Session 3 | Session 4 |
|---|---|---|---|---|
| $f_t - s_t$ | −31.8 | −20.4 | −1.7 | −3.9 |
| $\lvert f_t - s_t \rvert$ | 34.0 | 24.4 | 8.0 | 8.3 |
| $\sigma(f_t - s_t)$ | 34.0 | 20.1 | 12.9 | 10.0 |
| $\mathrm{Count}(f_{t+1} \neq f_t)$ | 5 | 3 | 4 | 5 |
| $\lvert f_{t+1} - f_t \rvert$ for $f_{t+1} \neq f_t$ | 3.6 | 9.0 | 13.8 | 8.0 |

The second pattern is that there is no relationship between the level of efficiency or capacity provision with the level, absolute value, or variance of the difference between spot and future prices. Session 1, in which capacity investment is close to optimal and efficiency is relatively high, has the greatest discrepancy between spot and future prices, and the greatest variability of the difference. On the other hand, session four, the other market with close to optimal investment and welfare levels, has a below-average discrepancy between spot and forward prices with below-average variability. Thus, overall efficiency is not correlated with the predictive power of the forward market for subsequent spot prices.

The third pattern is that the LTFTR markets tend to be characterized by modestly more frequent but smaller prices changes in the more efficiently operating markets. These relatively smooth price patterns in forward prices in sessions 1 and 4 may increase the credibility of the LTFTR prices in the view of the network operator in guiding his investment decisions.

## 8.6 Conclusion

In this chapter we have reported the results of four experimental sessions in which a network operator uses LTFTR to establish a forward market for her capacity. The experimental environment contains some distinctive features of energy markets. Demand exhibits an increasing trend, but varies unpredictably. Network users have more information about future demand than the network operator. A small number of network users have some market power in both the spot and LTFTR markets. The network operator is subject to cap regulation.

The data show that market behavior varies considerably among different sessions, despite the fact the underlying structure is the same in every way, except for the identity of the individuals participating, who are randomly drawn from the same subject pool. The four sessions we have conducted yield four different scenarios about how spot prices, LTFTR prices, investment, and welfare can interact. In session 1, capacity moves along a trajectory close to the optimum. However, spot prices are much greater than forward prices for most of the session, and thus forward prices give biased signals about future demand. Session 2 is also characterized by spot prices that are much greater than forward prices, and convergence of forward prices to levels close to those observed under the benchmark simulation. However, capacity remains consistently short of the optimal level, and this exerts a considerable, negative effect on efficiency. Here, the network operator may interpret the low forward prices as an indicator of low future demand and not invest sufficiently. Session 3 has forward prices more or less in line with spot prices, and thus forward prices are good predictors of spot prices. Although both prices are high relative to an optimal scenario, they do not induce a sufficient level of investment, and capacity and welfare remain low. Here, the high prices appear to arise as a response to low capacity, but the network operator fails to respond to the price signals. In session 4, forward and spot prices track each other fairly closely, exhibiting a decreasing trend over time. Despite the trend, capacity investments are made consistently over the time horizon and high welfare levels result. The decreasing prices over time appear to be a response to increasing capacity.

These results suggest that what can be expected if LTFTRs are implemented depend on several factors. How the network operator interprets the data from the LTFTR market is very important. If he takes high or increasing prices as a willingness to pay for an increase in capacity, he may invest more in response. On the other hand, if the LTFTR market is viewed as being driven by speculative demand unconnected to the underlying commodity, the operator may not respond to the price signals it generates. If the operator fails to correct for strategic underrevelation of demand in the LTFTR market, he may believe that demand is likely to decline in the future and withhold investment in response. On the other hand, if he fully corrects for this and interprets prices in that context, strategic behavior will not affect investment decisions.

## References


Abbink K, Brandts J, McDaniel T (2003) Asymmetric demand information in uniform and discriminatory call auctions: an experimental analysis motivated by electricity markets. J Regul Econ 23:125–144

Allaz B, Vila JL (1993) Cournot competition, forward markets and efficiency. J Econ Theory 59:1–16

Alsemgeest P, Noussair C, Olson M (1998) Experimental comparisons of auctions under single and multi-unit demand. Econ Inq 36(1):87–97

Brandts J, Pezanis-Christou P, Schram A (2008) Competition with forward contracts: a laboratory analysis motivated by electricity market design. Econ J 118:192–214

Bushnell J, Stoft S (1996) Electric grid investment under a contract network regime. J Regul Econ 10:61–79

Bushnell J, Stoft S (1997) Improving private incentives for electric grid investment. Resour Energy Econ 19(1–2):85–108

Cambini C, Rondi L (2010) Incentive regulation and investment: evidence from European energy utilities. J Regul Econ 38:1–26

Denton M, Rassenti S, Smith V (2001) Spot market mechanism design and competitivity issues in electric power. J Econ Behav Organ 44:435–453

Draaisma T, Noussair C (1997) Optimal bidding in a uniform price auction with multi-unit demand. Econ Lett 56(2):157–162

Fischbacher U (2007) z-tree: Zurich toolbox for ready-made economic experiments. Exp Econ 10(2):171–178
Henze B, Noussair CN, Willems B (2012) Regulation of network infrastructure investments: an experimental evaluation. J Regul Econ 42(1):1–38
Hogan W (1992) Contract networks for electric power transmission. J Regul Econ 4(3):221–242
Hogan W, Rosellón J, Vogelsang I (2010) Toward a combined merchant-regulatory mechanism for electricity transmission expansion. J Regul Econ 38(2):113–143
Holmberg P (2011) Strategic forward contracting in the wholesale electricity market. Energy J 31:169–202
Joskow P, Tirole J (2000) Transmission rights and market power on electric power networks. Rand J Econ 31(3):450–487
Kench B (2004) Let's get physical! or financial? A study of electricity transmission rights. J Regul Econ 25:187–214
Kiesling L (2005) Using economic experiments to test electricity policy. Electr J 18:43–50
Kiesling L, Wilson B (2007) An experimental analysis of the effects of automated mitigation procedures on investment and prices in wholesale electricity markets. J Regul Econ 31:313–334
Kristiansen T, Rosellón J (2006) A merchant mechanism for electricity transmission expansion. J Regul Econ 29:167–193
Krogmeier J, Menkhaus D, Phillips O, Schmitz J (1997) An experimental economics approach to analyzing price discovery in forward and spot markets. J Agric Appl Econ 29:327–336
Le Coq C, Orzen H (2006) Do forward markets enhance competition? Experimental evidence. J Econ Behav Organ 61:415–431
Levitt S, List J (2007) What do laboratory experiments measuring social preferences tell us about the real world? J Econ Perspect 21(2):153–174
Normann HT, Riccuti R (2009) Laboratory experiments for economic policy-making. J Econ Surv 23(3):407–432
Noussair C, Porter D (1992) Allocating priority with auctions: an experimental analysis. J Econ Behav Organ 19(2):169–195
Phillips O, Menkhaus D, Krogmeier D (2001) Laboratory behavior in spot and forward auction markets. Exp Econ 4:243–256
Rassenti S, Smith V, Wilson B (2003a) Controlling market power and price spikes in electricity networks: demand-side bidding. Proc Natl Acad Sci 100:2998–3003
Rassenti S, Smith V, Wilson B (2003b) Discriminatory price auctions in electricity markets: low volatility at the expense of high price levels. J Regul Econ 23:109–123
Smith V (1994) Economics in the laboratory. J Econ Perspect 8(1):113–131
Staropoli C, Jullien C (2006) Using laboratory experiments to design efficient market institutions: the case of wholesale electricity markets. Ann Public Coop Econ 77:555–577
Vogelsang I (2010) Incentive regulation, investments and technological change. CESifo working paper series no. 2964
Vossler C, Mount T, Thomas R, Zimmerman R (2009) An experimental investigation of soft price caps in uniform price auction markets for wholesale electricity. J Regul Econ 36:44–59
Williamson D, Jullien C, Kiesling L, Staropoli C (2006) Investment incentives and market power: an experimental analysis. Economic analysis group discussion paper, United States Department of Justice

# Chapter 9
# FTR Properties: Advantages and Disadvantages

**Richard Benjamin**

## 9.1 Introduction

While financial transmission rights (FTRs) were developed as a hedge for locational price risk,[1] their advocates envision them as a multifaceted tool, providing revenue sufficiency for contracts for differences, distributing the merchandizing surplus an independent system operator (ISO) or regional transmission operator (RTO) accrues in market operations, and providing a price signal for transmission developers.

While several economists have addressed the question of whether allocating incremental FTRs to developers will induce efficient grid expansion,[2] the issue of FTR allocation for the existing grid has basically flown under the radar. Economists generally argue that opening the electricity sector to competition will increase efficiency and thus decrease costs. While costs have indeed fallen,[3] retail electricity prices have not followed.[4] And while the exercise of market power has been well documented, both in the U.S. and the U.K electricity market, another, more subtle factor may be propping up retail rates as well. The rules for FTR distribution for the existing grid, FTR market settlement, and the treatment of FTRs in rate cases all have important implications for retail rates. Seemingly innocuous decisions may

---

[1] See Hogan (1992).

[2] See, e.g., Oren et al. (1995), Wu et al. (1996), Bushnell and Stoft (1996a, b, 1997), Hogan (1998), Barmack et al. (2003), Brunekreeft (2004), Calviou et al. (2004), Keller and Wild (2004), Bogorad and Huang (2005), and Joskow and Tirole (2005).

[3] See, e.g. Fabrizio et al. (2007) and Knittel (2002).

[4] See, e.g. Apt (2005) and Taber et al. (2006). This conclusion in not unanimous, however (see, e.g. Joskow 2006).

R. Benjamin (✉)
Round Table Group, Inc, 1074 Springhill Ct, Gambrills, MD 21054, USA
e-mail: dr_rbenj@yahoo.com

J. Rosellón and T. Kristiansen (eds.), *Financial Transmission Rights*,
Lecture Notes in Energy 7, DOI 10.1007/978-1-4471-4787-9_9,

have helped to inflate retail rates in restructured states above those in their traditional counterparts.

A second basic issue that has gone mostly unnoticed is the difference between wholesale electricity market settlements in theory and in practice, and the implications of this difference for the hedging characteristics of locational marginal pricing (LMP). While FTRs serve as a perfect hedge against transmission congestion in theory, the same is not true in practice when load is not settled at the LMP.

Finally, as Siddiqui et al. (2005) have shown, FTR markets are themselves flawed. Siddiqui et al. (2005) found that FTR market participants were systematically unsuccessful at hedging larger risk exposures. This paper studies the implications of allocation of FTRs for the existing grid and RTO FTR market rules on retail rates (i.e., the distributional aspects of FTR allocation), FTR hedging properties, and FTR inefficiencies. Section 9.2 provides a brief review of transmission pricing. Section 9.3 offers background on FTR allocation, both for the extant grid and grid additions. Section 9.4 then examines FTR allocation using a two-node model. It makes the point that FTR allocation has important distributional implications. In particular, it shows that FTR allocation is an important determinant of the ability of restructured markets to hold down the retail price of electricity to consumers. Section 9.5 examines the hedging qualities of FTRs in a three-node model. It shows that RTO FTR practices create a divergence between the theoretical result of perfect hedging and FTR hedging in practice. It also shows, through a counterexample, that even in theory, FTRs cannot universally serve as a perfect congestion hedge. Section 9.6 offers a further discussion of the hedging aspect of FTRs. Section 9.7 presents data demonstrating the magnitude of FTR cost-inflating factors in United States RTOs and ISOs. Section 9.8 presents observations on FTR properties. Section 9.9 offers suggestions for further research and concludes.

## 9.2 Transmission Pricing

In order to do justice to any discussion of the properties of FTRs, one must first discuss transmission pricing, emphasizing LMPs, as developed by Schweppe et al. (1988).

Hsu (1997) divides the overall costs for a transmission network into four major components: returns and depreciation of capital equipment, operation and maintenance to ensure the network is robust, losses incurred in transmitting power, and opportunity costs of system constraints. He adds that marginal cost pricing of transmission services defines the impact on the overall system costs when one additional megawatt is injected or withdrawn at some node. According to Hsu (1997), these costs include two major components: marginal losses throughout the network and the opportunity costs of not being able to move cheaper power due to transmission line congestion. Hsu (1997) argues that under an ideal marketplace, transmission service charges should equal the short-run marginal costs of providing that service. This is the standard argument that locational marginal prices should include a congestion component. The overwhelming majority of energy economists seemingly agree with the interpretation of congestion charges as opportunity costs.[5] Rosellón (2003) agrees that there is a general consensus regarding the marginal cost of electricity transmission usage.

Oren et al. (1995) stands in sharp contrast, however. This work counters that the opportunity cost component is based on an improper analogy to transportation costs and arbitrage theory. The authors state that the idea is that if the good is priced at level $p_A$ at location $A$ then the price at any other location $B$ cannot exceed $p_A$ plus the cost of transportation from $A$ to $B$. Oren et al. (1995) argue that marginal transmission losses can be interpreted as the equivalent of a transportation cost, and that in the absence of such losses, nodal price differences would reflect no physical transmission costs. Nodal price differences, however, reflect the welfare gain from relieving the congestion between nodes $A$ and $B$. The authors argue that the transportation analogy is misplaced; because in electricity networks there is no active competition among transmission operators to carry electrons over their wires. In electricity networks, transmission constraints and their pricing are determined by the action and judgments of grid operators rather than by the decentralized decision making of transmission companies and their clients. Oren et al. (1995) conclude that, as a consequence, a better analogy to the differences in nodal prices is an externality tax imposed by a network operator. They further argue that nodal price differentials are not appropriate for allocating congestion rents across the network, and thus an alternative mechanism to allocate network congestion rents has to be designed. The authors do acknowledge, though, that locational prices equal the marginal valuation of net benefits at different locations, and thus provide the right incentives for consumption and generation decisions, both in the short run and the long run.

## 9.3 Background on FTRs and FTR Allocation

Hogan (1992) developed FTRs as a tool for allocating scarce transmission capacity ("the congested highway"). He argues that defining FTRs as the right to locational price differences, (the sum of the loss and congestion components) between busses provided correct short-run incentives for transmission system use. In the short run, a holder of FTRs should be indifferent between physically delivering power between two nodes or financial compensation if loop flow or system contingencies prevent physical delivery. He sees this tradeoff as the key to providing complementary

[5] See, e.g. Borenstein et al. (2000), Brunekreeft (2004), Bushnell and Stoft (1997), Chao and Peck (1996), Green (1997), Hogan (1992), Joskow and Tirole (2000), Kristiansen (2005a), Rosellón (2003), and Rotger and Felder (2001).

long-term transmission capacity rights for new generation. An FTR holder can honor any long-term delivery commitment by either physical delivery or using FTR revenue to purchase power at the point of delivery, thus guaranteeing the economic viability of such transactions and solving the problem of loop flow preventing physical delivery of generation under contract. Hogan's mechanism envisions a two-part tariff for transmission usage, with fixed charges for long-term transmission access, and short-run congestion charges.

Since Hogan's seminal work, different variants of FTRs have been proposed. Hogan's original proposal has since been labeled "point-to-point FTR obligations." Chao and Peck (1996, 1997) propose flowgate FTRs. Flowgate FTRs are constraint-by-constraint hedges that convey the right to collect payments based on the shadow price associated with a particular transmission constraint (flowgate) (Kristiansen 2005a). The other determining factor for FTR type is obligation versus option. An obligation FTR compels payment for price differences, where an option FTR gives the holder the option to receive the price difference, which the holder will use provided the (directional) price difference is positive. Since obligation-type FTRs are the most common in practice and the most closely scrutinized in the literature, this chapter focuses on this type.

Kristiansen (2005a) differentiates between FTRs allocated for grid expansions and for the extant grid. He notes that they can be given to those who invest in transmission line or to load-serving entities (LSEs) and others who pay fixed cost transmission rates, either through direct allocation or through an auction process in which the LSE is allocated auction revenue rights (ARRs) that can be used to purchase FTRs. Kristiansen (2005a) states that FTRs for existing transmission capacity can be allocated based on existing transmission rights or agreements (historical and entitlements), auctioned off, or so that their benefits offset the redistribution of economic rents arising from tariff reforms, *inter alia*.

## 9.4 Distributional Aspects of FTR Allocation

As Benjamin (2010) notes, researchers studying FTRs generally take allocation as given, thus ignoring the implications of FTR allocation on the marketplace. This section looks at this issue in detail in the contexts of a two-node and a three-node model.

The two-node model is the most straightforward way to examine the distributional aspects of FTR allocation. Denote as $i$ and $j$ two nodes connected by a single transmission line. Let the first node represent a generation pocket, connected with a load pocket by a single transmission line, which we assume to be congested (day-ahead), so as to create a difference in day-ahead prices at the load and generation pockets, denoted $p_i$ and $p_j$, respectively, with $p_j > p_i$. For simplicity we also assume a single generator at each node, producing quantities $q_i$ and $q_j$, with $q = q_i + q_j$. Assume further that there is no load at node $i$ (Fig. 9.1).

Denote the proportion of load covered by long-term bilateral contracts as $\alpha$, so that the proportion not covered under contract is $1 - \alpha$. As per Hogan, FTRs

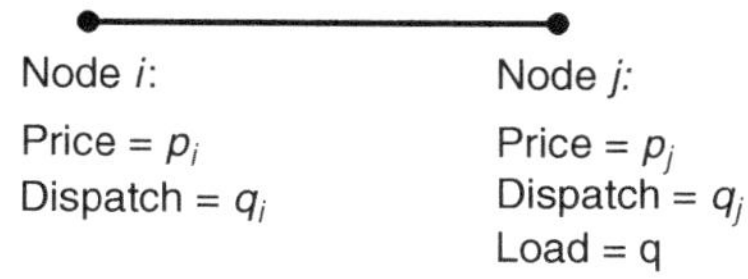


**Fig. 9.1** The two-node model

**Table 9.1** Settlements for non-contract power when the RTO allocates FTRs to generator $i$

| | Settlements | | |
|---|---|---|---|
| Entity | Energy | FTRs | Net |
| Generator $i$ | $(1-\alpha)p_iq_i$ | $(1-\alpha)\ (p_j-p_i)\ q_i$ | $(1-\alpha)p_jq_i$ |
| Generator $j$ | $(1-\alpha)p_jq_j$ | | $(1-\alpha)p_jq_j$ |
| LSE $j$ | $-(1-\alpha)p_jq$ | | $-(1-\alpha)p_jq$ |

**Table 9.2** Settlements for non-contract power when the RTO allocates FTRs to the LSE

| | Settlements | | |
|---|---|---|---|
| Entity | Energy | FTRs | Net |
| Generator $i$ | $(1-\alpha)p_iq_i$ | | $(1-\alpha)p_iq_i$ |
| Generator $j$ | $(1-\alpha)p_jq_j$ | | $(1-\alpha)p_jq_j$ |
| LSE $j$ | $-(1-\alpha)p_jq$ | $(1-\alpha)\ (p_j-p_i)\ q_i$ | $-\{[(1-\alpha)\ (p_iq_i+p_jq_j)]\}$ |

provide revenue sufficiency for the proportion of output covered by long-term contract. For that not covered by contracts, though, FTR allocation has important distributional implications. First let us assume that the RTO allocates FTRs from $i$ to $j$ to generator $i$. Settlements for this case are shown in Table 9.1, below:

Now let us examine the difference when the RTO allocates the FTRs to the LSE serving load at $j$ (Table 9.2).

We may graph these results as follows (Fig. 9.2).

Area (a) represents energy payments received by generator ($i$), while areas (a) plus (b) represent energy payments received by generator ($j$). Area (b) represents total FTR revenue.

When the RTO allocates FTRs to generator $i$, the LSE serving node $j$ ends up paying the load-pocket price for *all* energy procured, regardless of the source (That is, the total of Areas (a) through (c), or $(1-\alpha)q$), for total energy consumption. However, Under cost-based regulation, load-pocket energy procurement costs would be only

$$(1-\alpha)\big(p_iq_i+p_jq_j\big); \tag{9.1}$$

that is, the sum of Areas (a) and (c).

Let us assume that the market is in long-run equilibrium, so that bids reflect embedded cost. Then the difference between the net amount the LSE pays in the

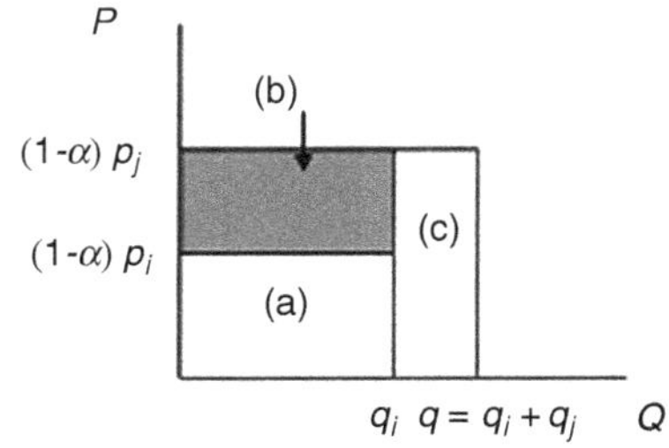


**Fig. 9.2** Settlements for power not under long-term contract

market when the RTO allocates the FTRs to it and the amount it pays when the RTO allocates FTRs to generators is the same as the difference between the cost that the LSE would pay for producing that power itself (i.e., if it were a vertically integrated utility) versus the amount it has to pay for that power in the competitive market, when it is not allocated the FTRs. We may thus calculate the cost inflation factor for the LSE purchasing power in the competitive market (relative to the cost of power procurement for the vertically-integrated utility), as

$$\frac{(1-\alpha)\left(p_j - p_i\right) q_i}{q}, \tag{9.2}$$

or Area $(b)/q$.

The argument that allocating FTRs to generator $i$ inflates the cost of electricity for consumers merits further discussion. In the traditional marketplace, the vertically-integrated utility (VIU) incurs a cost of

$$TC = c_i q_i + c_j q_j \tag{9.3}$$

to serve the load pocket, where $TC$ is total cost, and $c_k$ is embedded cost of generation, $k = i, j$. In the United States, many economists have argued that the primary goal of restructuring is to reduce retail electricity rates, that is, to decrease $p_k$ below $c_k$ sufficiently to make deregulation cost-effective.[6] However, as argued above, FTR "misallocation" adds another factor inflating the cost of power to consumers in competitive electricity markets.

The question remains as to whether there remain any dynamic arguments for allocating FTRs to generator $i$. That is, will allocating FTRs to generator $i$ facilitate attainment of the long-run equilibrium? Firstly, the desirable long-run equilibrium

[6] See, e.g. Fabrizio et al. (2007) and Joskow (2006). Deregulation may decrease prices by providing incentives for existing plants to improve their performance, and by providing price signals to new generating capacity and grid expansions (although, as mentioned above, the latter point is controversial). Joskow notes that restructuring in the U.K. was driven by the ideological commitment of the Thatcher government to competition as an alternative to regulated monopoly (p. 2), while the primary political selling point for competition in the United States was that it would benefit consumers by leading to lower costs and lower prices.

where generation at node $i$ earns a normal return is in no way contingent on this generator receiving FTRs. By the standard argument, short-run prices at $i$ will induce generation to enter or exit up to the point where all node $i$ generators receive zero economic profit. Let us denote the price corresponding to normal economic profit as $p_N$. That is, $p_N$ is equal to long-run average total cost of generation at node $i$,

$$p_N = LRAC_i. \tag{9.4}$$

If generators at node $i$ do not receive an allocation of FTRs, then, by the standard argument, price at node $i$ will gravitate toward $p_N$ in the long run.

Now let us assume that node $i$ generators are allocated FTRs. The most straightforward method of demonstrating the long-run equilibrium is to assume that there is now load at both nodes.[7] Let us assume for sake of simplicity, that generator $i$ is able to sell a constant amount of output, regardless of the amount of entry. Denote the amount generator $i$ sells at node $i$ as $q_i$, and the amount sold at $j$ as $q_j$. Next, assume that we are in the original long-run equilibrium, with $p_i = p_N$. Generator $i$'s total revenue is composed of two parts: (1) revenue from energy market settlements, and (2) revenue from FTR settlements, as shown in (9.5)

$$TR = p_i q + \left(p_j - p_i\right) q_j = p_i q_i + p_j q_j \tag{9.5}$$

where $TR$ is total revenue. Since $p_j > p_i = p_N$, generator $i$ is making positive economic profits. This will encourage other firms to enter until economic profits are again equal to zero. Since quantities are assumed to be unchanged, entry must occur until

$$p_i = \frac{p_N q_i - \left(p_j - p_N\right) q_j}{q_i} \tag{9.6}$$

That is to say, the node $i$ price must fall sufficiently to dissipate all FTR revenue earned. In this equilibrium, there is no price distortion, because LSEs pay generation at $i$ only enough for them to receive a normal return. The informational assumptions required for this equilibrium to obtain, however, are fantastic. That is, while it is difficult enough for a potential entrant to estimate the profitability of entry based on rapidly fluctuating electricity prices, asking the entrant to simultaneously gauge the profitability of future FTR revenues complicates the decision drastically. Therefore, by Occam's Razord,[8] it would be counterproductive to allocate FTRs to node $i$ generators in response to long-run equilibrium concerns.

---

[7] Otherwise, we would have to let entry decrease the amount of power sold by each generator, resulting in inefficient excess capacity.

[8] Occam's razor, (the law of parsimony, economy or succinctness), is a principle that generally recommends that, from among competing hypotheses, selecting the one that makes the fewest new

Note, though, that the analysis above does not preclude the possibility of merchant transmission investment being financed by incremental FTRs. Fundamentally, the issues of FTR allocation for the existing grid and for grid additions remain conceptually separate. Additionally, let us note that retail customers will ultimately pay for grid expansion, regardless of whether transmission additions are built by merchants and financed by FTR revenues, or by load-serving entities and financed through retail-rate adders. In the first case, LSEs pay congestion charges, either by buying FTRs to hedge congestion or paying congestion charges directly. In the second case, transmission expansions are amortized in retail rates. Thus, the only potential difference in retail-rate impacts is the risk that FTR revenues exceed the cost of the project. Numerous works, however, indicate that exactly the opposite will be the case because FTR revenues cannot be expected to fully-finance new transmission projects.[9] Thus, merchant transmission stands on its own merits, independent of the discussion in this paper.

By the above argument, then, the amount by which FTR allocation inflates the wholesale price of electricity relative to cost-based regulation depends on (1) to which party the FTRs are allocated, (2) the portion of electricity under long-term contract, (3) the amount of electricity imported into load pockets, and (4) the price difference between load-pocket and unconstrained generation. Let us defer discussion of point (1) briefly. Point (2) adds an additional argument for encouraging long-term contracting in the marketplace. Papers such as Blumsack et al. (2006), Rothkopf (2007), and Lave et al. (2007a, b) argue that maximizing the amount of capacity under long-term (and particularly "life-of-the-plant") contracts increase the competitiveness of wholesale electricity markets. Here, we find that in addition to any competitiveness issues, proliferation of long-term contracts will decrease any inflation of procurement cost for LSEs who are not allocated FTRs for load-pocket transactions and either have to pay the spot price for these transactions or purchase FTRs in the secondary market to hedge their spot-price exposure. Point (3) demonstrates that while increased transmission into a load pocket can bring more low-cost power into the area, load-pocket consumers will not benefit unless their LSE is allocated FTRs for such transactions or the transmission expansion reduces/eliminates the difference in LMPs. Correspondingly, Point (4) notes that the severity of the price distortion depends on the relative efficiency of generation in the load pocket to unconstrained generation. The older, less efficient the generation in the load pocket, the greater the distortion.

The distributional impact of FTR allocation ultimately depends on state regulators' treatment of FTR revenues. Suppose the state allows the LSE to keep all FTR revenues as profit, instead of crediting the amount against retail rates. The only distributional question regarding FTR allocation is whether generator $i$'s stockholders (when the generator receives the FTRs) or the LSE's stockholders

---

assumptions usually provides the correct one, and that the simplest explanation will be the most plausible until evidence is presented to prove it false. (thank you, Wikipedia)

[9] See fn. 1, above.

benefit.[10] In either case, retail rates will be inflated, as per (9.2). When the state rebates FTR revenues against retail rates, the redistributional results are telling. If the RTO allocates FTRs to LSEs who are required to credit FTR revenues against electricity-procurement costs, then the LSE's customers will benefit from lower retail rates. Otherwise, Generator *i*'s stockholders once again benefit. Given that support for electricity restructuring in the United States has extended as far as the consumer's energy bill, distributional concerns call for allocating FTRs to LSEs.

Making the assumptions that energy and capacity markets are competitive yields two additional results

**Result 1** *When all load in a two-node model is covered by FTRs, if the RTO allocates FTRs to the LSE serving the load pocket, then the LSE's cost of procuring wholesale energy is simply the cost of electricity generation.*

Result 1 follows from the argument that bids in the electricity markets will reflect embedded costs in long-run equilibrium, whether the transaction occurs in the spot market or the bilateral contract market. In this case, retail rates will fall provided the restructured company is more efficient than its traditional VIU counterpart.

When FTRs are allocated to generator *i*, however, load-pocket consumers pay the marginal price of (load-pocket) electricity production for all electricity consumed. As argued above, this result may act to inflate the retail price of electricity in restructured markets. In traditional markets, consumers pay the average of the embedded cost of all electricity produced, but in restructured markets, this is the case only if FTRs are allocated to LSEs. Section 9.7 examines the amount that retail rates have been inflated by all FTR market imperfections, as data limitations frustrate the effort to disentangle the separate effects.[11]

**Result 2** *In the perfectly competitive two-node model, as above, allocation of FTRs to the LSE serving the load pocket aligns the private and social incentives for transmission expansion, provided that the state regulatory agency allows the LSE to keep all of the cost savings attributable to the transmission expansion.*

Result 2 holds because, as per Leautier (2001) and Joskow and Tirole (2005), the social benefit from transmission expansion in the perfectly competitive market is equal to the redispatch cost savings attributable to the new line. This redispatch cost savings is also the LSE's benefit from building the line, provided the state allows the LSE to keep this savings.

Result 2 starts with the basic proposition that transmission expansion allows the substitution of less-expensive for more-expensive generation, reducing redispatch

---

[10] See Benjamin (2008), though, for a discussion of mechanisms made possible when load-pocket LSEs retain these revenues.

[11] Returning to the discussion of Sect. 9.3, allocation of auction revenue rights (ARRs) to FTR holders complicates matters further by introducing disparity between payments to LSEs and congestion revenues. Notice that the law of one price still applies for sale of electricity at each node, but see Lave et al. (2004), pp. 17–18 for an argument against paying the market-clearing price to all generation.

costs; and that this redispatch-cost savings is the value added of transmission expansion.[12] Next, it recognizes that the traditional investor-owned utility (IOU) serves the dual role of LSE and builder of the extant transmission system. Because the IOU serves these two roles, it reduces its own cost of procuring power for its retail customers when it builds new transmission. Because the IOU bears the full (social) cost of building new transmission (ignoring environmental externalities),[13] if it reaps the full benefit of transmission expansion, it will necessarily make socially optimal decisions if it is allowed to reap the full benefit of transmission expansion. Note that this is a sufficient, but not necessary condition for optimality. The state regulator may decide to decrease retail rates in this case, provided that it leaves the utility with at least a normal return to investment.

Result 2, of course, is not robust to increasing complexity of the transmission system. In meshed networks with loop flow, there will be many beneficiaries of transmission expansion, not simply a single LSE and its customers. Thus, remunerating transmission projects based on redispatch costs savings is no longer a simple exercise, but is fraught with the problem of potential free-riding.[14] This result does fit in nicely with Benjamin (2008), however, in that it provides further insight into the economics of load-pocket management. This result also adds explicit theoretical justification for the argument that transmission projects that alter nodal prices should not be done on a merchant basis. As noted already, again referencing Joskow and Tirole (2005) and Leautier (2001), the social justification for such projects is the redispatch cost savings they create, rather than incremental FTRs.

At first blush, Results 1 and 2 seem to yield conflicting recommendations regarding distribution of FTR revenue accruing to the LSE (Result 1 suggesting that it be refunded to retail ratepayers, so as to equate retail rates in restructured markets with those in traditional markets, Result 2 suggesting that the LSE keep these revenues). However, they are no more than variations on a theme, with Result 1 suggesting that incremental transmission be financed *separately in retail rates*, while Result 2 would have it financed directly *through FTR revenues*. Such a choice is ultimately in the hands of the regulator. Given that transmission expansion often alters nodal prices, regulators would be wise to finance transmission expansions in retail rates, as opposed to incremental FTR allocation.[15]

---

[12] Of course, this proposition also ignores the reliability-enhancing character of transmission, which is of great value as well.

[13] Although the transmission financing literature ignores questions such as scenic and environmental impacts of new transmission lines, NIMBY has a strong impact on transmission siting decisions, complicating actual transmission siting decisions.

[14] But see Benjamin (2007) for thoughts on how to award redispatch cost savings to transmission builders in meshed networks.

[15] Indeed, FERC has taken this tack in Order 679, "Promoting Transmission Investment Through Pricing Reform," 113 FERC 61,182.

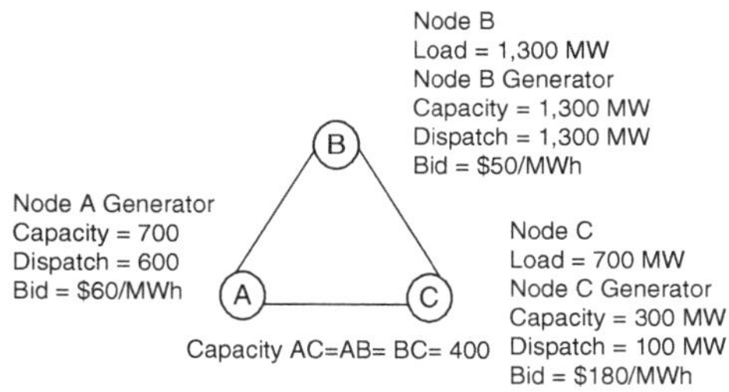


**Fig. 9.3** Three-node load pocket diagram

## 9.5 Hedging Aspects of FTRs Under Load Aggregation

Questions regarding the hedging characteristics of FTRs are generally relegated to situations where contingencies limit actual transmission capability below expected network capability, so that the merchandizing surplus will not fully finance allocated FTRs. Such inquiries, however, assume that load is settled on a nodal basis, when, in fact, it generally is not. For example, load in PJM is settled on a zonal basis, which has also been the plan in California under MRTU since it was MD02.[16] And the "perfect" hedge provided by FTRs is not robust to changes in settlements, as we will see below.

We illustrate this point using a three-node network, as it provides a richer set of results than does the two-node framework. First we examine a load pocket in a three-node network. While the RTO will dispatch some load-pocket generation for voltage support, in this example we motivate the dispatch of local generation as necessary to serve load as well. Denote the three nodes as A, B, and C, and assume that B and C both have local load, and that A, B, and C are all generation nodes. Assume all lines have equal impedance, so that 2/3 of the power generated at node A will flow on line AC, with 1/3 flowing on lines AB and BC, while 1/3 of the power generated at node B will flow on lines AB and AC, with 2/3 flowing on line BC. Let all lines have a capacity of 400 MWs. Finally, let the loads at nodes B and C be 1,300 and 700 MWs, and capacities at nodes A, B, and C be 700; 1,300; and 200 MWs, respectively. The diagram for the example follows (Fig. 9.3):

In this case, 400 MW of the 600 MW of generation at node A will flow on line AC, making the latter a binding constraint. With 200 MW flowing on lines AB and BC, neither is binding. As there is excess generation capacity at nodes A and C, node A and node C LMPs are the corresponding bid values of $60/MWh and $180/MWh, respectively. Since there is no excess generation at node B, an increment of load at node B would be met by an additional one-half MW from nodes

[16] That is, Market Redesign and Technology Upgrade and Market Redesign 2002.

**Table 9.3** Settlements for non-contract power when the RTO allocates FTRs to the buyer

| | Day-ahead energy and settlements | | | | | | |
|---|---|---|---|---|---|---|---|
| Entity | Energy (MWh) | LMP ($/MWh) | Load-weighted LMP ($/MWh) | Energy payment ($) | FTRs ($) | Net payment ($) | Net energy price ($/MWh) |
| Gen. A | 600 | 60 | n/a | 36,000 | 0 | 36,000 | 60 |
| Gen. B | 1,300 | 120 | n/a | 156,000 | 0 | 156,000 | 120 |
| Gen. C | 100 | 180 | n/a | 18,000 | 0 | 18,000 | 180 |
| Load B | 1,300 | 120 | 141 | −183,300 | 0 | −183,300 | 141 |
| Load C | 700 | 180 | 141 | −98,700 | 72,000 | −26,700 | 38.14 |

A and C (in order to satisfy the constraint on line AC), so the LMP at node B is simply ½(60 + 180) = 120.

Due mainly to political constraints, RTOs generally settle load on a weighted-average basis. A common concern for municipal utilities located in a transmission-constrained area is that because they are generally small, their service area fits entirely inside the constrained area. They would face high prices if load were settled on a nodal basis. Such is not a concern under weighted-average settlements, because their price simply becomes the zonal average price.

Let us also assume that B and C are the only nodes in the load-aggregation zone. The weighted-average price for load settlements is then

$$\frac{(1,300 \cdot 120) + (700 \cdot 180)}{2,000} = \$141/\mathrm{MWh}.$$

Now assume that load at node C receives the full allotment of FTRs with source of node A that sink at node C (AC FTRs).[17] The value of each FTR is equal to the difference in nodal prices between node C and node A, or

$$\$180 - \$60 = \$120/\mathrm{MWh}.$$

Since the LSE serving load at C holds 600 MWs of AC FTRs, The LSE receives

$$\$120 \times 60 = \$72,000/\mathrm{hour}$$

in FTR settlements. Total settlements are then as follows (Table. 9.3):

We may examine the hedging impacts of FTR allocation with respect to load at both nodes B and C. First, note that FTRs cannot hedge node-B load against

[17] Allocating 600 MWs of AC FTRs to load at node C is consistent with PJM's practice of distributing ARRs according to historical usage patterns (as long as we make the simplistic assumption that node B consumption and output have historically been equal).

congestion charges, because congestion creates a difference between the bid-price and LMP at a single node, and FTRs do not hedge against such an eventuality.[18] The best that node-B load can hope for is to convince the RTO that it should be allocated a share of the line AC FTRs, but this is contrary to the standard practice of allocating FTRs according to historical usage patterns.

The second point of note is that average pricing of load can over-hedge load-pocket consumption. This is the case in our example if the LSE at node C is allocated all of the AC FTRs. Assuming that the load pocket will have the highest nodal price, the average price paid by the node C LSE will be lower than the node C LMP, creating a subsidy of (LMP(C)—Weighted Average Price) per MWh. If load at nodes B and C are served by the same entity, this is a wash, but it is precisely because not all load in a zone *is* served by the same entity that RTOs employ nodal pricing, so this is a concern.

## 9.6 Further Thoughts on Hedging

A central strength of FTRs remains that properly defined FTRs serve as the perfect hedge for contracts for differences. Bushnell and Stoft (1997) show that FTRs supply revenue adequacy for CFDs, allowing for a fully hedged, fixed-price contract between traders. Indeed, the literature on managing electricity market risks echoes Bushnell and Stoft's analysis. Consider Deng and Oren (2006):

> A 1-MW bilateral transaction between two points in a transmission network is charged (or credited) the nodal price difference between the point of withdrawal and the point of injection. At the same time (assuming that transmission rights are fully funded), a 1 MW FTR between two points is an entitlement (or obligation) for the difference between the nodal prices at the withdrawal node and the injection node. Thus regardless of how the system is dispatched, a 1 MW FTR between two nodes is a perfect hedge against the uncertain congestion charge between the same two nodes.[19]

We may demonstrate this result borrowing Bushnell and Stoft's (1997) (Table 9.4).

As one may readily see, absent FTRs, the supplier will not receive the full contract price for power produced, but rather the contract price minus the nodal price difference $(p_j - p_i)$.

This analysis, however, begs a further question: What impact does the introduction of LMP and FTRs have on the contract market? That is, were bilateral market participants better off before, or after the introduction of LMP and FTRs? To

[18] Further, one cannot expect long-term contracting to rectify the situation, because generators have no incentive to sign long-term contracts for anything less than the expected LMP at node B, as the California crisis made abundantly clear.

[19] Deng and Oren (2006), pp. 950–951. See also Liu and Wu (2007), Sarkar and Kharparde (2008), and Yu et al. (2010).

**Table 9.4** Using FTRs to provide revenue adequacy for bilateral trades

| Contract or market | Payment | |
|---|---|---|
| | Supplier at node $i$ | demander at node $j$ |
| Spot market | $p_i \cdot q$ | $-p_j \cdot q$ |
| CFD for $q$ at $P_c$ | $(P_c - p_j) \cdot q$ | $-(P_c - p_j) \cdot q$ |
| FTR for $q$ from $i$ to $j$ | $(p_j - p_i) \cdot q$ | |
| Total | $P_c \cdot q$ | $P_c \cdot q$ |

answer this question, let us assume that the FTR market once again satisfies the efficient market hypothesis, so that $P_{FTR} = E(p_j - p_i)$. If we further assume that the long-term contract market is competitive, then in a restructured market, the contract price, $P_c$, will equal the competitive contract price, $P_c^*$, where the competitive contract price yields normal economic profit to the marginal generator, plus the price of an FTR, or

$$P_c = P_c^* + E(p_j - p_i) \tag{9.7}$$

Provided that the node $i$ LSE receives the all of revenue from FTR sales from node $i$ to node $j$ (generally through auction revenue rights, see e.g. PJM (2009), Joskow (2005), and Kristiansen (2005b, 2008)), both the seller and the buyer receive/pay the expected net price, $P_c^*$.[20] However, to the extent that FTR auction results contain a stochastic component and the parties are risk averse, then *both* the buyer and seller are made *worse off* by the introduction of LMP and FTRs!

Likewise, to the extent that market imperfections persist in the FTR market,[21] the introduction of LMP and FTRs produces lasting inefficiency in the long-term contract market.

This observation brings us back to Oren et al.'s (1995) contention that nodal price differences do not reflect opportunity cost because transmission operators do not compete with each other to carry electrons over their wires, but rather reflect an externality tax imposed by the system operator. Under this interpretation, electricity transactions under long-term contract arguably deserve a "tax-break," as they help to limit electricity spot-market price volatility. That is to say, there is nothing to stop the system operator from simply settling power under contract at the contract price, while eliminating the FTRs corresponding to the contracted energy. As argued above, doing so would both increase market efficiency and make parties to bilateral contracts better off. Admittedly, this would make existing FTR markets even thinner, but as per Benjamin (2010), the system operator might simply allocate FTRs to LSEs to cover actual power transactions. This would serve the dual

[20] The seller/buyer receives/pays $P_c = P_c^* + E(p_j - p_i)$ according to the energy contract and pays/receives $E(p_j - p_i)$ in the FTR auction.

[21] See Deng et al. (2010).

purposes of (1) further encouraging generators to sign long-term power contracts, and (2) give LSEs leverage in the long-term contract market, helping push down the price of power in imperfectly functioning long-term electricity markets towards the generator's embedded cost, improving the efficiency of restructured electricity markets.

## 9.7 FTR Cost Inflation

This section measures the net costs flowing to market participants after accounting for revenues FTRs provide to hedge their transactions. It starts with an analysis at the RTO level, then proceeds to the LSE level. At the RTO level, net costs associated with FTRs consist of the costs of running the FTR markets themselves (FTR administration fees), the difference between congestion charges incurred in settlements and revenues collected through FTRs and ARRs,[22] legal settlements the RTO pays stemming from FTR market disputes, any construction costs incurred in establishing FTR or long-term FTR facilities, and FTR defaults. These figures are shown for ISO-NE, NYISO, and PJM in Table 9.5 below.[23] Appendix A elucidates on the data sources for these values.

From 2006 to 2008, these values ranged from $244 to $625 million. As the data show, FTR market imperfections result in hundreds of millions of dollars of expenses paid by ratepayers yearly. To put these figures into perspective at an RTO level, I compare them with total costs of RTO operations for these three years in Table 9.6.

NYISO exhibits the greatest FTR market issues, with costs ranging from 149 % to 358 % of RTO operating costs from 2006 to 2009. This fact is mainly attributable to NYISO's practice of fully-funding FTRs. NYISO has been revenue insufficient in both the day-ahead and real-time markets, due to (1) transmission line deratings spurred by thunderstorm alerts (TSRs), and, particularly in 2008, (2) circuitous transactions—that is, fictional contract paths which exacerbate congestion and system operations. In the first 7 months of 2008, circuitous transaction scheduling around Lake Erie caused hundreds of millions of dollars in FTR underfunding.

The problems in NYISO, as well as participant defaults in PJM helped fuel FTR cost inflation from 46 % of total RTO operating costs in 2006 to *116 %* of RTO operating costs in 2008. These numbers provide additional support to economists such as Apt (2005), Blumsack et al. (2006), Lave et al. (2004, 2007a, b), Morey

[22] Each of the RTOs allocates ARRs to market participants, generally based on historical system usage. ARRs give their holders claims to the revenues collected in the FTR auctions held by RTOs. ARR holders may then either keep the ARRs or translate them into FTRs, through processes that differ from RTO-to-RTO. The difference between congestion and total ARR and FTR payments is known as "unhedged congestion."

[23] Data for the Midwest ISO and the California ISO did not provide comparable estimates, and are not included.

**Table 9.5** FTR Cost inflation factors (in thousands of dollars)

| RTO | Year | FTR admin. fees | Unhedged congestion charges | Legal settle-ments | Long-term FTRs | FTR defaults | Totals |
|---|---|---|---|---|---|---|---|
| ISO NE | 2006 | 270 | 7,423 | n/a | n/a | n/a | 7,693 |
| NYISO | 2006 | 2,337 | 211,000 | n/a | n/a | n/a | 213,337 |
| PJM | 2006 | 2,092 | 21,024 | n/a | n/a | n/a | 23,116 |
| Totals | 2006 | 4,699 | 239,447 | 0 | 0 | 0 | 244,146 |
| ISO-NE | 2007 | 315 | 7,334 | n/a | 719 | n/a | 8,368 |
| NYISO | 2007 | 2,074 | 252,000 | 1,542 | n/a | n/a | 255,616 |
| PJM | 2007 | 10,204 | 28,036 | n/a | n/a | 26,303 | 64,543 |
| Totals | 2007 | 12,593 | 287,370 | 1,542 | 719 | 26,303 | 328,527 |
| ISO-NE | 2008 | 476 | 8,627 | n/a | 960 | n/a | 10,062 |
| NYISO | 2008 | 2,648 | 504,000 | n/a | n/a | n/a | 506,648 |
| PJM | 2008 | 10,538 | 52,249 | n/a | n/a | 45,943 | 108,730 |
| Totals | 2008 | 13,661 | 564,876 | 0 | 960 | 45,943 | 625,440 |

**Table 9.6** FTR costs as a percentage of RTO operating costs

| RTO | Year | Net FTR costs ($ thousands) | RTO operating costs ($ thousands) | FTR costs as a percentage of RTO operating costs |
|---|---|---|---|---|
| ISO-NE | 2006 | 7,693 | 114,938 | 6.69 |
| NYISO | 2006 | 213,337 | 142,945 | 149.24 |
| PJM | 2006 | 23,116 | 274,536 | 8.42 |
| Totals | 2006 | 244,146 | 532,419 | 45.86 |
| ISO-NE | 2007 | 8,368 | 119,278 | 7.02 |
| NYISO | 2007 | 255,616 | 147,545 | 173.25 |
| PJM | 2007 | 64,543 | 281,194 | 22.95 |
| Totals | 2007 | 328,527 | 548,017 | 59.95 |
| ISO-NE | 2008 | 10,062 | 120,571 | 8.35 |
| NYISO | 2008 | 506,648 | 141,395 | 358.32 |
| PJM | 2008 | 108,730 | 277,895 | 39.13 |
| Totals | 2008 | 625,440 | 539,861 | 115.85 |

et al. (2005), Morrison (2005), and Rothkopf (2007) calling for changes in deregulated electricity markets.

At the LSE level, the ideal way to measure the retail-rate impact of FTRs would be to extract data from LSE retail-rate filings. However, these filings almost universally do not present this level of detail. Given this data limitation, I use FERC Form No. 1 data to estimate unhedged congestion for all entities operating in RTO or ISO markets and making FERC Form No. 1 filings for the years 2006

**Table 9.7** FTR overhedging

| year | Positive net transmission rights | Net transmission rights as a percentage of gross transactions | Retail rate impact ($/MWh) |
|---|---|---|---|
| 2006 | $203,052,352 | 7.793 | −0.741 |
| 2007 | $240,146,073 | 8.269 | −0.811 |
| 2008 | $360,401,276 | 8.482 | −1.014 |

through 2008.[24] FTR data is spotty at best for years before 2006, dictating the initial year of the data set. Statistics used to calculate unhedged congestion is found on page 397 the utility's Form No. 1 filing, as shown in Appendix B. These values appear under various categories such as transmission rights, congestion, auction revenue rights, transmission rights-sales, and transmission rights-purchases. Data was also gathered for total sales to ultimate consumers (Form No. 1, page 300, line 10), as well as gross transactions. Gross transactions are computed as the sum of absolute values of net purchases (account 555) and net sales (account 447), as found on Form No. 1, page 397.[25] As one cannot determine, *a priori*, whether the utility is using FTRs as a hedge for sales or purchases, it is prudent to simply include both transaction categories. Figures for net transmission rights are obtained from Form No. 1, page 397 as well.

The data of Appendix B confirm the prediction of distributional impacts of FTR allocation. Of the 21 utilities listed, ten were underhedged in each of the 3 years, while six were overhedged for each of the years for which congestion data was available.[26] FTRs therefore systematically overhedge some utilities, while underhedging others. To estimate the retail-rate impact of FTR revenue surpluses and shortages, I divided net transmission rights by MWhs in total sales to ultimate consumers. Estimates range from pennies per MWh to $5.35/MWh for Central Vermont in 2007. To estimate the scale of FTR under/overfunding per transaction undertaken by utilities in RTO markets, I divided net FTRs by gross transactions, in dollars. Appendix B lists this information as well.

Consistent with the data obtained at the RTO level, LSE-level data also shows that the rate impact of FTR market imperfections has increased over the past 3 years. Summary figures are given in Tables 9.6, 9.7, and 9.8 below.

Whereas this conclusion was largely dependent upon the NYISO market at the RTO level, it is independent of NYISO at the LSE level, because, as per footnote 28, firms operating solely in this market are omitted. In both cases, the data

[24] The FERC requires all major electric utilities to make Form No. 1 Filings. FERC defines "major" as having (1) one million MWh or more; (2) 100 MWh of annual sales for resale; (3) 500 MWh of annual power exchange delivered; or (4) 500 MWh of annual wheeling for others (deliveries plus losses). Because NYISO uplifts a large portion of congestion costs, I omit firms operating solely in this market.

[25] The accounting convention used on page 397 is to list debits as *positive* figures, and credits as *negative* values.

[26] Congestion data for Wisconsin Power was not provided for 2006. One might also interpret these differences as containing a component due to risk aversion of the various market participants.

**Table 9.8** FTR underhedging

| Year | Negative net transmission rights | Net transmission rights as a percentage of gross transactions | Retail rate impact ($/MWh) |
|---|---|---|---|
| 2006 | $124,454,187 | 6.083 | 0.285 |
| 2007 | $172,250,364 | 5.981 | 0.406 |
| 2008 | $209,070,380 | 9.446 | 0.579 |

demonstrates that FTR markets are not maturing, or, at least, their kinks are getting not smaller, but rather larger over time. The distributional aspect of FTR settlements is sizeable, with average rate deflation of over 1 dollar per MWh for those who benefited, and costing an average of $0.58/MWh for those whose costs were inflated in 2008.

## 9.8 Observations

This section comments on the ability of FTRs to serve the four functions they have been proposed to serve: (1) providing a hedge for nodal price differences, (2) providing revenue sufficiency for contracts for differences (CFDs), (3) distributing the merchandizing surplus an ISO or RTO accrues in market operations, and (4) providing a price signal for transmission and generation developers.

Let us examine each of these functions in turn.

1. Hedging. As Sect. 9.4 shows, once load aggregation enters the picture, FTRs are no longer the perfect hedge as envisioned in theory. The political constraints that have served to thwart load settlement at LMPs have seen to that. Further, though, the hedging properties of FTRs were developed in the context of long-term bilateral contracts. As we have seen, when these contracts are not in place, FTRs serve as a hedge to only the holder, with important distributional consequences.
2. Distributing the Merchandizing Surplus. While FTRs still serve to distribute the RTO's merchandizing surplus, the choice of FTRs to distribute the merchandizing surplus is arbitrary. Any number of mechanisms might serve this function. And as the analysis of Sect. 9.7 shows, FTRs constitute a quite expensive method of serving this function.
3. FTRs as a Price Signal. The ability of FTRs to signal transmission and generation development has come into question, and rightly so. Both experience and even theory show that FTR value serves only as a very imprecise signal of need for new investment. To be fair, though, the difference spoken of between congestion rent and congestion (redispatch) cost arises only because transmission investment is lumpy. On the margin, the two are the same. To the

economist, for whom marginal analysis is king, the assumption that FTRs and LMPs should provide the correct price signals is only natural.

Finally, note that nothing in this analysis precludes FTRs as an instrument for funding transmission grid expansions. Allocating incremental FTRs to a grid developer does not alter the fundamental recommendation that FTRs should match physical trades, as long as incremental FTRs are simultaneously feasible. RTOs which issue FTRs for grid additions would necessarily run minimal FTR markets, as parties transacting over incremental lines wish to hedge congestion costs. However, as per FERC Order 679, as well as numerous works on the subject, a sea change has already taken place with respect to funding new transmission through incremental FTR allocation.[27]

## 9.9 Conclusions

This paper examines the properties of FTRs, delving into their advantages and disadvantages. Using a two-node model it finds that if the ISO allocates FTRs to the LSE serving the load pocket, then the LSE's cost of procuring wholesale energy is simply the cost of electricity generation and that allocation of FTRs to the LSE serving the load pocket aligns the private and social incentives for transmission expansion, provided that the state regulatory agency allows the LSE to keep the cost savings. The paper argues that the first result is more relevant, though, because the second result breaks down when congestion reduction alters nodal prices. The paper then shows that under zonal pricing of load, FTRs no longer serve as a perfect hedge against congestion costs, as well as showing that, even in principle, FTRs do not necessarily serve as a perfect hedge for congestion. The work goes on to examine the magnitude of distributional consequences and inefficiencies caused by FTR allocation and FTR market imperfections. It shows that the magnitudes are great, mounting to hundreds of millions of dollars per year, with average distributional affects in the range of −$1 to + $0.5/MWh. It further calls into question the notion of FTRs as a hedge, arguing that while FTRs may serve as a "perfect hedge" for transmission congestion, this does not accord with the standard definition of a hedge as an instrument to hedge the variability of a firm's profit. Finally, the paper argues that while FTRs serve wonderfully as a complement to contracts for differences in providing revenue sufficiency for contracts for differences, their success in serving other of their proposed functions is lacking. Based on these observations, and the current state of restructured electricity markets, the paper concludes that RTOs should undertake a far less ambitious FTR program, limiting them to their hedging function (with trading in secondary markets limited to hedging purposes for

[27] But see Benjamin (2011), Gans and King (2000), Hogan et al. (2010), Rosellón (2003), Rosellón et al. (2011), Rosellón and Weigt (2011), and Vogelsang (2001), inter alia, for further thoughts on this matter.

transmission expansions financed by FTRs). The paper argues that allocating FTRs to LSEs while carving out energy served under long-term contracts will boost the negotiating position of LSEs in the long-term contract market, bringing the price of contract power closer to a generator's embedded cost while simultaneously reducing the cost of energy LSEs procure in the real-time market.

## Appendix A: Data Sources

FTR administration charges are the expense the RTO incurs in running the transmission rights market and are found on p. 322 of the RTO's FERC Form 1 filing (available at http://www.ferc.gov/docs-filing/forms.asp). I use these figures for all RTOs except PJM. PJM's FTR administration value is ambiguous, because one may measure it using the "Transmission Rights Market Facilitation" entry on p. 322, or the "Schedule 9–2" entries, found on p. 302. The former measures the *cost* PJM incurs in running its FTR markets. For 2007 and 2008 these values are $1,582,839 and $1,581,491, respectively. The latter measures the *revenue* that PJM collects from FTR administration fees. This revenue stream has two parts. The first is to FTR holders based on FTR megawatts and hours each FTR is in effect. For January 1, 2008 through December 31, 2010, the rate for this fee is $0.0027/MWh, subject to quarterly refund for revenue over-collection. The second is a charge to FTR auction participants based on the number of hours associated with each FTR obligation bid submitted in an FTR auction. The values for these two categories are given on p. 322 as two separate Schedule 9–2 entries. I use these values for PJM's FTR Administration charges because they are the amounts paid by LSEs (and other market participants).

The second category, unhedged congestion osts, is an estimate based on data found in the RTO's state of the market reports for 2007 and 2008. Because data supplied varies from RTO to RTO, differing methods of calculation are unavoidable and different interpretations are possible.

Let us start with PJM. Since June 1, 2003, PJM has allocated ARRs to network service and long-term, firm point-to-point transmission service customers. These customers may take their allocated ARRs or the underlying FTRs through a self-scheduling process. The PJM market monitoring unit (MMU) argues for measuring the effectiveness of ARRs and FTRs as a hedge against congestion by comparing the revenue received by ARR and FTR holders with the congestion across the corresponding paths. That is, it adds total payouts of ARR and FTR holders and subtracts the amount FTR holders paid at auction to determine ARR plus FTR payouts. It then compares this value with total congestion charges on the underlying transmission paths. Table 9.4 lists these amounts that the PJM MMU has computed for the 2006–2007 and 2007–2008 planning periods.

ISO-NE's annual markets report lists both day-ahead and real-time congestion charges and total revenue generated in FTR auctions. Therefore one might use either day-ahead or real-time congestion charges minus total auction revenue to estimate unhedged congestion in ISO-NE. I use day-ahead congestion charges minus FTR auction revenue to approximate unhedged congestion. Day-ahead congestion charges are the values that LSEs pay for congestion as calculated in the day-ahead energy market. Differences between real-time dispatch and the day-ahead schedule can cause real-time congestion charges to differ from day-ahead values. While there are two basic causes of this difference: (1) difference between load forecasts and actual load, and (2) generation or transmission outages/derating, the latter is the more important, so I will focus on it. Because ISO-NE settles the real-time market on deviations from the day-ahead market, real-time congestion can be either positive or negative.[28] In recent years, real-time congestion in ISO-NE has been negative due to transmission outages/deratings. Because load is settled on a day-ahead basis, outages/deratings will not affect the amount of congestion payments by ISO-NE LSEs, so I approximate unhedged congestion based on day-ahead, instead of real-time congestion figures. The estimate is imprecise because deviations between forecasted and actual load do occur. LSEs are "fully-hedged" if the revenue they collect in FTR auctions is equal to or greater than this value. Thus, unhedged congestion is approximated by any positive difference between congestion costs minus rights to auction revenues (ARRs). One may find the requisite data at ISO-NE 2008 Annual Markets Report, pp. 72–74 and ISO-NE 2007 Annual Markets Report, pp. 124, 129.

The *2008 State of the Market Report* for the NYISO does not list unhedged congestion charges as such. But one may approximate the impact of FTR[29] market imperfections in NYISO as day-ahead market plus balancing market congestion revenue shortfalls since NYISO fully funds FTRs. When revenue shortfalls occur in NYISO FTR markets, NYISO makes up the difference through uplift charges to load. Thus every dollar shortfall translates into a dollar increase in charges to load. In 2006, day-ahead and real-time congestion revenue shortfalls were $40 million and $171 million, for a total of $211 million. In 2007, day-ahead and real-time congestion revenue shortfalls were $93 million and $159 million, respectively, for a total of $252 million. The respective figures for 2008 were $179 million and $325 million, for a total of $504 million. The marked increase in these values in 2008 was due largely to market manipulation, in the form of circuitous transaction scheduling around Lake Erie in the first 7 months of the year.[30]

---

[28] Real-time congestion is positive if real-time dispatch changes to allow more power to flow over transmission lines, some of which are congestion. It is negative if, say, a transmission outage or derating allows less power to flow over transmission lines.

[29] Called transmission congestion contracts, or TCCs in NYISO.

[30] For further information, see *New York ISO 2008 State of the Market Report, Section II.*

Of the other three categories, two, legal settlements and FTR defaults are non-recurring expenses associated with litigation and non-payment of FTR settlements, respectively. The third, "long-term FTRs" is construction expenses (apparently associated with a new facility to house a long-term FTR market command center). All of this data is found in the RTO annual market reports mentioned above.

## Appendix B: FTR Information by Major Utility

| Company | Year | Gross transactions ($1,000) | Total sales to ultimate consumers (MWh) | Net transmission rights ($1,000) | Net transmission rights as a percentage of gross transactions | Estimated retail rate impact ($/MWh) |
|---|---|---|---|---|---|---|
| ALLETE, Inc. | 2006 | 70,028 | 9,078 | 2,774 | 3.962 | 0.306 |
| | 2007 | 137,512 | 9,001 | 349 | 0.254 | 0.039 |
| | 2008 | 84,432 | 9,138 | 256 | 0.303 | 0.028 |
| Appalachian Power Company | 2006 | 37,937 | 30,328 | −1,289 | −3.397 | −0.042 |
| | 2007 | 238,847 | 33,875 | 2,876 | 1.204 | 0.085 |
| | 2008 | 74,775 | 34,210 | −676 | −0.904 | −0.020 |
| Atlantic City Electric Company | 2006 | 309,142 | 9,931 | −13,058 | −4.224 | −1.315 |
| | 2007 | 259,767 | 10,187 | −3,652 | −1.406 | −0.358 |
| | 2008 | 297,238 | 10,089 | −9,175 | −3.087 | −0.909 |
| Central Vermont Public Service Corporation | 2006 | 22,608 | 2,284 | 931 | 4.119 | 0.408 |
| | 2007 | 19,672 | 2,320 | 12,420 | 63.138 | 5.353 |
| | 2008 | 26,014 | 2,259 | −2,109 | −8.107 | −0.934 |
| Columbus Southern Power Company | 2006 | 24,171 | 19,567 | −863 | −3.571 | −0.044 |
| | 2007 | 136,463 | 22,009 | 1,281 | 0.939 | 0.058 |
| | 2008 | 41,885 | 22,206 | −733 | −1.749 | −0.033 |
| Connecticut Light & Power Company | 2006 | 230,529 | 23,638 | 301 | 0.131 | 0.013 |
| | 2007 | 239,656 | 24,032 | 16,734 | 6.982 | 0.696 |
| | 2008 | 296,607 | 23,145 | 1,482 | 0.500 | 0.064 |
| Dayton Power & Light Company | 2006 | 123,059 | 14,767 | −3,907 | −3.175 | −0.265 |
| | 2007 | 181,162 | 15,234 | −6,073 | −3.352 | −0.399 |
| | 2008 | 191,910 | 14,932 | −7,753 | −4.040 | −0.519 |
| Delmarva Power & Light Company | 2006 | 13,607 | 13,479 | 14,271 | 104.879 | 1.059 |
| | 2007 | 40,730 | 13,685 | 26,530 | 65.137 | 1.939 |
| | 2008 | 34,100 | 13,016 | 34,100 | 100.000 | 2.620 |
| Duke Energy Indiana, Inc. | 2006 | 174,074 | 28,592 | 13,686 | 7.862 | 0.479 |
| | 2007 | 274,658 | 29,734 | 37,685 | 13.721 | 1.267 |
| | 2008 | 324,148 | 28,548 | 16,594 | 5.119 | 0.581 |
| Duke Energy Kentucky, Inc. | 2006 | 39,078 | 3,884 | 198 | 0.506 | 0.051 |
| | 2007 | 78,321 | 4,142 | 883 | 1.127 | 0.213 |
| | 2008 | 75,771 | 4,041 | 663 | 0.875 | 0.164 |
| Pacific Gas & Electric Company | 2006 | 94,875 | 84,421 | 2,061 | 2.173 | 0.024 |
| | 2007 | 135,839 | 86,313 | −1,744 | −1.284 | −0.020 |
| | 2008 | 204,414 | 88,269 | 19,065 | 9.326 | 0.216 |
| Potomac Electric Power Company | 2006 | 30,095 | 26,488 | 23,688 | 78.709 | 0.894 |
| | 2007 | 23,650 | 27,451 | 38,552 | 163.008 | 1.404 |
| | 2008 | 64,325 | 26,863 | 37,674 | 58.568 | 1.402 |

(continued)

| Company | Year | Gross transactions ($1,000) | Total sales to ultimate consumers (MWh) | Net transmission rights ($1,000) | Net transmission rights as a percentage of gross transactions | Estimated retail rate impact ($/MWh) |
|---|---|---|---|---|---|---|
| Public Service Company of New Hampshire | 2006 | 74,479 | 8,034 | −2,899 | −3.892 | −0.361 |
| | 2007 | 81,837 | 8,132 | −2,735 | −3.342 | −0.336 |
| | 2008 | 109,935 | 7,926 | −3,070 | −2.793 | −0.387 |
| Public Service Electric & Gas Company | 2006 | 46,149 | 43,678 | 242 | 0.524 | 0.006 |
| | 2007 | 65,525 | 44,709 | 5,082 | 7.756 | 0.114 |
| | 2008 | 56,484 | 43,734 | 4,554 | 8.062 | 0.104 |
| San Diego Gas & Electric Company | 2006 | 20,662 | 9,508 | 2,295 | 11.107 | 0.241 |
| | 2007 | 36,912 | 10,087 | 2,425 | 6.570 | 0.240 |
| | 2008 | 94,017 | 12,320 | 4,063 | 4.322 | 0.330 |
| Southern California Edison Company | 2006 | 314,249 | 88,729 | 44,584 | 14.187 | 0.502 |
| | 2007 | 305,066 | 88,805 | 18,235 | 5.977 | 0.205 |
| | 2008 | 338,565 | 89,809 | 79,434 | 23.462 | 0.884 |
| UGI Utilities, Inc. | 2006 | N/A | 1,030 | 118 | N/A | 0.114 |
| | 2007 | 2,466 | 1,016 | 2,512 | 101.870 | 2.473 |
| | 2008 | 1,706 | 1,003 | 430 | 25.192 | 0.429 |
| Upper Peninsula Power Company | 2006 | 18,974 | 800 | −152 | −0.800 | −0.190 |
| | 2007 | 20,000 | 861 | 409 | 2.046 | 0.475 |
| | 2008 | 15,893 | 848 | −2,289 | −14.400 | −2.699 |
| Virginia Electric & Power Company | 2006 | 651,585 | 76,149 | −144,694 | −22.207 | −1.900 |
| | 2007 | 894,392 | 79,892 | −114,281 | −12.777 | −1.430 |
| | 2008 | 1,404,890 | 78,664 | −271,778 | −19.345 | −3.455 |
| Wisconsin Power & Light Company | 2006 | 204,678 | 28,189 | 0 | 0.000 | 0.000 |
| | 2007 | 125,467 | 10,852 | −15,565 | −12.406 | −1.434 |
| | 2008 | 185,604 | 10,529 | −8,084 | −4.355 | −0.768 |
| Wisconsin Public Service Corporation | 2006 | 180,812 | 10,580 | −26,163 | −14.470 | −2.473 |
| | 2007 | 259,946 | 11,106 | −22,694 | −8.730 | −2.043 |
| | 2008 | 207,869 | 10,892 | −11,866 | −5.708 | −1.089 |

# References


Apt J (2005) Competition has not lowered U.S. industrial electricity prices. Electr J 18(2):52–61

Barmack M, Griffes P, Kahn E, Oren S (2003) Performance incentives for transmission. Electr J 16 (3):9–22

Benjamin R (2007) Principles for interregional transmission expansion. Electr J 20(8):36–47

Benjamin R (2008) Load-pocket management in restructured electricity markets: a hybrid model. Electr J 21(6):51–59

Benjamin R (2010) A further inquiry into FTR properties. Energy Policy 38(7):3547–3556

Benjamin R (2011) Tweaking merchant transmission expansion(again). Round table group. Available from author by request

Blumsack S, Apt J, Lave L (2006) Lessons from the failure of U.S. electricity restructuring. Electr J 19(2):15–32

Bogorad C, Huang W (2005) Long-term rights for new resources: a crucial missing ingreent in RTO markets. Electr J 18(7):11–24

Borenstein S, Bushnell J, Stoft S (2000) The competitive effect of transmission capacity in a deregulated electricity industry. Rand J Econ 31(2):294–325

Brunekreeft G (2004) Market-based investment in electricity transmission networks: controllable flow. Utilities Pol 15(4):269–281
Bushnell J, Stoft S (1996a) Electric grid investment under a contract network regime. J Regul Econ 10(1):61–79
Bushnell J, Stoft S (1996b) Grid investment: can a market do the job. Electr J 9(1):74–79
Bushnell J, Stoft S (1997) Improving private incentives for electric grid investment. Resour Energy Econ 19(1–2):85–108
Calviou M, Kleindorfer P, Paravalos M (2004) Creating value through transmission in an era of RTOs. In: Paper presented at the electricity transmission in deregulated markets conference, Carnegie-Mellon University, 15–16 Dec 2004. Available at http://www.ece.cmu.edu/~electriconf/2004/Kleindorfer_CKP-CMU-Paper-D021.pdf
Chao H, Peck S (1996) A market mechanism for electric power transmission. J Regul Econ 10 (1):25–59
Chao H, Peck S (1997) An institutional design for an electricity contract market with central dispatch. Energy J 18(1):85–110
Deng SJ, Oren S (2006) Electricity derivatives and risk management. Energy 31(6–7):940–953
Deng SJ, Oren S, Meliopoulos S (2010) The inherent inefficiency of simultaneously feasible financial transmission rights auctions. Energy Econ 32(4):779–785
Fabrizio K, Rose N, Wolfram C (2007) Do markets reduce costs? Assessing the impact of regulatory restructuring on US electric generation efficiency. Am Econ Rev 97(4):1250–1277
Gans J, King S (2000) Options for electricity transmission regulation in Australia. Aust Econ Rev 33(2):145–160
Green R (1997) Electricity transmission pricing: an international comparison. Utilities Pol 6 (3):177–184
Hogan W (1992) Contract networks for electric power transmission. J Regul Econ 4(3):211–242
Hogan W (1998) Transmission investment and competitive electricity markets. Harvard University. Available at http://www.hks.harvard.edu/fs/whogan/trn98.pdf
Hogan W, Rosellón J, Vogelsang I (2010) Toward a combined merchant-regulatory mechanism for electricity transmission expansion. J Regul Econ 38(2):113–143
Hsu M (1997) An introduction to the pricing of electric power transmission. Utilities Pol 6 (3):257–270
Joskow P (2005) Transmission policy in the United States. Utilities Pol 13(2):95–115
Joskow P (2006) Markets for power in the United States: an interim assessment. Energy J 27 (1):1–36
Joskow P, Tirole J (2000) Transmission rights and market power on electric power networks. Rand J Econ 31(3):450–487
Joskow P, Tirole J (2005) Merchant transmission investment. J Ind Econ 53(2):233–264
Keller K, Wild J (2004) Long-term investment in electricity: a trade-off between co-ordination and competition. Utilities Pol 12(4):243–251
Knittel C (2002) Alternative regulatory methods and firm efficiency: stochastic frontier evidence from the U.S. electricity industry. Rev Econ Stat 84(3):530–540
Kristiansen T (2005a) Financial transmission rights: experiences and prospects. Kema 0000000ments/infraday/2005/papers/kristiansen_Financial_Transmission_Rights_Experiences_and_Prospects.pdf
Kristiansen T (2005b) Markets for financial transmission rights. Energy Stud Rev 13(1):25–74
Kristiansen T (2008) Allocation of long-term financial transmission rights for transmission expansion. Eur J Oper Res 184(3):1122–1139
Lave L, Apt J, Blumsack S (2004) Rethinking electricity deregulation. Electr J 17(8):11–26
Lave L, Apt J, Blumsack S (2007a) Deregulation/restructuring part I: reregulation will not fix the problems. Electr J 20(8):9–22
Lave L, Apt J, Blumsack S (2007b) Deregulation/restructuring part II: where do we go from here. Electr J 20(9):10–23

Leautier T (2001) Transmission constraints and imperfect markets for power. J Regul Econ 19 (1):27–54
Liu M, Wu F (2007) Risk management in a competitive electricity market. Int J Electr Power Energy Syst 29(9):690–697
Morey M, Eakin K, Kirsch L (2005) RTOs and electricity restructuring: the chasm between promise and practice. Electr J 18(1):31–51
Morrison J (2005) The clash of industry visions. Electr J 18:14–30
Oren S, Spiller P, Varaiya P, Wu F (1995) Nodal prices and transmission rights: a critical appraisal. Electr J 8(3):24–35
PJM Interconnection, L.L.C. (2009) PJM manual 06 financial transmission rights. Available at http://pjm.com/~/media/documents/manuals/m06.ashx
Rosellón J (2003) Different approaches towards electricity transmission expansion. Rev Netw Econ 2(3):238–269
Rosellón J, Weigt H (2011) A dynamic incentive mechanism for transmission expansion in electricity networks—theory, modeling and application. Energy J 32(1):119–148
Rosellón J, Myslíková Z, Zenón E (2011) Incentives for transmission investment in the PJM electricity market: FTRs or regulation (or both?). Utilities Pol 19(1):3–13
Rotger J, Felder F (2001) Promoting efficient transmission investment: the role of the market in expanding transmission infrastructure. Transenergie Report. Available at http://www.hks.harvard.edu/hepg/Papers/Promoting%20Efficient%20Transmission%20Investment%20(Rotger-Felder%20Nov%202001).pdf
Rothkopf M (2007) Dealing with failed deregulation: what would price C. Watts do? Electr J 20 (7):10–16
Sarkar V, Kharparde SA (2008) A comprehensive assessment of the evolution of financial transmission rights. IEEE Trans Power Syst 23(4):1783–1795
Schweppe F, Caramanis M, Tabors R, Bohn R (1988) Spot pricing of electricity. Kluwer, Dordrecht
Siddiqui S, Bartholomew E, Marnay C, Oren S (2005) Efficiency of the New York independent system operator market for transmission congestion contracts. Manag Finance 31(6):1–45
Taber J, Chapman D, Mount T (2006) Examining the effects of deregulation on retail electricity prices. Department of Applied Economics and Management, Cornell University, working paper 2005–14. Available at http://dyson.cornell.edu/research/researchpdf/wp/2005/Cornell_Dyson_wp0514.pdf
Vogelsang I (2001) Price regulation for independent transmission companies. J Regul Econ 20 (2):141–165
Wu F, Varaiya P, Spiller P, Oren S (1996) Folk theorems on transmission access: proofs and counterexamples. J Regul Econ 10(1):5–23
Yu N, Somani A, Tesfatsion L (2010) Financial risk management in restructured wholesale power markets. Lowa State University. Available at http://www2.econ.iastate.edu/tesfatsi/FinRiskTutorial.IEEEPESGM2010.pdf

# Chapter 10
# FTRs and Revenue Adequacy

**Guillermo Bautista Alderete**

## 10.1 Introduction

With the implementation of electricity markets, the provision of transmission rights has become a natural extension. Transmission rights can be of the flowgate or financial types and have been developed to manage the risk posed by volatile congestion costs. A financial transmission right (FTR) is an instrument to hedge source-to-sink congestion and entitles its holder the right – or obligation- to collect a payment when congestion arises in the energy market. The basic definition of an FTR consists of a source and a sink node that identify the point-to-point direction of the right, a MW award that is constant for the full life term of the instrument, a time of use for which the instrument is settled and a life term which identifies the period over which the instrument is valid. Although the definition is point-to-point, FTRs are not necessarily limited to be defined between individual nodes, load zones or trading hubs are aggregated nodes widely used for FTRs.

Nowadays, most of FTR markets offer obligation type,[1] for which the holder has either the right to collect a payment when congestion occurs in the energy market or the obligation to pay when the congestion in the energy market is in the opposite direction to the FTR definition. The payment or charge is computed as the price differential between the sink and source nodes times its MW award. An FTR option, in contrast, provides only the upside benefit to its holder since there is no charge to the holder when congestion is in the opposite direction of the FTR (Lyons and Fraser 2000). Since

The ideas expressed in this chapter are solely those of the author and do not represent any official position from the California ISO.

[1] The Pennsylvania, New Jersey, Maryland (PJM) market offers both obligation and option types of FTRs.

G.B. Alderete (✉)
California Independent System Operator, Folsom, CA 95630, USA
e-mail: gbalderete@caiso.com

J. Rosellón and T. Kristiansen (eds.), *Financial Transmission Rights*, Lecture Notes in Energy 7, DOI 10.1007/978-1-4471-4787-9_10,

FTRs are only financial instruments, the payments or charges are independent of the actual use of the transmission system by their holders; this separation provides efficiency by not interfering with the optimal operation of the system. FTRs hedge only congestion costs, even though there have been several theoretical proposals to have instruments to also hedge losses (Rudkevich and Bagnall 2005).

Markets using locational marginal prices, mainly throughout the United States, have put in place mechanisms to provide participants with financial transmission rights (Alsac et al. 2004). Such mechanisms are usually allocation or auction processes run by an Independent System Operator (ISO). Regardless of the means to acquire FTRs, the hedging properties of the instruments are the same. After the initial release, FTRs can be traded bilaterally. The definition of the processes to release financial transmission rights is mainly driven by the specific needs of a given market. Regardless of the means to release FTRs, an ISO needs to limit the overall amount of FTRs that can be feasibly released. A simultaneous feasibility test is the underlying process to determine the amount of FTR awards. When FTRs are modelled in the release processes, such as auctions, the source and the sink used to define every FTR represents bilateral trades for which injections and withdrawals of power determine the power flow contributions in the transmission system. Thus, any set of FTRs that can be released has to be a feasible power flow, in which no transmission constraints are violated. The transmission system used in the release processes represents as close as possible the transmission system and configuration that will be used later in the energy market. Since the release process is usually driven by an optimization engine, the optimal solution or set of feasible FTRs is determined by considering simultaneously all FTRs. Thus, the optimal set of FTRs is necessarily simultaneously feasible as FTRs will provide counter-flows to each other. By using a simultaneous feasibility test to determine the optimal set of FTRs to be awarded, revenue adequacy can be ensured. Revenue adequacy is the condition in which sufficient money from the forward energy market is collected to cover all FTR payments over a given period of time. Revenue adequacy is the subject of analysis throughout this chapter.

## 10.2 Nomenclature

This section introduces the notation and acronyms used throughout this chapter.

### *10.2.1 Acronyms*

| | |
|---|---|
| ATC | Available Transfer Capability |
| AC | Alternate Current |
| CAISO | California Independent System Operator |

(continued)

| | |
|---|---|
| CRR | Congestion Revenue Right |
| DC | Direct Current |
| LMP | Locational Marginal Price |
| FTR | Financial Transmission Right |
| OTC | Operating Transfer Capability |

### 10.2.2 Symbols

| | |
|---|---|
| $\Phi$, $\Phi_{p.u.}$ | Revenue adequacy in dollars or per unit, respectively |
| $\lambda_{t,i}$ | Locational marginal price at node $i$ for hour $t$ |
| $\lambda_s$ | Locational marginal price at the slack node |
| $d_{t,i}$ | Demand at node $i$ in interval $t$ |
| $g_{t,i}$ | Generation at node $i$ in interval $t$ |
| $\tau_{jl}$ | Financial transmission right award from node $j$ to node $l$ |
| $i$, $l$ | Index for nodes in the system |
| $t$ | Index for hour |
| T | Set of hours for a given accounting period, say, a calendar month |
| I | Set of nodes in a transmission system |
| $\pi_v$ | Share of revenue gap for participant $v$ |
| $\tilde{d}_v$ | Average demand of participant $v$ |
| $z_k^{FTR}$ | Equivalent power flow on constraint $k$ due to injections from FTRs |
| $\bar{z}_k$ | Upper limit for transmission constraint $k$ |
| $\Phi_k$ | Revenue adequacy on transmission constraint $k$ |
| $\tilde{\tau}_i$ | Equivalent power injection from all FTRs at node $i$ |
| $S_{i,k}$ | Sensitivity factor for transmission constraint $k$ due to power injection at node $i$ |
| $\mu_k$ | Shadow price associated with transmission constraint $k$ |
| $\underline{\gamma}$, $\bar{\gamma}$ | Dual variables associated with generation limits |
| $\underline{\varsigma}$, $\bar{\varsigma}$ | Dual variables associated with demand limits |
| $\alpha, \beta$ | Bid factors for supply and demand |
| $G$ | Susceptance matrix |
| $H$ | Reactance matrix |
| $\delta$ | Vector of nodal angles |

## 10.3 Simultaneous Feasibility Test

The release of FTRs is done through an optimization process that determines the optimal set of FTRs that are simultaneously feasible. This is implemented using a network model as similar as possible to the model used in the forward energy market, including both the transmission configuration and constraints. The transmission constraints are related to physical and operational limitations such as thermal limits, and voltage and stability. Contingencies[2] and nomograms that define simultaneous path limits need also to be modeled. All these constraints are well known as security constraints.

Since FTRs hedge only congestion, this justifies to certain extent not modeling losses and reactive power in the FTR market. In general, DC-based models for FTRs markets are used among ISOs; one exception is the New York market that uses an AC model for FTRs (Alsac et al. 2004). Among other considerations, the DC model assumes that there is sufficient voltage support throughout the system and thus the voltage profile can be maintained at 1 p.u.[3] Therefore, the reactive power component can be disregarded from the full network formulation (Wood and Wollenberg 1996). However, the energy market may include losses and reactive power and this may create a gap. In the actual transmission system, limits are defined for full apparent power. If DC models are used in the FTR market then the capacity that the reactive power would use in the transmission elements need be considered. Also transmission limits may need to get adjusted to represent voltage constraints of the transmission system. A simple but crude way to account for losses and even reactive power is to reduce system- or element-wise the transmission limits by a given percentage.

Linearity is the main advantage of using the DC model which allows modeling the point-to-point FTRs as balanced trades, with MW injections equals to MW withdrawals. One implicit consideration of the simultaneous feasibility test is that all FTRs are exercised; this is required because counter-flows are part of the feasibility while determining the set of FTRs. The simultaneous feasibility test is implicit in the clearing process of the FTR markets. The FTR auctions, for instance, are simply a process to find a solution of a security constrained optimal power flow where the objective function is to maximize the utility function of FTR bidders while enforcing the base and contingency (typically n-1 type) transmission constraints. Such constraints are usually linearized power flows. In DC-based models with linear (or even quadratic) objective function and linear constraints, linearity implies that the feasible set is convex and, thus, any solution found will be a global solution (Nocedal and Wright 1999). This is the theoretical foundation to ensure revenue adequacy for FTRs under the premise that the same network model is used also in the forward energy market (Hogan 1992). With a DC model, obviously, there is the implication that lack of realism could also lead to issues with revenue adequacy; this aspect, however, is difficult to quantify. If an AC model is used in the FTR market, revenue adequacy cannot be guaranteed due to the non convexities (Lesieutre and Hiskens 2005). Since the control area of an ISO is

---

[2] The n-1 contingency is usually the most common security constraint used in the operation of a power systems and accounts for the loss of a single generator or transmission element in the system.

[3] P.u. stands for per unit and is used in power systems studies to handle calculations with different voltage levels.

typically connected to neighboring control areas, loop flows may also create revenue adequacy issues.

## 10.4 How to Measure Revenue Adequacy

Revenue adequacy is inherently a financial notion related to an electricity market; since it materializes in the settlements of a market, it is defined in monetary terms as the balance of the money inflows (congestion rents) and outflows (FTR payments) of the FTR account,

$$\Phi = \sum_{t \in \mathrm{T}} \left\{ \sum_{i \in \mathrm{I}} \lambda_{t,i}\left(d_{t,i} - g_{t,i}\right) - \sum_{(j,l) \in \mathrm{I}} \tau_{jl}\left(\lambda_{t,l} - \lambda_{t,j}\right) \right\} \tag{10.1}$$

The monetary value, however, is not a good indicator of the size of the market and, thus, it may not be a reference of the extent of adequacy. Instead, a widely used metric of revenue adequacy among ISOs is the revenue adequacy as the proportion of available congestion rents to the amount of FTR payments,

$$\Phi_{p.u.} = \frac{\sum_{t \in \mathrm{T}} \sum_{i \in \mathrm{I}} \lambda_{t,i}\left(d_{t,i} - g_{t,i}\right)}{\sum_{t \in \mathrm{T}} \sum_{(j,l) \in \mathrm{I}} \tau_{jl}\left(\lambda_{t,l} - \lambda_{t,j}\right)} \tag{10.2}$$

Depending on the specific features of the market, revenue adequacy may be composed of other elements such as exemptions for existing transmission, auction revenues and reimbursed FTR payments. The goal of a market operator is obviously to attain a revenue adequacy of 100 % (zero dollar gap). This ideal value represents the best tradeoff between the number of FTRs awarded to market participants and the ability to fully fund such FTRs; a value smaller than 100 % indicates a revenue inadequacy. A sustained revenue adequacy greater than 100 % (an excess of money) may also be an indication of an inefficiency due to having, for instance, a process that is too restrictive or conservative in releasing FTRs, and depriving participants of more hedging opportunities.

The engineering notion of revenue adequacy can be simply stated as the requirement of allocating as much transmission capacity in the form of financial transmission right as transmission capacity made available in the energy market. This notion can be quantified in terms of available transfer capability (ATC).[4] Although the engineering notion of revenue adequacy can also be defined in terms of the total

[4] Operating Transfer Capability (OTC) refers to the nominal values of transmission constraints. Once any reserved capacity is discounted, one may refer to the Available Transfer Capacibility (ATC) which is the transmission capacity that is made available to the market.

available transfer capability (MW), this metric would turn out a futile exercise since it does not have a meaningful interpretation of the overall market condition.

## 10.5 Causes of Revenue Inadequacy

The theoretical concept to ensure revenue adequacy relies on two factors: (1) use of the same shift factors, and (2) the enforcement of the same constraints in both the FTR and the forward energy market. In real-life markets, however, it is not possible to satisfy such conditions all the time due to the inherent changing nature of a power system. In a mature market, the set of transmission constraints may be well defined and remain relatively constant over time, reducing the mismatch of constraints enforced between markets and reducing, consequently, the room for inadequacies. During the evolution of a relatively new market, in contrast, transmission constraints may be developing. This, compounded with the inherent timing of running the FTR market well ahead of the energy market (in some markets there may be leading times of a few months), may result in some constraints not enforced in the FTR market, creating the potential for revenue inadequacies.

A more typical cause, however, are derates of transmission elements. Transmission limits may change due to system conditions, including operational needs and weather. Derates, which are the fact of modifying the normal rate of a transmission constraint, happen all the time in a transmission system. The limits used in the FTR market cannot fully account for derates that will happen because derates will be known until close to real-time operation of the system. There may be only a few planned derates known by the time the FTR is run, but other derates, such as forced derates, will take place in the last moment. This typically leads to overestimate the transmission capacity released as FTRs.

The use of different shift factors is typically the main cause of revenue shortfalls and is mainly driven by outages. Similar to derates, some planned outages may be known by the time the FTR market is run, but many other outages, typically forced outages, cannot be known until real-time operations. Outages change the system configuration and result in different shift factors. Thus, if FTRs were released with certain configuration and then the energy market runs with some last-moment outages, there is an increase possibility of revenue shortfalls. For this reason, one of the main concerns among ISOs is how to account for outages in the network model used to run the FTR markets. Empirical results show that using the nominal OTC values in the FTR market usually results in an over allocation of FTRs since any outages or derates will further constraint the energy market, resulting in less congestion rents to fund FTRs.

Figure 10.1 shows the daily revenue adequacy for CRRs during 2010 in the California ISO market.[5] The bars in blue stand for the daily revenue adequacy,

[5] In the California ISO markets, FTRs are named Congestion Revenue Rights (CRRs).

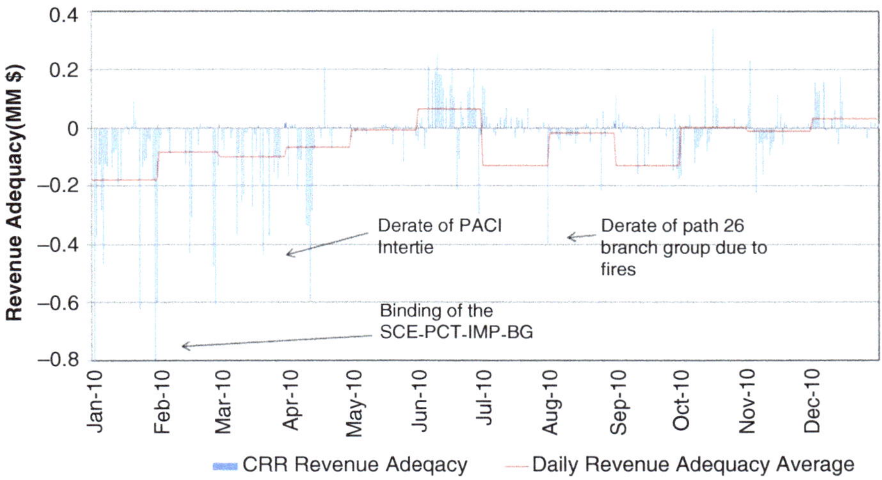


**Fig. 10.1** Trend of daily revenue adequacy in 2010 in the California ISO market

while the line in red shows the average daily revenue adequacy over a calendar month. Positive values stand for revenue surplus while negative values stand for deficiencies. For illustration purposes, large revenue shortfalls are a reference to system events. Derates in two major transmission paths resulted in large revenue shortfalls, while the enforcement of a new transmission constraint also drove revenue deficiencies. In the CRR market, the limits used to model such constraints could not account for all derates that happened in actual operation of the system and many of them were not known by the time the CRR market was run. With both the energy and the CRR markets being recently new, compounded with the CRR market running a few months ahead, a new transmission constraint was introduced in the market after the CRR market was run, resulting in a more restrictive energy market with less congestion rents available to fund all the amount of CRRs that were awarded without having such a constraint.

## 10.6 Allocation of Revenue Adequacy Gaps

Revenue adequacy is one of the main items for CRRs monitored and studied by ISOs since this is the primary indication of the overall condition of the process to release transmission rights. Indeed, depending on the specific design, revenue inadequacy is also the concern of various market players, since revenue shortfalls may be socialized among certain participants of the market. For instance, if full funding for FTRs is guaranteed then any revenue shortfall will be offset by some means. Usually, some market participants, such as load serving entities or transmission owners, will be charged the revenue shortfall. If the revenue shortfall is systematic and the charge is not spread to all participants, however, this becomes a transfer of wealth from one subset of participants to another. On the other hand, this same set of participants that are charged to cover for revenue shortfall may be the same participants to be paid when a revenue surplus exists. One approach to allocate any shortfall or surplus of revenue, say, to load serving entities, is by using a pro-rata approach

$$\pi_v = \frac{\tilde{d}_v}{\sum_{v \in V} \tilde{d}_v} \Phi \tag{10.3}$$

where $\pi_v$ is the payment uplift for each market participant subject to the revenue gap. This payment can be calculated over a period, as short as an hour, but typically over a longer accounting period such as a month. Depending on the case, $\tilde{d}_v$ can represent hourly average cleared demand (or measured load) or average energy over a specific accounting period for participant $v$.

A second alternative to attain revenue adequacy is a pro-rata adjustment of FTR payments. That effectively reduces the FTR payments among all FTR holders in a proportion such that FTRs payments are covered up to the amount of available congestion rents. The revenue adequacy ratio defined by Expression (10.2) can be used as the pro-rate value to adjust all FTR payments. This alternative, however, introduces certain degree of uncertainty as CRRs no longer hedge up to its nominal value. There may also be the approach of rolling revenue gaps over accounting periods. If a current month, for instance, has a revenue inadequacy, the difference may be rolled over the upcoming periods expecting that future months have a surplus. In the California ISO, for instance, CRRs are fully funded and any revenue gaps – surpluses or shortfalls- are covered through uplifts to measured demand. In the New York ISO, in contrast, any revenue gap is allocated to transmission owners. In PJM and MISO markets, FTRs are rather scaled pro-rata.

## 10.7 Practical Issues of Revenue Adequacy

Since revenue adequacy is a metric obtained from the FTR settlements, revenue adequacy is naturally calculated as a market-wise metric. When multiple transmission elements are congested at the same time, it is not possible to accurately identify the root cause of revenue gaps – shortfalls or surpluses – from values available in settlements. This is so because settlements are calculated on a nodal basis, which represents the overall impact on prices of congestion arising from multiple constraints. Fundamentally, revenue adequacy is no more than releasing as much transmission capacity in the FTR market as capacity is made available in the forward energy market. Revenue adequacy can be evaluated at its most fundamental element, which is on a transmission constraint basis. Conceptually, each transmission constraint needs to be revenue adequate, making the overall market adequate as well. In real-life markets, however, it is not possible to attain this concept given all the factors impacting the conditions of the energy market. Usually, some transmission constraints experience revenue shortfalls and others have revenue surpluses; overall they partially offset each other and the overall revenue adequacy condition will mask that effect. The engineering notion of revenue adequacy, using the concept of available transfer capability, yields the explicit identification of the revenue gaps at its most fundamental level: by each transmission element.

The following derivation is based on the well known linearized DC power flows of a locational marginal pricing scheme, which was implemented in the California ISO markets to identify the root causes of revenue inadequacies (CAISO 2010). This approach exploits the linearity and superposition of DC power flows and identifies the relationship between the financial and engineering notion of revenue adequacy. The DC optimal power flow is ubiquitous in the technical literature and is next presented for completeness of the description of this chapter. This model takes into account only congestion since it is a lossless representation of the transmission system.

Let us consider a DC OPF in its simplest form where supply and demand are cleared using an LMP scheme for one single trading interval (Bautista Alderete 2010),

$$\begin{array}{ll} \min & \alpha^T g - \beta^T d \\ & G\delta = g - d \\ & H\delta \leq \bar{z} \\ s.t. & \underline{g} \leq g \leq \bar{g} \\ & \underline{d} \leq d \leq \bar{d} \end{array} \tag{10.4}$$

where $\alpha,\ \beta$ are the transposed vectors of cost and benefit bid parameters; $g,\ d$ are the vectors of generation and demand, respectively; $G,\ H$ are the Susceptance and Reactance matrices, while $\bar{z}$ is the vector of transmission constraint limits. The vector of nodal angles is defined by the symbol $\delta$. The objective function is to minimize the social cost as defined by the supply and demand bids, the first constraint stands for the nodal power balance, the second constraint stands for the transmission limits, while the last two constraints stand for the lower and upper limits of bids. The Lagrangian function for this minimization problem yields the following expression,

$$\begin{aligned} L(g,d,\delta,\lambda,\mu) = {} & \alpha^T g - \beta^T d - \lambda^T(G\delta - g + d) - \mu^T(H\delta - \bar{z}) \\ & - \underline{\gamma}\left(\underline{g} - g\right) - \bar{\gamma}(g - \bar{g}) - \underline{\varsigma}(\underline{d} - d) - \bar{\varsigma}(d - \bar{d}) \end{aligned} \tag{10.5}$$

where $\lambda,\ \mu,\ \underline{\gamma},\ \bar{\gamma},\ \underline{\varsigma},\ \bar{\varsigma}$ are the dual variables (multipliers) associated to each constraint of the DC OPF. In particular, the variables $\lambda,\ \mu$ associated with the power balance and the flow limit constraints, respectively, are used as prices. The variable $\lambda$ stands for the locational marginal prices at each node $i$, while the variable $\mu$ stands for the price of each transmission constraint.

The first order optimality conditions of this Lagrangian yields an equilibrium point. Since this is a linear programming problem, a solution for the Lagrangian yields also an optimal point for the market (Nocedal 1999). For the sake of this derivation, let us consider only the optimality conditions with respect to the variable of nodal angles $\delta$, i.e.

$$G^T\lambda + H^T\mu = 0 \tag{10.6}$$

If this expression is multiplied by the vector of nodal angles, one obtains

$$\lambda^T G\delta + \mu^T H\delta = 0 \tag{10.7}$$

Introducing the power balance constraint and the transmission limits from the DC OPF (10.4), the following expression is obtained:

$$\lambda^T(g - d) + \mu^T\bar{z} = 0 \tag{10.8}$$

Expanding the vectors in its scalar elements and rearranging terms yields

$$\sum_{i\in \mathrm{I}} \lambda_i(d_i - g_i) = \sum_{i\in \mathrm{K}} \mu_k\bar{z}_k \tag{10.9}$$

In this lossless transmission system, the locational marginal prices have the energy and congestion components only. Since the energy component is unique across all nodes of the system, locational price differentials are only due to congestion. If congestion arises, then congestion rents exist and are calculated as the difference between charges to demand and payments to supply as defined by the left hand side term of Expression (10.9). This is how the actual settlement of forward energy markets is done. The right hand side of Expression (10.9) provides the equivalence of the congestion rents in terms of transmission constraints. This term is the basic relationship needed to identify the root causes of revenue inadequacy since congestion is fundamentally accrued over each congested element and represented at each node of the system.

The composition of a locational marginal price for a lossless power system is ubiquitous in the technical literature and is defined by two components: marginal energy and congestion prices

$$\lambda_i = \lambda_s + \sum_{k\in \mathrm{K}} S_{i,k}\mu_k \tag{10.10}$$

where $\lambda_s$ is the price at the slack node and represents the marginal energy component. This composition is derived from the fact that congestion accrued on a transmission constraint is distributed and priced throughout the system's nodes according to the flow impact of each node to the given transmission constraint, as defined by the shift factors $S_{i,k}$.

In the calculation of FTR payments, the settlement process of an ISO computes the overall payments for all FTRs based on the price differential of LMPs between the two given nodes used to define each FTR,

$$\psi = \sum_{(j,l)\in \mathrm{I}} \tau_{jl}\left(\lambda_l - \lambda_j\right) \tag{10.11}$$

Since this is usually based on DC power flows, Expression (10.11) can be arranged using the relationship of shift factors and shadow prices of transmission constraints given in Expression (10.10) to yield the following relationship

$$\psi = \sum_{i\in \mathrm{I}} \tilde{\tau}_i \lambda_i = \sum_{i\in \mathrm{I}} \tilde{\tau}_i \left(\sum_{k\in \mathrm{K}} S_{i,k}\mu_k\right) = \sum_{k\in \mathrm{K}} \mu_k \left(\sum_{i\in \mathrm{I}} S_{i,k}\tilde{\tau}_i\right) \tag{10.12}$$

where $\tilde{\tau}_i$ is the net nodal injections calculated as the algebraic summation of all MW quantities from all FTRs having node $i$ in its definition of either source or sink. The term within parenthesis is the product of shift factors times nodal injections from FTRs and stands for the equivalent power flow estimated from nodal injections of FTRs. These are the equivalent power flows if the FTR injections and withdrawals from sources and sinks were actually materialized with the shift factors of the energy market. This is simply the equivalent MW value of the FTR payments, and is also the counterpart of the power flows in the energy market needed to estimate revenue adequacy, i.e.

$$\psi = \sum_{k\in \mathrm{K}} \mu_k \left(z_k^{FTR}\right) \tag{10.13}$$

Both expressions (10.9) and (10.13) can be expanded to account for multiple trading intervals by just adding the sub-index $t$, and then be used to substitute the corresponding terms in Expression (10.1) in order to give rise to an alternate expression for revenue adequacy

$$\Phi = \sum_{t\in \mathrm{T}} \left\{ \sum_{k\in \mathrm{K}} \mu_{t,k}\bar{z}_{t,k} - \sum_{k\in \mathrm{K}} \mu_{t,k} z_{t,k}^{FTR} \right\} = \sum_{t\in \mathrm{T}} \sum_{k\in \mathrm{K}} \mu_{t,k}\left(\bar{z}_{t,k} - z_{t,k}^{FTR}\right) \tag{10.14}$$

This expression provides a natural interpretation of revenue adequacy in an engineering context, by pricing any gap between the transmission capacity made available in the FTR and energy markets, accruing over all transmission constraints and all intervals of the accounting period. Obviously, this also allows estimating revenue adequacy of each transmission constraint over an accounting period, i.e.

$$\Phi_k = \sum_{t \in \mathrm{T}} \mu_{t,k} \left( \bar{z}_{t,k} - z_{t,k}^{FTR} \right) \tag{10.15}$$

This expression provides with a formal definition of several intuitive features of revenue adequacy that are not possible to visualize with Expression (10.1) used in the actual settlements of markets. This expression also uses the actual transmission limits $\bar{z}_k$ for each transmission constraint in the energy market while the equivalent power flows $z_k^{FTR}$ from FTRs are not explicitly bounded. Instead, the FTR power flows are bounded implicitly by the amount of FTRs released in the FTR markets, which in turn were bounded by the transmission limits in the FTR market. The estimated power flows from FTRs usually differ from the transmission limits used in the FTR market because (1) not all constraints in the FTR may be binding and (2) the power flows used the shift factors from the energy market which internalizes all the changes occurred in the energy market in comparison to the FTR market.

The shadow prices of transmission constraints, $\mu$, which are a by-product of the optimization process to clear the energy market, are nonzero prices that indicate the associated cost of relaxing a transmission constraint. If a transmission constraint is not binding, usually the shadow price will be zero and, consequently, the gap of transmission capacity between the energy and the FTR markets has no impact at all in revenue adequacy, even if there was an over allocation of transmission capacity in the FTR market for this transmission constraint.

The overall revenue adequacy is composed of the revenue adequacy of each transmission constraint and Expression (10.15) explicitly allows such a calculation. This is not possible with Expression (10.1) as the impact of each transmission constraint has been decomposed and aggregated at each node in what becomes the marginal congestion component. This feature becomes relevant to identify the root cause of revenue gaps because each transmission element can be analyzed independently. System-wise revenue adequacy masks the offsetting between revenue surpluses and deficiencies accrued among different transmission elements; this becomes more pronounced in meshed network where multiple transmission constraints are congested simultaneously. For instance, in the California ISO market for CRRs (CAISO 2010), there was a revenue surplus of $66,000 in October 2010, which represents about 100 % of revenue adequacy given the size of the congestion market of $7 million. This value, however, does not provide insights whether surpluses in some constraints offset shortfalls in others. With the metric of revenue adequacy per constraint, it can be observed that indeed there was systemic revenue shortfall in the IPPADLN branch group[6] which was fully offset by the revenue surplus on the IID-SDGE branch group, as observed in Fig. 10.2. This metric helped unveil a mismatch between the capacity released in the CRR

[6] Branch groups, transmission corridors or inter-tie constraints are just different types of transmission constraints used in the California ISO markets. Such elements are usually identified with acronyms or names related to the area of their physical location.

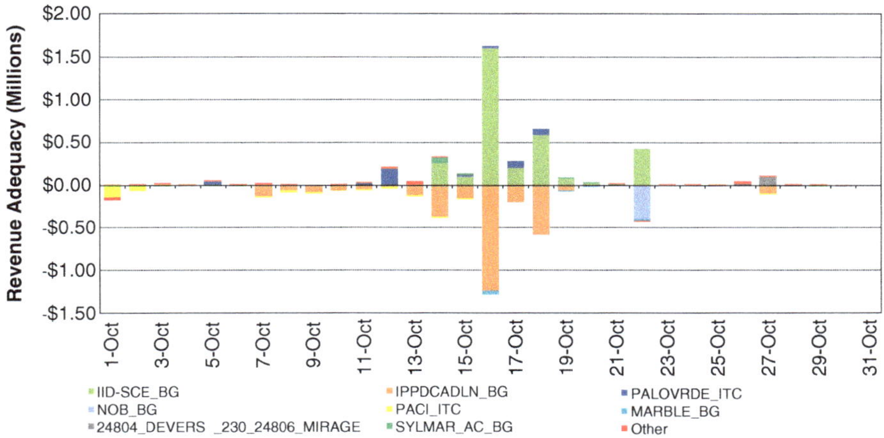


**Fig. 10.2** Daily revenue adequacy in August 2010 in the California ISO. Revenue adequacy is organized by transmission element

market and the capacity released in the energy market due to the enforcement of different constraints in the energy market.

The underlying condition to ensure revenue adequacy is that the same amount of transmission capacity is made available in both the energy and the FTR markets. This requires that (a) the same transmission constraints are enforced in both markets and (b) the same transmission configuration (as defined by the shift factors) is used. From expression (10.15) it is clear that if a transmission constraint is not enforced in the FTR market but is enforced in the energy market, there is a likelihood that revenue deficiencies will occur since the gap of transmission is inherently bounded to be negative. This obviously has its root in the fact that the constraint is not being enforced in the FTR market and, hence, the release of transmission capacity for FTRs over that specific transmission constraint is unlimited.

## 10.8 Revenue Neutrality

Using the engineering notion of revenue adequacy, the ratio of revenue adequacy for transmission constraint $k$ can be stated as follows

$$\Phi_{p.u._k} = \frac{\sum\limits_{t \in \mathrm{T}} \mu_{t,k} \bar{z}_{t,k}}{\sum\limits_{t \in \mathrm{T}} \mu_{t,k} z_{t,k}^{FTR}} \tag{10.16}$$

For the ideal condition of a revenue adequacy of 100 %, the left hand side term of Expression (10.16) has a value of unity and yields

$$\sum_{t \in \mathrm{T}} \mu_{t,k} z_{t,k}^{FTR} = \sum_{t \in \mathrm{T}} \mu_{t,k} \bar{z}_{t,k} \tag{10.17}$$

This relationship is quite intuitive and stands for the requirement that the money paid to FTRs over constraint $k$ matches the money collected in rents over the same constraint $k$. In real life markets, the ATC in the energy market $\bar{z}_{t,k}$ may change as much as every hour. Similarly, the power flows computed from the nodal injections from FTRs, $z_{t,k}^{FTR}$, will change as often as hourly due to changes in the shift factor values used in the calculation of the power flows. Such changes are mainly due transmission system outages, switching or different operating points. From a point of view of the FTR market, however, the amount of FTRs released is a constant and unique value for the full accounting period. For instance, a monthly auction will release FTRs that are defined for the same MW amount for the whole month. Therefore, from a revenue adequacy point of view, the power flow defined by the FTRs over a transmission constraint can be seen as constant value and derived directly from Expression (10.17)

$$z_{t,k}^{FTR} = \frac{\sum_{t \in \mathrm{T}} \mu_{t,k} \bar{z}_{t,k}}{\sum_{t \in \mathrm{T}} \mu_{t,k}} \tag{10.18}$$

It is worth noticing that this MW value obtained from Expression (10.17) is based solely on the outcome of the energy market, namely shadow prices of the transmission constraints and ATC available, which only have an impact when constraints are binding. This calculation can be done only after the fact once the accounting period is complete; i.e., it is based on historical performance of the energy market. This value for revenue neutrality indicates what would have been the ideal transmission capacity released in the FTR market to attain a revenue adequacy of 100 % once the energy market materialized and, therefore, represents the ideal transmission limit for constraint $k$ to be enforced in the FTR market. Obviously, there is no guarantee that historical values are a close representation of future occurrences. Nonetheless, this metric provides the means to analyze the pattern of revenue adequacy on specific and problematic transmission constraints and can help find out deeper issues such as persistent over allocation of FTRs on specific constraints due, for instance, to modeling of outages or derates.

For instance, the California ISO has been able to develop the CRR revenue neutrality metric. Figure 10.3 shows the revenue neutrality point for one specific transmission path (Paloverde Intertie) in the California ISO control area. Such point is estimated based on the market outcomes during the full calendar season of 2010 for on-peak time of use. This plot also shows the OTC duration curve of the intertie with the line in blue, while the dotted line in green is the average CRR amount in MW released in such a constraint. The duration curve is built by sorting from the highest to the lowest MW value. The curve reads by referencing what percentage of

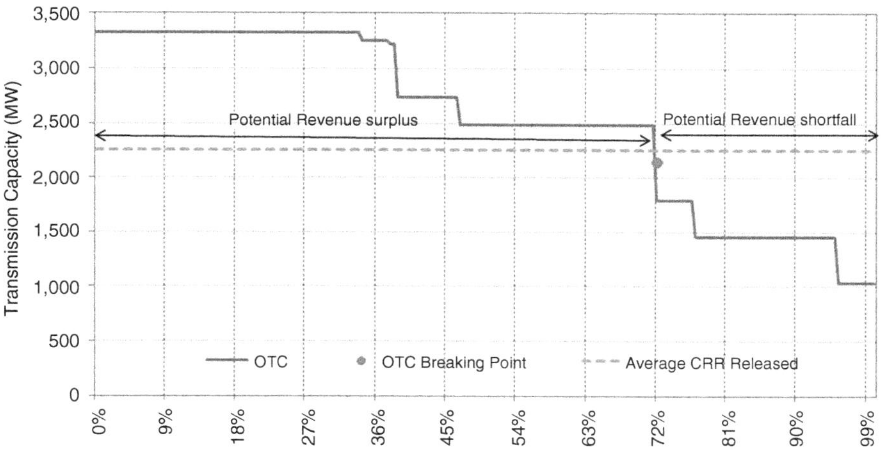


**Fig. 10.3** Duration curve for Operating Transfer Capability of Palo Verde Intertie for calendar season 1, 2010

the time the OTC value has been higher than a given MW value. For instance, about 34 % of the time the OTC value of this intertie was the nominal value (no derates).

This OTC value includes the capacity on the path associated with encumbered rights which is capacity reserved and not made available in the markets, neither for CRRs nor for energy. If such reserved capacity is discounted, then the remaining capacity is the well-known ATC. In this case, the breakeven point is slightly below the CRR value which means that on this intertie and for the given period there was a slight revenue surplus, implying the amount of transmission capacity over this transmission capacity released in the CRR markets was adequate. This metric and its analysis give a reference about the right amount of transmission capacity needed to be released to attain revenue neutrality, and therefore, can be a straight indicator of systemic issues about the modeling of the transmission system in the CRR market to help improve revenue adequacy.

The duration curve goes from the nominal value on the left hand side through lower values on the right hand side; the variations result from derates happening during the actual operation of the system. As can be observed, about 5 % of the time this intertie was heavily derated to about 1,000 MW, less than a third of its nominal value. On one hand, when the system elements are derated to such extent, congestion on such elements will lead inevitably to large revenue deficiencies. On the other hand, when elements are at nominal values or lightly derated, revenue surplus could be observed. From the point of view of the FTR market, the break-even point for revenue adequacy represents a MW value at which revenue surpluses collected in some intervals will balance the revenue shortfalls of other intervals so that over all the accounting period they offset each other.

## 10.9 Factors to Consider for Revenue Adequacy

In its more general terms, revenue adequacy is the fact that there is enough money from the energy market to pay all FTRs holdings. The fundamental concept, however, is more specific. It refers that given a market outcome, say, for one hour or one trading day, the congestion rents from the energy market will be sufficient to pay the FTR holdings for that same hour or day. In a broader context, different markets have adopted variations of the concept for revenue adequacy. These variations respond to other market design aspects and specific needs. The following are some variations that can be considered when determining what factors to use for revenue adequacy.

- Revenue adequacy on an hourly basis over an accounting period. The main factor in revenue adequacy is the inherent system changes, such as derates and outages that may alter the transmission configuration and limits. Such changes may lead to revenue shortfalls in some hours, while other hours can accrue revenue surpluses. Since the forward market, such as the day-ahead market, usually has time intervals of an hour and FTRs usually are settled only at the day-ahead prices, the smallest interval for which revenue adequacy can be calculated is an hour. With changes happening from hour to hour, however, calculation of hourly revenue adequacy may become unnecessary. For instance, if revenue gaps – surpluses or shortfalls- are allocated to demand and let us assume that there are shortfalls in 1 h and surplus in the next one, such hourly settlements are going to implicitly offset each other. Revenue adequacy is usually settled over a full accounting period such as a calendar month or season. In this way, revenue adequacy is a more reflective metric of the FTR process rather than a by-product of the dynamic system conditions over short periods. Naturally, the accounting period can span over the life term of FTRs. In other instances, revenue shortfalls or surpluses can be rolled over from period to period and allowed to offset.
- Day-ahead market versus real-time market. Settlements for FTRs are usually based on day-ahead prices. Since LMP-based markets usually rely on a two-step settlement (day-ahead and real-time), congestion rents may also arise in the real-time market and, thus, can also be put in the funds to pay FTRs. Sometimes, the moneys from settling the real-time market congestion, however, can be negative and this effectively reduces the funds available to pay FTRs.
- Congestion versus losses. FTRs are designed to hedge congestion arising from the day-ahead market. Although some academic concepts for financial instruments to also hedge losses have been proposed, currently only FTRs for congestion are available among markets. On the other hand, an LMP-based market will also have losses rents similar to congestion rents. Depending on the market design, such losses rents may also be included in the funds to pay for FTRs; in this context, losses rents serve as a buffer against revenue deficiencies.

- Auction rents. Depending on the specifics of the design, in instances where there are FTR auctions, the FTR auction revenues may also be used to fund FTR payments.
- Existing transmission rights. Based on contractual arrangements, markets may have to accommodate existing transmission rights. Such rights are exempt from congestion. This means that congestion rents from the day-ahead market need to be reduced to account for such exemptions, and this effectively account for the existing transmission rights.
- Reimbursements of FTR payments. It is well known that the ownership of FTRs may be an extra incentive for profit seeking opportunities (Joskow and Tirole 2000). Markets usually have a process to screen and identify instances where FTR payments may have been increased due to participants' actions in the energy market. For instance, virtual bidding may increase the value of certain FTRs. In such instances, the portion of the FTR value that fails to pass certain test is not paid (or the participant is required to reimburse that quantity depending on the settlements configuration). The money from this process can also be used to fund the FTRs payments, or from another point of view, these proceeds effectively reduce the overall FTR payments used in the determination of revenue adequacy.

## 10.10 Final Remarks

Given the fact that revenue adequacy is an indication of the health of the processes to release FTRs; a shortfall indicates that too many FTRs were released. It is important to understand the intricacies of the process to identify potential mechanism to control revenue adequacy. The main problem is the uncertainty associated with outages and derates. Processes to release FTRs, such as auctions and allocations, usually rely on deterministic approaches, where the system transmission configuration and transmission limits are defined a priori. This is further compounded with the timing for running the FTRs processes well ahead of the energy market, which in some instances can amount to a few months ahead. Ideally, one could derate the OTC of each transmission constraint to a level that represents what historically has been available. Since this approach would rely on historical performance to account for future releases, there is no guarantee that previous performance would occur. Nonetheless, this approach somehow would be more conservative than just ignoring the likelihood of derates and outages and use the nominal OTC, which in turn would lead to revenue shortfall more frequently since there is no room for any change happening in the operation of the system. When there insufficient historical data, it is more plausible to identify constraints that systematically drive revenue deficiencies and they can be target more specifically. When the metric to identify individual revenue adequacy of transmission constraints is not available, the simplest approach can be to derate by a given factor the entire set of transmission constraints. The drawback of this system-wise derate is that revenue shortfalls may be concentrated in certain regions of the system and some regions would be funding deficiencies from others, affecting certain market participants.

## References

Alsac O, Bright JM, Brignone S (2004) The rights to fight price volatility. IEEE Power Energy Mag 2:47–57

Bautista Alderete G (2010) Competition in electricity markets: modelling and economics. Ed. VDM Verlag, Germany

CAISO (2010) Annual market performance report. Folsom, USA. www.caiso.com. Accessed December 2010

Hogan W (1992) Contract networks for electric power transmission. J Regul Econ 4:211–242

Joskow P, Tirole J (2000) Transmission rights and market power on electric power networks. RAND J Econ 31:450–487

Lesieutre BC, Hiskens A (2005) Convexity of the set of feasible injections and revenue adequacy in FTR markets. IEEE Trans Power Syst 20:1790–1798

Lyons K, Fraser H (2000) An introduction to financial transmission rights. Electr J 13:31–37

Nocedal J, Wright S (1999) Numerical optimization, Springer series in operations research. Springer, New York

Rudkevich A, Bagnall J (2005) Loss hedging rights: a final piece in the LMP puzzle. In: 38th PHICSS, Hawai, pp 1–7

Wood AJ, Wollenberg BF (1996) Power generation, operation and control. Wiley, New York

# Chapter 11
# Trading FTRs: Real Life Challenges

**Jose Arce**

## 11.1 Introduction

The problem of trading FTRs can be understood as one of decision making under uncertainties, where the boundary conditions are set by the laws of physics that govern the electric power flows. Under this setup, a typical FTR desk has to deal not only with standard roles of trading financial products, but also with technical ones of power analytics. Building and operating a successful FTR business is a complex enterprise, with multiple factors to consider. Additionally, the still exotic nature of the product makes standard solutions from the trading industry difficult to use. Accordingly, this chapter describes some of the challenges we currently face while trading FTRs in the US, covering three aspects of the business.

The first one deals with the process of building an FTR portfolio and executing the trade (Sect. 11.2). The idea is to go over the different steps mentioning standard practices and most relevant challenges, which are described in the subsections: Data, Analysis, Portfolio Construction, and Trade Execution. The second one (Sect. 11.3) covers alternatives for managing risk and the role played by the FTR desk. Also here, the goal is to describe current situation and open issues, which are elaborated in the sub-sections: Managing Current Exposure, Risk Management, Interaction with Other Desks, and Profile of the "FTR Trader". The third one (Sect. 11.4) mentions a potential evolution of the FTR business. A brief description of alternative scenarios is mentioned in the sub-section: Next Steps. Finally, this chapter concludes with a summary of challenges we encounter in the real life operation of an FTR business.

---

J. Arce (✉)
Maple Analytics, 590 Broadway #2, Somerville, MA 02145, USA
e-mail: jose.arce@mapleanalytics.com

J. Rosellón and T. Kristiansen (eds.), *Financial Transmission Rights*,
Lecture Notes in Energy 7, DOI 10.1007/978-1-4471-4787-9_11,

## 11.2 Building an FTR Portfolio and Executing Trade

### *11.2.1 Data*

Currently in the US there are six markets where it is possible to trade FTRs: PJM, MISO, ISONE, NYISO, CAISO, and ERCOT (The ISO/RTO Council 2012). The general concepts are the same in all of them; however, there are differences in implementation. The first barrier faced in dealing with FTRs is the lack of standards in producing and publishing relevant market data. This problem has implications in three sub-problems: gathering, normalizing, and storing in database. The first one deals with identifying the best places to collect and implementing systematic processes to capture data. The second one relates to the most laborious task, normalizing the data which includes, among other things, mapping different names to the same physical element and with the same format. This task cannot be fully automated, requiring laborious manual intervention. And finally, the third one refers to the efficient storage of data in master database. The main source of raw data comes from the ISOs which can be classified as indicated in Table 11.1

There is an additional set of data coming from ISOs' meetings (committees, subcommittees, task forces, working groups, etc.) which provide very valuable information. In small to mid-size companies, the task of following these meetings is performed by the FTR desk, however this additional function is difficult to accomplish properly, considering the number of activities to cover. Large organizations, on the other hand, have Market Affairs teams or Regulatory Policy teams dedicated to this function. However, due to the technical details involved, it is difficult for them to identify exactly what may be valuable for different areas of the company. Sometimes, companies complement their coverage subscribing to services provided by Market Specialists/Consultants.

In general, the topics discussed in these meetings are relevant to the FTR business, however, on specific instances they are critical to understand or value substantial changes in the market. Companies that can translate this type of information into trading signals have a clear advantage. Definition of new interfaces, retirement of reliability must run units, implementation of special protection schemes on active binding constraint, redefinition of load pockets, derates on critical facilities, are few examples of topics presented in some of the mentioned meetings and that generally impact the market.

Unfortunately, it is usually difficult to automate the identification and collection of relevant information from written documents (or voice records). State of the art software that can interpret text/voice, like the ones used in equities trading (RavenPack 2012), should facilitate this task.

In addition, there are services provided by third parties that help having a better picture of the market dynamics. Some of them are listed in Table 11.2.

Clearly, the objective in this initial step is to concentrate, normalize, and store all these diverse data in an efficient manner. However, implementing and managing this task is very challenging.

**Table 11.1** Data from ISOs

| | | |
|---|---|---|
| | LMPs | DA/RT/5 min RT prices |
| Day Ahead (DA) and Real Time (RT) markets | Congestion | DA/RT/5 min RT binding constraints DA/RT/5 min RT transmission shadow prices |
| FTR market | FTR auction results | Inventory of FTRs |
| | | Binding constraints |
| | | Prices |
| Network representation | Transmission system | Network model |
| | | Operating procedures |
| | | Monitor elements |
| | | Contingent elements |
| | LMP/FTR models | Nodes available for trading (CPNodes) |
| | | Hubs, Aggregated |
| | | Interfaces, Flowgates, Nomograms |
| | CPNode changes | New CPNodes |
| | | Terminated CPNodes |
| Operation | DA/RT realization | Load |
| | | Inter-tie flows |
| | | Wind power |
| | Weather | Temperature |
| | | Thunderstorm alerts (TSA) |
| | Transmission outages | Active |
| | | Scheduled |
| Operation | Historical bidding data | Generation bids |
| | | FTR bids |
| Planning | Transmission/ generation | Transmission upgrades |
| | | Generation queues |
| | | Retirements |

An operation covering PJM and MISO which is evaluating to build an infrastructure to manage 2 years of data would have to consider for instance the following requirements:

- *Dimensionality:* building database with around 210 tables, three billion records, and 400 GB of disk space
- *Dispersion of data sources:* maintaining 15 web data collectors (scrapers)
- *Lack of standards:* mapping and normalizing 50,000 records

The scale of this problem is equivalent to the one managed by a leader mobile telecom operator serving four million customers.

## 11.2.2 Analysis

The next step is to process the data looking for trading signals. Here we consider two alternative approaches, one based on fundamental analysis, and the other based on quantitative analysis.

**Table 11.2** Data from third parties

| | Historical prices |
|---|---|
| Normalized market data | FTR inventories |
| Generation status | RT production |
| | Outages |
| Flows data | RT power flows |
| Market intelligence | Price forecasts |
| | Congestion forecasts |
| Policy | Policy meeting reports |
| | Environmental issues |
| | FERC filings |
| | Geopolitics |
| Weather | Temperature forecast |
| | Seasonal forecasts |
| | Storms |
| Wind | Wind profile |
| | Wind forecast |
| | Wind power production |
| Water | Reservoir levels |
| | Precipitations |
| | Snow pack levels |
| Fuels | Inventories |
| Over the counter (OTC) | Prices for tradable products |

The fundamental based approach relies on the fundamentals of power systems to explain the occurrence of congestion. There are different options to perform this analysis but all of them share the principle of linking congestion events with particular scenarios of supply, demand, transmission, and operation of the system. AC and DC Power Flow models, Optimal Power Flow analysis, Unit Commitment Security Constrained OPF simulations, are some of the concepts or methodologies commonly used to perform this task (Wood and Wollenberg 1996). Typically, this analysis is performed using some of the standard software products available in the industry (e.g. PSSE, PowerWorld Simulator, DAYZER, SCOPE, GEMAPS).

The quantitative based approach relies on principles of statistical analysis to process large amounts of data to identify overall trends. There are different alternatives to perform this task but all of them share the same idea of objectively identifying trends or patterns out of noisy data. Linear and Non-Linear Regressions, Data Mining, Time Series Analysis, Principal Component Analysis, are some of the concepts or methodologies commonly used to perform this task (Nisbet et al. 2009). In this case, the analysis is done using proprietary models written in technical-oriented programming languages (e.g. Matlab, Mathematica, R, C#, Java)

For both approaches, the process follows the sequence indicated in Fig. 11.1.

The objective in this step, independently of the approach, is to obtain trading signals. In terms of FTRs, trading signals refer to bullish or bearish views on congestion. However, if the inputs for the quantitative approach are prices, then trading signals could be source-sink paths.

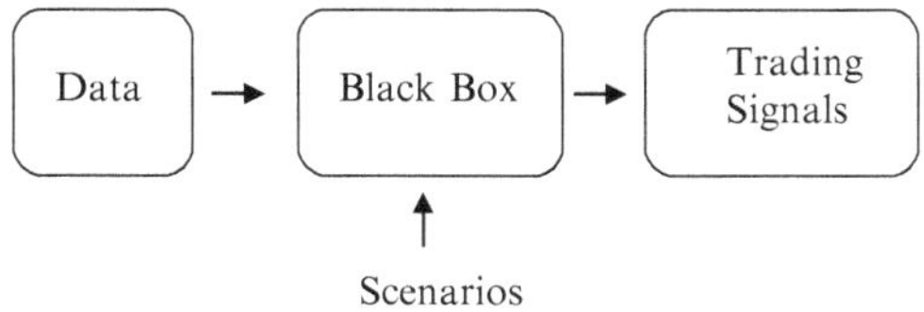


**Fig. 11.1** Analysis flowchart

**Table 11.3** Drivers

| Seasonality | Periodic patterns such as summer weather triggering specific BCs |
|---|---|
| Wind | Speed and persistence above threshold creating oversupply scenarios and congesting weak links |
| Thunderstorms | System operates in conservative manner when TSA is declared, adding pressure to the transmission system (N-2 secure instead of N-1 secure) |
| Fundamental changes | Upgrades in the grid reduce/eliminate historical BCs, and sometimes shift the problem to new BCs |
| Outages | Short-term generation/transmission outages creating localized congestion |
| Chronic problems | Devices operating close to their limit almost permanently |

To finish, it is necessary to quantify the relevance of the simulated signals (ranking), which are obtained comparing expectation (edge) relative to dispersion (conviction).

Some of the challenges include:

- *Confidence in data:* unfortunately, it is not rare to observe changes in relevant published information after the auction is closed, invalidating the simulated signals.
- *Technology barrier:* building and processing complex simulations (e.g. Unit Commitment Security Constrained OPF runs) for large systems is still beyond most operations' technical capabilities.

### *11.2.3 Portfolio Construction*

In context of the standard optimization problem solved by the ISOs, congestion refers to transmission binding constraints (BCs), which are specified as monitor element (monitor) and contingent element (contingency) (Schweppe et al. 1998). Not all BCs share the same drivers, therefore, they have different behaviors. Some of these drivers are listed in Table 11.3. In order to be systematic in this classification, it is necessary to quantify these behaviors (e.g. using higher moments).

One of the main differences between FTRs and other financial products is that the selection of the contract to trade, source-sink path (path), is a decision variable. Depending on the strategy, it may be even more relevant selecting a path than pricing it.

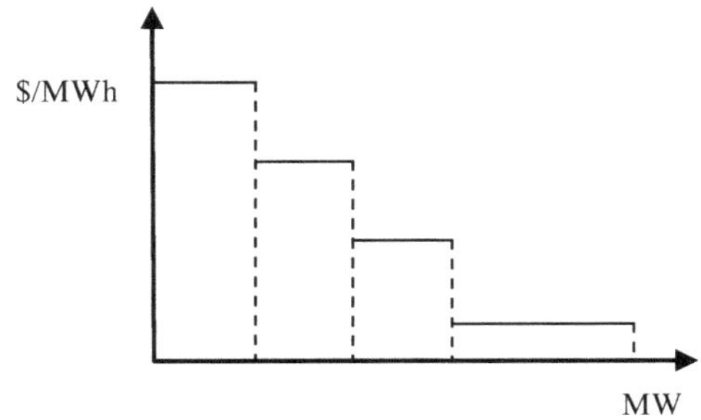


**Fig. 11.2** Step function bid curve

Here, it is necessary to remark that a path is impacted by active BCs with positive or negative contribution depending on its exposure. Accordingly, a long exposure refers to paths that have positive correlation to a target BC, receiving positive revenues when the BC is active. Bullish views tend be expressed by paths with long exposure to the associated BC. On the other hand, a short exposure refers to paths that have negative correlation to a target BC, receiving negative revenues when the BC is active. Bearish views tend to be expressed by paths with short exposure to the related BC. Following the same logic, a long-short exposure refers to paths that have positive correlation to one target BC and negative correlation to a different one. Combining bullish and bearish views can be expressed by paths with long-short exposures. In addition, counterflow is a particular case of exposure to a BC different from the desired one. In general the expression refers to adverse congestion that produces revenues with the opposite sign to the expected when initiated the trade.

In markets with limited number of CPNodes, it is difficult to select source-sink pairs that have exposure to a single or dominant BC. Here there are two side effects to consider, one is the cost of paying for undesired BCs, and the other is counterflow risk. In the case of Obligation FTRs, the second issue is maybe FTR Traders' most feared risk. To address this problem, some ISOs have implemented Option FTRs (Pameshwaran and Muthuraman 2009).

The construction of the bidding curve requires definition of Auction, Period, Source, Sink, Time of Use (On Peak, Off Peak), Trade Type (Buy, Sell), Hedge Type (Obligation, Option), Price, and Volume. To simplify the pricing for different periods, it is usual to work in $/MWh terms and then convert it to $/MWPeriod before submitting. In this problem, Price, Volume and Shape of the Bidding Curve are the key decision variables. Figures 11.2 and 11.3 describe the accepted formats for bidding curves.

An interesting characteristic of FTRs is that price (P) and quantity (Q) are unknown before executing the trade. We only have control over maximum price to pay/minimum price to receive, and maximum volume to clear. So, FTR auction simulators are built to evaluate contingent performances of the working portfolio. Adjustments in bidding curves are made until differences between simulated and target portfolios are acceptable. Finally, we obtain the portfolio to be submitted in the FTR auction.

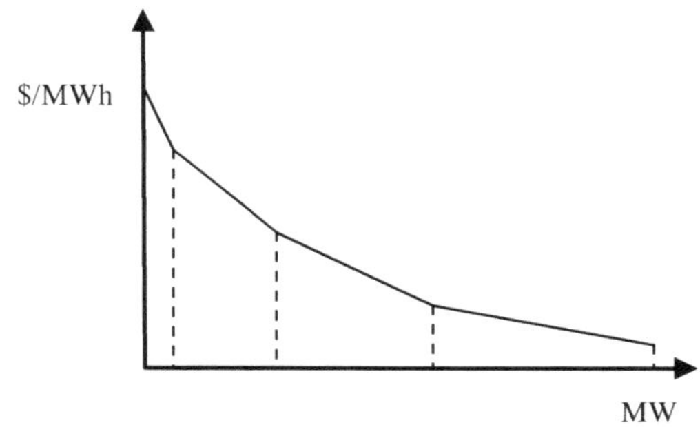


**Fig. 11.3** Piece-wise bid curve

Some notable challenges include:

- *Counterflow risk:* limited number of CPNodes available for trading Obligation FTRs (in some ISOs) makes difficult selecting source-sink paths with limited counterflow exposure. Furthermore, not all ISOs have Option FTRs.
- *P and Q are unknown:* acquiring FTRs from auctions adds a new layer of complexity in the risk taking process. The uncertainty associated with price and quantity results in getting a cleared portfolio different from the targeted one.

### *11.2.4 Trade Execution*

The trade execution is a simple process; however, there are some requirements to satisfy and validations to perform. The first requirement is related to collateral to support FTR bids. Here again, each ISO has different level of collateralization requirement according to its credit policy, but all of them share the principle that to participate in the FTR auction, a market participants has to have sufficient capital.

Then, the CPNodes used in the different paths have to be valid for the particular FTR auction we plan to submit. For example, CPNodes valid for prompt auctions may not be necessarily valid for non-prompt auctions. Sometimes, during early stages of the portfolio construction the valid CPNode list for the next auction is not available, therefore it is a good practice to implement a CPNode validation step.

Furthermore, it is necessary to convert the target portfolio to the accepted format (xml files). Although this formatting process is not complex, the cost paid for mistakes here can be enormous. For example, changing sources for sinks automatically converts long exposures into short exposures (or vice versa), or using the wrong number of hours for a given period changes the bidding prices.

The submission is implemented electronically, through secure sites, uploading xml files manually or through programmatic interfaces. As mentioned before, the cost of operational mistakes in this step could be high. Therefore, a prudent step is to validate that the submitted portfolio is exactly the portfolio we wanted to submit.

After this final validation, the trading execution is concluded. Auction results, in general, are published within 10 days.

Some relevant challenges in FTR trade execution are:

- *Prone to costly mistakes:* it is common to have several auctions overlapping during the same period of time. For small/mid-size operations, in particular, this issue creates substantial pressure when controlling and validating different portfolios/auctions. In large operations, on the other hand, this problem is reduced; however, distractions from crowded trading floors work against them too. Compounded by the fact of dealing with an almost illiquid product, mistakes in trading execution could be just too high to bear.
- *Execution infrastructure:* building a robust infrastructure is critical to mitigate execution risk. However, here also the lack of standards creates some frictions that require special treatments.

## 11.3 Managing Risk and the Role Played by the FTR Desk

### *11.3.1 Managing Current Exposure*

Currently, and depending on the ISO, it is possible to trade FTRs from 3 years to 1 month forward (Long-Term: 1–3 years, Annual: 1 year, Balance of the Year: less than 1 year, Monthly: 1 month).

This temporal discretization goes in line with different market needs. The power market evolves over time, so does our trading signals, convictions and target portfolios. So, it is common to start accumulating core positions in Long-Term Auctions and/or Annual Auctions and then adjusting the portfolio in Monthly/ Balance of the Year Auctions. The last opportunity we have for implementing this strategy is during the Monthly Auction just before delivery.

However, because of the few opportunities we have to trade (in comparison with other financial products), it is very difficult to arrive to delivery with a balanced portfolio. An alternative to improve this situation is to trade FTRs in secondary markets. Most ISOs have implemented an environment for this purpose. Unfortunately, participation has been minor. On the other hand, attempts to build a bilateral market for predefined paths have gained some interests. However, the reality is that most FTR paths target idiosyncratic factors which are difficult to match in bilateral trades, limiting the attractiveness of the concept.

During delivery, the FTR portfolio is subject to DA congestion. In case we prefer to get exposed to RT congestion, then it is possible to do it using some of the daily DA-RT swaps available in the market. The most common DA-RT product is Virtual Bidding (VB) (Metin et al. 2010), which is a contract specified by hour and CPNode. There are two products, INC that settles as the difference in LMPs between DA and RT, and DEC that settles as the difference in LMPs between RT and DA.

The strategy requires to INC at source and DEC at sink of the FTR we want to get exposure from RT market. This strategy is easy to implement however transaction cost can be significant. Furthermore, there is volumetric risk when clearing unbalanced portfolio results in net long/short exposure to absolute LMP instead of desired locational spread. Additionally there is spread risk which refers to price taker strategy predisposed to unlimited DA congestion cost.

To address these risks, some ISOs have implemented a product that trades balanced spreads (e.g. Up to Congestion contracts in PJM), which settles as the difference in LMPs between RT and DA (PJM 2012). In this case the strategy requires using the same source and sink of the FTR we want to get exposure from RT market. In case they are not available, then use some proxy CPNodes at the expense of getting different BCs' exposure. The main advantage here is that this product solves both volumetric and spread risks.

There are two other aspects of relevance about these DA-RT contracts. The first one relates to Transaction Costs, which sometimes can be significant. Furthermore, these costs are known after the fact, turning it difficult to incorporate properly in the trading strategy. Moreover, this friction limits the success of its original goal of improving convergence between DA and RT markets.

The second one is more controversial, and is related to the impact that this activity has in DA results. These strategies, Virtual Injections and Virtual Withdraws, create additional power flows in the DA solution, and consequently affect DA congestion. Furthermore, in the case of using proxy CPNodes, DA congestion may diverge from the expected based on fundamentals (phantom congestion). However, the most problematic issue arrives when strategic bidding creates DA congestion on purpose to increase FTR revenues (or the value of any other contract that settles on DA prices). Strict monitoring is necessary to identify and mitigate these behaviors.

In the OTC market, the alternatives for proper portfolio rebalancing are even more limited. Even though there are products that have good liquidity (ICE 2012), the main limitation is the weak correlation between these OTC products and FTR portfolios. A reason for this observation is that the typical factors that explain most dynamics in OTC products are less relevant for FTR portfolios. On the contrary, specific FTR paths provide complementary value to OTC portfolios.

Based on these reasons, the concept of active portfolio management to optimize current exposure is difficult to implement in the case of FTRs.

Some of the current challenges include:

- *Liquidity:* opportunities to trade FTRs are few, in general once a month with limited volume in reconfiguration auctions. Furthermore, the secondary market has not developed as anticipated, and the OTC products are poorly correlated with FTRs.
- *Transaction costs:* during delivery, there is a possibility to move part of the DA exposure to RT, however transaction costs for VB (including operating reserve charges, volumetric and spread risks) have worked against this strategy.

### 11.3.2 Risk Management

It is necessary to consider the different components of risk involved in the whole process before, during, and after the auction. Before the auction, the main risks come from inaccuracies in data, assumptions, models, and/or their usage in Analysis and Portfolio Construction steps. Procedures to control this type of risk include performing quality control of data, validating models and hypothesis, and verifying most recent published information. Within the risk management process, these risks tend to be part of Operational Risk and Modeling Risk considerations.

During the auction, the main risks come from operational mistakes and/or technology failures during Trade Execution. Procedures to control these risks include submitting preliminary portfolios to test own infrastructure/technology, performing validation of submitted portfolio against target portfolio, and building and testing technology back-up infrastructure. Within the risk management process, these risks tend to be part of the Operational Risk and Execution Risk concerns.

After the auction, the main risks come from the impact of realized congestion on cleared portfolio. Here, it is important to recognize two levels of realizations. On one hand, a normal range, where congestion is related to drivers such as weather events, unexpected outages, over/under-commitment, etc., which tend to produce transitory patterns. On the other hand, an extraordinary range, where congestion is related to a permanent pattern change. Furthermore, and primarily due to non-storage nature of electricity (wholesale level) and operational constraints (ramping, operating procedures, localized inflexibility due to outages), it is observed that tail events are a lot more common in FTRs than in other energy products (i.e. leptokurtosis) (Adamson et al. 2010). Within the risk management process, these risks tend to be part of the Market Risk, Liquidity Risk, and Credit Risk concerns. Additionally, Underfunding Risk and Default Risk require special considerations.

The standard risk management role includes periodic evaluation of Value at Risk (VaR), which tends to provide a good indication of risk involved in a portfolio for a normal range of realizations. To complement this metric, some forms of Stress Testing and Concentration Analysis are also performed looking for risk associated with realizations in the extraordinary range. These evaluations are part of the Market Risk assessment and are described below.

- *VaR:* the maximum loss that will not be exceeded with a given probability (confidence level) over a given period of time. In general, a simulative model is created, using historical congestion realizations adjusted by seasonality and giving more weights to more recent realizations. This approach is very flexible and easy to implement, however it ignores congestion patterns not present in the historical sample.
- *Stress Testing:* this analysis is performed for specific scenarios looking for extreme realizations. The key issues are, calculating net exposure for different BCs, and defining under which circumstances the portfolio is exposed to counterflow.

- *Concentration Analysis:* the basic approach is to use a risk aggregator to convert source-sink paths in net exposure (MW) for each branch monitored in DA market.

Sometimes to complement these three metrics, the risk management function also calculates Conditional Value at Risk (CVaR) that is more sensitive to the shape of the loss distribution in the tail of the distribution (Uryasev 2001).

With this information, the risk manager evaluates Liquidity Risk. The basic idea is to make sure that the company has allocated enough capital to the FTR account to pay invoices. Given the uncertainties and assumptions involved in different calculations, a conservative approach is to keep liquid funds to pay invoices equals to a multiple of the current VaR.

Additionally, on a daily basis, a common metric used by different roles within the organization is the Profit and Loss (PL) report. Standard reports include Year to Date PL, Month to Date PL, and Today's PL. Sometimes Inception to Date PL is also included. Here it is important to clarify the difference between Realized PL and Marked to Market PL.

- *Realized PL:* results from calculating the difference between DA revenues and FTR auction cost. This PL is replicable by anyone (FTR inventories and DA prices are public information).
- *Marked to Market PL:* results from calculating the difference between future DA revenues (represented by a market quote) and the corresponding FTR auction cost. The challenge here is that there is no liquid forward market for FTRs. As a result, it is common to use models to estimate future revenues adjusted by a liquidity factor. In this case, this model-driven PL is more difficult to replicate and may create disagreements.

These PL refer to gross values, therefore to obtain the net PL it is necessary to include in the calculation Underfunding, Defaults, and Fees/Adjustments.

- *Underfunding:* in some ISOs FTR is not a fully funded contract, therefore PL has to be adjusted by this factor. Basically, if the transmission capacity sold ahead of time in the auction is more than the available transmission capacity during delivery, then the ISO does not collect enough revenues to pay its obligations. This problem is not minor, and is currently a topic of debate.
- *Defaults:* in case of default events, the ISO socializes the incurred losses among market participants proportional to their participation in the different markets administered by the ISO (even participants with no FTR positions share part of the default cost).
- *Fees and Adjustments:* there are some administrative fees per bid and cleared position as well as adjustments in case of corrections in prices or other factors that require proper considerations.

In context of bilateral contracts, the concept of credit risk deals with credit exposure and credit quality associated with counterparties; where credit exposure refers to the magnitude of the risk and credit quality refers to the likelihood of the risk. In the case of FTRs, where there is no specific counterparty besides the ISO, the concept of credit risk is adjusted to include Underfunding and socialized Defaults.

Additionally, the risk involved with policy changes is not minor, also, very difficult to quantify. An alternative approach to deal with regulatory risk is to have an active participation in the different policy meetings relevant to the business. However, as mentioned in section on data (Sect. 11.2.1), this task is not easy to address effectively.

Moreover, some return on risk metrics (e.g. Sharpe ratio) can mislead the risk the portfolio is running if it is not analyzed properly. As explained in section Managing Current Exposure (Sect. 11.3.1), most FTR paths are accumulated in Long-Term/ Annual Auctions, therefore setting portfolio's performance until delivery. A good Sharpe ratio could just reflect that a dominant position acquired in Long-Term Auction is suddenly in the money due to a particular congestion pattern, but does not say much about the other "sleeping" paths.

Finally, and given the specific characteristics of FTRs, it is beneficial to also include some risk management practices used for Alternative Investments, for example similar to the ones described in (Jorion 2009).

Some of the remaining challenges on FTR risk management include:

- *Underfunding:* this issue is nowadays a serious concern, at the extreme of making some trading strategies unprofitable. Furthermore, the problem is even adding risk to standard hedges that are not working as designed.
- *MtM models:* MtM PL is a metric generally requested not only by groups within the company but also by investors. However, its value can be challenged, creating additional burden to the desk.
- *Path dependence:* portfolio performance is strongly dependent on the FTR paths locked in during Long-Term/Annual Auctions, therefore simplistic performance metrics could underestimate the portfolio's risk.
- *Choosing proper risk management approach:* standard models for quantifying risk do not necessarily apply to FTRs. Furthermore, even if risk is properly quantified, nature of product makes difficult to rebalance the portfolio. Therefore, some risk management approaches used for Alternative Investments may be a good complement.

### 11.3.3 Interaction with Other Desks

Originally, with a single price per control area (or power pool), the focus of transmission analysis was primarily concentrated on inter-ties. However, the arrival of locational pricing shifted the focus to the transmission system within control areas. The immediate reaction has been to allocate more resources to transmission analysis, and then build an FTR desk. Currently, there are multiple players participating in the FTR business such as investment banks, hedge funds, private equity shops, proprietary desks, global energy companies, merchant power plants, municipalities, utilities, cooperatives, service providers, etc.

**Table 11.4** Management, researchers/analysts, and risk takers roles

| | | |
|---|---|---|
| **Management** | | |
| Head of trading | In charge of the whole risk taking process | |
| Portfolio manager | In charge of particular risk taking desk | |
| **Researchers/analysts** | | |
| Strategists | Provide analytics and research to risk takers, converting data into trading signals | |
| Meteorologists | Supply weather forecasts and different reports on temperature, wind, hurricanes, precipitation | |
| Data/IT | In control of data gathering | |
| | Managing storage space in database | |
| | Server maintenance in context of a 24 h operation | |
| | Backup process | |
| Market affairs | Communicate relevant information from different meetings | |
| | Quantify impact of policy changes | |
| **Risk takers** | | |
| OTC | Term (directional) | Trades long-term dynamics, directional power contracts, highly correlated to fuel prices, overlapping interests with natural gas desks |
| | Term (heat rates) | Trades long-term dynamics, relative value contracts (power prices/fuel prices), idiosyncratic to the power business |
| | Options | Trades medium-term/short-term dynamics, still an exotic desk in most operations, limited liquidity beyond short-term horizons |
| | Basis | Trades medium-term dynamics, locational spread contracts, similar to FTR if traded within the same ISO, additional component of supply stack function if traded between different ISOs |
| OTC | Cash | Trades short-term dynamics, directional power contracts, highly correlated to fundamental drivers |
| ISOs | FTR | Trades medium-term/long-term dynamics, congestion specific contract |
| | VB | Trades short-term dynamics, directional or locational spreads |
| | Up to congestion | Trades short-term dynamics, locational spreads |
| Physical | RT | Trades according to physical power needs, 24 h operation |
| Origination/sales | Structured products | Trades long-term dynamics, satisfying customized deals |
| Exchanges | Quants | Trades very short-term dynamics, state of the art technology-driven business |

Independently of the type of player, it is common to have Management, Researchers/Analysts, Risk Takers, and Back Office personnel. The arrival of the FTR desk creates an interesting dynamics, in particular with the Power Desks and Back Office roles. A brief description of these different roles and the link with the FTR desk is described in Tables 11.4 and 11.5.

The strong link between FTR desk and these roles comes from the current relevance that congestion has in power prices. Therefore, it is observed that the

**Table 11.5** Back Office roles

| Back Office | |
|---|---|
| Risk management | Measures and manages everything related with risk |
| Settlement | Reconciles PL (realized and MtM PL), including underfunding, fees, and adjustments |
| Accounting | Monitors liquidity situation and implements budgeting plan for different needs |
| Compliance/ legal | Guarantees compliance with company and market requirements<br>Providing legal support and interpretation of different regulations |
| Human resources | In charge of recruiting needs (critical role) |

desk is a permanent provider of congestion views for different scenarios and time horizons. In particular, it is highly requested when a new congestion pattern arrives in the market. Consequently, nowadays the FTR desk plays a central function within the Power business.

In this case also, the arrival of this new product impacted these teams. In particular, its exotic nature has forced the FTR desk to be creative to explain its business and to be flexible to adapt to standard company's requirements.

Some main challenges on the interaction of desks include:

- *Diverse interests:* the difficulty comes not only from satisfying multiple and sometimes conflicting interests but also from explaining nature of FTR business to diverse audiences.
- *Integration:* even though the relevance of the FTR desk in the trading floor has increased, its true value that comes from a full integration has been difficult to materialize.

### *11.3.4 Profile of the "FTR Trader"*

The traditional trading business separates roles among IT, Data, Analytics, and Trading. However, in the case of the FTR business, these roles tend to be self-contained within the FTR desk. Therefore, the "FTR Trader" performs tasks beyond the standard ones. Accordingly, this new profile requires proficiency according to the ones presented in Table 11.6.

Clearly, it is difficult to find candidates who score high in these four skills. Therefore, a more realistic proposal is to build a team with members complementing each other. The recruiting effort is not minor, on one hand the pool of experienced talent is not big (FTR is still a niche), and on the other hand the job itself is very demanding. There is consensus among recruiters that there are only three true job interview questions (Bradt 2011), which in terms of the FTR business refer to:

1. *Can you do the job*? This question is the one generally addressed in the interviews, where technical skills and specific knowledge (i.e. Transmission,

**Table 11.6** Skills according to new trading profile

| | |
|---|---|
| Transmission | Be capable to analyze complex dynamics and identify the right trading signals, skills in general associated with formal education in engineering, physics, or mathematics |
| Risk taking | Be able to convert systematically trading signals in profitable trading strategies, properly quantifying opportunities and risks, skills in general obtained with formal education in economics or finance |
| IT/data | Be proficient to design and implement sophisticated and scalable IT infrastructure according to the needs of a data-intensive 24 h operation, skills gained not only with formal education in computer science but also with experience in real life implementation |
| Interpersonal | Be flexible to accommodate challenging schedules and demanding projects, be able to adjust to emotional swings associated with financial outcomes, and finally (and may be most important) be able to work well within a team |

Risk Taking, IT/Data) are evaluated. Moreover, the answers can be quantified properly and comparison among candidates is easier.

2. *Will you love the job*? This one refers to comparing expectation with reality of the open position. Most of the time the "FTR Trader" has to deal with tasks that can be considered tedious and sometime even repetitive/boring but in the end result critical to the overall success (e.g. normalizing data, reading long reports, analyzing power flow cases). It is very important to communicate this reality to the candidate looking for honest feedbacks.
3. *Can we tolerate working with you*? Sometimes also known as "The Airport Test", this question focus on the candidate's interpersonal skills and how well he/she fits within the existing team's working culture.

Finally, after building the FTR team and working together for 1 or 2 years, the desk starts to consolidate.

The main two challenges presented in this section are:

- *Recruiting:* the pool of experienced talent is not big enough to satisfy current hiring needs. Moreover, recruiting out of school requires substantial investment in training and coaching.
- *Building and consolidating:* finding the right candidates is only part of the challenge, it is even more difficult to keep them long enough to consolidate the business. Consolidation is a process that takes time, unfortunately many companies are not patient enough to make it a reality.

## 11.4 Potential Evolution of the FTR Business

### *11.4.1 Next Steps*

A natural evolution should occur to both the product FTR and the FTR desk. The first one would require addressing some of the issues indentified in this chapter, in particular underfunding and liquidity. The second one would require institutionalizing the whole trading process. This will be even more necessary if additional areas within the US and/or other countries decide to implement LMPs and FTRs.

Also, a good integration between FTR and Structured Products desks providing liquidity beyond the time horizon covered by FTR auctions would be necessary. Tolling agreements, customized deals, load serving contracts, are some of the transactions that require hedging basis risk. Nowadays, this is difficult to achieve considering the limited quotes beyond liquid hubs. That is where FTR desks should appear in the process pricing competitively illiquid locations and working close by Structured Products desks implementing these multipart deals.

Besides, a better interaction with state of the art Quant desks would add complementary skills to this technology intensive business. As time evolves it is becoming more evident of the critical role played by technology in a more globalized business environment.

Here, some of the challenges include:

- *Evolution and consolidation:* the real challenge in the next years would be for the current FTR desks to adjust fast enough to a more global and sophisticated trading environment, and for the FTR concept to consolidate as a liquid financial instrument.
- *Expanding beyond the US:* attempts to transition towards full LMPs and FTRs in some countries have not evolved beyond initial discussions.

## 11.5 Conclusions

In the last 10 years, the FTR business has evolved substantially, with more markets to trade and more sophisticated FTR operations. During the early days, traders with their own spreadsheets and simplistic models participated in the market. Nowadays, there are several teams of researchers approaching the problem in a more quantitative manner, running highly sophisticated trading platforms, turning FTRs in a technology driven business.

Moreover, the low correlation between FTRs and global financial markets has made this product very appealing. This fact has attracted the interest from financial institutions and a diverse set of investors. Furthermore, over time, it is expected that the area covered by LMPs and FTRs be sizable enough to allow even more attractive business opportunities.

However, there are still multiple challenges to address before realizing the full value associated with the concepts of LMPs and FTRs. Some of them, as seen from the proprietary trading side, have been discussed in this chapter and are summarized as follows:

- *Data*: The volume, dispersion of sources, and lack of standards makes the data management problem the first obstacle to pass. The scale of this problem requires highly sophisticated solutions. However, normalizing data also involves tedious manual intervention.

- *Analysis:* Independently of the approach, fundamental-based or quantitative-based, it is critical to have reliable data. Unfortunately, it is not rare to observe changes in relevant published information after the auction is closed, invalidating the simulated signals. Furthermore, building and processing complex simulations for large systems is still beyond most operations' technical capabilities.
- *Portfolio Construction:* The limited number of CPNodes available for trading Obligation FTRs (in some ISOs) makes difficult to select source-sink paths with limited counterflow risk. Option FTRs present an interesting solution to this problem, but unfortunately only two ISOs offer the product and not for all CPNodes. Furthermore, acquiring FTRs from auctions adds a new layer of complexity in the risk taking process. The uncertainty associated with price and quantity results in obtaining a cleared portfolio different from the original targeted portfolio.
- *Trade Execution:* The reality of having several auction deadlines overlapping during the same period of time, distractions from crowded trading floors, pressure of dealing with an almost illiquid product, and compounded by the nature of electronic execution, results in a process that is naturally prone to costly mistakes.
- *Managing Current Exposure:* Comparing with other financial products, the opportunities to trade FTRs are very few, in general once a month with limited volume in reconfiguration auctions. Furthermore, the secondary market concept has not developed as anticipated, and the OTC products are poorly correlated with FTRs. As a result, and for most practical terms, a portfolio of FTRs is considered illiquid. During delivery, there is a possibility to move part of the DA exposure to RT; however, transaction costs (including operating reserve charges, volumetric and spread risks) have worked against this strategy. Based on these reasons, the concept of active portfolio management to optimize current exposure is difficult to implement.
- *Risk Management:* Currently, underfunding is a hot issue. The severity of this problem turns some trading strategies unprofitable. Besides this problem, the standard models for quantifying risk do not apply necessarily to FTRs. Moreover, even if risk is properly quantified, the nature of this product makes difficult to rebalance the portfolio. Therefore, some risk management approaches used for Alternative Investments may be a good complement.
- *Interaction with other Desks:* The strong link between FTR and the different Power Desks comes from the current relevance that congestion has in power prices. Therefore, the FTR desk is a permanent provider of congestion views for different scenarios and time horizons. In particular, it is highly requested when a new congestion pattern arrives in the market. Consequently, nowadays the desk plays a central function within the Power business. Also, the arrival of this new product impacted Back Office as well. In particular, its exotic nature has forced the FTR desk to be creative to explain its business. Summarizing, the relevance

of the FTR desk in the trading floor has increased, however, its true value that comes from a full integration has been difficult to materialize.

- *Profile of the "FTR Trader":* The traditional trading business separates roles among IT, data, analytics, and trading. However, in the case of the FTR business, these roles tend to be self-contained within the FTR desk. Therefore, the "FTR Trader" performs tasks beyond the standard ones. Accordingly, this new profile requires proficiency in transmission, risk taking, and IT/data. Additionally, on the interpersonal side, he/she has to be able to tolerate the always demanding trading environment. Besides the difficulty in recruiting the right candidates, the business consolidation is a process that takes time.
- *Next Steps:* The real challenge in the following years would be for the current FTR desks to adjust fast enough to a more global and sophisticated trading environment, and for the FTR concept to consolidate as a liquid financial instrument.

**Acknowledgements** The Author would like to thank Dr. Carlos Larisson from UNSJ and Dr. Fernando Olsina from CONICET and UNSJ for reviewing early versions of this chapter and for their constructive suggestions. However, the Author is the only responsible for errors that may have occurred.

## References

Adamson S, Noe T, Parker G (2010) Efficiency of financial transmission rights in centralized coordinated auctions. Energy Econ 32:771–778

Bradt G (2011) Top executive recruiters agree there are only three true job interview questions. http://www.forbes.com/sites/georgebradt/2011/04/27/top-executive-recruiters-agree-there-are-only-three-key-job-interview-questions. Accessed 6 June 2012

ICE (2012) OTC electricity contracts. https://www.theice.com/otc_electricity.jhtml. Accessed 6 June 2012

Jorion P (2009) Risk management for alternative investments. http://merage.uci.edu/~jorion/varseminar/Jorion-CAIA-isk_Management.pdf. Accessed 6 June 2012

Metin C, Attila H, Philip QH (2010) Virtual bidding: the good, the bad, and the ugly. Electr J 23:16–25

Nisbet R, Elder J, Miner G (2009) Handbook of statistical analysis & data mining applications. Academic, Amsterdam

Pameshwaran V, Muthuraman K (2009) FTR-option formulation and pricing. Electr Power Syst Res 79(7):1164–1170

PJM (2012) Two settlement virtual bidding and transactions. http://www.pjm.com/training/~/media/training/core-curriculum/ip-transactions-201/transactions-201-two-settlement.ashx. Accessed 6 June 2012

RavenPack (2012) RavenPack News Analytics. http://ravenpack.com. Accessed 6 June 2012

Schweppe F, Caramanis M, Tabors R et al (1998) Spot pricing of electricity. Kluwer, Boston

The ISO/RTO Council (2012) IRC Documents. www.isorto.org. Accessed 6 June 2012

Uryasev S (2001) Conditional value at risk (CVaR): algorithms and applications. http://www-iam.mathematik.hu-berlin.de/~romisch/SP01/Uryasev.pdf. Accessed 6 June 2012

Wood A, Wollenberg B (1996) Power generation operation and control. Wiley, New York

# Chapter 12
# Participation and Efficiency in the New York Financial Transmission Rights Markets

**Seabron Adamson and Geoffrey Parker**

## 12.1 Introduction

As many authors have observed, the allocation of scarce transmission capacity presents a major market design challenge. The electric power system is subject to generation and transmission technology constraints that make it difficult to define tradable property rights for physical transmission. This difficulty has led economists to instead create markets for financial transmission rights (FTRs) settled against the congestion price component of locational marginal prices (LMPs) (Hogan 1992). This market structure has been increasingly adopted in the United States and other countries.

While there has been a substantial literature on the relative attractiveness of FTR markets over other market design, there has been significantly less empirical analysis of how these markets have performed in practice. In this chapter, we trace the operation of the one of the earliest FTR markets, operated by the New York Independent System Operator (NYISO). In particular, we present new analysis showing how the mix of firms that have participated in the NYISO FTR markets has changed over time. We also summarize the econometric analysis of Adamson et al. (2010) on FTR market efficiency and learning over time.

## 12.2 Transmission Congestion Pricing in New York

NYISO, along with the Pennsylvania, New Jersey, Maryland Interconnection (PJM), was one of the first LMP markets in the United States and has conducted periodic FTR auctions since 1999. The NYISO publishes day-ahead and real-time

S. Adamson (✉) • G. Parker
Tulane Energy Institute, Tulane University, New Orleans, LA 70118, USA
e-mail: sadamson@tulane.edu; gparker@tulane.edu

J. Rosellón and T. Kristiansen (eds.), *Financial Transmission Rights*,
Lecture Notes in Energy 7, DOI 10.1007/978-1-4471-4787-9_12,

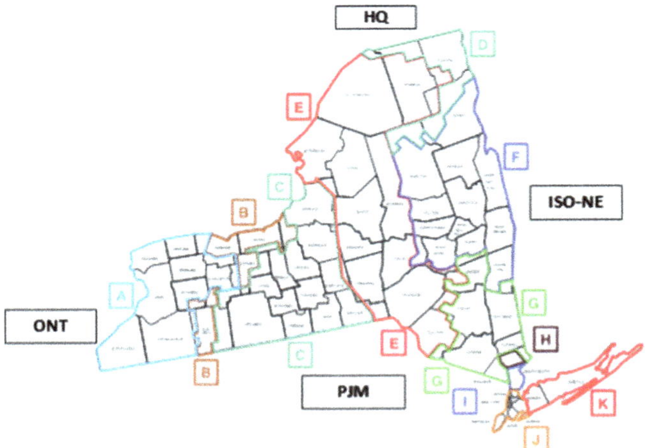


**Fig. 12.1** NYISO load zones (Source: NYISO)

**Table 12.1** Zones and import zone names in NYISO

| Zone | Zone name | Zone | Zone name |
|---|---|---|---|
| A | West | I | Dunwoodie |
| B | Genesee | J | New York City |
| C | Central | K | Long Island |
| D | North | HQ | Hydro Quebec |
| E | Mohawk Valley | PJM | PJM |
| F | Capital | IMO | Ontario |
| G | Hudson Valley | ISONE | New England |
| H | Millwood | | |

LMPs at numerous points across New York State's power grid, which has a complex interconnected topology. These LMPs include a congestion price component reflecting the impact of transmission constraints.

Under the NYISO market design generators are considered to generate at their bus, while loads are considered to consume in a load zone. The NYISO grid is divided into 11 load zones – labeled "A" to "K" as shown in Fig. 12.1 below – plus 4 import zones that are used to price imports and exports to and from the neighboring PJM and ISO-New England markets in the US and the Ontario (IESO) and Hydro Quebec (HQ) markets in Canada.

Prices are denoted in dollars per megawatt-hour. For example, a generator which produces 100 MW for an hour at a specific node x within Zone A will be paid 100 times the node x price for that hour while a load at a specific node z of 10 MW in Zone J will pay 10 times the local price for that hour (Table 12.1).

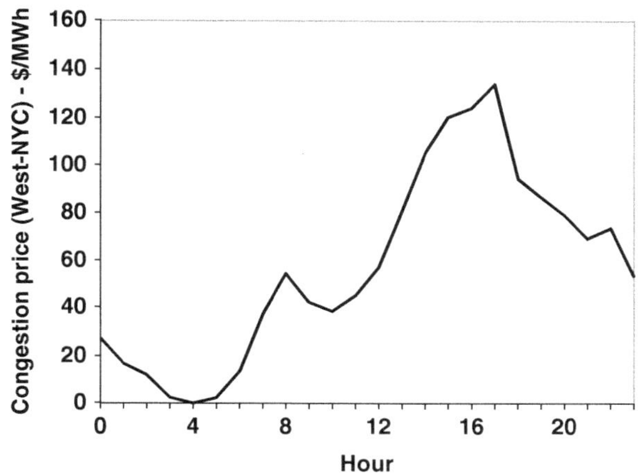


**Fig. 12.2** Example of NYISO congestion prices over a day (Source: Adamson et al. 2010)

## 12.3 New York Financial Transmission Rights Markets

Although this spot market pricing system is effective at addressing the realities of power flow on an interconnected grid, on its own it poses substantial financial risks for both generators and users of power. As can be seen in Fig. 12.2, there can be substantial congestion price volatility across a single day. This example shows the hourly congestion charge (per MWh) in each hour for a hypothetical bilateral transaction between the West Zone (Zone A) and New York City (Zone J) for 1 day in early July 2008.

Given the magnitude and volatility of congestion prices in an LMP market, a method is needed to hedge the price risks posed by spot power prices that vary from location to location and by hour. In response to this problem, Hogan (1992) proposed a system of financial hedging contracts designed to mitigate the component of this risk associated with congestion. These financial hedging contracts – fundamentally similar to financial swaps – pay the owner of the congestion contract the quantity (in MW) times the congestion price difference between a specified Point of Injection (PoI) and Point of Withdrawal (PoW) for each hour in the term of the contract. These FTRs are called "transmission congestion contracts" or "TCCs" in the NYISO lexicon; we will use the more standard "financial transmission rights" term in this chapter. In the NYISO markets, FTRs play the role that ordinary point-to-point transmission rights play in physical market designs, although in this case they act solely as financial swaps and have no direct effect on system operations.

For example, a monthly FTR might be defined with a PoI of Albany and a PoW of New York City. For each hour in the month, the FTR holder is paid the difference between the NYC and Albany congestion prices. FTR payments over an hour (or longer periods) can be negative – an FTR is an obligation to pay the sum of congestion price differences even if this sum is negative.

NYISO has conducted periodic FTR auctions since 1999. Market participants include utilities, marketers, generators and financial firms such as banks and hedge funds. In New York, FTRs have been sold for varying durations – ranging from 1 month to 2 years. As described above, a 1-month FTR is the right to hourly differences between congestion prices at two specified locations for the period of a calendar month. Since the FTR is defined as an obligation, and not an option, it may have a negative value, in which case a reverse auction is used to allocate it. Both positive and negative FTRs are allocated in the same auction. An auction of FTRs covering a month is conducted early in the preceding month, so that a FTR covering the month of November, for example, will be auctioned in early October.

## 12.4 Participation in New York FTR Auctions

NYISO publishes extensive data on its FTR auctions; this information specifically identifies the market participant that was awarded the FTR, the contract duration, the price paid, and the POI/POW pair that defines the FTR.[1] Note that the dataset identifies only FTRs awarded, but does not identify bidding firms that did not win in the auction.

In the first New York auctions, FTRs were generally of short duration, with a term of less than or equal to 6 months. In 2001 and 2002, more longer-term (e.g. 2 years) FTRs were offered, but this trend has since reversed and more recently 1 year and shorter FTRs have become the norm, as shown in Fig. 12.3.

The NYISO dataset also includes data on grandfathered FTRs. These FTRs were awarded to market participants in the early days of NYISO operations to replace pre-existing physical transmission rights in the grid, before market opening. Many New York utilities had such rights, some of which were of very long duration. Under the NYISO tariff, these holders of existing transmission rights had the option to convert them into FTRs and many did so. As these FTRs were not awarded in the auctions, and represented existing transmission rights in the grid, these have been excluded from our analysis.

Using the NYISO data, it is possible to examine trends in the number and POI/POW locations of FTRs awarded and to classify the market participants awarded FTRs. FTR market participants have been divided into five classes for this analysis:

- ***Utilities:*** This category includes New York investor-owned utilities, state agencies that serve loads in NYISO (such as the New York Power Authority) and a number of smaller municipal utilities. Out-of-state utilities acquiring FTRs in the NYISO auctions – which would typically be done by a competitive marketing group – are not included in this category.
- ***Generators/marketers:*** This category includes the major NYISO generators, out-of-state utilities selling power into NYISO, and the power marketing firms, many of which are part of combined generation/marketing firms.

[1] http://www.nyiso.com/public/about_nyiso/understanding_the_markets/financial_markets/. Accessed 17 Sept 2011.

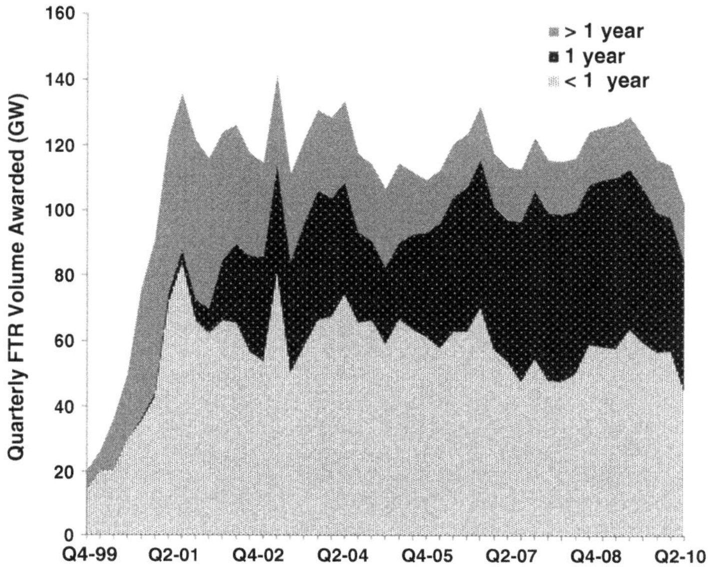


**Fig. 12.3** Quarterly FTR awards by contract duration (Source: NYISO)

- ***Retailers:*** The competitive retail sector in New York consists of firms which primarily market electricity directly to individual end-use customers.
- ***Banks:*** The major Wall Street investment banks, through various proprietary and commodity trading desks, are active in the NYSIO FTR markets.
- ***Funds:*** This category includes non-bank hedge funds and trading groups. Many of the funds most active in the NYISO market are specialized entities; some of which that focus almost entirely on the FTR markets in NYISO and other U.S. markets.

It is not possible, using this data, to classify neatly those FTRs acquired for "speculation" versus "hedging" purposes. Some generalizations, however, can be made. Utility and retailer FTR purchases, given the nature of these firms, have most likely been made to hedge congestion risk. For example, a New York City utility or retailer that had a purchase contract with a generator upstate, but had load obligations downstate, would be exposed to risk in the congestion component of LMPs; this could be hedged using FTRs. At the opposite extreme, hedge funds and other specialized trading groups generally do not have offsetting load exposures and their FTR purchases most likely represent allocations of purely speculative capital.

The FTRs purchased by generators/marketers and the bank trading desks cannot be classified a priori as being for hedging or speculative purposes. These entities both engage in speculative trading but also have extensive portfolios of power positions that FTRs can help to hedge. For example, an upstate generator could sell power under a contract to a downstate customer fixing the price at the customer's

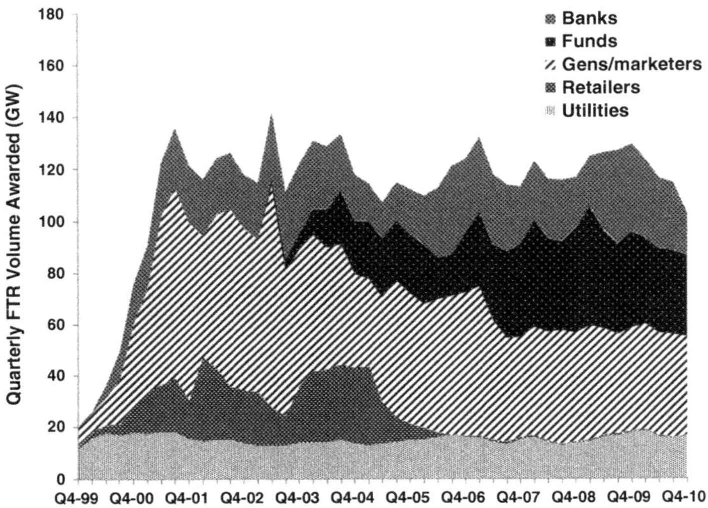


**Fig. 12.4** FTR awards by participant type and quarter (Source: NYISO)

location; an FTR could then be used to hedge the congestion component of the basis risk. Similarly, a bank trading desk may enter into a swap position with a customer in one zone but have some of the risk offset by a corresponding purchase in another zone. Again this risk could be managed using FTRs. Overall however, both the marketers and investment banks are known to allocate significant amounts of speculative capital to FTR trading and at least a significant fraction of these total volumes likely represent speculative transactions.

Figure 12.4 shows the total volume of FTRs awarded (in gigawatts) in NYISO by quarter, broken down by category of market participant. The volume of FTRs awarded by NYISO grew quickly in 2000 and 2001, and has remained largely stable ever since.

The primary trend apparent in Fig. 12.4 is the increasing importance of financial sector firms (banks and funds) over time. These two classes of market participants were of minimal significance in the early days of the NYISO FTR markets but now represent approximately half of all FTR volumes. Conversely, retailers were important in the 2000–2005 period, but are no longer significant FTR market participants, reflecting perhaps the state of the competitive retail market in New York. The share of FTRs awarded to utilities has remained relatively constant over the period.

The most congested major interfaces in the NYISO system are those that cross into the downstate New York City and Long Island zones (Zones J and K in Fig. 12.1). For FTRs with a POI or POW in Zones J and K, a similar pattern emerges in Fig. 12.5 in terms of market participation, with a somewhat higher share of financial sector FTRs awarded to funds in comparison to investment banks. Utilities received a larger share of these FTRs, reflecting perhaps their interest in hedging risks associated with power purchase contracts upstate.

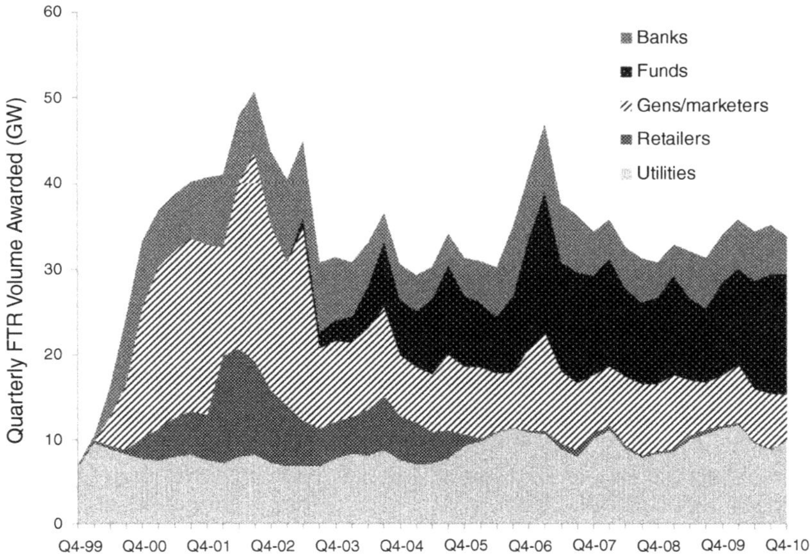


**Fig. 12.5** FTR awards involving New York City/Long Island zones (Source: NYISO)

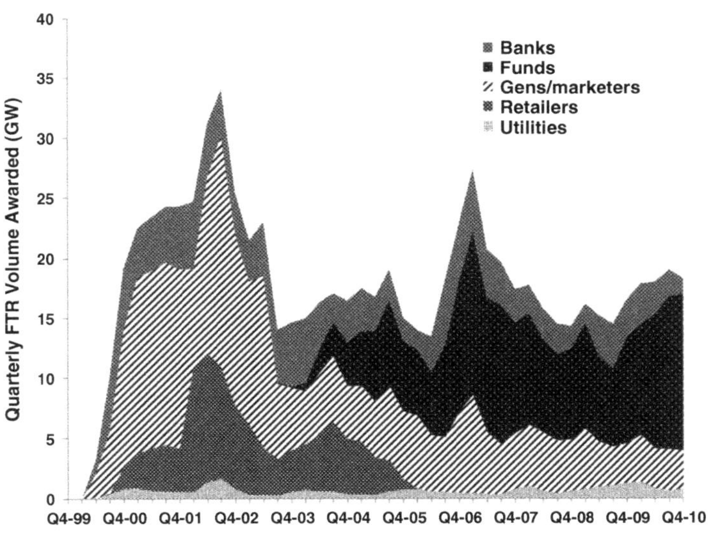


**Fig. 12.6** FTRs awarded solely within New York City/Long Island Zones (Source: NYISO)

The trend of increasing share of FTRs awarded to financial sector firms is even stronger for FTRs that have both a POI and POW within Zones J and K (New York City and Long Island), as shown in Fig. 12.6. Few of these FTRs have been acquired by utilities, and since 2007 the majority of FTRs within NYC/LI have been awarded to specialist funds. The investment banks have played a smaller role in this component of the FTR auctions. The funds' focus on zones J and K may not be surprising given that there appear to be participation and informational costs unique to the NYC/LI market that have prevented transaction profits from being eliminated (Adamson et al. 2010).

## 12.5 Efficiency of New York FTR Auctions

FTRs settled against day-ahead locational congestion prices allows congestion price risks to be hedged, while allowing the system operator to centrally commit and dispatch all generation units while meeting transmission security constraints. The FTR-based market design thus allows market participants to hedge price risk while allowing the system to maintain least cost unit commitment and dispatch.

In LMP-based markets, such as New York, longer-term contracts (including FTRs) are effectively financial hedges settled against spot prices. In the case of FTRs, these are spot congestion price differences. Examining the efficiency of FTR markets in an LMP-based design such as that of NYISO therefore provides some insights into the longer-term allocative efficiency of the whole market.

Several authors have examined FTR market efficiency, in many cases relying on NYISO data. An early analysis concluded that the NYISO FTR market was highly inefficient in its early operations, circa 2000–2001 (Siddiqui et al. 2005). Their analysis examined only four auctions in the early years of the market (and is hence based on only four independent data points). Adamson and Englander also suggested that NYISO FTR auctions were initially highly inefficient, although efficiency did improve somewhat over time (Adamson and Englander 2005).

A recent paper documented a significant divergence between spot and forward prices for 1-month TCCs in the NYISO in 2006 and 2007, finding that forward prices exceeded spot in 2006 and spot exceeded forward in 2007 (Hadsell and Shawky 2009). As the authors of the paper themselves point out, the dependence of realized congestion charges on large low frequency shocks (e.g., 2005s Hurricanes Katrina and Rita) makes estimating the expected profit from forward contracts using a short time series of observations problematic. Adamson, Noe, and Parker analyzed a much larger and richer data set and the results of their analysis are summarized below (Adamson et al. 2010).

From a more theoretical perspective, Deng, Oren, and Meliopoulos postulated that the inherent design of these FTR auctions, rather than limits on price discovery and information flows, may lead to inefficiency (Deng et al. 2004). In their model, FTR auction clearing prices will differ from expected FTR payoffs, even if bidders have perfect foresight, depending on the quantity of bids in the auction, due to the simultaneous feasibility constraints imposed in the FTR auction design.

Below, we discuss the econometric models Adamson et al. (2010) used to test hypotheses about FTR market efficiency and whether including a dynamic component in a learning model helps to improve the model fit. We then describe the data set they used to estimate model parameters and their summary statistics.

## 12.6 Econometric Models to Test Efficiency and Learning

Learning has been studied by economists, perhaps most famously, in the analysis of airplane manufacturing costs conducted by Wright (1936). Argote provides a comprehensive review of learning models and econometric specifications (Argote 1999). Most of this analysis has been performed in log-linear models with the underlying relation of a time variable to capture learning effects. However, in this case neither realized spot prices nor forward FTR prices need must be positive. Therefore, it is difficult to apply the standard log-linear learning framework to FTR markets. Thus, Adamson et al. (2010) analyzed two econometric specifications that do not require commitment to the unbiased forward rate hypothesis and do not require positive prices.

Their base model is the classic joint hypothesis test for bias and efficiency in a forward market (Engel 1996).

$$S_t = \beta_0 + \beta_1 F_t + \mu \tag{12.1}$$

$S_t$ is the spot price in period t (in this case, the sum of realized congestion rents), $F_t$ is the forward price for delivery in period t (in this case, the price paid for the FTR in the auction), and $\mu$ is an error term. If the market is efficient, then the intercept $\beta_0$ will not differ systematically from zero and the constant term $\beta_1$ will not differ systematically from one.

The second, dynamic model is specified as:

$$S_t = \beta_0 + \beta_{01}/(1 + t) + \beta_1 F_t + \beta_{11} F_t/(1 + t) + \mu \tag{12.2}$$

The dynamic model relates spot prices to forward prices through a constant linear relation ($\beta_{01}$) subject to diminishing bias over time. This model also allows the linear relation itself ($\beta_{11}$) to vary over time so that the model approaches a long-run equilibrium value. Learning is indicated by non-zero coefficients for these dynamic effects. The joint hypothesis test of H0: $\beta_0 = 0$ and $\beta_1 = 1$ can be used to examine the long run efficiency of the market.

## 12.7 New York FTR Auction Data

To test the base and dynamic models discussed above requires data on the forward FTR prices, and the realized spot congestion prices. This section describes the operations of the New York FTR markets in more detail and how forward and spot prices for FTRs are calculated.

Adamson et al. (2010) analyzed a large data set of all NYISO 1-month FTR auctions over the period from September 2000 through June 2006.[2] There were 2,250 unique PoI/PoW (source/sink) combinations in this data set, between both points and zones within the NYISO control area. Each set of monthly results often included prices for multiple contracts with the same source and sink zone.[3]

The spot congestion prices are subject to many of the same shocks and hence are not independent. Therefore robust regression models were used to verify model significance and correct standard errors (Huber 1964; White 1980).

Adamson et al. split their data set into four groups by contract type and geography. First, their data set was separated by "positive" FTRs – those for which a positive price was paid by the winning bidder in the auction – and "negative" or "counterflow" FTRs, where the auction price is negative.[4] The efficiency of positive and negative contract auctions was found to be quite different so analysis was done separately on positive and negatively priced contracts.

Adamson et al. also analyzed the New York City/Long Island region (Zones J and K in Fig. 12.1) separately from the others. Congestion within these two zones is qualitatively and quantitatively different from elsewhere in the NYISO, owing to the very high load and generation density of the transmission system within this region, especially during summer periods, and a complex pattern of voltage as well as thermal constraints creating transmission congestion.[5] Thus, the analysis was split into four major groups: (1) positive contracts not solely within zones J and K, (2) negative contracts not solely within zones J and K, (3) positive contracts solely within zones J and K, and (4) negative contracts solely within zones J and K.

Table 12.2 shows the summary statistics for the time series data divided into these four groups. FTR spot and forward prices are very fat tailed, with many more extreme observations than one would expect from a normal distribution with a similar variance (Corrado and Su 1996).

[2] These data sets includes Day-Ahead congestion prices, TCC auction bids and TCC auction results for over 9,000 FTRs as obtained from the NYISO website.

[3] The "source zone" is the zone in which the POI is located and "sink zone" is the zone in which the corresponding POW for the FTR is located.

[4] For a counterflow FTR, the winning bidder is paid to take the FTR but has the obligation to pay congestion rents to the TSO. Counterflow FTRs are sold in the same auctions as positive FTRs.

[5] Significant parts of the New York City transmission grid are operated to a higher reliability standard than the rest of the New York market: using an N-2 criterion rather than the usual N-1 standard (NYISO 2008).

**Table 12.2** Summary statistics for positive/negative FTRs by group

| | | | | | | |
|---|---|---|---|---|---|---|
| Group 1: Positive FTR contracts | | | | | | |
| *Crossing outside zones J & K* | *Mean* | *Std. dev.* | *Kurtosis* | *Min* | *Max* | *N* |
| Spot price (MW-month) | $626 | $1,887 | 28 | −$7,351 | $19,618 | 2,719 |
| Forward price (MW-month) | $653 | $1,693 | 34 | $0 | $22,520 | 2,719 |
| Spot – forward | −$28 | $1,360 | 44 | −$19,688 | $11,226 | 2,719 |
| Quantity (MW-month) | 26 | 66 | 139 | 0 | 1,160 | 2,719 |
| Transaction profit $ | $4,039 | $210,638 | 1,585 | −$2,807,776 | $9,568,143 | 2,719 |
| Group 2: Negative FTR contracts | | | | | | |
| *Crossing outside zones J & K* | *Mean* | *Std. dev.* | *Kurtosis* | *Min* | *Max* | *N* |
| Spot price (MW-month) | −$659 | $2,543 | 28 | −$19,894 | $3,703 | 2,992 |
| Forward price (MW-month) | −$808 | $2,415 | 25 | −$24,597 | $0 | 2,992 |
| Spot – forward | $148 | $1,489 | 46 | −$11,254 | $21,847 | 2,992 |
| Quantity (MW-month) | 26 | 60 | 126 | 0 | 1,147 | 2,992 |
| Transaction profit $ | $3,060 | $150,326 | 305 | −$1,658,063 | $4,343,408 | 2,992 |
| Group 3: Positive FTR contracts | | | | | | |
| *Solely within zones J & K* | *Mean* | *Std. dev.* | *Kurtosis* | *Min* | *Max* | *N* |
| Spot price (MW-month) | $1,706 | $3,221 | 20 | −$11,495 | $36,852 | 1,923 |
| Forward price (MW-month) | $1,061 | $1,484 | 12 | $0 | $12,500 | 1,923 |
| Spot – forward | $645 | $2,917 | 15 | −$11,882 | $29,358 | 1,923 |
| Quantity (MW-month) | 14 | 28 | 116 | 0 | 564 | 1,923 |
| Transaction profit $ | $10,962 | $78,746 | 228 | −$849,082 | $1,956,745 | 1,923 |
| Group 4: Negative FTR contracts | | | | | | |
| *Solely within zones J & K* | *Mean* | *Std. dev.* | *Kurtosis* | *Min* | *Max* | *N* |
| Spot price (MW-month) | −$1,076 | $2,961 | 14 | −$26,511 | $11,500 | 1,625 |
| Forward price (MW-month) | −$1,701 | $2,102 | 11 | −$21,889 | $0 | 1,625 |
| Spot – forward | $625 | $2,644 | 12 | −$19,565 | $21,776 | 1,625 |
| Quantity (MW-month) | 14 | 23 | 24 | 0 | 220 | 1,625 |
| Transaction profit $ | $4,489 | $84,511 | 128 | −$1,207,868 | $1,432,660 | 1,625 |

Source: Adamson et al. (2010)

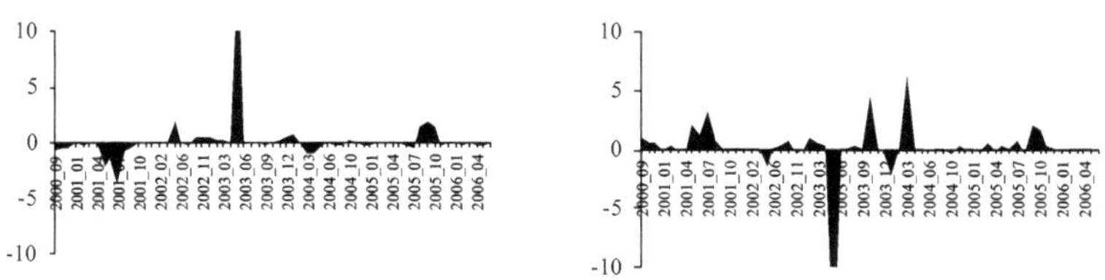

Group 1: Total Transaction Profits ($M) by month (Positive Contracts outside NYC/Long Island)

Group 2: Total Transaction Profits ($M) by month (Negative Contracts outside NYC/Long Island)

**Fig. 12.7** Total transaction profits by month ($M) for groups 1&2 (Source: Adamson et al. 2010)

Figure 12.7 presents transactions profits for contracts that cross outside the New York City/Long Island market for positive and negative contracts. Figure 12.8 details transactions profits for contracts solely within the New York market for both positive and negative contracts.

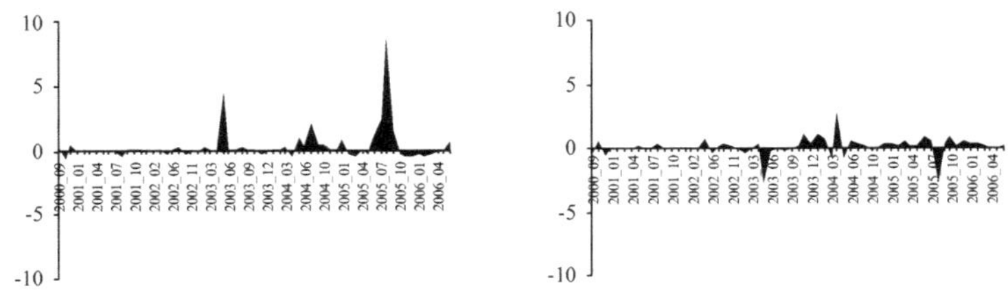


Group 3: TotalTransaction Profits ($M) by month (Positive Contracts within NYC/Long Island)
Group 4: Total Transaction Profits ($M) by month (Negative Contracts within NYC/Long Island)

**Fig. 12.8** Total transaction profits by month ($M) for groups 3&4 (Source: Adamson et al. 2010)

**Table 12.3** Regression results for base and dynamic models – groups 1&2

| | Group 1: Positive | | Group 2: Negative | |
|---|---|---|---|---|
| Model | Base | Dynamic | Base | Dynamic |
| Dep variable: | S | S | S | S |
| ($B_0$) Constant | 104.8*** | 94.6*** | 38.4 | 30.9 |
| | (30.9) | (24.7) | (22.4) | (20.2) |
| ($B_1$) Forward | 0.798*** | 0.919*** | 0.864*** | 0.944*** |
| | (0.057) | (0.042) | (0.041) | (0.036) |
| ($B_{01}$) 1/(1 + t) | | −471 | | 552 |
| | | (392) | | (296) |
| ($B_{11}$) Forward/(1 + t) | | −1.75*** | | −1.76*** |
| | | (0.20) | | (0.15) |
| Wald test | [6.93]** | [7.45]*** | [16.4]*** | [3.72]* |
| ($B_0 = 0$ & $B_1 = 1$) | | | | |
| N | 2,719 | 2,719 | 2,992 | 2,992 |
| Robust F statistic | 196 | 164 | 436 | 238 |
| $R^2$ | 0.510 | 0.560 | 0.674 | 0.701 |

* p<0.05, ** p,0.01, *** p<0.001. Source: Adamson et al. (2010)

The left hand panel of Fig. 12.7 shows that initially transaction profits were negative for positive contracts not entirely within New York City and Long Island. After this initial period of about 12.5 years, transactions profits were on average non-negative. The right hand panel of Fig. 12.7 presents transactions profits on negative contracts not entirely within New York City/Long Island. Early transaction profits were positive, followed by a final period in which transaction profits were small in absolute size.

The left hand panel of Fig. 12.8 depicts the transactions profit on positive contracts entirely inside the New York City/Long Island zones. Initially profits were small in absolute magnitude. However, toward the end of the sample period, very large positive profits were realized, the largest profit spike being associated with Hurricanes Katrina and Rita in 2005, which created major shocks in US natural gas markets and hence power prices. The right hand panel of Fig. 12.8 shows that for negative contracts entirely in the New York City/Long Island zones the absolute

**Table 12.4** Regression results for base and dynamic models – groups 3&4

| | Group 3: Positive | | Group 4: Negative | |
|---|---|---|---|---|
| Model | Base | Dynamic | Base | Dynamic |
| Dep variable: | S | S | S | S |
| ($B_0$) Constant | 725.4*** | 684.7*** | 116.5 | 206.3* |
| | (86.9) | (101.5) | (90.3) | (93.7) |
| ($B_1$) Forward | 0.924*** | 1.170*** | 0.701*** | 0.819*** |
| | (0.090) | (0.126) | (0.064) | (0.075) |
| ($B_{01}$) $1/(1 + t)$ | | 3,020** | | −4,199*** |
| | | (1,065) | | (737) |
| ($B_{11}$) Forward/$(1 + t)$ | | −9.73*** | | −5.36*** |
| | | (1.81) | | (1.41) |
| Wald test | [66]*** | [68]*** | [49]*** | [17]*** |
| ($B_0 = 0$ & $B_1 = 1$) | | | | |
| N | 1,923 | 1,923 | 1,625 | 1,625 |
| Robust F statistic | 105 | 48 | 119 | 45 |
| $R^2$ | 0.181 | 0.193 | 0.248 | 0.252 |

* p<0.05, ** p,0.01, *** p<0.001. Source: Adamson et al. (2010)

**Table 12.5** Expected long run spot – forward price differences

| $/MW-month | Group 1 | Group 2 | Group 3 | Group 4 |
|---|---|---|---|---|
| Mean forward price (in year 6) | $598 | −$960 | $1,131 | −$2,033 |
| Expected spot price in long run | $644 | −$876 | $2,007 | −$1,458 |
| Dynamic regression model standard error | $1,258 | $1,392 | $2,894 | $2,562 |
| Expected long run spot – forward price | $46 | $84 | $876 | $575 |

Source: Adamson et al. (2010)

variability of contract profit was smaller. On average profits were positive throughout the study period.

Tables 12.3 and 12.4 summarize results for the base and dynamic models for each of the groups.

The results in Table 12.4 below indicate that the market for contracts solely within the New York City/Long Island sample (zones J and K) was less efficient than that for contracts that are outside New York City/Long Island. For positive contracts, the constant ($\beta_0$) was significantly above zero for both the static and dynamic model. For negative contracts, the coefficient on forward price was significantly less than one, leading to high positive expected spot – forward price differences.

Table 12.5 presents expected spot – forward price differences (per MW-month) that are calculated using the parameters from the dynamic model. A "representative" contract price is modeled using the mean forward price seen in the last 12 months of the data set.

Expected spot – forward prices are positive for all four groups, but are much larger for contracts that are within the New York City and Long Island zones. However, the corresponding standard errors are much larger than the expected profits in all cases and are especially large for groups 3 and 4, indicating a high likelihood of negative profit on any given transaction.

The implication of this table is that expected profits from participating in the FTR market are positive, but are highly variable, indicating that many market participants realize negative returns.

## 12.8 Conclusions

This chapter has presented descriptive data on the entities that have participated in NYISO FTR auctions and how the efficiency of these auctions has changed over time. The analysis shows that direct load-serving entities such as utilities and competitive retailers have purchased a relatively small fraction of FTRs auction, although they may have benefitted indirectly from energy price hedges sold to them by generators and marketers (who were major FTR purchasers) in their load zone. The most noteworthy aspect of the participation analysis has been the rapid rise in importance of financial institutions (including bank trading desks and specialist funds) in the New York FTR markets.

The importance of these financial sector entities in the NYISO FTR markets is especially pronounced for FTRs with a POI and POW solely within the New York City and Long Island zones. This may reflect the fact that this market appears to be less efficient from an economic perspective and hence trading profits on average may be larger. We have previously hypothesized that the costly modeling systems and staff required to analyze this complex transmission system may limit the willingness of firms to participate given the overall small size of the market, helping preserve positive expected transaction profits over time.

From a broader market design perspective, the results of the analysis of Adamson et al. (2010) are encouraging. Confirming the results of earlier analyses, the initial efficiency of the FTR auctions was relatively low, although it improved quickly over time, consistent with rapid learning by market participants. This suggests that the overall forward-looking allocative efficiency of these FTR market designs is generally robust.

Analysis of FTR auction data should allow a range of other research questions to be addressed. The NYISO FTR market was one of the first to begin operations, but subsequently several others have started in the United States. It may be hypothesized that initial efficiency would be higher, or learning more rapid, in these later markets, given that many of the same firms participate. The rich level of firm-level data should allow hypotheses of firm entry and exit to be tested using FTR market data.

## References

Adamson S, Englander S (2005) Efficiency of New York transmission congestion contract auctions. In: Proceedings of the 38th Hawaiian international conference on system sciences, HICSS-38, Big Island

Adamson S, Noe T, Parker G (2010) Efficiency of financial transmission rights markets in centrally coordinated periodic auctions. Energy Econ 32:771–778

Argote L (1999) Organizational learning: creating, retaining and transferring knowledge. Kluwer, Norwell

Corrado CJ, Su T (1996) Skewness and Kurtosis in S&P 500 index returns implied by option prices. J Financ Res 19:175–192

Deng S, Oren S, Meliopoulos S (2004) The inherent inefficiency of the point-to-point congestion revenue right auction. In: Proceedings of the 37th Hawaii international conference on systems sciences, Big Island

Engel C (1996) The forward discount anomaly and the risk premium: a survey of recent evidence. J Empir Finance 3:123–192

Hadsell L, Shawky H (2009) Efficiency and profit in the NYISO transmission congestion contract market. Electr J 22:47–57

Hogan W (1992) Contract networks for electric power transmission. J Regul Econ 4:211–242

Huber P (1964) Robust estimation of a location parameter. Ann Math Stat 35:73–101

New York ISO (2008) Transmission and dispatching operation manual. Available from www.nyiso.com. Accessed October 23, 2012

SiddiquI A, Bartholomew E, Marnay C, Oren S (2005) On the efficiency of the New York independent system operator market for transmission congestion contracts. J Manag Finance 31:1–45

White H (1980) A heteroskedasticity-consistent covariance matrix estimator and a direct test for heteroskedasticity. Econometrica 48:817–838

Wright TP (1936) Learning curve. J Aeronaut Sci 3:122–128

# Chapter 13
# Experience with FTRs and Related Concepts in Australia and New Zealand

**E. Grant Read and Peter R. Jackson**

## 13.1 Introduction

As one of the first Full Nodal Pricing (FNP) electricity markets, New Zealand was also one of the first places where FTR concepts were developed and considered for implementation, actually as early as 1989. Ironically, though, it is only now, after more than two decades of discussion, that a limited FTR market seems likely to be actually implemented. This long delay may be partly attributed to failures in the regulatory process, but it also reflects the special circumstances facing the small hydro-dominated New Zealand market, in which a relatively small group of vertically integrated participants compete over a fairly sparse network, in which losses and reserve support requirements play a more important role than line transfer limits, per se. Thus there has been considerable debate over whether classical FTR concepts are really suitable. We discuss several variant proposals, one of which is moving toward implementation by 2012.

Although geographically close, the Australian market developed along very different lines from the New Zealand market, both before and after reform. The market design is zonal, not nodal, creating quite different hedging requirements for participants, and raising quite different design issues. Congestion still occurs in such a market, and still affects both dispatch and pricing outcomes, but the nature of that impact depends significantly on whether the constraints involved are inter-regional, intra-regional, or indeed trans-regional. Once again, variations on classical FTR concepts have been developed to deal with these situations, both inter-regional and intra-regional. We describe the simplified inter-regional hedging arrangement currently available, which employs a less precise mathematical representation of transmission system realities than has been normal elsewhere, and was designed to facilitate integration with financial markets. We also describe a generalised hedging

---

E.G. Read (✉) • P.R. Jackson
University of Canterbury, Private Bag 4800, Christchurch 8140, New Zealand
e-mail: Grant.Read@Canterbury.ac.nz

J. Rosellón and T. Kristiansen (eds.), *Financial Transmission Rights*,
Lecture Notes in Energy 7, DOI 10.1007/978-1-4471-4787-9_13,

framework that has been developed, but not implemented, to address that limitation, as well as deal with intra-regional and trans-regional hedging and contracting issues that arise in zonal markets.

## 13.2 New Zealand Experience to Date

The first "FTR" concept discussed for implementation in New Zealand was not based on the classic work of Hogan (1992). Read (1989) proposed a "shareholding" concept, defined on a line-by-line basis, with participants receiving proportional shares of the rents, rather than a fixed MW allocations. Thus, in some respects, it anticipated the financial version of "Flow Gate Right" (FGR) concept later proposed by Chao et al. (2000), and the "constraint based right" concept of Biggar (2006). In other respects it was similar to the proportional rental share bundles auctioned under the Australian SRA concept, discussed later.

This proposal was intended to reduce the potential distortion arising in Schweppe's nodal pricing proposal when the actions of a small group of users influence the price differential. And it was supposed to incentivise optimal transmission investment, by protecting those investors paying for a share of the line, as discussed by Read (1997a). Such proportional shareholdings have the advantage of always being revenue adequate, but they do not match the hedging requirement of a participant wanting to trade a fixed MW quantity of power across the network, who would have to identify and acquire a whole bundle of such rights, ideally covering all lines over which any of its trade might flow, in order to be fully hedged. Thus, while the shareholding concept could have been developed into an FGR like regime, it was only applied to simple situations involving specific assets, and has not featured strongly in subsequent discussions on meeting hedging requirements in the interconnected AC system.

Instead, the original design for the New Zealand electricity market, as described by Read (1997a), proposed the introduction of FTRs in the classic form of Hogan (1992). Implementation of that aspect of the market design was deferred, though, and subsequently become problematic. Read (1997b) notes that the sector had no regulator and, once the centralised publicly owned organisations were dismembered and/or privatised, there was no way of reaching consensus agreement on such matters, or enforcing any agreement that might be reached. In the early days, allocation of rents also seemed less urgent than other matters, partly because there was little congestion and significant rents seldom arose.[1]

In the interim, rents were simply allocated back to those parties paying for various parts of the transmission system. Parties paying for dedicated "connection assets" receive all the rents collected on them but the aggregate rent involved is small. Since they control the flow on those assets they are then indifferent as to whether the flow

[1] This lack of congestion is partly an illusion. In a small market, a few major users can effectively prevent a line reaching its upper flow limit, and without FTRs they are motivated to do so.

actually reaches its upper limit, or is controlled to some slightly lower level. Inter-island HVDC transmission costs are allocated to South Island generators, and they currently receive corresponding shares of the inter-island rents. This means that those generators are approximately hedged against inter-island price differentials arising in wet years, when they are exporting excess power to the North Island. But it also means that those generators, who are all vertically integrated generator/retailers are approximately hedged against inter-island price differentials arising in dry years, when they are importing power from the North Island to meet South Island load commitments. For the remainder of the system, though, transmission pricing gradually evolved away from the market-based regime proposed by Read (1997a) towards a situation in which most participants simply pay a "postage stamp" charge covering their share of the "inter-connected" intra-island AC transmission system costs, based on a measure of peak load. Consequently, rents are also implicitly allocated in proportion to peak load.

While, of itself, allocating rents in that way might be considered distortionary, it may also be seen as reducing any distortion inherent in the allocation of transmission costs. The parties paying transmission charges are either major users, or distribution businesses, who generally pass the rents through to retailers operating in their distribution areas. The formulae used differ by area, and a few smaller distribution companies do not pass rents through at all. But, in principle, this provides all loads, everywhere, with a proportional share of the rent an all AC system lines. And that effectively provides a proportional hedge (roughly an FTR) from a notional island Generation Weighted Average Price (GWAP) hub to a notional island Load Weighted Average Price (LWAP) hub. Thus nationally diversified retailers, serving a proportional share of load at each node, are effectively as fully hedged against intra-island price differentials as they can be, given network capacity. Meanwhile, smaller regional participants effectively hedge their position physically, by developing retail markets close to their generation, and vice versa. Thus many parties see little to be gained by more formal hedging arrangements.

When Transpower (2002) proposed, and developed the software for, a comprehensive FTR market it was strongly opposed by most of the industry. In part, that opposition stemmed from parties resisting the loss of revenue streams to which they had become accustomed. Some participants saw little to gain, and others could not be expected to welcome any regime that allowed competitors to access their "captive" local markets more easily. But most participants also argued that the proposal was far more complex than the situation demanded, and would impose unjustifiable costs on parties who would be obliged to interface with the market, disadvantaging smaller participants in particular. The work reported in Pritchard and Philpott (2005) also raised fears that parties in a position to exercise market power in the spot market, and thus effectively control the ultimate payout of FTRs, would be able to play more complex strategic games in the FTR market, and potentially purchase FTRs that actually strengthened market power in some retail markets. Participants interested in penetrating regional markets where they might be exposed to incumbent market power were concerned by this, and further concerned that Transpower's proposal to only offer FTRs a year or two ahead would not provide sufficient protection to justify developing a long term regional retail position, or industrial supply contract. Thus many parties felt that some kind of regionalised variation on the existing rental allocation would be simpler, cheaper to implement, less prone to manipulation, and might actually serve their hedging requirements better.

Responding to these concerns, Read (2002) proposed a hybrid regime. This attempted to meet long term concerns over retail competition by making an automatic allocation of shares in regional rental streams to loads, via whatever entities were serving those loads. But it also proposed that those implicit hedging positions be treated as a form of "Auction Revenue Right" (ARR), as discussed by Sarkar and Khaparde (2008), and employed in the PJM and New England ISO markets, for example. Thus they would be effectively convertible into explicit FTRs, to be traded in a formal market perhaps up to a year ahead. Those recommendations were endorsed by the Government, which issued a policy statement to the then newly formed Electricity Commission, but little progress was actually made until quite recently.

## 13.3 Current New Zealand Developments

In recent years the locational hedging situation has received more attention, partly because congestion rents have increased. But there has still been significant debate as to whether the expense of establishing locational hedging arrangements was really justified and, if so, what the arrangements should be. As noted above, many parties already have implicit hedging arrangements, which suit their circumstances well enough. There was concern, though, that, while major cities were reasonably well served by competing retailers, the lack of any formal inter-regional hedging made it difficult for North Island generators to compete in the South Island, and some smaller regional markets had quite limited retail choice. So the focus has been on the potential to increase competition by reducing the risk for out-of-region retailers to operate in regional markets.

Potential economic gains have been reduced, though, by recent moves to re-allocate assets between state-owned generators which have increased competition in the South Island, and the potential gain from better competition in smaller regional markets is not all that large, either. Nor has there been a clear consensus that FTRs, per se, are the best way of dealing with the problem. Some regard them as just too expensive to implement, and too complex for smaller niche participants to deal with. Others still consider that FTRs may actually worsen incumbent market power in smaller regional markets, or even at the inter-island level.

For all these reasons, significant effort was expended on developing an alternative "Locational Rental Allocation" (LRA) regime that would automatically assign rents to loads in a way that mimicked the likely outcome of an efficient FTR market regime. The preferred regime, as described by Read (2009), was in fact a hybrid FTR/LRA regime that would effectively hedge all loads in each island to that island's Generation Weighted Average (GWAP) hub prices, with inter-island FTRs traded between GWAP hubs. By construction (in a lossless system), the intra-island rents are just sufficient to hedge between GWAP and the Load Weighted Average Price (LWAP), and hence to hedge all loads to GWAP. The SPD market-clearing engine described by Alvey et al. (1998) includes a piece-wise linear representation of line losses, implying a hedgeable loss rental component, and a non-hedgeable loss cost component.[2] Since many parties argued that the LRA should not cover loss-induced differentials at all (and in order to facilitate analytical comparisons) the LRA regime was formulated in terms of explicitly allocating congestion rents on binding constraints, using participation factors determined by converting MCE constraints into the generic form of (13.1) below.

As Read (2009) notes, simply assigning rents in exact proportion to real time loads would actually undo the effect of locational marginal pricing, effectively creating a regional pricing regime for loads.[3] Some parties have actually supported that approach, moving the New Zealand market design closer to that in Australia, or more exactly Singapore, where generators face nodal prices, but loads face a uniform price across the whole island. That would obviously remove any intra-regional hedging problem for loads, and allow generators to trade hedges at the GWAP hubs. But the general consensus is that regional pricing would unacceptably compromise efficient economic signalling. Thus it was proposed to base locational rental allocations on some measure other than the actual load in the trading interval, such as historic average load share for similar periods. This implies some compromise to hedging effectiveness, but places a load that is allocated rents for a fixed MW quantity in much the same situation as a load that is holding an FTR for the same quantity.[4] Thus it largely preserves the locational signalling advantages of nodal pricing, for operational purposes, because any incremental consumption faces the full nodal price.[5] Long run locational signalling would still be compromised because, on average, the LRA regime would still hedge loads to the chosen reference hub price, in this case GWAP, without requiring any payment for that hedge.[6] But this was only considered

---

[2] The residual market settlement surplus must be all loss rents, and the residual nodal price differences must be loss-induced. Since the piece-wise linearization represents an underlying quadratic loss function, these two components are approximately equal, so the remaining surplus should cover about half the loss-induced price differences.

[3] If the local price is $P_n$, the effective price is just $P_n - (P_n - GWAP) = GWAP$ as in the IDMA representation of a zonal market, discussed in Sect. 13.6 below.

[4] Efficient signalling is still compromised to the extent that increasing consumption in one period increases rental allocations in future periods, though. Consideration was given to excluding periods in which congestion occurred from the historical load calculation, but this makes little difference if congestion is infrequent. Distortion of pervasive loss-induced differentials is more difficult to deal with, and some versions of this proposal excluded them entirely.

[5] Read also notes that these dynamic signals are of limited relevance to small users, who do not actually face spot prices, and argues that LRA actually improves on the status quo for large loads. With no locational hedging available, they face a disproportionate second order signal to avoid causing congestion, since the resultant high local prices would apply to their entire load. They are thus incentivised to reduce consumption in favour of smaller loads which, being oblivious of the second order considerations, effectively see the nodal price as a pure SRMC signal.

[6] By way of contrast, if participants have to purchase FTRs they effectively pay the average locational price differentials, in the FTR purchase price.

to be a major problem for new large electricity-intensive loads, which seem unlikely for the foreseeable future.

In theory, it may be argued that a comprehensive FTR regime might produce a better outcome, in terms of both hedging and economic signalling. But the advantage is not actually clear, even apart from issues of complexity and expense, or concerns about FTR market gaming. FTRs are perfectly suited to hedging the position of parties wishing to trade a fixed MW amount for a fixed period from one node to another. But very little of the power sold by New Zealand retailers is traded on that basis. To compete in regional markets, a stand-alone retailer faces the hedging problem of matching a continuously variable load pattern, across many nodes, with a set of energy hedges, probably bought at a major trading hub. But a typical New Zealand retailer also has some generation plant, and might well be interested in FTRs from its generation nodes to a trading hub. Much of that capacity is hydro, though, with highly variable output, so the pattern of FTRs required would be constantly changing. The implied hedging requirements would be difficult to match well unless FTRs were traded for quite short time intervals, and no small retailer has the resources to deal with that kind of complexity. Conversely, it seems easy to devise allocation formulae that match hedge quantities to loads at least as closely as seems likely under a regime in which FTRs might only be traded in monthly blocks, and between only a few participants.

Thus many participants felt that their needs would actually be better met by the LRA regime than by even a comprehensive FTR regime. That was not a uniform consensus, though. Some parties would prefer to wait for a more comprehensive intra-island FTR regime, and Read (2009) points out that the LRA regime is not without its own problems. First, it could create even worse localised market power problems than an FTR market by automatically allocating FTR-like rental streams to incumbents who may already have market power. Second, existing regional participants are rightly concerned that they would face major risks if, having based their business on locating generation near load, their load was now hedged to an island hub while their generation was not. These are not reasons for rejecting an LRA based design, per se because both problems could be overcome by applying LRA to net, rather than gross loads. Equivalently, LRA could be applied to generation as well as consumption. This would focus energy trading on the LRA hubs, and could provide an effective compromise between the FNP/FTR paradigm, and the zonal paradigm employed in Australia, for example. But there would be significant cost in switching to such a market design, now that participants have established market positions based on the status quo. Thus, while development of intra-island hedging is still on the agenda, the LRA proposal is not being implemented at this time.

What is being implemented, though, is the inter-island FTR component of the hybrid proposal, which most (but not all) agree will be beneficial. NZEA (2010) discusses the background, and describes the original proposal, which has since been modified by NZEA (2011a). Originally it was planned to hedge between island GWAP hubs, because these would have desirable properties in terms of long term stability, and form a suitable reference point for any future LRA regime.[7] Thus encouraging trading at those hub prices seemed desirable. However it has been decided to use the existing trading hubs at the major North Island load centre of Auckland (Otahuhu, or OTA), and the major South Island generation centre in the Waitaki valley (Benmore, or BEN). So the planned "inter-island" FTR will not just hedge price differentials across the inter-island HVDC link, but also across much of the North Island AC system. Even this modest development is not without its conceptual difficulties and debates, though.

First, this is a hydro dominated system, with strong South–north flows in a wet year, but similarly strong North–south flows in a dry year. This "tidal flow" situation also implies that a conventional "obligation-inclusive" FTR will not fit the hedging requirements of many parties. In a wet year, a typical South Island-based generator-retailer will expect to be exporting power to the North Island, and want to hedge that trade. In a dry year, the same party may expect to be importing power from the North Island, and want to hedge that trade. Since inflow fluctuations are relatively unpredictable in New Zealand, and reservoirs are relatively small, the situation can change rapidly, and a party wanting to hedge more than a few months ahead is likely to want to cover both situations. But that cannot be done using any combination of obligation-inclusive FTRs. By buying an obligation-inclusive South–north FTR, a South Island generator would implicitly be committing to have that amount of power available to send north, even in a dry year, and that is exactly the kind of commitment that a hydro generator cannot afford to take on. Aversion to downside risk may well imply that even a net exporter would offer a negative price for export FTRs on that basis. Thus it has been decided that the market will include option FTRs from the beginning.

Second, there has been significant debate as to whether hedging should only cover the congestion component of differentials or also include either the loss rental component, or the entire inter-hub price differential (including loss costs as well). While some have argued that loss costs and/or rentals are relatively small and stable, and need not be covered, this is not actually the case. Hume (2009) reports that average transmission system loses are only 3.7 %, but Fig. 9 in that paper implies instantaneous marginal losses as high as 45 %, from one end of the transmission system to the other, and 35 % between OTA and BEN, with the direction of the differential reversing due to tidal flows. Monthly average loss differentials are lower, but loss-induced price differentials also reflect the fact that losses must be (implicitly) bought in from a highly volatile spot market, as may be seen from Fig. 13.1 in NZEA (2011b).

Figure 13.1 above shows that, while the average loss-induced price differential may be only \$1.28/MWh, from south to north, it swings from +\$12/MWh down to −\$58/MWh over the period sampled, and accounts for a significant proportion of the volatility in inter-island price differentials. After much debate, it has been decided that inter-island FTRs will cover the full price differential, including the loss cost and

[7] By construction (in a lossless system), the intra-island rents are just sufficient to hedge between GWAP and the Load Weighted Average Price (LWAP), and hence to hedge all loads to GWAP.

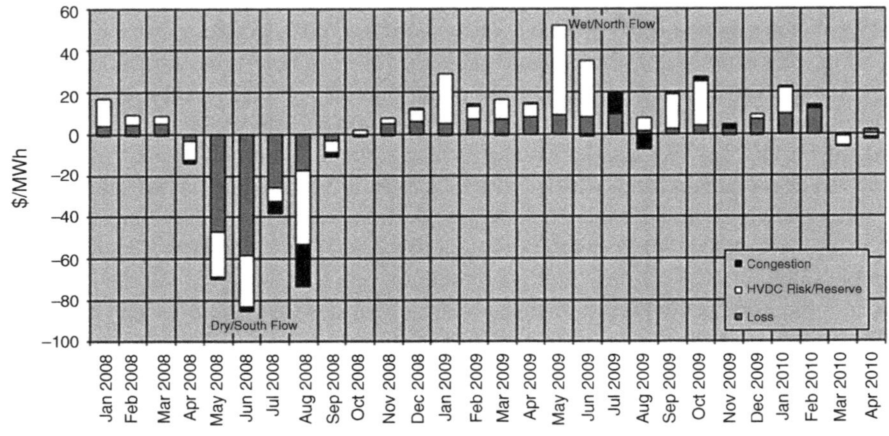


**Fig. 13.1** OTA-BEN price differential components, by month (Source: NZEA 2011b)

rental components.[8] This allows an obligation-inclusive FTR to be expressed as a symmetrical "swap" contract similar to those which participants might enter into between themselves, thus hopefully facilitating integration with the energy hedge market trading regime, and boosting its liquidity.

Third, inter-island HVDC flows are often not constrained by a lack of link capacity, per se, but by a relative shortage of (economic) instantaneous reserve to cover link failure, in the receiving island. This is a major issue in a small system. At low to moderate transfer levels the HVDC can normally cover its own reserve requirements, by flows "failing-over" from one pole to the other,[9] leaving a residual requirement that is often less than the size of the largest unit operating in the receiving island. Beyond that "self support" transfer level, though, the energy/reserve co-optimisation model described by Read (2010) implies that the inter-island price differential will be set by the marginal cost of reserve provision, that is the reserve market clearing price, plus marginal losses, even when no "congestion" occurs. Figure 13.1 shows that this accounts for a major part of the risk faced by inter-island traders, and cannot be ignored by the hedging regime.

Covering price differentials due to losses and reserve support requirements has major implications for FTR revenue adequacy. In both cases, there is not really a binding capacity limit on which rents are generated, but rather an interaction between a supply curve for loss/reserve, and a demand curve implied by the inter-island difference in energy price offers. In both cases, the rent will be less than that required to

---

[8] Note that alternatives, such as covering only congestion costs, or loss rents but not loss costs, would require us to calculate what those components actually were. But this could be done, as for the LRA proposal.

[9] The high inter-island differentials shown in Fig. 13.1 are partly due to this facility being unavailable, with one pole out of service. Commissioning of a replacement will restore that facility, but also greatly increase capacity. So, while inter-island flow limits will occur less often, the proportion caused by congestion will most likely fall too.

hedge the observed flow, because loss and/or reserve support costs for that flow must be covered. The loss cost is well defined by the need to buy energy to cover a piece-wise linear loss curve, but the reserve support situation is less clear. The rent generated by the inter-island price differential is not attributable to any constraint in the transmission system, and not necessarily available for hedging trades across the transmission system. With energy/reserve co-optimisation all reserve suppliers are paid the market clearing price. If every MW of transfer, above the self support level, required a MW of reserve support to be paid for from the inter-island settlement account, then market participants would logically be looking to the reserve suppliers (rather than to the transmission system owner) to provide hedging for those flows, out of the rents implied by the difference between the market clearing reserve price, and the offered supply curve (or actual cost of provision.)[10]

In practice, the situation is much less severe. First, there will be no reserve–induced price differential so long as flows are below the self-support level. Second, the inter-island transfer account should at least be assigned the rents corresponding to its self-support level when flows do exceed it. Third, the costs of reserve provision are not all charged to the party setting the reserve requirement, but shared between all parties whose potential breakdown is covered by the reserve. In any case, it could be argued that the cost of HVDC reserve support should be recovered from the industry in the same way as other transmission system costs. On that basis, the current proposal, actually avoids the problem of paying for reserve support by not deducting them from the inter-island settlement surplus.

A revenue adequacy issue remains, though, because the market settlement surplus will not cover actual loss costs. Perhaps more importantly, loss/reserve effects imply price differentials in every period, so some payout will always be required on the full FTR volume, even when flows are below that volume, and rents are only generated on that lower flow.[11] The fact that these factors increase average payout requirements is not really a problem, because the extra premium that participants are expected to pay for loss/reserve-inclusive FTRs should more than cover the extra payments required, on average. FTR purchasers are effectively pre-purchasing a volatile stream of loss/reserve-induced energy costs that, if the FTRs did not cover them, would eventually have to be made up from the spot market. The FTR issuer does face increased risk, though.

Read and Miller (2011) argue that the FTR provider could cover its risk by using the auction proceeds to contract forward for losses and/or reserve support to cover the FTR volume issued, while Philpott (2011) suggests using a more general set of "unbalanced FTRs" (including loss contracts). If contracts for both loss and reserve support were bid into the FTR market-clearing auction, we would get a trade-off, now, between the forward demand for FTRs and loss/reserve support contracts, mirroring the future spot

[10] This situation differs from that for losses, where rents implied by the piece-wise linear "loss requirement curve" remain in the market settlement surplus.

[11] By way of contrast, the classic revenue adequacy result rests on the assumption that price differentials only arise when a constraint binds, at which time it will generate enough rent to cover any FTR up to the binding limit.

market trade-off between the demand for inter-island transfer and the cost of losses and reserve support. Studies conducted for the NZEA have confirmed that such contracts could increase the volume of FTRs that can be offered without undue risk of scaling due to revenue adequacy problems. Concerns have been raised about having the FTR manager take an active role in other hedging markets, though. So, initially, it is planned to scale back FTR offerings to levels at which revenue adequacy should not normally be a problem, and then also scale back FTR payouts if necessary.

Finally, the issue arises as to what rent can legitimately be taken to support FTRs that are only available between a limited set of hubs. Of course, standard revenue adequacy results also apply to this limited FTR offering, and it has been argued that all network rents should be left in the FTR support pool, in order to maximise FTR firmness. But it is also clear that line capacity over which flows between the FTR hubs cannot pass could be removed from the network without affecting the FTR revenue adequacy test. It is not obvious why rents generated on the South Island AC system, for example, should be used to support an inter-island hedging product for trades that do not utilise any part of the South Island AC network. These rents also provide an approximate form of hedging with respect to intra-regional price differences, in their current form, and would be required to form the basis of any future intra-regional hedging regime. Thus, while all parties will receive what are effectively ARRs corresponding to their share of the rents used to support FTRs, some parties currently receiving rents generated on capacity not required to support FTRs wish to continue receiving those rents, pending further intra-regional hedging developments.

Thus only rents generated on what might be termed the "FTR support grid" will be available to support FTRs, with the remainder being passed though to existing recipients, as described by Miller and Read (2011). The FTR support grid does not consist of a set of assets but, for each line, we take the rent generated by flows lying between the maximum flows implied by possible inter-hub flow patterns in the forward and reverse directions. Congestion rents are only generated when the line is fully utilised in the actual dispatch, and the FTR support rent is taken to be the FTR support capacity times the shadow price on line capacity. A similar calculation is performed for "security constraints" involving multiple line flow variables. The calculation is extended to apportion the loss rent implicit in the market settlement surplus between inter-island and intra-island pools. That loss rental is generated by the shadow prices on the loss tranches of the piece-wise linear representation. The market clearing software does not report those shadow prices, but they can be inferred from the solution, because the marginal cost of having to utilise a higher loss tranche is just the cost of buying in more power at nodal energy prices. Thus it is possible to perform a line by line calculation of loss rents generated, and to partition them using the same set of extreme FTR flows used to partition congestion rents.[12]

[12] Binding line limits and (on a much smaller scale) loss tranche limits both generate similar pricing effects, inducing (positive or negative) rents to be collected on all lines in all loops in which that line is involved. But what matters, for revenue adequacy of the inter-island FTR, is to partition rents according to flows on lines where rent is generated, not where it is collected.

While NZEA (2011a) describes the Code changes required to implement the arrangements above, the Code focuses more on securing the rental streams required to develop the market, than on the market design itself. Many details remain to be determined by the FTR market manager, in consultation with the Electricity Authority and participants. Thus the final shape of the market may depend significantly on the choice of FTR manager. To some, it seems natural that the FTR market manager should be associated with the transmission system owner and/or system operator, since they are experts on the capabilities of the transmission system. To others, it seems more natural for the FTR market manager to be associated with an existing operator of hedge markets, since they are experts on the design of financial products, and better placed to integrate FTRs into the overall financial market framework.[13] Whatever manager is chosen, though, the market will initially provide Southward and Northward obligation-inclusive and option FTRs, in monthly blocks, over a time horizon progressively extending out to 2 years. Further developments could include adding further hubs, reserve/loss support contracting,[14] facilitation of secondary trading, or differentiation of hedge products by degree of "balance" (as suggested by Philpott (2011)) or target "firmness" (as suggested by Read and Miller (2011)). But such developments are contingent on the success of the initial market arrangements, which has yet to be demonstrated.

## 13.4 The Australian Market Context

"The Australian National Electricity Market" (NEM) was initially operated by the National Electricity Market Management Company (NEMMCO), whose operations were subsequently merged into the Australian Electricity Market Operator (AEMO). The Australian Electricity Market Commission (AEMC) is responsible for policy development. Although developed soon after, using a version of the same software, the Australian market has quite a different structure from that in New Zealand, being organised into regions corresponding to the States. The reasons for this are not just historical or political. Each state is centred on a major population centre, and there has historically been little development, and hence network complexity, close to state

[13] This is an important issue in a small market, where fragmentation of trading platforms increases the difficulty of achieving desirable liquidity on any one platform. Conversely, Fig. 13.1 suggests that accurate modeling of possible congestion limits probably has less practical impact on revenue adequacy than dealing with the loss and reserve costs issues. Read and Miller (2011) point out that, for a small number of hubs, the FTR feasible region could be represented as the set of all possible convex combinations of the extreme inter-hub flow patterns used to determine the rents available for FTR support. (For the initial two hubs, this only involves the maximum forward/reverse flows between them). Given a set of extreme flows determined by the System Operator, the FTR manager could actually clear the FTR market without any direct knowledge of the transmission system at all.

[14] Possibly via an integrated auction somewhat similar to that proposed by O'Neill et al. (2002).

borders.[15] The NEM initially covered the states of South Australia (SA), Victoria (VIC), New South Wales (NSW) and Queensland (QLD), although the last was not physically interconnected for some years after market start. Tasmania (TAS) subsequently joined the NEM, once it was connected to the mainland by HVDC cable.

Market prices are calculated every 5 min, but summed to form half hourly prices. All participants in a particular region face essentially the same "regional reference price", which is calculated as the marginal cost of supply to the major regional load centre, plus or minus a node-specific intra-regional loss factor, which is fixed annually. Commercially, the NEM operates as if there were "notional interconnectors" linking these regional reference nodes into a "tree" structure, with no loops.[16] A piece-wise linear loss function is calculated for each notional interconnector, by varying injection/extraction at its source/sink under typical conditions.[17]

The physical transmission network does not exactly match the structure assumed for commercial trading purposes, though. Thus NEMDE[18] imposes constraints to ensure that generator dispatch is actually feasible, given the loads and network capacity available. But NEMDE cannot impose constraints on line flow variables, because it does not contain a nodal model of the network, and hence of inter-nodal flows. Instead, off-line studies have been used to create a large library containing several thousand constraints that might need to be applied under particular load and network conditions. In canonical LP form, the "Left Hand Side" (LHS) of each constraint is an algebraic sum of interconnector flow and generator output terms, each weighted by what we will call a "Constraint Participation Factor" (*CPF*), while the RHS is a constant determined by the load/network conditions at the time.[19]

$$\sum_{i \in Variables_k} CPF_{ik}{}^* x_i \quad \leq RHS_k = - \sum_{i \in Constants_k} CPF_{ik}{}^* x_i \tag{13.1}$$

These constraints are thus like the "generic" or "security" constraints often overlaid on the basic network constraint structure in nodal models such as that of Alvey et al. (1998). Initially, NEMDE constraints were expressed in a variety of ways, often inherited from pre-market regional system operators. But CRA (2003a) showed that, while a wide variety of constraint forms may achieve the same physical dispatch

[15] One major exception relates to the Snowy Mountains hydro-electric development, which lies in New South Wales, but close to the Victorian border, and which until recently formed a region of its own (SNY).

[16] South Australia, Victoria, New South Wales and Queensland form a chain, with the island of Tasmania linked to Victoria.

[17] Thus zero cross-border flow does not generally imply minimum losses, or zero marginal loss.

[18] NEMDE is now the NEM market clearing engine, replacing a version of SPD.

[19] We have expressed that RHS constant as a linear combination of terms, and multiplied by −1, so as to facilitate discussion of a general pricing/hedging framework in Sect. 13.6.

outcome, the correct pricing outcome will only result if all constraints are consistently "oriented" toward regional reference nodes.[20] Thus, constraints derived in other ways have been progressively "re-oriented".[21] Once expressed in this form, these constraints play a fundamental role in providing a consistent theoretical framework for both intra-regional "access" and inter-regional hedging in a zonal market. Before discussing that framework, though, we describe current hedging arrangements in Australia.

## 13.5 The Current Australian Locational Hedging Framework

In a sense, intra-regional hedging is a non-issue in a zonal market. Provided a generator is dispatched, it faces no price risk if contracting to sell at its own regional reference price, because it will be paid that price, adjusted by a known loss factor. This is most often the case, because regional networks have been developed in such a way as to make intra-regional congestion relatively rare. When it does occur, though, it is obviously not possible to dispatch all generators that wish to be dispatched at the regional reference price. Thus they face a volume risk that may be described as creating a "firm access problem". Despite years of debate, work to resolve that problem is still ongoing. But some of the approaches to resolve it are discussed below because they effectively amount to exposing the participants involved to a form of nodal pricing, and issuing them with instruments which, in combination with the regional pricing arrangements, are effectively FTRs from their location to their regional reference node.

The basic theoretical framework of inter-regional hedging in Australia was established in a series of papers prepared by Putnam Hayes and Bartlett, including PHB (1997). As an independent market/system operator, with no transmission asset base, NEMMCO did not consider it appropriate to enter the business of issuing, or operating markets in, financial instruments of any kind, including FTRs. Conversely, by that stage, financial intermediaries were already developing some expertise in creating financial instruments, or deals, to meet the hedging needs of participants. Without access to the market settlements surplus, those deals were partially under-pinned by offsetting swaps by parties wishing to trade in opposite directions, but also involved some risk-taking by the issuing parties themselves. As a result, some parties actually opposed the introduction of an FTR-type regime, because they believed that the evidence showed that the sector's inter-regional hedging requirements could be met without access to the market settlements surplus. This theme has re-surfaced in recent New Zealand debates, and it has been an ongoing challenge to convince some

[20] In brief, this means that the participation factors referred to above must correspond to the increase in the constraint LHS (e.g. the flow over a constrained line) if a notional 1 MW flow were sent from the generator in question to the regional reference node. For simple line flow limits, these *CPF*s are just PTDF's using the regional reference node as "swing bus".

[21] This can be done by using regional energy balance equation(s) to substitute out for injection at the regional reference node(s).

that, without such access, a swap market must either be perfectly balanced (thus supporting a net inter-regional transfer of zero), or expose issuers to significant risk.

Having established that point, though, the design philosophy was not to "establish an FTR market", per se, but to make the "Inter-Regional Settlements Residue" (IRSR)[22] available to competing providers of inter-regional hedging products. It was deemed inappropriate, in this relatively small national market, to divert liquidity away from existing financial markets, or to create a competing "financial institution". So the goal was to foster liquidity in existing markets, and allow existing financial institutions to use their financial market expertise to provide participants with integrated hedging products, based on the IRSR, swaps or other instruments, to best cover the risks they faced in those markets.

To operationalize hedging based on the IRSR, proportional bundles of inter-regional rents are defined, such as "x% of the rents from NSW to Victoria, when flow is in that direction" and auctioned as non-firm units in the Settlement Residue Auction (SRA). The rents are determined by subtracting the value of export flows, at the exporting region's reference price, from the value of import at the importing region's reference price. Import/export flows are calculated from modeled interconnector flows, using (annually) fixed proportions to apportion interconnector losses to each regional reference node. Thus the calculation accounts for the cost of interconnector losses, as well as the pricing impact of marginal losses, and all constraints affecting inter-regional price differentials. Inasmuch as they are proportional, these IRSR bundles are somewhat similar to the "shareholdings" of Read (1989), or the Auction Revenue Rights available in some US markets. If the transmission system really did match the NEM market design, with no loops linking regions, they would also be similar to (financial) "Flow Gate Rights" (FGRs) of the type discussed by Chao et al. (2000). In reality, though, two complications arise.

First, the Australian market allows Market Network Service Providers (MNSPs) to develop and operate "controllable" links whose capacity is offered into the market to transport power between regions, at a price. Such MNSPs will operate when (after adjustment for MNSP losses) capacity is offered at less than the inter-regional price difference, and will collect rents when doing so. That rent does not form part of the IRSR, though, and is only available for hedging purposes if the MNSP owners choose to make it available to participants through their own market arrangements.[23] Three such (HVDC) links were actually developed, and two of those operated, to some extent, in parallel to "regulated" interconnectors joining NEM regions. But Mountain and Swier (2003) report that they have struggled to compete, and the only MNSP still operating is the HVDC link between Tasmania and Victoria, which was built at the instigation of, and is effectively controlled by, the dominant state-owned generator in Tasmania. We understand that, by default, the dominant generator currently retains the

[22] Initially known as "Inter-Regional Settlements Surplus" (IRSS), but later changed to "Inter-Regional Settlements Residue" (IRSR) because it is not always positive.

[23] The SRA auction makes no provision for inclusion of MNSP rentals in the auction process.

entire IRSR (and thus effectively FTRs) for both import and export over that link. Tasmanian market arrangements are currently under review, though.

Second, inter-regional flows will typically be limited by constraints that are more complex than simple bounds on a single interconnector flow. And that means that the flow between any pair of regions can be effectively limited by any generator output, load, or other inter-regional flow that appears in any binding constraint in which it appears. CRA (2003b) analysed the pricing impacts of such constraints and concluded that they can create a significant misalignment between physical flows, which NEMDE optimises for actual network conditions, and the flows that might appear to be optimal in the simplified NEM market model. And that means that the rents collected on particular links may not match the hedging requirements of participants particularly well, either.[24]

The auction design incorporates an extended price discovery process, in which no party can "corner the market". Initially, 25 % of the rent bundles for a particular quarter were released four quarters in advance, then another 25 % three quarters in advance, and so on.[25] That process has since been progressively extended to include 12 quarterly auctions spanning 3 years, each releasing 1/12th of the available units, defined in terms of a notional interconnector capacity.[26] Bids for SRA units may be submitted for each particular connector/quarter or, within the same quarterly auction, as linked bids defining a set of interconnector/quarter combinations.[27] The SRA process is open to standalone generators, integrated generator-retailers, and traders, with each group accounting for about one third of purchases. There are 30 parties currently registered, and 17 active in recent auctions, with current annual turnover of approximately $120 m.

The SRA process commenced in 1999, and there have been various modifications to interconnector specifications since then, due to upgrades to interconnector capacity,

[24] In extreme cases power may actually flow across a border in a direction opposite to the price difference, causing the IRSR to be negative. Originally, negative residues were offset against positive residues within the same weekly sub-period, so SRA units could have negative values. Recently this has been changed so that SRA units always have positive payouts, with any revenue shortfall due to counter-price flow being effectively subtracted from the auction proceeds passed through to TNSPs.

[25] The proceeds of these auctions are deemed to belong to the parties providing the underlying network capacity. However, the regulatory regime operates in such a way that those parties effectively have no financial interest in the auction or IRSR outcomes, and are not therefore incentivized to either provide, or withhold, capacity. Read (2008) suggests a regime that would partially expose transmission providers to the outcome, with the aim of incentivizing maximum economic capacity availability, but that suggestion has not been pursued further.

[26] In theory, there is provision to carry unsold units forward to the following auction, but this does not happen because units have unambiguously positive value, and always sell. Since the auction process makes no provision for resale of units, the quantity available in each auction is constant for each interconnector/direction.

[27] Linked bids that are accepted are charged the sum of the individual component prices that comprise the bid. But we understand this facility is seldom, if ever, used.

**Table 13.1** Settlement residue auction statistics (Source: AEMO 2010, 2011)[a]

| | | Payout: unit cost ratio | | | |
|---|---|---|---|---|---|
| Quarters | SRA share | Average | Standard deviation | Hedge effectiveness[b] | |
| Q1 | 45.93 % | 1.42 | 0.60 | – | – |
| Q2 | 14.77 % | 1.79 | 2.31 | – | – |
| Q3 | 17.57 % | 0.99 | 0.51 | – | – |
| Q4 | 21.72 % | 2.31 | 2.49 | – | – |
| Interconnectors | | | | Beta | $R^2$ |
| SAVIC | 3.03 % | 0.99 | 1.52 | 0.167 | 0.077 |
| VICSA | 22.55 % | 1.42 | 1.58 | 0.517 | 0.991 |
| VICNSW | 32.69 % | 2.00 | 3.21 | 0.329 | 0.943 |
| NSWVIC | 20.69 % | 0.84 | 1.18 | 0.206 | 0.713 |
| NSWQLD | 3.68 % | 1.09 | 3.11 | 0.202 | 0.871 |
| QLDNSW | 17.36 % | 2.13 | 2.82 | 0.489 | 0.579 |
| Aggregate | 100.00 % | 1.59 | | | |

[a]The data here is aggregated for both directions, over the entire market history, and includes changes in interconnector capacity, and unit definition. For the years over which the market included a SNY region, the VI-SNY and SNY-NSW interconnector results have been aggregated to represent a notional VIC-NSW interconnector

[b]Although the SRA's fractional share of rents is linear, $R^2$ has been calibrated using the total sum of squares from an affine approximation of the relationship

absorption of previously MNSP links, and changes to the NEM regional structure.[28] There have also been two periods during which the characteristics of SRA units changed: First, as a result of modification of the charging regime to re-balance rents between generators in the SNY region and the SNY-NSW interconnector during the 2005–2007 *CSP/CSC* trial described later; Second, to eliminate negative settlement residues, after 2010. Both changes were motivated by a desire to increase SRA unit firmness, and that should have thus increased their value, but neither has (yet) operated for long enough to form a judgement about actual impact on auction values. We are, however, able to reach some broad conclusions from the full set of SRA proceeds and payouts, as in Table 13.1. As expected, both total and per unit values vary significantly between interconnectors, and between quarters. There is a strong seasonal pattern, with most value arising in Q4 (Oct–Dec), and particularly Q1 (Jan–Mar), when Australia's summer peak in electricity demand occurs, with highly variable returns as a result of "summer" events occurring in late Q4, or early Q2.

These auction results, and residues, align well with expectations based on the physical characteristics of the system. Alignment with theoretical expectations relating to risk management, is another issue. Assuming that auction participants are risk averse, and that SRA units assist market participants by reducing the risks they would otherwise face as a result of inter-regional trading activities, we would expect to see SRA units sold at a premium to their expected value. At least, since speculators are allowed to trade, we would expect competition to set a floor on unit prices not too far

[28] Queensland was added in 2001, shortly after the scheme commenced, and the Snowy region incorporated into New South Wales in 2008.

below their expected payout value. But the data in Table 13.1 does not really support that hypothesis. Rather than paying a premium for risk, in aggregate purchasers of SRA units have only paid 63 % of their actual value, as it has turned out.

This result is not explained by the time value of money. Even when units are "purchased" 3 years in advance, the auction payment is not required until the quarter to which the units actually apply. Nor does it seem to be explained by learning effects. Even if the market only involved pure speculators, we might expect that the price discovery process would tend to converge, so as to better reflect actual settlement residues as more information becomes available in later auctions, with lower prices reflecting significant uncertainty for units auctioned well in advance, and later auctions clearing at something closer to actual payouts. But detailed examination of the data reveals no such trend. At a broader level, we might also expect that discounts would have reduced over the years, as the market learned the true value of units. But this is not the case, either. The market appeared to be in approximate equilibrium after 3 years, with cumulative revenue approximately matching cumulative payouts, to that point. But there appears to have been a consistent discrepancy since then.

It is often suggested that SRA units are being sold at a discount because they are "not firm". AEMO (2011) attributes this lack of firmness to network outages, implicit limitations on interconnector flow arising from the dispatch process, and flow reversals within pricing periods. Loss effects also have a pervasive impact, as discussed below. Based on accumulated quarterly price differences reported in AEMO (2010), settlement residues could only support FTRS for about 39 % of nominal link capacity on a firm basis over the period considered.[29] But this does not really explain the large observed discount to expectation. Figure 13.2 reveals that, while speculators face a significant probability of making a loss on units for any particular interconnector/quarter, the probability of making a loss on investment in a spread of interconnector auctions is only 47 %, and this is strongly outweighed by the size of the profits made when rents are high.

From an electricity market participant's perspective, though, the issue is not the variability of the SRA unit payout, but the correlation of that payout with their inter-regional trading risks. Even though electricity market participants would obviously pay a higher premium for firmer instruments, theory suggests that they should be prepared to pay some premium for any instrument whose payout is positively correlated with their inter-regional trading risks. We understand, anecdotally, that some participants in the SRA process do behave in ways that suggest they are hedging changes in their portfolio trading positions. But the fact that SRA units are consistently sold at significant discounts suggests that NEM participants do not see the SRA as providing a particularly effective risk management tool. Our analysis of the data suggests that the apparent "lack of firmness" may be over-stated, though.

Figure 13.3 plots actual settlement residue against a full hedging requirement, as defined by accumulated inter-regional price differentials, both on a per MW basis, for

[29] Even if the transmission system was 100 % firm, rents should only cover half the (loss-induced) inter-regional differential in periods with no congestion.

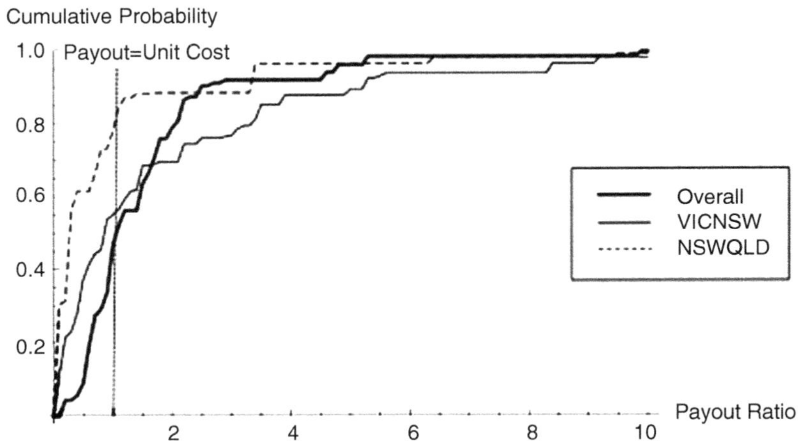


**Fig. 13.2** Cumulative payout ratio distribution for selected interconnectors (Source: AEMO 2011). This figure has been constructed by weighting interconnector/quarter results by the $ value of auctioned units, so as to represent consistent pdfs of the payout ratio per $ invested

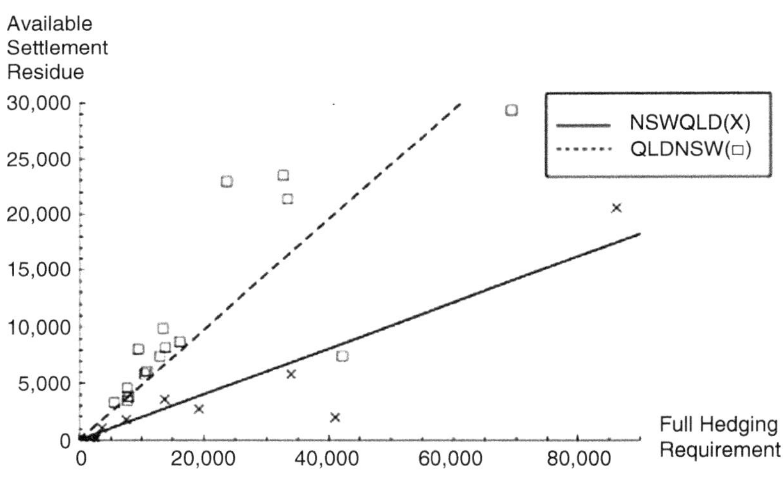


**Fig. 13.3** Hedging performance and firmness for selected interconnectors (Source: AEMO 2010, 2011)

some sample interconnectors, and shows "best fit" rays through that data.[30] Table 13.1 reports $\beta$, the estimated ratio of the settlement residue to full hedging requirements, and $R^2$, the proportion of variation explained by the rays fitted, for all interconnectors. Ideally, the correlation would be perfect, and this data would form diagonal lines with $\beta = 1$, and $R^2 = 1$. In reality $\beta$, is significantly less than unity for all interconnectors,

[30] Better fits could be obtained for lines that did not pass through the origin. But such lines do not represent the hedging available from an auction of proportional rental shares.

reflecting an inability for the SRA process to fully hedge the nominal interconnector capacity. The degree of firmness varies significantly though.

For example, the VIC-SA interconnector has $\beta = 0.517$, and $R^2 = 0.991$, in that direction. Thus it can only support hedging of approximately 52 % of its nominal interconnector capacity. This may be due to loss effects,[31] or it may be that effective inter-regional capacity is really much less than cross-border capacity, once intra-regional flow requirements are accounted for. But the hedging it does provide on that 52 % is actually quite "firm". Thus a participant can quite effectively cover its trading exposure by buying two SRA units for every MW of exposure. The fact that the hedging available is (fairly) firm, but that the quantity available is only half the nominal capacity, should really drive SRA unit prices up, not down.

In the reverse direction, though, (i.e. for SA-VIC) we only have $\beta = 0.167$, and $R^2 = 0.077$. Thus this link cannot reliably support hedging on even 17 % of nominal interconnector capacity in that direction. This kind of situation typically occurs where interconnector flows must compete with each other, or with variable intra-regional flow requirements, for capacity on congested lines. Thus it is not possible to support hedging up to the nominal capacity of each interconnector, simultaneously, because the transmission system cannot simultaneously support flows up to that level, and the proportion assigned to any particular interconnector depends on the dispatch of the day. Proposals to improve hedging performance in that situation are discussed in the next section.

## 13.6 Possible Developments in Australia

The NEM design, and the SRA process, would match reality if the interconnectors really did link regions into an unlooped tree, and congestion could only occur on those interconnectors. But that is not the case, and the market must somehow deal with the real congestion issues, which affect market outcomes in ways that are not always obvious. Three situations have caused particular concern: The inability of the market to provide generators with "firm access" to sell all their output at their own regional reference node; The way in which network constraints interact to cause IRSR accounts to be non-firm even when the underlying transmission capacity is firm; And the way in which "counter-price flows" can make the IRSR negative.

These issues have been considered more than once, and another review is currently underway (AEMC (2011)). It has been suggested that some of these issues could be dealt with by devices such as constrained on/off payments, or network support contracts, which are already provided for in the NEM code, but seldom used. Another

[31] Quadratic losses imply that rents will only cover half the (loss-induced) inter-regional differentials on the actual flow, when flow is unconstrained. This may not matter much because differentials are typically low in such situations. Unlike the congested situation, though, flow volume in such periods is essentially what participants, in aggregate, have decided they want it to be. Thus the ratio reported here will be under-stated because true "hedging requirements" may be significantly less than nominal interconnector capacity, and maybe close to actual flows.

possibility is that all of these problems could be resolved, using FTRs in a nodal pricing framework. In fact, the original NEM design included a rule that required new regions to be created once congestion on any constraint reached some quite low threshold. If applied, this would have progressively moved the market toward something like nodal pricing. It never has been applied, though, partly because it was shown that a large number of regions might have to be created to capture the pricing impacts of a single constraint arising in a loop. Thus, in a study conducted for the Ministerial Council on Energy (MCE), CRA (2004a, c) considered a range of alternative proposals, including full nodal pricing, and "generator nodal pricing" (GNP), under which nodal prices would be averaged into zones for loads.

Naturally, such proposals will always be opposed by generators whose price is expected to drop, and by loads whose prices are expected to rise. But it should also be said that successive studies referenced in AEMC (2008) suggest that intra-regional congestion has only had a modest impact on efficiency (perhaps 0.5 %), making it difficult to justify radical change. In that context CRA (2004c) proposed an intermediate alternative, which allowed participants to be selectively exposed to a form of nodal pricing, but provided with FTR-like instruments to hedge the risk implied by that exposure. That framework was based on the earlier proposal of CRA (2003b) for NEMMCO, to use a combination of price and contract mechanisms to incentivise participants to support an agreed level of interconnector flow, while, at the same time, firming up the corresponding IRSR account. Read (2008) later generalised the framework for the AEMC so as to include load, network capacity, and ancillary service providers, as well as interconnectors and generators. Before discussing its potential application to various NEM situations, we will describe the framework in this general form.

In NEMDE, only the generator and interconnector terms appear on the LHS of any generic constraint, with load, ancillary service and intra-regional network capacity terms combining to form the RHS constant, as in (1).[32] The generalised framework accepts that formulation, but introduces a conceptual distinction between terms (and hence parties) that are "exposed" to the price impact of the constraint, and therefore face price risk, and terms that are "protected" by an implicitly assigned "dispatch matching" hedge. In general, there is no reason why variable terms may not be "protected", or fixed terms "exposed". So, denoting "protected RHS capacity" as *PRHSC*, we can re-arrange equation (13.1) as:

$$\sum_{i \in EXPOSED_k} CPF_{ik}{}^{*}x_i \quad \leq PRHSC_k = - \sum_{i \in PROTECTED_k} CPF_{ik}{}^{*}x_i \tag{13.2}$$

Given a Constraint Shadow Price (*CSP*) for each NEMDE equation, each market design can be expressed in terms of exposing a particular subset of the parties represented in constraint form (13.1) to the *CSP* on a particular subset of constraints.

[32] In reality, other terms, such as the inertia of generating units assumed or observed to be running, may affect the RHS, but that complication will be ignored.

For example, if we expose a participant (generator or load) at node *i* to the pricing impact of constraints $k = 1,\ldots K$, they will pay, or be paid, this *CSP* component in addition to the regional reference price they already face for their net injection. This effectively creates an Adjusted Nodal Price (*ANP*) of:

$$ANP_i = RRP - \sum_k CPF_{ik}{}^* CSP_k \tag{13.3}$$

Charging these prices to affected participants thus creates a limited form of nodal pricing, in the context of a zonal market. It provides affected participants with the same price signals they would see in a nodal market, at least with respect to adjusting operations so as to respect the limits implied by key constraints. But it also creates a hedging problem for those participants, if they wish to trade power with other parties facing the regional reference price. If only generators are exposed, that would include all intra-regional loads, since they are implicitly hedged to the regional reference node, and inter-regional parties who purchase (IRSR based) inter-regional FTRs, to transport power away from that node. So we introduce a Constraint Rental Right (*CRR*) which gives the holder the right to call on the rents generated by a portion of the *PRHSC* of the corresponding constraint, in form (13.2). Before accounting for any energy contracts, the net market exposure of a participant injecting $INJ_i$, while holding a bundle of *CRRs* for constraints $k = 1,\ldots K$, would be[33]:

$$NetExp_i = INJ_i{}^*(RRP - \sum_k (CPF_{ik}{}^* CSP_k)) + \sum_k (CRR_{ik}{}^* CSP_k) \tag{13.4}$$

If all parties, including transmission and ancillary service providers, were exposed to *CSP*, all would appear on the LHS, and $PRHSC = 0$, implying a net rental pool of zero. In fact the total pool of rent available to support hedging is zero in all cases. But each market design partitions that total pool differently, between implicit and explicit hedging pools. The rental pool available for explicit hedging of exposed participants is always determined by the *PRHSC*. But we can think of the protected terms making up *PRHSC*, as being implicitly hedged, by what Read (2008) called Implicit Dispatch Matching Allocation (IDMA). The market does not work this way, but it is as if protected participants are always assigned, ex post, a *CRR* that exactly matches their dispatch position, and so face no exposure to the pricing impact of this constraint at all. But, since the entire hedging pool sums to zero, this hedging cannot have positive value, in aggregate, to both "protected" and "exposed" parties. The aggregate rent available to be paid to (or paid by) exposed parties is just the net available after protecting the protected parties from constraint pricing effects. For example, in a nodal market, both load and generation terms are exposed to the pricing impact of all constraints, and the *ANP* defined above is just the nodal price. Transmission line

[33] We have also defined them in fixed MW capacity terms, but assume that revenue adequacy is dealt with by scaling, thus creating a problem with "firmness".

capacity terms, on the other hand, are typically protected, in the sense that the transmission service provider does not ultimately receive the rents generated on those constraints, and so is not exposed to variability in the rent. This protection, which has a negative expected value for the transmission service provider, allows the rents to be used to support FTR hedging which is as firm as the transmission capacity limits themselves.

*CRR* is defined in terms of constraint RHS units, but dividing by $CPF_{ik}$ gives an equivalent Participant Rental Right (*PRR*) defined in terms of MW injection at node *i*. If a participant holds the same *PRR* quantity for each constraint in which it is exposed, then minor re-arrangement shows that the *PRR* quantity is effectively sold at *RRP*, with the remainder sold at *ANP*. That is:

$$NetExp_i = PRR_i{}^{*}RRP + (INJ_i - PRR_i)^{*}ANP_i \tag{13.5}$$

In other words, the bundle of *CRR*s for all constraints in which a nodal injection is involved effectively forms an FTR (for *PRR* MW) from that node to its reference node. Similarly, a consistent bundle of *CRRs* for all constraints in which an interconnector flow is involved effectively forms an FTR from its source to its sink. Together, these FTRs form a complete hedging structure for the system, but how firm can they be?

In its simplest form, classic FTR theory states that the rental pool available for hedging is just the rents on transmission capacity constraints, including any weighted sum of line limits. If transmission is not exposed to *CSP* then these appear as part of the *PRHSC*, and the rent available for generator/load hedging is just the rents generated by the transmission system. So the FTR pool is as firm as the *PRHSC* defined by the transmission system capacity. But ancillary service terms may appear in transmission constraints, too, if ancillary service support can allow increased interconnector flows. This may be implicit in the calculation of the constraint RHS, or explicit in the kind of co-optimised market discussed by Read (2010). If those services are paid for by some other means, e.g. as part of a regulated transmission system cost recovery regime, they can be considered protected, and thus provide part of the *PRHSC* from which rent is available for hedging by exposed participants. This will typically increase the FTR hedging pool, but that pool will then be only as firm as the transmission capacity, adjusted for whatever ancillary service support happens to be provided.

In a zonal market, by way of contrast, load and generation terms are not exposed to the pricing impact of constraints, but are implicitly hedged to their own regional reference prices. This means that the firmness of hedging available for interconnector flows, which are the only terms exposed to *CSP* in that market design, is significantly compromised. The *PRHSC* of each constraint in which an interconnector is involved will depend on the pattern of load, generation and ancillary service levels, for each party involved in that constraint, irrespective of whether these are treated as constants determining the RHS of that LP constraint in each NEMDE LP run, or variables to be optimised on the LHS. And this gives rise to what has been described in Australia as "disorderly bidding". For example, generators involved in binding constraints which would, in a nodal market, imply a low nodal price, will bid very low so as to be dispatched at a high level, knowing that they will be paid their own regional reference price for whatever quantity they can be dispatched for. Equivalently, they know that, simply by being dispatched, they will implicitly be granted (retrospective) *CRR* transmission rights to cover that dispatch quantity. This obviously creates a conflict with inter-regional flows, which do not share that privilege, but must instead purchase what are effectively bundles of *CRRs*, representing whatever constraint capacity has not been taken up by protected parties, from the IRSR pool.

These effects contribute significantly to the "lack of firmness" discussed in the previous section, and can be characterised as creating a need for "interconnector support". In that context, CRA (2003b) proposed to selectively expose participants in key constraints to the pricing impact of those constraints, and then contract with then for "support" services. *CSP* was referred to then as a Constraint Support Price, while *CRRs* were expressed in the form of Constraint Support Contracts (*CSCs*). *CSCs* could be thought of as equivalent to a (financial) FGR with respect to a particular constraint. As with any such right, a generator facing the *CSP*, in addition to its regional reference price, would have first order incentives to increase or decrease generation in ways that reduce congestion on that constraint. Risk aversion, and localised market power,[34] mean that a party holding a *CSC* also has second order incentives to move dispatch toward the *CSC* level.[35] If the *CSC* level is set appropriately, this has the physical effect of "supporting" interconnector flows, as well as the financial effect of creating a firmer inter-regional settlement pool for hedging purposes. Effectively, the *CSC* quantity becomes protected, and becomes part of the *PRSHC*. Similar incentives apply if other parties, including transmission and/or ancillary service providers, are exposed to congestion pricing, and contracted in a similar way.

In this context, it was envisaged that the overall deal required to induce parties to behave in ways which increased net interconnector capacity would generally have negative expected value for those parties, relative to the status quo, and so would need to be paid for. In some cases exposure to *CSP* would have a negative impact, while the *CSC* had a positive value. In other cases it could be the other way around. But increased revenue should be available to make such payments, to the extent that parties buying IRSR bundles value the increase in firm interconnector capacity. And generators contracted in this way would also receive "firm access", to deliver the *CSC* quantities through the constraints to which they are exposed to the regional reference node. Thus *CSC*s are effectively financial FGRs with respect to particular constraints. If all generators were exposed to *CSP* for all constraints, though, we would effectively have GNP.

*CSC*s would have positive or negative value to a generator, depending on whether its *ANP* was typically above or below the regional reference price. But their firmness

[34] Which can be very important in some of these constrained situations where a single generator may act as a "gatekeeper" determining effective interconnector flow capacity.

[35] *CSC*s held with respect to various constraints might imply different preferred generation levels, and conflicting second order incentives, if those constraints bind simultaneously. But participants are free to target a dispatch level that makes an appropriate trade-off, given the relative prices involved on any occasion.

would still be limited, so long as loads were not exposed to constraint pricing. Thus, in a GNP regime, the rents available to support hedging with respect to a transmission constraint are not as firm as the transmission capacity of that constraint, but as variable as the transmission capacity minus the weighted sum of load levels defining the *PRHSC*. Although that may not be very firm, it is really the upper limit on firm hedging that can be provided for generators in such a market. Firmer hedging can be provided to generators, but only at the expense of exposing loads to constraint pricing, and requiring them, too, to protect themselves explicitly with FGRs or FTRs.

Exposing loads to constraint pricing effects has not found general favour in Australia, but the framework has been extended to deal with issues arising out of interactions between interconnector flows. In reality, power does not flow from one region to another over a single piece of wire. Cross border flows may occur at several points, and interconnector flows between several regions may be involved in some of the same loop flow constraints. Under the status quo, the IRSR pool available for each notional interconnector is effectively determined by the flow NEMDE dispatches on that interconnector, which is determined by the dispatch position generators using that interconnector are able to achieve. Conversely, the rental pool available to support hedging on each interconnector, with respect to a constraint, is the RHS of that constraint minus all ancillary service, load, generation, and other interconnector terms involved in that constraint, at whatever dispatch level they happen to achieve. Understandably, that can imply significant variability in the IRSR bundle available to each interconnector.

CRA (2004b) proposed to resolve this problem by assigning each interconnector a *CSC* corresponding to a defined share of the constraint RHS capacity. Suppose *CSC*s were assigned to proportionally partition *PRHSC* capacity between multiple interconnectors involved in a common constraint. Then rents collected on any interconnector from flows above its *CSC* level would be paid into a common pool, and re-assigned to interconnectors whose flows were below their *CSC* levels. The effect would be to make the IRSR pool available to each interconnector involved as firm as the *PRHSC*. Under the status quo, generator terms are protected, so the *PRHSC* available for interconnector hedging would still vary significantly, as generator dispatch varied. But if all generator terms were exposed to constraint pricing effects the joint hedging pool available for interconnector and generator hedging would be as firm as the RHS capacity of the NEMDE constraint (i.e. the transmission capacity adjusted for load and ancillary service terms.)

Biggar (2006) proposed a more comprehensive framework in which *CRR*s would be defined, and traded, for every possible constraint. Read (2008) describes that approach in terms of the *CSP*/*CRR* framework, but argues that it would be impractical since there were, at that time, over 12,500 possible constraints in the NEM.[36] Thus an FNP/FTR framework was considered preferable, if a comprehensive NEM wide locational pricing/hedging regime was thought to be worthwhile. What CRA

[36] In an appendix, CRA also argues that there are conceptual errors in the Biggar paper, and in its critique of earlier work by CRA.

(2004c) actually recommended, though, was to leave the NEM regional market structure largely intact, while using the *CSP/CRR* approach to deal with situations arising around congestion on critical constraints which might not be severe enough, or expected to persist for long enough, to justify a change to regional boundaries. When applied in this way, the *CSP/CRR* framework can be compared with a "financial" (rather than physical) application of the FGR framework of Chao et al. (2000). But it was developed in the context of a zonal market, and thus relates to an old proposal of Stoft (1998). When applied to inter-regional flows, it is similar to the approach adopted in the Texas market before introduction of full nodal pricing, as described by Baldick (2003), or the regime proposed for Europe by Pérez-Arriaga and Olmos (2005). If generator terms are included, as above, it deals with the kind of situation identified by Baldick, in which generators classified as being in a particular region may actually have significantly different participation factors in particular constraints.

In reality, while these proposals received a reasonable level of support, and similar proposals are again under discussion, none of these applications is current in the NEM. The major problem identified by CRA (2004c) was the need to determine how capacity on congested flow gates would be allocated between generation and interconnector flows, between different generators or interconnector flows, and between new participants and old. This has proved a significant barrier to widespread implementation of the regime, given the many vested interests involved. From 2005, though (an approximation to) the *CSP/CSC* approach was employed in a large scale trial to deal with constraint problems around the former SNY market region.[37] While generally successful, the trial was eventually discontinued when the SNY market region was amalgamated into the NSW market region, thus "resolving" the boundary flow issues by moving the market design in the opposite direction from the gradual proliferation of regions envisaged in the initial rules. In 2009, a rule was introduced to deal with one of the most obvious mis-pricing problems by simply truncating any negative IRSR at a low positive value, should it persist, on average, for more than a week. [38]

But it would be fair to say that, in practice, the inter-related problems of inter-regional boundary definition, firm inter-regional hedging, and firm intra-regional access, remain unresolved. The situation is again being reviewed but, so far, one can only conclude that the acknowledged deficiencies of the current framework are not (yet) causing enough economic inefficiency to justify the transaction costs of establishing a consensus in favour of any particular proposal for a better regime. In that context, though, the *CSP/CRR* approach at least provides a consistent general

[37] The constraints involve a trade-off between generation in the SNY region and both VIC-SNY, SNY-NSW interconnector flows. Initially, the trial did not involve the VIC-SNY interconnector, so *CSP/CSC* transfers were only made between SNY generation and the SNY-NSW interconnector, within an aggregate rental pool that was less firm than the transmission capacity. This was subsequently modified to eliminate negative residues on the VIC-SNY interconnector.

[38] This may be seen as a very limited application of the *CSP/CRR* framework, in which the transmission system provider, who ultimately receives the SRA proceeds, effectively guarantees that the net interconnector capacity available to support flows in each direction will at least be non-negative.

framework within which a wide variety of market designs, involving elements of both FGR and FTR paradigms, can be described, compared, and potentially implemented.

## 13.7 Conclusions

Experience with FTRs, and related concepts, has been mixed, and success limited, in both Australia and New Zealand. These two markets are actually very different in structure, with one pricing power at every node, and the other over quite broad regions. In both cases, the industry has evolved to accommodate itself to the lack of effective hedging arrangements. And that means that there can be significant resistance to change from participants whose modus operandi has been developed to exploit some particular aspects of the status quo arrangements. Thus the pressure for change has often come more from regulatory institutions, which have only gradually gained strength over the years.

While the SRA process has been moderately successful in providing inter-regional hedging for Australia, the mismatch between the market architecture and the physical network configuration makes this hedging less firm than it could be, and it seems the market participants do not value it highly. Those problems might be overcome by applying some version of the *CSP/CRR* framework, or perhaps by moving to full nodal pricing, but it is far from universally accepted that the NPV benefits of introducing either regime would be positive. For New Zealand, the "inter-island" FTR regime now planned will provide a level of inter-regional hedging comparable to that in Australia. But participants there will still face significant intra-regional price risk. Thus, although recent studies have not predicted overwhelmingly positive NPV benefits, further development may be expected. On the other hand, we might suggest that, in both cases, the cost incurred by all concerned in continuing a debate which has already lasted for more than a decade may account for a significant part of the projected "implementation cost".

**Acknowledgement** The authors wish to thank the EA (New Zealand) and AEMO (Australia) for provision of data, and permission to publish.

## References [39]

AEMC (2008) Congestion management review: final report (on AEMC website)

AEMC (2011) Transmission framework review, directions paper (on AEMC website). http://www.aemc.gov.au/Market-Reviews/Open/Transmission-Frameworks-Review.html

[39] All URLs accessed 1 Oct 2011. "NZEA website" is http://www.ea.govt.nz/our-work/programmes/priority-projects/locational-hedges/, but will be archived in future. "MCE website" is http://www.mce.gov.au/emr/elec_trans/archive.html. "AEMC website" is http://www.aemc.gov.au/Market-Reviews/Completed/Congestion-Management-Review.html. Older reports available on request from the corresponding author.

AEMO (2010) Submission to AEMC transmission frameworks review issues paper (on AEMC Website)
AEMO (2011) Settlement residue auction results. At http://www.aemo.com.au/electricityops/results.html
Alvey T, Goodwin D, Ma X, Streiffert D, Sun D (1998) A security-constrained bid-clearing system for the New Zealand wholesale electricity market. IEEE Trans Power Syst 13(2):340–346
Baldick R (2003) Shift factors in ERCOT congestion pricing. At http://www.ece.utexas.edu/~baldick/papers/shiftfactors.pdf
Biggar D (2006) Solving the pricing and hedging problems in the NEM using constraint-based residues. Report to the AEMC (on AEMC website)
Chao HP, Peck S, Oren S, Wilson R (2000) Flow-based transmission rights and congestion management. Electr J 13(8):38–58
CRA (2003a) Constraint orientation: principles and pricing implications. Report to NEMMCO
CRA (2003b) Dealing with NEM interconnector congestion: a conceptual framework. Report to NEMMCO
CRA (2004a) NEM regional boundary issues: theoretical framework. Report to MCE (on MCE website)
CRA (2004b) NEM interconnector congestion: dealing with interconnector interactions. Report to NEMMCO (on MCE website)
CRA (2004c) NEM transmission region boundary structure. Report to MCE (on MCE website)
Hogan WW (1992) Contract networks for electric power transmission. J Regul Econ 4(3):211–242
Hume D (2009) Losses in the New Zealand power system. In: New Zealand EEA conference, Christchurch
Miller RT, Read EG (2011) The EA's 2-hub FTR market. In: EPOC winter workshop, Auckland (expanded version on NZEA website)
Mountain B, Swier G (2003) Entrepreneurial interconnectors and transmission planning in Australia. Electr J 16(2):66–76
NZEA (2010) Managing locational price risk: proposal. Consultation paper at: http://www.ea.govt.nz/document/11101/download/our-work/consultations/priority-projects/lpr-management-proposal/
NZEA (2011a) Managing locational price risk: amendments to code (on NZEA website)
NZEA (2011b) Financial transmission rights: market outline (on NZEA website)
O'Neill RP, Helman U, Hobbs BF, Stewart WR, Rothkopf MH (2002) A joint energy and transmission rights auction: proposal and properties. IEEE Trans Power Syst 17(4):1058–1067
Pérez-Arriaga IJ, Olmos L (2005) A plausible congestion management scheme for the internal electricity market of the European Union. Utilit Policy 13(2):117–134
PHB (1997) Inter-regional hedging in the Australian National Electricity Market: theoretical framework. Report to NEMMCO
Philpott AB (2011) On revenue adequacy in electricity pool markets with line losses and reserve constraints. Appendix to report on locational price risk management, Stochastic Optimization Limited report to the EA (on NZEA website)
Pritchard G, Philpott AB (2005) On financial transmission rights and market power. Decis Support Syst 40(3–4):507–515
Read EG (1989) Pricing and operation of transmission services: long run aspects. In: Turner A (ed) Principles for pricing electricity transmission. Transpower, New Zealand. http://www.hks.harvard.edu/hepg/rlib_rp_transmission_flowgate.html
Read EG (1997a) Transmission pricing in New Zealand. Utilit Policy 6(3):227–236
Read EG (1997b) Electricity sector reform in New Zealand: lessons from the last decade. Pac Asia J Energy 7(2):175–191
Read EG (2002) Financial transmission rights for New Zealand: issues and alternatives. EGR Consulting Ltd, report to the New Zealand Ministry of Economic Development. http://www.med.govt.nz/templates/MultipageDocumentTOC____20696.aspx
Read, EG (2008) Network congestion and wholesale electricity pricing in the Australian National Electricity Market: an analytical framework for describing options, EGR Consulting Ltd, report to AEMC (on AEMC website)

Read EG (2009) Locational hedging options for New Zealand: issues and options. EGR Consulting Ltd, report to NZEA. http://www.ea.govt.nz/document/2426/download/our-work/consultations/priority-projects/locational-price/
Read EG (2010) Co-optimization of energy and ancillary service markets. In: Rebennack S, Pardalos PM, Pereira MVF, Iliadis NA (eds) Handbook of power systems, vol I. Springer, Berlin/Heidelberg, pp 307–327
Read EG, Miller RT (2011) Possible FTR developments. In: EPOC winter workshop, Auckland (expanded version on NZEA website)
Sarkar V, Khaparde SA (2008) A comprehensive assessment of the evolution of financial transmission rights. IEEE Trans Power Syst 23(4):1783–1795
Stoft S (1998) Congestion pricing with fewer prices than zones. Electr J 11(4):23–31
Transpower (2002) FTR design as implemented. At http://www.transpower.co.nz/ftr-design-2002

# Chapter 14
# Transmission Rights in the European Market Coupling System: An Analysis of Current Proposals

**Gauthier de Maere d'Aertrycke and Yves Smeers**

## 14.1 Introduction

Regulation EC No714/2009 on "… conditions for access to the network for cross-border exchanges in electricity …" (see European Commission 2009) and the Framework Guidelines on Capacity Allocation and Congestion Management (ACER 2011) formally introduced transmission rights in the European Electricity System. None of these documents really explain what these transmission rights should be but the Framework Guidelines are slightly more explicit than the Regulation: transmission rights can be physical or financial; physical rights should be options subject to a use it or sell it clause (UIOSI[1]); and financial rights can be options or obligations. Further details will come with the grid codes that Transmission System Operators (TSOs) are preparing. The Framework Guidelines do not really elaborate on the market design that must accommodate these rights. They simply mention that *"TSOs implement capacity allocation in the day-ahead market on the basis of implicit auctions … based on the marginal pricing principle"*. Both Market Splitting (MS), which is now well established in the Nordic power market and Market Coupling (MC), which is emerging as the European reference system outside of the Nordic countries satisfy these conditions. The US experience shows that the transmission rights and market design are closely intertwined and that one cannot discuss the former without referring to the latter. We follow suit and discuss the extent to which transmission rights can be meaningfully implemented in Market Coupling. Market Coupling can be seen

[1] Physical rights must be nominated ("Use It") before the opening of the day-ahead market in order to be used. Otherwise ("Or Sell It") they are automatically sold back to the day-ahead market at the price that will come out from that market.

G. de Maere d'Aertrycke
FEEM, Corso Magenta 63, Milan 20123, Italy

Y. Smeers (✉)
CORE, Voie du Roman Pays 34, L1.03.01, Louvain-la-Neuve 1348, Belgium
e-mail: yves.smeers@uclouvain.be

J. Rosellón and T. Kristiansen (eds.), *Financial Transmission Rights*,
Lecture Notes in Energy 7, DOI 10.1007/978-1-4471-4787-9_14,

as a very simplified version of nodal pricing (replacing nodes by zones and hoping that the rest applies). It is thus convenient to discuss transmission rights in Market Coupling keeping the nodal system in background. Chapter 3 of this book (Oren 2012) offers an in depth discussion of transmission rights in the nodal pricing model: we continuously (most often implicitly) refer to this chapter during the discussion. Much of the analysis of congestion management in nodal pricing was constructed on examples of two and three nodes grids. It is thus also reasonable to follow that approach and reason on two and three zone (not node) systems that we construct from a six node (not zone) network. The rest of this introduction gives a brief survey of the literature on Market Coupling and the structure of the paper.

Nodal pricing has been elaborated during the restructuring of the US electricity system (see HEPG website (http://www.hks.harvard.edu/hepg/) and the list of research papers thereof). A summary is presented in Chap. 1 of the book to which we refer the reader. Market Coupling has been extensively discussed in the so-called "Florence Forum" that is driving the thinking on the completion of the internal electricity market in Europe (see the website of the Directorate General Energy at http://ec.europa.eu/energy/gas_electricity/electricity/forum_electricity_florence_en.htm). The Forum produced many informal presentations that can be found on its website. Transmission System Operators (TSOs) (see the website of ENTSO-E, formerly ETSO at https://www.entsoe.eu/resources/publications/) and regulators (see the website of ACER at http://www.acer.europa.eu/portal/page/ACER_HOME/Public_Docs) also produced many papers on the subject. More technical documents are available in studies initiated by the Commission (see the website of the Directorate General Energy under the heading "Studies" at http://ec.europa.eu/energy/studies/index_en.htm). In contrast, the academic literature on Market Coupling is more limited. Buglione et al. (2009) offers an extensive and very pedagogical presentation of many aspects of the problems posed by the internal electricity market including a discussion of transmission rights and market coupling. The paper also contains an extensive list of academic and non academic literature. Glachant (2010) provides a quite readable paper that summarizes the important ideas underlying Market Coupling and gives references. Van Vyve (2011) offers a rigorous formal presentation of Market Coupling and its relation to nodal pricing. Janssen et al. (2011) and Wobben (2009) analyze European transmission rights as financial instruments. The improved efficiency of the power system and the resulting price convergence that it entailed attracted a lot of attention (De Jong et al. 2007; Dijkgraaf and Janssen 2009; Huisman and Kilic 2011; Kurziden 2010; Parisio and Bosco 2008; Pellini 2011; Zachmann 2008).

Most of these academic presentations refer to the current implementation that makes the assumption that the electricity grid can be described by a set of zones (today countries) connected by interconnections described by their sole transfer capacities (the Transfer Capacity or TC model). Neglecting bloc bids (which represent machine indivisibilities) for the sake of brevity in this paper, Market Coupling determines the price in the different zones by solving a welfare maximization problem subject to these transfer capacities. The success of Market Coupling in bringing about a closer integration of European electricity markets is remarkable given the simplicity and very approximate nature of the underpunning model. This success is also recognized in the reports of the European Commission on the progress in the internal electricity and gas markets (e.g. European Commission 2011).

The representation of an interconnection by its sole transfer capacity requires two simplifications: one can neglect Kirchhoff's second law and one can aggregate lines between different pairs of nodes connecting two zones into one interconnection between these zones. TSOs who introduced that model had advised very early (ETSO 2001) about its strange properties. They explain that transfer capacities have none of the usual algebraic properties of day-to-day life (adding and subtracting), which also rule flows on individual lines (adding and netting). These difficulties induced the search for a new model, named as "flow based" (FB) that was first proposed in ETSO-EuroPEX (2004) and later elaborated more extensively in ETSO-EuroPEX (2009). The flow-based version of Market Coupling also solves a welfare maximization problem but it replaces the TC representation of the grid by PTDF like relations: these are common in the nodal system but are here applied to a zone-to-zone representation of the grid. The FB model is more technical than its TC predecessor and less developed in the literature or in the slides of the Florence Forum. TSOs have produced important documents on that system (Amprion et al. 2011a, b) that we refer to in this paper. As for the TC model, academic studies have examined the efficiency gains accruing from the FB model. They generally adopt a simplified view of the FB representation of the grid that they assume to be pure flowgate (see Oren 2012). This contrasts with the real FB model that goes through an aggregation of the real network to construct the PTDFs of the zone-to-zone model. This aggregation is one of the themes of this paper.

The demand for transmission rights originated from stakeholders, whether traders or large industrial consumers. Papers from the Florence Forum on the subject are mainly descriptive and do not go into technical discussions of feasibility (e.g. ETSO 2006; EFET 2008). Duthaler and Finger (2008) seem to be the first ones to look at the problem from a more formal point of view. A recent in depth study by Booz & Co (2012) recommends going to the nodal system for granting transmission rights. This chapter falls into this more analytical category: it explores the feasibility of constructing firm transmission rights on the basis of the TC and FB models.

In conclusion, the literature on Market Coupling is more limited and much more informal than the one that drove the discussions of the nodal and flowgate systems in the USA. It is also more limited and less technical than the literature of market splitting that is an older sibling of Market Coupling. We contribute to that literature in the following way. Following the tradition in congestion management we analyze the principles of transmission rights discussed in EU regulatory and legal European texts on the basis of two and three zone models that we construct in Sect. 14.2. We examine the properties of transfer capacities in these models with reference to a "superposition property" that has been important for creating portfolios of tradable transmission rights in the nodal system and underlies legal requirements such as netting (Sect. 14.3). We show that these basic properties are not verified in general (as was already recognized in ESTO 2001). Our analysis goes into some depth in the notion of Generation Shift Key (GSK) that is crucial in the FB model but remains largely ignored in the literature (except by TSOs who introduce the notion but do not explore its properties). We then explain how congestion charges in the day-ahead market of Market Coupling (there is no real-time market in Market Coupling) increase transmission risk compared to a nodal system and are more difficult to hedge with transmission rights (Sect. 14.4). We then go more in detail in that discussion and explain that these shortcomings dramatically complicate the "simultaneous feasibility" requirement that is central to firmness in the Financial Transmission Rights of the nodal system. We finally come back to the description of the transmission rights proposed by the Framework Guidelines (Sect. 14.5) and show that financial rights would not satisfy the "simultaneous feasibility"[2] criterion and that physical rights suffer from well known shortcomings that justified the abandonment of the contract path model. Whether financial or physical, the rights foreseen by the Framework Guidelines are thus unlikely to be firm or if so will be unduly restricted. We conclude with a pessimistic view on development of a liquid market of transmission rights.

## 14.2 A Primer on Market Coupling

Market Coupling is a zonal model that assigns responsibilities on energy and transmission to different entities while organizing some interaction between them. As in many implementations of nodal pricing, the energy market in Market Coupling is a combination of a centralized market organized around Power Exchanges (PXs) and a decentralized market of bilateral contracts. PXs receive energy bids in each zone and clear their respective intra-zonal market in day-ahead assuming that it is free from transmission constraints. Bilateral contracts, whether domestic or international are concluded outside of the PXs and are thus not part of this market clearing.

Transmission System Operators (TSOs) deal with the transport of electricity. Transmission is conventionally unlimited inside a zone (physical limitations need to be handled by counter-trading) but zones are linked by capacitated interconnections. Market Coupling adopts the EU common but contestable (Duthaler et al. 2008 and Duthaler and Finger 2009) assumption that transmission limitations occur at the border between countries. TSOs therefore provide a simplified view of the grid consisting of zones (today identical to countries) connected by capacitated interconnections. PXs (more specifically EpexSpot and Apx Belpex) clear this inter-zone market in day-ahead on behalf of the PXs using the interconnections and taking account of the clearing of the domestic market already achieved by the PXs. These intra- and inter-zone clearings of the energy market take place in day-ahead; there is no real-time energy market in Market Coupling but intraday markets (between day-ahead and real-time) are in development inspired by the Nordic ELBAS market. It is recognized that a

[2] Simultaneous feasibility is a property that requires that the set of transmission rights, whether financial or physical (but this notion was introduced for financial rights), be physically feasible (that is satisfy Kirchoff's first and second laws) for the real grid. We come back to that question later.

grid composed of infinite intra-zone capacities and limited inter-zone interconnections cannot represent the real physical system; TSOs may therefore have to proceed to intra and inter-zonal counter-trading[3] in order to restore the feasibility of the grid after the clearing of the energy market. Counter-trading does not give rise to tradable transmission rights and is therefore not discussed in this paper. Our focus here is the extent to which a zonal system based on this organization of the day-ahead market can provide the background for firm and tradable transmission rights.

Transmission rights mainly developed in the restructured US systems in the form of hedges of congestion charges (Financial Transmission Right). Congestion charges in the nodal system, whether computed from node-to-node transactions or on particular links (flowgate), are intimately related to the solution of an Optimal Power Flow (OPF) that simultaneously clears the real-time energy and transmission markets. Market Coupling does not rely on an OPF to clear a real-time energy market but the algorithm (Djabali et al. 2010) used in the day-ahead market has a flavor of a simplified OPF at least if we exclude technically difficult issues such as "bloc bids". It thus makes sense to reason on transmission rights allowed by Market Coupling in the EU market by analogy to those offered in the restructured nodal systems.

As mentioned above, Market Coupling does not have a real-time market but is in the process of implementing an intraday market. All our discussion refers to the existing day-ahead market. This implies that we implicitly transpose considerations made for the real-time US markets to the day-ahead European market. The real-time market is seen as the real spot market in several electricity organizations but Market Coupling does not comply with that philosophy and considers (so far and before the full deployment of intraday markets) the day-ahead market as the spot market.

### 14.2.1 From Nodal to Zonal Models

Market Coupling is based on a zonal decomposition of the electricity market. It was first implemented in November 2006 on the so-called "trilateral market" consisting of Belgium, France and The Netherlands and it expanded to include Germany in November 2010. Further linkage with the Nordic system (known as Nordpool and consisting of Denmark, Finland, Norway and Sweden) is in progress and extensions to the Southern Peninsulas (Iberia and Italy) are foreseen. We reason in this paper on the basis of two and three zone illustrative examples that we construct on the basis of Chao and Peck (1998) six-node example (see Fig. 14.1). This test problem can be summarized as follows:

[3] Counter-trading is an operation whereby TSOs buy adjustment injections and withdrawals of power at different nodes to generate counter-flows on congested line. A counter-trading operation may require an expensive plant to ramp up and a cheap plant to ramp down. These operations are also called "out of merit" because they violate the economic order of plant operations. Counter-trading is the responsibility of the TSOs and does not involve PXs; it must be planned on the basis of the real characteristics of the grid and not of the simplified model provided by the TSOs to the PXs.

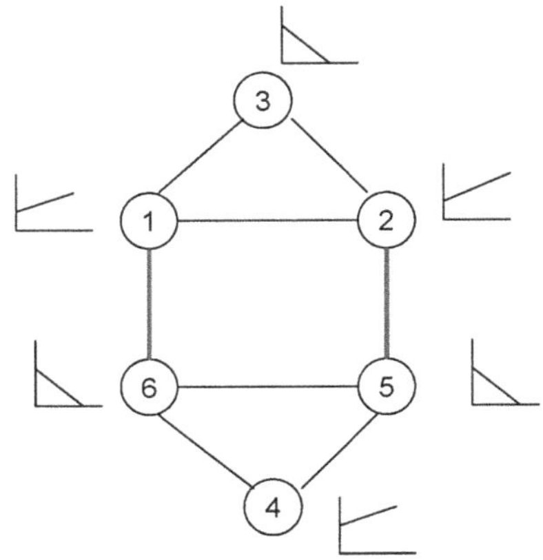


**Fig. 14.1** The six nodes, eight lines, test problem

**Table 14.1** Demand and cost functions

| Node | Function type | Function |
|---|---|---|
| 1 | Marginal cost | 10 + 0.05q |
| 2 | Marginal cost | 15 + 0.05q |
| 3 | Inverse demand | 37.5 − 0.05q |
| 4 | Marginal cost | 42.5 + 0.025q |
| 5 | Inverse demand | 75 − 0.1q |
| 6 | Inverse demand | 80 − 0.1q |

Source: Chao and Peck (1998)

**Table 14.2** PTDFs in the six node example

| Power (1 MW) injected at node | Power flow on link 1 → 6 (MW) | Power flow on link 2 → 5 (MW) |
|---|---|---|
| 1 | 0.625 | 0.375 |
| 2 | 0.5 | 0.5 |
| 3 | 0.5625 | 0.4375 |
| 4 | 0.0625 | −0.0625 |
| 5 | 0.125 | −0.125 |
| 6 (hub) | 0 | 0 |

There are three generators respectively located at nodes 1, 2 and 4 with linear (affine) marginal cost curves. Three demand nodes with linear (affine) demand functions are located at nodes 3, 5 and 6. Two lines, namely 1–6 and 2–5 of impedance 2 are capacity constrained respectively at 200 and 250 MW. The other lines are unconstrained and their impedances are 1. The marginal cost curves and demand functions are documented in Table 14.1. The PTDFs of lines 1–6 and 2–5 are given in Table 14.2. Real time nodal prices are obtained by solving an OPF and reported in Table 14.3. The difference of nodal prices between two nodes (e.g. nodes 1 and 6) is the congestion charge of a node-to-node transmission service. The dual variable of the line constraint (e.g. line 1–6) in the solution of the OPF is the congestion charge on the "flowgate" (see Oren 2012) defined by these lines. Differences of nodal prices can be

**Table 14.3** Demand, generation and power prices of the nodal pricing model

| Node | Demand (MWh) | Generation (MWh) | Prices (€/MWh) |
|---|---|---|---|
| 1 | | 300 | 25 |
| 2 | | 300 | 30 |
| 3 | 200 | | 27.5 |
| 4 | | 200 | 47.5 |
| 5 | 300 | | 45 |
| 6 | 300 | | 50 |

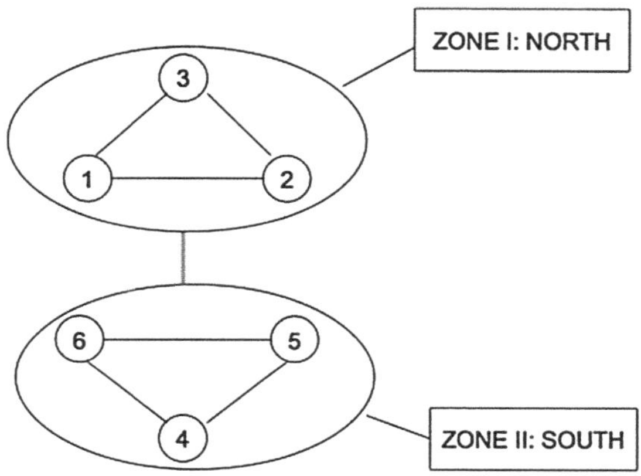


**Fig. 14.2** The two zone example

constructed as PTDF weighted sums of the dual variables of constrained lines (which are the real-time market prices of line services in the US flowgate system). We use the six node example to construct two zonal models respectively with two and three zones.

### *14.2.2 A Two Zone Model*

Consider Fig. 14.2 where nodes 1, 2 and 3 have been aggregated into a Northern zone, the three other nodes 4, 5 and 6 being aggregated in a Southern zone.

The two lines 1–6 and 2–5 form the single interconnector between the zones. This two zone grid can be viewed as part of the 2006 trilateral market between Belgium, France and the Netherlands after grouping Belgium and the Netherlands (see Fig. 14.3). The cost and demand data of the example reported in Table 14.3 are obviously unrelated to these countries but the radial topology of Fig. 14.2 offers some resemblance to the trilateral market of Fig. 14.3 that can be useful.

Market Coupling organizes trading between the Northern and Southern zones through a market-clearing model whose principle is as follows (we describe a very simplified version of Market Coupling and refer the to Djabali et al. (2010) and Van Vyve (2011) for further details).

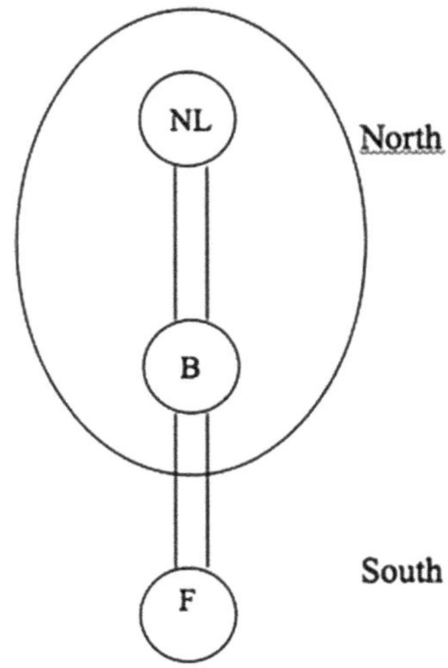


**Fig. 14.3** The two zone network as a reduction of the trilateral B-F-NL market

**Table 14.4** Intra zonal equilibrium before cross border trade

| |
|---|
| **Northern zone** |
| Global supply (horizontal addition of supply functions at nodes 1 and 2) |
| $p_N = 12.5 + 0.025\ Q_N$ |
| Global demand |
| $p_N = 37.5 - 0.05\ Q_N$ |
| Equilibrium |
| $p_N = 20.83,\ q_1 = 216.6,\ q_2 = 116.6$ |
| $q_1 + q_2 = q_N = q_3 = 333.3$ |
| **Southern zone** |
| Global supply (node 4) |
| $p_S = 42.5 + 0.025\ Q_S$ |
| Global demand (horizontal addition of demand functions at node 5 and 6) |
| $p_S = 77.5 - 0.05\ Q_S$ |
| Equilibrium |
| $p_S = 54.16,\ q_5 = 258.33,\ q_6 = 208.33$ |
| $q_5 + q_6 = q_S = q_4 = 466.6$ |

#### 14.2.2.1 Clearing Zonal Markets

Recalling the Market Coupling assumption that intra-zonal transmission systems are unconstrained, we suppose that there is one Power Exchange in each zone (here Northern and Southern PXs) that clears the energy market and finds the equilibrium between supply and demand in that zone. One can verify that all generators are active and that nodal demands are positive at the intra-zone equilibrium. Equilibrium prices and quantities in that equilibrium are given in Table 14.4.

**Table 14.5** Impact of a unit export from North to South on the intra-zone equilibria

**Northern zone**

Global supply with export

$p_N = 12.5 + 0.025\ (Q_N + Ex)$

Global demand

$p_N = 37.5 - 0.05\ Q_N$

Start from Ex = 0 and suppose a unit export $\Delta Ex = 1$

$\Delta p_N = 0.0166,\ \Delta q_1 + \Delta Ex_1 = 0.33,\ \Delta q_2 + \Delta Ex_2 = 0.33$

$\Delta Q_N + \Delta Ex = 0.66,\ \Delta Q_N + \Delta q_3 = -0.33$

Note that one cannot separate $\Delta Q_i + \Delta Ex_i$, i = 1, 2

**Southern zone**

Global supply (node 4)

$p_S = 42.5 + 0.025\ Q_S$

Global demand

$p_S = 77.5 - 0.05(Q_S + Ex)$

Variation of the equilibrium for $\Delta Ex = 1$

$\Delta p_S = -0.0166,\ \Delta q_4 = -0.666$

$\Delta q_S + \Delta Ex = 0.333$

Note that one cannot separate $\Delta Q_i + \Delta Ex_i$, i = 5, 6

**Table 14.6** Import and export functions of the Northern and Southern zones

**Northern zone**

Global supply

$p_N = 12.5 + 0.025\ (Q_N + Ex)$

Global demand

$p_N = 37.5 - 0.05\ Q_N$

Eliminating $Q_N$ we get

$p_N = \frac{1}{3}\ (62.5 + 0.05Ex)$

**Southern zone**

Global supply

$p_S = 42.5 + 0.025\ Q_S$

Global demand

$p_S = 77.5 - 0.05(Q_S + Ex)$

Eliminating $Q_S$ we get

$p_S = \frac{1}{3}(162.5 - 0.05Ex)$

The price difference between the two zones (of the order of 35.5) suggests that trading should take place from North to South. This is what Market Coupling organizes using a model that bears some resemblance to an OPF as we discuss now.

#### 14.2.2.2 Coupling the Northern and Southern Markets

Consider a unit export from North to South. The principle of Market Coupling is to consider the export as a parameterized demand (here noted Ex for Export) in the Northern zone and as a parameterized supply (the same Ex) in the Southern zone. The equilibrium in each zone is then modified as stated in Table 14.5 that also reports the price and generation changes in the two zones. Because marginal cost and demand curves are linear in this example, these coefficients can be used to construct linear

**Table 14.7** The "OPF" like model of the inter zone market clearing of the two zone model

| |
|---|
| $\text{Max} \int_0^{Ex} p_S(e)dE - \int_0^{Ex} p_N(e)de$ |
| s.t. "$E$ is feasible for the interconnection" |

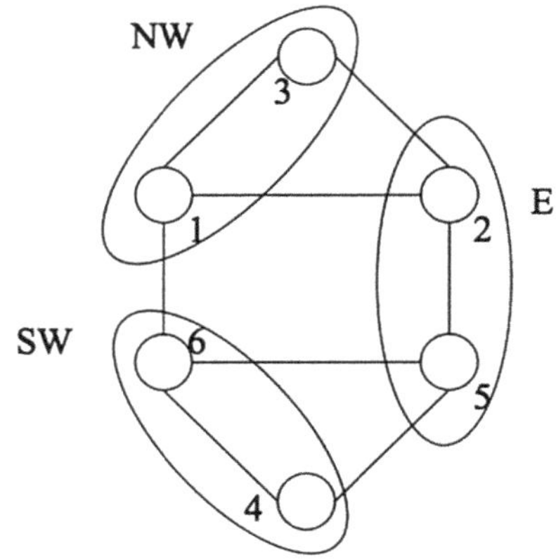


**Fig. 14.4** The three zone model

export/import curves in both zones as depicted in Table 14.6. It is then possible to set up a welfare maximization problem very similar to an OPF on a two zone grid provided we have a representation of that grid. This is done in a stylized way in Table 14.7 where the grid is provisionally represented by the statement "Ex is feasible for the interconnection". We discuss the construction of the grid and the implied transmission services after the presentation of the three zone model.

### *14.2.3 Three-Zone Model*

Three node models have been instrumental in analysis of congestion management; Oren (2012) uses a three-node model in his chapter on Financial Transmission Rights. The following constructs a three zone model using the six node example. Suppose a partitioning of the nodes in three zones respectively noted East (E) consisting of nodes 2 and 5, North West (NW) comprising nodes 1 and 3, and South West (SW) formed by nodes 4 and 5. This is depicted in Fig. 14.4. Each zone contains one generator and one customer so that one can clear each intra-zone energy market. In contrast with the two zone model, zone E now also contains a domestic transmission constraint namely line 2–5. This will require some particular treatment as explained below. Last the three zones are linked by interconnections composed of one or two lines. The interconnection between the two Western zones is limited to the sole capacity constrained line (1–6). The other interconnections consist of unconstrained lines: NW-E is formed by lines 3–2 and 1–2 while SW-E consists of lines 6–5 and 4–5. This three zone model can again be viewed as a simplified representation of the Pentalateral market that went live in November

**Table 14.8** Intra-zone equilibrium in the three zone model before cross border trade

| |
|---|
| **NW zone** |
| Supply (node 1): $p_{NW} = 10 + 0.05\ q_{NW}$ |
| Demand (node 3): $p_{NW} = 37.5 - 0.05\ q_{NW}$ |
| Equilibrium: $p_{NW} = 23.75$, $q_{NW} = 275$ |
| **SW zone** |
| Supply (node 4): $p_{SW} = 42.5 + 0.025\ q_{SW}$ |
| Demand (node 6): $p_{SW} = 80 - 0.1\ q_{SW}$ |
| Equilibrium: $p_S = 50$, $q_{SW} = 300$ |
| **E zone** |
| Supply (node 2): $p_E = 15 + 0.05\ q_E$ |
| Demand (node 5): $p_E = 75 - 0.1\ q_E$ |
| Equilibrium: $p_E = 35$, $q_E = 400$ |

**Table 14.9** Impact of a unit export from each zone on the intra-zone equilibrium

| |
|---|
| **NW zone** |
| Supply: $p_{NW} = 10 + 0.05\ (q_{NW} + Ex_{NW})$ |
| Demand: $p_{NW} = 37.5 - 0.05\ q_{NW}$ |
| Unit export: $\Delta Ex_{NW} = 1$ |
| $\Delta p_{NW} = 0.025$, $\Delta q_{NW} + \Delta Ex_{NW} = 0.5$, $\Delta q_{NW} = -0.5$ |
| Generation increase: 0.5, demand decrease: 0.5 |
| **SW zone** |
| Supply: $p_{SW} = 42.5 + 0.025\ (q_{SW} + Ex_{SW})$ |
| Demand: $p_{SW} = 80 - 0.1\ q_{SW}$ |
| Unit export: $\Delta Ex_{SW} = 1$ |
| $\Delta p_{SW} = 0.02$, $\Delta q_{SW} + \Delta Ex_{SW} = 0.8$, $\Delta q_{SW} = -2$ |
| Generation increase: 0.8, demand decrease: 0.2 |
| Equilibrium: $p_S = 50$, $q_{SW} = 300$ |
| **E zone** |
| Supply: $p_E = 15 + 0.05\ (q_E + Ex_E)$ |
| Demand: $p_E = 75 - 0.1\ q_E$ |
| Unit export: $\Delta Ex_E = 1$ |
| $\Delta p_E = 0.0333$, $\Delta q_E + \Delta Ex_E = 0.666$, $\Delta q_E = -0.0333$ |
| Generation increase: 0.8, demand decrease: 0.2 |

2010 where SW and E respectively represent France and Germany while NW integrates Belgium, Luxemburg and The Netherlands. As in the two zone model the cost and demand data of the example are unrelated to those of the Pentalateral market but the topology of Fig. 14.4 offers some resemblance with the real system that can usefully be kept in mind.

Proceeding as in the two node grid and neglecting the domestic transmission constraint in zone E for the time being, we find market clearing quantities and prices in each zone as depicted in Table 14.8.

Supposing a unit export from each zone, we can again compute the price variations and the changes of generation and demand that result from that export in the three zones. These are listed in Table 14.9. As before the linearity of the supply and demand curves allows one to use these coefficients to construct linear import/export curves in each zone. These are listed in Table 14.10. Assuming a grid model to represent export/import between zones the clearing of the cross border

**Table 14.10** Impact of a unit export from each zone on the intra-zone equilibrium

| |
|---|
| **NW zone** |
| Supply: $p_{NW} = 10 + 0.05\ (q_{NW} + Ex_{NW})$ |
| Demand: $p_{NW} = 37.5$ |
| Eliminating $q_{NW}$ we get $p_{NW}\ (Ex_{NW}) = 23.75 - 0.025\ Ex_{NW}$ |
| **SW zone** |
| Supply: $p_{SW} = 42.5 + 0.025\ (q_{SW} + Ex_{SW})$ |
| Demand: $p_{SW} = 80 - 0.1\ q_{SW}$ |
| Eliminating $q_{SW}$ we get $p_{SW}\ (Ex_{SW}) = 50 - 0.02\ Ex_{SW}$ |
| **E zone** |
| Supply: $p_E = 15 + 0.05\ (q_E + Ex_E)$ |
| Demand: $p_E = 75 - 0.1\ q_E$ |
| Eliminating $Q_E$ we get $p_E(Ex_E) = 35 - 0.033\ Ex_E$ |

**Table 14.11** The "OPF" like model of inter-zone market clearing of the three zone model

$$\text{Min} \int_0^{Ex_{NW}} p_{NW}(e_{NW})de_{NW} + \int_0^{Ex_{SW}} p_{SW}(e_{SW})de_{SW} + \int_0^{Ex_E} p_E(e_E)de_E$$

$$\text{s.t. } Ex_{NW} + Ex_{SW} + Ex_E = 0$$

$$(Ex_{NW},\ Ex_{SW},Ex_E\ )\ \textit{is feasible for the grid}$$

market can be obtained by an OPF like model that is given in Table 14.11 where we have left the detail of the grid for the next section.

## 14.3 Transmission Services in Market Coupling

Market Coupling is a zonal system where zones are today identical to countries. In contrast with the USA, energy and transmission markets remain separated in the EU: PXs clear the energy market in each zone and provide the import/export curve of the zones; TSOs remove intrazonal congestion and provide the representation of the grid used for clearing the inter-zone energy market. Two representations of the grid are relevant for the discussion: the so-called Transfer Capacity model (TC) is currently in operation; a new Flow Based model (FB) is being developed. Whatever the model, representing the grid by zones and interconnections between them can only approximate reality. TSOs may thus need to conduct both inter and intra-zone counter trading services to remove overflows that result from these simplifications. Counter-trading adds to the balancing operations that take place in real-time and are not part of the day-ahead energy market. None of these involve tradable transmission rights relevant for the energy market; they are thus not discussed in this paper.

Transmission services offered on this zonal model will differ from those in nodal systems. We first consider the TC model that is implemented today and then turn to the FB model that is meant to be the reference system for the future. We conduct the discussion on the two and three zone examples.

## *14.3.1 The Two Zone Model*

Consider first the two-zone model depicted in Fig. 14.1. Section 14.2.2.1 shows that exports should normally flow from North to South. A single interconnection links the Northern and Southern zones. We examine how this interconnection compares to the standard flowgate (line) of the nodal model.

### 14.3.1.1 The TC Model

Computing Interconnection Characteristics

The PTDF of the Northern zone to the interconnection is obviously equal to 1: all exports from North must go through this single interconnection and similarly for exports from South. The only remaining relevant question is the capacity of this interconnection.

TSOs publish interconnection capacities but do not guarantee them before day-ahead. TSOs also do not explain today their method of calculation but will have to do so in the network code as required by the Framework Guidelines. One should not expect too much from this publication though: transfer capacities of interconnectors composed of several lines cannot be determined unambiguously in a meshed network and are always conventional figures obtained by some heuristics. Rious et al. (2008) illustrate a possible approach: they provide a "worst case" analysis that we apply as follows.

Suppose a unit export from North to South. First assume that domestic transactions have no impact on the flows on the interconnection (this is corrected later in the section). The most effective use of the interconnection occurs for a transaction between nodes 3 and 4 where Kirchoff's laws equally separate the physical flow between the two lines 1–6 and 2–5. The maximal export is then 400 MW. In contrast the minimal use of the interconnection lines occurs for a transaction between nodes 1 and 6. Most of the flow (1/0.625 where 0.625 is the PTDF of node 1 on line (1–6)) is directed to the 200 MW line 1–6 while the rest flows on line 2–5 of capacity 250. Because the distribution of transactions between North and South is not known beforehand, the TC is obtained by selecting the most unfavorable pattern, which is the transaction from 1 to 6. This gives a flow of 200/0.625 or 320 MW. The interconnection can accommodate all flow patterns resulting from an export of 320 from North into South.

The 320 MW value must be corrected for the impact of pre cross-border trade on the interconnection lines. Following Rious et al. (2008) we suppose that zonal TSOs rely on some intra-zone trade scenarios to compute the flow on the interconnection lines before cross-border trade. In order to fix ideas we assume that the flows at equilibrium before cross-border trade described in Table 14.4 reflect this practice. Taking node 6 as a hub, it is easy to verify (see Appendix 14.1) that these domestic transactions imply net (rounded) flows of 3 MW and −3 MW on lines 1–6 and 2–5 respectively. The TC available for North South commercial transactions is then reduced by that amount and becomes 317 (320 − 3).

TCs differ by direction. Consider now the TC from South to North. The TSOs can realistically assume that only transactions from 4 (the generator in South) to 3 (the demand node in North) are possible. Flows then distribute equally on lines 1–6 and 2–5, which gives the most efficient use of the interconnection. The TC from South to North is then 403 (400 + 3) when there is no contingency.

Contingency analysis is important for determining firm transmission services. Suppose a failure of one of these two lines, for instance 1–6. The capacity of the interconnection is then 250, which is the capacity of the remaining line (2–5). The transmission capacity drops to zero when both lines are down.

### Superposing Transactions in the TC Model

The capability to superpose transactions in the representation of the grid is a key element for developing portfolios of tradable transmission rights in the nodal system (Oren 2012). Note that Regulation EC No. 714/009 also requires some sort of superposition property when stating "nominations of transmission rights in the opposite direction shall be netted" (Annex 1, paragraph 4.2). We analyze whether a superposition property also holds in the zonal system. We first examine whether transactions of opposite directions on the N-S interconnection can be netted (as requested by the Regulation) to determine the utilization of the interconnection. We then explore whether transactions can be added for determining the utilization of the interconnection. The conclusion is that these operations are not guaranteed in the TC model (see ETSO 2001 for a more extensive discussion).

#### *Subtracting (Netting) Transactions*

Assume two inter-zonal transactions of equal amount (e.g. 400 MW) but opposite direction. Suppose that the N to S transaction is between nodes 1 and 6 and the S to N transaction between nodes 4 and 3. Using the PTDFs one checks that the combination of these transactions leaves a residual flow of $400 * 0.125 = 50$ MW on line 1–6 and $-400 * 0.125 = -50$ MW on line 2–5. While the sum of the inter-zonal transactions is algebraically zero, the flows on the lines are not zero. The two transactions would have netted to zero line flows if they had taken place between the same nodes. We find that, in contrast with node-to-node transactions, zone-to-zone transactions cannot be netted on interconnections (aggregate flowgates) when their entry and exit points differ.

The lack of netting is inconsequential for the interconnection lines of 200 and 250 MW that can accommodate the residual flows of 50 and −50 MW. This is no longer true if the capacities of these lines were respectively 49 MW for line 1–6 (changed from 200 to 49) and 250 MW for line 2–5 (unchanged). Netting the two transactions to zero implies that the interconnection is not saturated while it is effectively saturated by line (1–6). This misjudgment can be corrected in two ways.

One is to permit some violation of the physical capacities of the lines in the day-ahead market and to correct that violation by counter-trading in real-time. The other possibility is not to allow netting and to only work with single direction transmission rights. There are then two TCs, one from North to South and the other from South to North. Transactions are then allocated according to the directional TCs. This excludes the N to S 400 MW transaction but permits the one from S to N. This is the solution adopted in the EU system where transmission rights are directional (see Djabali et al. 2010) for day-ahead transactions and EnBW (undated) for yearly and monthly transmission rights. This implies that trade can be restricted because of the sole market design without physical capacities or generators' market power having any responsibility for this restriction.

*Adding Transmission Rights*

A similar reasoning applies to the adding of transactions. Consider two N to S transactions of 200 MW from node 1 to node 6 and from node 2 to node 5 respectively. Their combination entails flows of 200 MW on lines 1–6 and 2–5 respectively or globally 400 MW. These two transactions are feasible for the real network but exceed the TC of the interconnection. One transaction will then be rejected. Here again trade is limited by the market design without physical capacities or generators' market power having any responsibility with this limitation.

### Long and Short Term TC

The Framework Guidelines require that long and short term computations of capacities be compatible. This seems difficult to achieve. A worst-case analysis indeed depends on the set of flow patterns that it considers. The larger this set, the lower the TC value. Long term transmission rights are based on a long term view of the TCs for which the set of plausible transactions is probably larger than in the day-ahead. TSOs are aware of that and will therefore restrict the computed long term TC through some rule of thumb in the hope of making it compatible with the short-term one. Long term and short term TCs are thus bound to be assessed differently even though the network and hence its physical possibilities in different contingencies are identical. Besides the difficulty of satisfying the compatibility requirement of the Framework Guidelines, this raises the question of how to maximize TCs while at the same time insuring the firmness of transmission rights. This is the central question addressed later in the paper.

### Conclusion

Transactions can be netted or added on lines (flowgates) in the nodal system: they satisfy the superposition property, at least if we assume constant PTDFs. This is not true for interconnections in the zonal grid. Node-to-node services in the nodal system are constructed from portfolio of flowgate services where adding and netting play a fundamental role. Because superposition of transactions on interconnection does not hold in the zonal system, it will also be missing on interconnections and hence in zone-to-zone services. The implication is that one cannot construct portfolios of transmission rights in the sense that one cannot verify that a set of zone-to-zone services that is found to be feasible for the zonal grid is effectively feasible for the real grid. If this verification cannot be done in day-ahead, it is unlikely that it can be done for transmission rights in the forward market.

#### 14.3.1.2 The Flow Based Model

The Flow-Based (FB) model is meant to offer a better representation of the grid for clearing the inter-zone energy market in day-ahead. It can alternatively also be used to increase the TCs in today inter-zone market-clearing as explained by TSOs in Amprion et al. (2011a, b). The FB model crucially relies on so-called "Generation Shift Keys" (GSK). The importance of these factors is generally overlooked in academic studies of the FB models even though they are essential elements of the model. According to TSOs *"It is the Generation Shift Key (GSK) that defines how a change in net position is mapped to the generation units in a bidding area"* (Amprion et al. 2011a). Section 5.7 of Amprion et al. 2011b describes the way different TSOs compute GSK: *"RTE puts its best effort to anticipate the best generation pattern for France in D + 2"*. The German TSOs only mention that they include *"power plants in GSK that are very quick and flexible"*. Elia gives the same type of information. Tennet's approach is still less informative: *"Tennet has no access to the merit order of units, however the list of units that appear in the GSK is evaluated by the operators on a daily basis for known outages"*. We come back to these statements when discussing the forward market. In the meantime and because these quotations of TSOs do not really constitute a description of how they compute GSKs, we introduce a notion of "perfect" GSKs that leads to an exact representation of the real grid. We then explain that real GSKs and hence the FB model can necessarily be "imperfect". We conclude with the implications for long term transmission rights.

TSOs usually describe the FB model in terms of bilateral transactions between PXs. Appendices of their reports (e.g. Amprion et al. 2011a, b) explain that it is more logical to conduct the discussion in terms of net positions of PXs (the outcome of market clearing). We adopt this more logical presentation and analyze the situation in terms of net positions on the exchanges.

##### "Perfect" GSK

Beginning with the clearing of the intra-zone energy markets before cross-border trade depicted in Table 14.4, we consider changes of generation and demand resulting from a 1 MW export from North to South (Table 14.5). Using the

PTDFs we compute the incremental utilization of lines (1–6) and (2–5) due to this export. Generator 4 originally produces 466.6 (Table 14.4) and decreases its output by 0.66 for each MW export. It will stop generating when the export is 700 MW. Demand at node 3 decreases by 0.33 for each 1 MW incremental export and is still positive (100 MW) at the 700 MW export level. The other generation and demand levels increase with export and hence remain positive. Because supply and demand curves are linear, the incremental generations and demands with respect to the no cross-border trade equilibrium are proportional to the export as long as it remains lower than 700 MW. Using the PTDFs one infers per unit incremental line flows as shown in Appendix 14.2: 1 MW export respectively leads to 0.5 MW and 0.5 MW additional flows on lines 1–6 and 2–5. We refer to these values as "zonal PTDFs" (ZPTDFs) of lines (1–6) and (2–5) because they represent the impact of a zonal export on these lines. The ZPTDFs assume that the energy markets clear in the two zones for these different export levels as in the philosophy of Market Coupling (see Djabali et al. 2010). One can easily verify that these ZPTDFs remain valid as long as the set of active generators and demands does not change, which is the case here for any export level not exceeding 700 MW.

Taking into account residual loop flows before cross-border trade (3 MW from North to South on line 1–6 and −3 MW in the opposite direction on line 2–5) one obtains that the 700 MW export from North to South leads to an utilization of lines 1–6 and 2–5 equal to 353 MW and 347 MW respectively. These flows are larger than the capacities of the lines of respectively 200 and 250 MW and hence are not feasible for the grid. Applying a standard reasoning in terms of DC load flow approximation, this suggests introducing two "ZPTDF" inequality constraints as in Table 14.11 to express the real limitation of export implied by the grid. These constraints have the usual PTDF form of the DC load flow approximation, but use ZPTDFs.

ZPTDFs are computed on the basis of both the original PTDFs (using node 6 as the hub) and of incremental demands and supplies induced by export as computed in Table 14.5. These are the GSKs mentioned in TSOs' documents: they indicate how generation (and here also demand) is modified as a result of the unit export from the Northern to the Southern zone. Because we also deal with demand in this computation we refer to these coefficients as Generation *and Demand* Shift Keys (GDSKs). Line capacities in the grid constraints of Table 14.12 are based the original data taking into account the flows on line (1–6) and (2–5) before cross-border trade.

One can alternatively consider an import from the Southern to the Northern zone. This transaction requires the expensive generator in node 4 to increase production to displace the cheap generators at nodes 1 and 2 and hence is unlikely to take place. This should not be of concern for the TSOs in charge of modeling the network. Using the same reasoning as before one can write the ZPTDF constraints of Table 14.13 that express the limitation of the export from the Southern zone.

Taking export from the Northern to the Southern zone while adapting the clearing of the zonal markets, one finds that the capacity of the N-S interconnection without contingencies ranges from −406 (−203/0.5) to 396 (197/0.5) where the

**Table 14.12** ZPTD constraints on lines (1–6) and (2–5) in the two zone model in the North South direction

| |
|---|
| $0.5\ Ex \leq 197$ |
| $0.5\ Ex \leq 253$ |

**Table 14.13** ZPTD constraints on lines (1–6) and (2–5) in the two zone model in the South–North South

| |
|---|
| $-203 \leq 0.5\ Ex$ |
| $-247 \leq 0.5\ Ex$ |

two bounds are reached because of saturation of line 1–6. Capacity in both directions is higher than the one computed with TCs. This improves cross-border trade compared to the TC model as announced by the proponents of the FB model (Amprion et al. 2011b).

### Superposing Transactions in the FB Model

Because the superposition of transactions is essential for the construction of portfolios of transmission rights, we again successively consider netting and adding transactions.

#### *Subtracting (Netting) Transactions*

Assume again two transactions of equal amount (e.g. 400 MW) but opposite direction. Because they are inter-zonal, their impact is accounted for through the PXs. This implies that their impact on lines (1–6) and (2–5) is computed using the GDSKs determined from market clearing. These coefficients are identical in the two directions (at least locally as we shall explain), which implies that these line flows cancel out. The two transactions can thus effectively be netted in the sense that N-S and S-N transactions that algebraically add up to zero imply flows on the interconnecting lines that are also zero.

#### *Adding Transmission Rights*

It is easy to verify that the same result applies when adding two N-S transactions of 197 MW (twice 197 MW or 394 is the maximal export capacity found in Table 14.11 in Sect. 14.2.1.1). The flows on the interconnection lines implied by the two transactions are twice the flows implied by one of these transactions. This suggests that adding and netting are both allowed, meaning that transactions are superposable and hence that the FB model meets at least in the day-ahead market the property of superposition mentioned in Oren (2012). The following shows that this conclusion is unfortunately too optimistic.

**Table 14.14** Impact of an incremental N-S export when plant 2 is at capacity

| |
|---|
| **Northern zone** |
| Global supply: $p_N = 10 + 0.05\ (q_N + Ex - 200)$ |
| Global demand: $p_N = 37.5 - 0.05\ q_N$ |
| Suppose a unit incremental export $\Delta Ex = 1$ |
| $\Delta p_N = 0.025$, $\Delta q_N + \Delta Ex = 0.5$ |
| $\Delta q_1 + \Delta q_N + \Delta Ex = 0.5, + \Delta q_2 = -0.5$ |
| **Southern zone** |
| Global supply: $p_S = 42.5 + 0.025\ Q_S$ |
| Global demand: $p_S = 77.5 - 0.05\ (Q_S + Ex)$ |
| Suppose a unit import $\Delta Ex = 1$ |
| $\Delta p_S = -0.01666$, $\Delta q_4 = -0.666$ |
| $\Delta q_5 + \Delta Ex_5 = 0.1666$, $\Delta q_6 + \Delta Ex_6 = 0.166$ |
| $\Delta q_S + \Delta Ex_S = 0.333$ |

## Non-linear GDSK

TSOs insist on the "linearity" of GSKs (Amprion et al. 2011b). The perfect GDSKs are constant, which in the terminology of Amprion et al. means that the effect of exports on generation (and demand in our case) is a linear function of these exports. The wording is different (constant vs. linear) but our "perfect GDSKs" meet the requirements of Amprion et al. (2011b). The real situation is more complex (even in this extremely simplified six node example) as we explain. Suppose the same data as before except that the plant located at node 2 has a limited generation capacity of 200 MW. This is larger than the generation of 183 MW of this plant when export is 198 MW. It is not possible for this plant to ramp up sufficiently to accommodate twice that export level. The plant will keep generating at its 200 MW limit when export becomes sufficiently large and plant 1 as well as demand at node 3 will adapt to increase export. This requires recomputing the GDSKs to reflect to the new generation pattern where plant 2 is at capacity. Table 14.14 reports the marginal impact of a unit export when machine 2 is at capacity.

Appendix 14.3 shows the computation of the new GDSKs that respectively amount to 0.8 and 0.2 for the lines (1–6) and (2–5). Computations show that an export of 250 MW leads to a generation of plant 2 just below 200 and to ZPTDFs of 0.5 and 0.5 for both lines; in contrast ZPTDFs become 0.8 and 0.2 for the fraction of the export exceeding 250 MW. The two 198 transactions are therefore not superposable because their combination implies a change of ZPTDFs.

The example shows that the model of the aggregate grid is ambiguous because the ZPTDFs change with export. This conclusion obviously extends to the case when the supply and demand curves are non-linear as happens if one introduces further constraints on the capacities, or non-linear marginal cost curves (the usual case) or demand, or a mixture of both. GDSKs will thus in general be non-constant, with the implication that the ZPTDFs of the aggregate network become non-linear. Non-linear ZPTDFs imply that transaction cannot be superposed. TSOs assume constant GSKs (Amprion et al. 2011b) but do not justify that assumption.

**Table 14.15** Impact of zonal exports in the three zone model (GDSK)

| |
|---|
| **NW zone** |
| Supply: $p_{NW} = 10 + 0.05\ (q_{NW} + Ex_{NW})$ |
| Demand: $p_{NW} = 37.5 - 0.05\ q_{NW}$ |
| Variation of the equilibrium for $\Delta Ex_{NW} = 1$ |
| $\Delta p_{NW} = 0.025$, $\Delta q_{NW} + \Delta Ex_{NW} = 0.5$, $\Delta q_{NW} = -0.5$ |
| Generation increase: 0.5, demand decrease: −0.5 |
| $\Delta q_1 + \Delta q_N + \Delta Ex = 0.5, + \Delta q_2 = -0.5$ |
| **SW zone** |
| Supply: $p_{SW} = 42.5 + 0.025\ (q_{SW} + Ex_{SW})$ |
| Demand: $p_{SW} = 80 - 0.1\ q_{SW}$ |
| Variation of the equilibrium for $\Delta Ex_{SW} = 1$ |
| $\Delta p_{SW} = 0.02$, $\Delta q_{SW} + \Delta Ex_{SW} = 0.8$, $\Delta q_{SW} = -0.2$ |
| Generation increase: 0.8, demand decrease: 0.2 |
| **E zone** |
| Supply: $p_{Ex} = 15 + 0.05\ (q_{Ex} + Ex_E)$ |
| Demand: $p_{Ex} = 75 - 0.1\ q_{Ex}$ |
| Variation of the equilibrium for $\Delta E_{Ex} = 1$ |
| $\Delta p_E = 0.033$, $\Delta q_{Ex} + \Delta E_{Ex} = 0.666$, $\Delta q_{Ex} = -0.333$ |
| Generation increase: 0.666, demand decrease: 0.333 |

### Imperfect FB Models

The superposition property of transactions in the FB model justified calling the GDSKs "perfect" in Sect. 14.3.1.2.1. This was unexpected good news: it is indeed known that the aggregation of a meshed grid by a reduction of the number of nodes or lines is never perfect. It is thus relevant to identify what entails this welcome but unexpected "perfection". Two linearity assumptions guarantee the result in the example. One is the linearity of the supply and demand functions of the six-node model. The second assumption comes from keeping the same set of active generators and demand as export increases until hitting the capacity constraints. These two linearity conditions imply that constant GDSKs measure the changes of generation and demand resulting from an export/import deviation from the intra-zonal equilibrium. These together with the PTDFs of the original grid give the changes of flows (ZPTDF) on the lines resulting from these exports and imports. ZPTDFs are true PTDFs of the aggregate network operating under Market Coupling but they are only valid locally and the range of this validity is a matter of empirical analysis. The scant information on the computation of GSKs provided by TSOs (Sect. 5.7 in Amprion et al. 2011b as recalled in Sect. 14.3.1.2) gives little evidence that GSKs are effectively constant in practice. Some additional comments of TSOs cast additional doubt on the matter. In Sect. 5.7.1 of the above report RTE states "*in order to avoid unwanted behavior of the GSK on major critical branches, RTE excludes some generating units from the GSK*". The statement has a flavor of discrimination that will probably be unnoticed by competition authorities. From the perspective of this paper, it signals that GSKs can be subject to ad hoc manipulation by TSOs of the type already mentioned for TCs.

Long and Short Term TC Model

ZPTDFs are constructed on the basis of GDSKs that are provided by the clearing of the day-ahead market. This raises the question as to how construct ZPTDFs representing the grid in the forward market. This relates to the distinction between implicit and explicit auctions in the EU language. Market Coupling is an implementation of implicit auctions where transmission rights are allocated implicitly by the double clearing of the intra-zonal and inter-zonal energy markets. This determines ZPTDFs that are locally valid in the day-ahead market or non-linear ZPTDFs that are valid everywhere but do not allow for superposing transactions by adding and netting. TSOs exclude non-linear GDSKs (Sect. 5.7 in Amprion et al. 2011b). These constructions may look a bit complicated but are well under control when the information from market clearing is on hand, that is in implicit auctions.

The situation is quite different in the forward market where transmission rights are allocated by so-called "explicit auctions", without reference to any energy market clearing. It is again useful to recall here that the Framework Guidelines require that the short and long term descriptions of the grid be compatible. This raises the question of computing compatible GDSKs for both the long and short terms, or in other words to reconcile the explicit and implicit auctions. This worry does not appear in the nodal system that relies on a single physical description of the grid both for the long and short terms without reference to the energy market. But Market Coupling requires adapting the method for computing daily GDSKs to monthly and yearly GDSKs. Transposing RTE statement recalled in Sect. 14.3.1.2 (which is the most explicit among TSOs quotes on the determination of GSKs), this would imply "the best effort to anticipate the best generation pattern for France in *one month/one year*". This is certainly more difficult than making the same anticipation "in D + 2" as in the day-ahead market. Tennet's statement (recalled in Sect. 14.3.1.2) is the most deceptive in that respect: how can one construct a year ahead merit order without even the knowledge of it in day-ahead?

A Final Note: Do Perfect GDSKs Lead to the Nodal System?

It is easy to verify that the clearing of the inter-zone model by Market Coupling using the computed transmission capacities between North and South is not identical to the result of the nodal system, reported in Table 14.3. Market coupling indeed imposes that all transactions in a zone clear at a single price. This constraint is absent from nodal pricing. The FB model therefore improves on the TC model but remains less efficient than the nodal system.

## *14.3.2 The Three Zone Model*

The discussion of the two zone network easily extends to three zones at the cost of additional complications for the TC model.

#### 14.3.2.1 The TC Model

TSOs will publish the details on the computation of transfer capacities in the grid codes. In the meantime we only mention in passing the difficulty of applying the worst-case analysis but do not elaborate. Consider computing TCs between the NW and SW zones. These are linked by an interconnection composed of a single line of capacity 200 MW. But 200 MW is not the TC between the two zones! The available capacity for transaction between NW and SW is indeed influenced by both the domestic transactions of zone E as well as by the transactions between the E and the NW and SW zones. TSOs must thus make worst-case assumptions both about the transactions that are completely external to their operations and about the transactions between domestic and other zones.

#### 14.3.2.2 The FB Model

ZPTDF for the Tree Zone Model

In contrast with the TC model, the analysis of the FB model for the two zone system easily extends to the three zone case. Table 14.15 reports the change of generations and demands resulting from a unit export in each zone while maintaining intra-zonal market clearing. These figures can then be used with the PTDFs to compute the ZPTDFs of lines 1–6 and 2–5 (keeping node 6 as the hub). These are documented in Appendix 14.4 and the results reported in Table 14.16.

The clearing of the energy market before cross-border trade implies flows on the interconnections. These are computed in Appendix 14.5 and respectively amount to 186 and 214 MW on lines 1–6 and 2–5. They are feasible for the grid (where capacities of these lines are respectively 200 and 250) but leave little additional room for cross-border trade.

Combining the GDSK and PTDFs one obtains a representation of the grid model in three relations. Two ZPTDF inequalities express constraints on line flows, a third balance equation sums zonal export to zero. These are stated in Table 14.17. This model has the same form as the PTDF representation of a three node system. The only difference is the use of ZPTDFs.

The Role of the Line 2–5 Constraint

The three-zone model comprises two capacitated lines but only one capacitated interconnection. The jurisprudence of European Institutions concentrates on limitations to cross-border trade caused by insufficient interconnections even though the impact of domestic limitations is progressively recognized as a factor for determining cross-border capacities (Duthaler et al. 2008; Duthaler and Finger 2008, 2009; ACER 2011). This raises the question of the treatment of the

**Table 14.16** ZPTDs of the three zone model

| | (1–6) | (2–5) |
|---|---|---|
| NW | 0.6 | 0.4 |
| SW | 0.05 | −0.05 |
| E | 0.375 | 0.291 |

**Table 14.17** The FB model of the three zone grid

$$-386\ (-200-186) \le 0.594\ E_{NW} + 0.05\ E_{SW} + 0.375\ E_E \le 14\ (200-86)$$
$$-464\ (-250-214) \le 0.406\ E_{NW} - 0.05\ E_{SW} + 0.292\ E_E \le 36\ (250-214)$$
$$E_{NW} + E_{SW} + E_E = 0$$

**Table 14.18** Two dimensional view of the FB model in the three zone grid

$$-386 \le 0.544\ E_{NW} + 0.325\ E_E \le 14$$
$$-464 \le 0.456\ E_{NW} + 0.342\ E_E \le 36$$

constrained line 2–5 in the model of Table 14.17. We first neglect this question and briefly discuss the proposal to use the FB model to construct improved Transfer Capacities (Amprion et al. 2011b).

## From the FB the TC Model?

The three zone model of Table 14.17 contains two ZPTDF inequality constraints and one balance equation each involving the PXs' export variables. Because export variables are unconstrained, the balance equation can be used to express one of the export variables as a function of the other two. Table 14.18 reports the model obtained after elimination of the SW export.

Amprion et al. (2011a, b) suggest using this reduced formulation to obtain improved (larger) TCs compared to the classic method. We illustrate the approach in Fig. 14.5 on a FB model consisting of the sole constraint on the interconnection NW-SW (line 1–6). The principle is to select a rectangle inside the space determined by the FB constraint. The advantage compared to the standard method is to avoid any "worst case" analysis and hence to increase the TCs.

The reasoning is certainly useful to determine transfer capacities for the day-ahead market. But other elements should come into play when discussing transmission rights. The selection of a rectangle (the TCs) inside the domain defined by the ZPTDF constraints reintroduces the discretion of the TSOs for constructing TCs. It again illustrates the conventional character of TCs whatever the method to determine them. Starting from ZPTDF constraints obtained with perfect GDSKs as we have done here, we first need to select the export variable to eliminate when going from the three-dimensional to the two-dimensional model. This implies that the TCs that represent access to different zones may have no relation with the physical

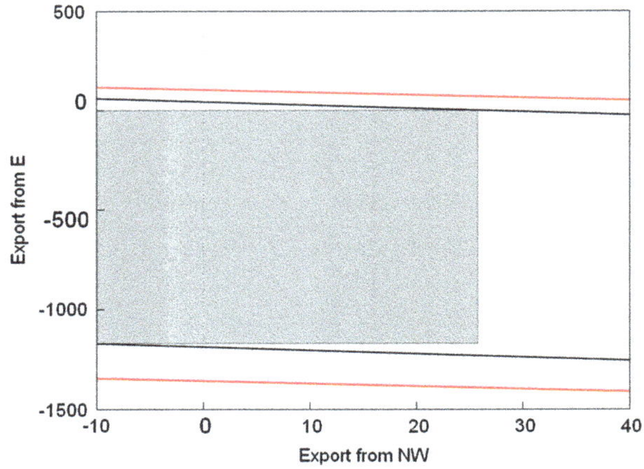


**Fig. 14.5** Construction of TC from ZPTDF constraints

capacities between these zones. We have here chosen to eliminate the SW export and obtained a graph in the sole NW and E exports. This gives a TC to and from zone E, which is in reality connected to the two other zones by unconstrained interconnection. This is in sharp variance with the common discourse of European and national authorities that systematically refer to the insufficient physical capacities between the zones to justify their conclusion of national relevant geographic markets. There is here no physical limitation between zone E and its neighbors but the physical limitation between NW and NS creates transfer capacities on zone E. Given this arbitrary selection of the eliminated export variable, there is also an infinite number of choices for allocating the possibilities given by the FB constraints into the two transfer capacities. The authors of the FB model explain that TSOs can take advantage of that flexibility to allocate the overall constraint between the two PX exports. This is certainly true but it also introduces ambiguity in the definition of the TCs that as we shall see later spoils any hope of producing the necessary mechanisms to guarantee firmness of transmission rights, or alternatively drastically reduces the scope of transmission rights for which firmness can be guaranteed.

#### Long and Short Term TC Model

Previous comments on possible discrepancies between day-ahead and the long term TC model equally apply to the FB model. The model of Table 14.18 embeds GDSK information collected from the clearing of the day-ahead market. There is no such information in the explicit auction of transmission rights in the forward market. Except for resorting to past observation, there is no objective rationale to select GDSKs for constructing the long-term possibilities of the grid.

## 14.4 Congestion in Market Coupling

### *14.4.1 Generalities*

The three node network has been the reference example in most discussions on congestion management in the nodal model. It is also used in Oren (2012) to present the fundamental notions of transmission rights in the nodal system. We similarly use a three zone model to analyze inter-zonal congestion in Market Coupling. By analogy with the nodal system, congestion charges can be measured as difference between zonal prices or as congestion charges on interconnections. These charges are the risks that one wishes to hedge through transmission rights and hence are the underlying of these rights. A first question is what influences this underlying. We successively discuss the question on the TC and FB models.

### *14.4.2 Congestion Charges in the TC Model*

Congestion can occur on line 1–6 which is the single capacitated interconnection in the three zone model. It can also occur on the 2–5 domestic capacitated line, which is not part of the interconnection. The implicit principle in the EU is that zonal TSOs handle domestic congestion by counter-trading or other remedies and that these congestion costs are socialized in access charges. Market clearing then consists in maximizing welfare computed on the PX import/export curves subject to the sole constraints on interconnections. We treat this case first and then turn to domestic congestion (here on line 2–5).

#### 14.4.2.1 Case 1: Market Coupling Does Not Recognize Domestic Congestion

Suppose that Market Coupling only recognizes congestion on interconnections, or in this example on line 1–6 between zones NW and SW. It may be surprising to note that this implies a single price for the three zones and zero congestion cost. Prices and quantities are identical to those occurring when there is no line limitation. The reason is that the TC model allows MC to redirect flows so as to avoid saturating the interconnection. Table 14.19 gives a TC feasible solution of Market Coupling. The reported flows are incompatible with Kirchhoff's law and hence with the possibilities of the real grid, but they are feasible for the TC representation of the grid.

These price and flow patterns differ from those of nodal pricing where a single constrained line implies three different prices in a network three node network. Dropping the PTDFs of the nodal model to arrive at the TC model therefore implies going from three to one price. The property is general: a single constrained line induces as many different prices as the number of nodes in the nodal system but

**Table 14.19** A TC feasible solution with one capacitated line (1–6)

| | Price | Generation (+) Demand (−) | Line flows in a TC solution | |
|---|---|---|---|---|
| 1 | 35 | +500 | | |
| 2 | 35 | +400 | 1–2 | 450 |
| 3 | 35 | −50 | 1–3 | 50 |
| 4 | 35 | 0 | 2–5 | 850 |
| 5 | 35 | −400 | | 450 |
| 6 | 35 | −450 | | 0 |

only one "copper plate" (obtained without congestion) price in the TC model. More surprisingly any single capacity located anywhere in the three zone network gives the same single price in the TC model! The constraint and its value are irrelevant to the outcome of Market Coupling in the TC model! There is no congestion cost and hence nothing to hedge. In contrast nodal prices are node specific and their pattern depends on the location and value of the constraint as one could expect. There is a congestion cost in the nodal model as there is in reality.

This paradoxical result sheds doubts on the TC model. Indeed combining the common assumptions that intra-zonal congestion does not interfere with inter-zonal congestion, with the principle that one can represent the interconnection by its sole capacity leads in this example to the patently absurd result that one can do away with congestion altogether. The TC model is wrong in this simple example and is thus unlikely to be right in the complexity of the real world. The following elaborates on this paradox by turning to the constrained domestic line 2–5.

#### 14.4.2.2 Case 2: Market Coupling Recognizes Domestic Congestion

The implicit assumption of the EU zonal system is that TSOs can manage domestic congestion to arrive at a single price in each zone. To do this in the three zone example, the E-TSO first checks whether the outcome of Market Coupling violates the possibilities of its grid. This is the case with the results of Table 14.19 where the flow on line 2–5 is equal to 416 MW and thus exceeds the 250 MW capacity of the line. The TSO then remedies the situation by counter-trading, that is, by reducing the flow from 416 to 250 by a counter flow from node 5 to 2. This counter flow would here amount to 166, which is significant and would cost a lot in access charges to the clients (generator and consumer) of the E zone. The TSO may thus want to find an alternative approach more acceptable for its jurisdiction.

This is what the Swedish TSO is reputed to have done in 2006 (DG Competition 2009 and Chauve et al. 2010). Because it could not accommodate the combination of domestic (to Swedish customers) and international (to Denmark) demands on its grid, it introduced a TC on the interconnection to Denmark. The TC was not a physical reality of the interconnection but a conventional number aimed at preventing domestic congestion by reducing the global use of the Swedish grid. It was a transfer from domestic to international congestion. Bjørndal et al. (2003) had mentioned the possibility of this practice well before 2006. We can adopt a similar approach here and introduce a TC of 250 MW (the capacity of line 2–5) on the NW-E interconnection; note that we introduce this fictitious capacity notwithstanding the fact that this interconnection is composed of non congested lines 1–2 and 3–2 (the congestion is domestic to zone E and has nothing to do with the interconnection)! The problem becomes feasible for the E-TSO and the new TC creates two-zonal prices in the system.

The Danes and the Swedes belong to the Nordpool system and both understand its functioning; the Danes found out and complained to European Competition Authorities (European Commission 2010). The practice has since been declared illegal and is also prohibited by the Framework Guidelines: TSO cannot move domestic congestion to the border by introducing fictitious TCs. The problem is that the practice is difficult to detect in a complex meshed network. The Swedish TSO denied any wrong doing but has since created domestic zones that will in any case prevent any misbehavior. Applying the same remedy in the three zone example implies creating subzones of the E zone, that is, separating the zone (2, 5) into two subzones (2) and (5). The impact of this separation is considerable. Market coupling of the three zone model now gives the same result as the North–South model (see Table 14.4) where generators 1 and 2 receive the same low price 20.83, which is also the price paid by consumer 3. In contrast, consumer 5, which was located in the same zone as generator 2 before the separation but is now in a different subzone, pays the much higher price 54.16 induced by generator 4. This solution better represents the physics and economics of the system but its application maybe controversial for the E zone.

#### 14.4.2.3 Summing Up

This discussion reveals that the TC model is largely conventional at least in a meshed network where TSOs need to remove domestic congestion to arrive at a single zonal price: whatever the value set for the TC on line 1–6, the outcome of Market Coupling is the same as if this value had been set at zero! This solution can also create domestic difficulties in zone E. One way out it to introduce a purely artificial TC; the other solution is to create additional zones, possibly with dramatic consequences for the operators of the initial zones. All this seems somewhat ad hoc and may explain that TSOs have so far limited the description of the TC calculation to very general considerations. This will be remedied with the publication of the grid codes but the clarification can only be in terms of procedures as nothing substantive can be said about the physical and economic reality of TCs. The result is that the congestion cost in the TC model is fundamentally arbitrary and importantly when it comes to transmission right subject to discretionary manipulation by TSOs. The question is then whether this constitutes a good underlying for hedging instruments.

The possibility to hedge congestion costs is necessary to conclude long-term contracts.[4] TSO announce TCs but not fully guarantee them before day-ahead. Congestion charges in the TC model therefore depend not only on the characteristic of the grid and on commercial activities but also on its management by TSOs. Dependence on short term ad hoc decisions cannot be avoided in the real time management of the electricity grid. But long term ad hoc decisions should be excluded at the design stage of the system. We shall formally argue in the next section that the very nature of TCs makes it impossible for TSOs to guarantee the firmness of the rights except by drastically reducing their scope.

The Nordic system, which is more radial than the one of Central West Europe, should be less vulnerable to this problem. But as the Denmark-Sweden event of 2006 revealed, even that system can suffer from meddling congestion management with grid physical characteristics. Changing zones in response to congestion is a congestion management technique that tries to rationalize TCs to better represent the use of capacitated lines. But it also mixes congestion management with grid physical realities and hence create the ambiguity of the TC signal that impacts on zonal prices and their difference (Bjørndal and Jørnsten 2001). The Nordic system responded to the problem by creating a decentralized market of Contracts for Differences (CfD) on zonal prices (Hagman and Bjorndalen 2011). The system does not involve the TSOs and hence dispenses them from providing firmness. This is probably unavoidable as changing zones modifies the topology of the grid on which zonal prices are computed and hence prevents firmness or reduces the scope of transmission rights. This CfD market is active but not very liquid in contrast with the Nordic energy market that is very liquid (NordREG 2010).

### *14.4.3 Congestion in the FB Model*

Turning to the FB model we examine whether it offers a more solid basis to congestion charges and their use as underlying of transmission rights. There is indeed an improvement, but its extent depends on the capability to determine constant GDSKs in the day-ahead and forward markets. The FB model effectively offers a realistic (possibly non-linear) representation of the capabilities of the grid in the day-ahead market when bidding information is available. But it fails to do so in the forward market when transmission rights need to be allocated. Moreover the practice of the TSOs referred to in Amprion et al. 2011a and 2011b, Sect. 5.7 suggests a somewhat ad hoc determination of GSKs. We discuss the day-ahead market in this

---

[4] There is a guarantee before day-ahead but it only holds if the security of the grid is not endangered. Needless to say the security of the grid in real-time depends on the set of transmission rights that have been allocated, as well as on the real-time conditions of the grid. A guarantee of transmission right that can be waived because of an initial misallocation of transmission rights by the TSO is not a guarantee. There is however an incentive not to unduly curtail transmission rights as TSOs are not sure to recover the compensation in their tariffs.

**Table 14.20** Market coupling in the FB model with congestion limited to interconnection

| |
|---|
| $\text{Min} \int_0^{E_{NW}} P_{NW}(E_{NW})dE_{NW} + \int_0^{E_{SW}} P_{SW}(E_{SW})dE_{SW} + \int_0^{E_E} P_E(E_E)dE_E$ |
| s.t. $E_{NW} + E_{SW} + E_E = 0$ |
| $-386\ (-200 - 186) \leq \text{APTDF}_{NW,1-6}E_{NW} + \text{APTDF}_{SW,1-6}E_{SW} + +\text{APTDF}_{E,1-6}E_E$ $\leq 24\ (200 - 186)$ |

section and treat the forward market in the next one. We successively take up the cases where Market Coupling only encompasses the single capacitated interconnection (1–6) and the one where it also treats the domestic line 2–5. These are referred to "critical branch" in the FB literature and we shall sometimes use that expression. We do not discuss the case where the FB model is used to construct improved TC models as this per se reintroduces discretionary decisions of the TSOs that spoil any hope of creating an adequate underlying for transmission rights.

#### 14.4.3.1 Case 1: Market Coupling Does Not Recognize Domestic Congestion

Applying the reasoning of Sects. 14.2.2.2 and 14.2.3, one sees that Market Coupling solves a problem analytically similar to the standard three node problem of the nodal system (see Oren 2012). This problem is restated in full in Table 14.20. It involves the ZPTDF constraint expressing the limitation on line (1–6) and the balance equation.

The solution is also structurally similar to the one of the nodal system in the sense that it consists of three different zonal prices. The situation is thus analogous to the one of a standard three-node grid provided the ZPTDFs are interpreted as ordinary PTDFs. We have seen that is this is the case as long as the GDSKs extracted from the clearing of the energy market remain constant.[5]

The congestion costs to be hedged are well defined in the sense that they depend on the technological characteristics of the grid and of the GDSKs, which in principle are only functions of the bids and offers in the day-ahead market. In more technical language zonal prices are "measurable" with the state of the system if one includes the GDSK in that state. The advantage compared to the TC model is that the discretionary decisions of the TSOs (which are not "measurable" with the state of the system) have now disappeared at least in "perfect" GDSKs. But the drawback compared to the nodal system is that the representation of the grid now also depends on commercial operations.

[5] The solution obviously differs from the nodal prices of the six-node system as MC imposes that the number of prices is at most equal to the number of zones. But this is taken for granted (and even desired) at the outset.

**Table 14.21** Market coupling in the FB model with congestions on lines (1–6) and (2–5) node explicit

| |
|---|
| $\text{Min} \int_0^{Ex_{NW}} P_{NW}(Ee_{NW})dEe_{NW} + \int_0^{Ex_{SW}} P_{SW}(e_{SW})de_{SW} + \int_0^{Ex_E} P_E(e_E)de_E$ |
| $Ex_{NW} + Ex_{SW} + Ex_E = 0$ |
| $-386\ (-200 - 186) \leq APTDF_{NW,1-6}\, Ex_{NW} + APTDF_{SW,1-6}\, Ex_{SW} + APTDF_{E,1-6}\, Ex_{1-6} \leq 14\ (200 - 186)$ |
| $-464\ (-250 - 214) \leq APTDF_{NW,2-5}\, Ex_{NW} + APTDF_{SW,2-5}\, Ex_{SW} + APTDF_{E,2-5}\, Ex_{2-5} \leq 36\ (250 - 214)$ |

**Table 14.22** Solution of the model 21

| | Net positions | Equilibrium prices | Equilibrium quantities |
|---|---|---|---|
| NW | 32.37 | 24.56 | 258.81 |
| SW | −21.27 | 49.6 | 304.2 |
| E | −11.11 | 34.63 | 403.70 |

#### 14.4.3.2 Case 2: Market Coupling Recognizes Domestic Congestion

Accommodating the sole interconnection (1–6) leaves the congestion of the domestic line (2–5) untreated. In the interest of brevity, we skip the discussion of countertrading for treating this domestic congestion and immediately move to a FB model that recognizes domestic congestion as proposed in Sect. 14.4.2.2. The market clearing problem solved by Market Coupling is stated in Table 14.21.

The solution of the model is given in Table 14.22. It leaves the domestic line unconstrained and implies a transfer from the low demand zone (NW) to the two other zones. Note that this is not a general outcome. Increasing the capacity of line 1–6 and decreasing the one of 2–5 will congest the sole domestic line, and not the interconnection. Whatever line is congested, there could be three zonal prices. It will thus be necessary to explain to stakeholders that there are three zonal prices even though there may be no congestion on the interconnections.

#### 14.4.3.3 Implication for Hedging

In contrast with the TC model the FB model does not meddle congestion management with the physical realities of the grid. It thus give unambiguous congestion charges when GDSKs are constant (commercial transactions that determine the GDSKs are observable). The result extends to non-linear GDSK but this is not considered by TSOs and hence not discussed here. The suggestion of the authors of the FB-model to use the PTDF constraints of the import/export possibilities to derive TC constraints would reintroduce arbitrariness in the process and hence be counterproductive for guaranteeing firmness of transmission rights. These good properties depend on whether GDKs are determined in a proper manner (constant, obtained from the PXs and without meddling of the TSOs). It remains to see whether these advantages of the FB model carry through to the forward market.

## 14.5 Hedging Congestion by Transmission Rights in Market Coupling

### *14.5.1 Problem Statement*

Consider a long term cross border bilateral transaction that one wants to financially hedge against congestion charges between the two zones; alternatively suppose that one wants to physically guarantee access of this transaction to the relevant zones. The question is whether the TC and FB models can provide adequate underlying for transmission rights. The preceding section elaborates on the better properties of the FB model. We therefore begin by discussing the extent to which transmission rights can be constructed on that model. A small subsection later adapts the discussion to the TC model.

The discussion of the preceding sections indicates that the FB zonal model completed with appropriate GDSKs (constant GDSK or export/imports dependent DGSKs determined from the PXs' clearings) gives a realistic representation of the possibilities of the grid and hence well-defined congestion charges in the day-ahead market. The model is also formally very close to the nodal model presented in Oren (2012). This suggests adapting the financial transmission rights of the nodal model to the zonal case. The preceding discussion also shows that this can be done under two conditions: first transaction should be superposable, second GDSKs should be available in the forward market: alternatively TSOs should be willing to bear the risk of changing GDSKs and ZPTDFs as North American ISOs bear the risk of changing PTDFs. Both conditions are important. First, superposition is important for agents to constitute portfolios of tradable rights. Second, TSOs are regulated companies and hence should not be forced to take on risks that the market design does not allow them to control. Specifically, while one can argue that ISOs in the nodal system can be responsible for the risk induced by varying PTDFs because these are characteristics of the grid (and hence ISOs should have some capital at risk for doing so), it seems more difficult to demand that TSOs should also be responsible for GDSKs that are essentially outcomes of the energy market and hence the result of PX operations at the border between the bilateral and organized markets. We explore the impact of these caveats to assess differences between Financial Transmission Rights in the nodal and zonal systems.

Transmission rights of the nodal system can be of the point-to-point or link (flowgate) type. The relative advantages of the two rights have been the subject of intense and deep discussions during several years in the USA. Oren (2012) briefly presents the two types of rights and summarizes these discussions. The US debate concluded in favor of point-to-point services. Oren (2012) analysis can be formally transposed from the nodal to the FB model but the question is whether a formal transposition is justified in substance. One can think of two criteria to assess the validity of this transposition. One criterion deals with the relation between node-to-node (zone-to-zone in FB parlance) and portfolios of flowgates (critical branch rights in FB parlance). It is possible under certain conditions to assemble flowgate rights to arrive at point-to-point rights in the nodal model; the criterion is whether one can similarly assemble critical branch rights into zone-to-zone rights. The second criterion is about firmness. One can under "simultaneous feasibility" guarantee the firmness of node-to-node rights in the nodal system. The question is whether a similar property holds in the zonal system.

We first show that an equivalence between zone-to-zone rights and a portfolio of rights on critical branches also holds in the zonal system, but note that the property diverges from what is stated in the Framework Guidelines. One can indeed in principle hedge zone-to-zone congestion costs by a portfolio of hedges on critical infrastructures even when GDSKs are significantly non-linear. Coming to the second criterion, we then explain that the main weakness of the transposition from nodal to zonal is the difficulty of foreseeing GDSKs in the forward market. Last, we show that the default makes it unlikely that firmness of transmission rights can be guaranteed for the different rights presented in Framework Guidelines, or alternatively that the set of rights that can be guaranteed will be a small fraction of what the grid allows.

## *14.5.2 Financial Rights in the Nodal and Zonal Systems*

### 14.5.2.1 On the Relation Between Zone-to-Zone and Critical Branch Rights

The relation between node-to-node congestion charge and flowgate value in the nodal system is well known. The difference of nodal prices is equal to a PTDF weighted sum of the value of the saturated lines. Both nodal prices and values of saturated flowgates are determined by the solution of the OPF. A hedge on point-to-point congestion can thus be obtained as a portfolio of hedges on the flowgate values. Different portfolios of hedging instruments can be offered in the nodal context and Oren (2012) discusses their respective advantages and drawbacks.

This relation can be formally transposed to the zonal system. Consider an adaptation of the model of Table 14.22 where the dependence of the ZPTDFs on import/export is made explicit (see Table 14.23).

The optimality conditions of that problem show that zonal prices are weighted sums of the values of the lines. This is similar to the nodal system except for one major difference. The weights are now a sum of ZPTDFs and derivatives of ZPTDFs with respect to export variables. The relation between zone-to-zone prices and critical line valuations thus follows the same principle as in the nodal system when GDSKs are constant. It is more complex when GDSKs are non-linear. Whether the non-linearities are significant enough to be taken into account is an empirical question that we cannot address here. We thus conclude that this equivalence is still warranted but is more complex to implement in practice. Oren (2012) comments on the advantages of node-to-node versus portfolio of flowgate rights

**Table 14.23** Market coupling in the FB model with congestions on lines (1–6) and (2–5) node explicit

$$Min \int_0^{Ex_{NW}} P_{NW}(e_{NW})dEx_{NW} + \int_0^{Ex_{SW}} P_{SW}(e_{SW})de_{SW} + \int_0^{Ex_E} P_E(e_E)de_E$$

$$Ex_{NW} + Ex_{SW} + Ex_E = 0$$

$$-386\ (-200 - 186) \le ZPTDF_{NW,1-6}(Ex_{NW}, Ex_{SW}, Ex_E)Ex_{NW} + ZPTDF_{SW,1-6}(Ex_{NW}, Ex_{SW}, Ex_E)E_{SW} + ZPTDF_{E,1-6}(Ex_{NW}, Ex_{SW}, Ex_E)E_E \le 14\ (200 - 186)$$

$$-464\ (-250 - 214) \le APTDF_{NW,2-5}(Ex_{NW}, Ex_{SW}, Ex_E)Ex_{NW} + ZPTDF_{SW,2-5}(Ex_{NW}, Ex_{SW}, Ex_E)E_{SW} + ZPTDF_{E,2-5}(Ex_{NW}, Ex_{SW}, Ex_E)Ex_E \le 36\ (250 - 214)$$

would be reinforced in a comparison of zone-to-zone and portfolios of critical branch rights.

It is useful to note that the Framework Guidelines adopt a TC oriented but FB incompatible view on that question. Suppose congestion for zone-to-zone services is defined as the difference between zonal prices (which is effectively what cross-border transactions pay). The Framework Guidelines define the value of an interconnection between these zones as the difference between the two zonal prices. This definition is correct for the TC model but incorrect in the FB case where the zone-to-zone congestion cost is a weighted sum of the value of all critical branches of the zone-to-zone network. Applying the Framework Guidelines definition to the nodal system would conclude that the value of a link is the difference of the prices of the nodes connected by that line, which is definitely erroneous.

#### 14.5.2.2 On Firmness of Zonal Financial Transmission Rights

The long-term (year-ahead and month-ahead) auctioning of node to node Financial Transmission Rights under constraints of "simultaneous feasibility" is central to the firmness of these rights in the nodal system (see Oren 2012). The question is whether such a central auctioning can be transposed to the zone-to-zone system.

Back to Superposition: In the Day Ahead and Forward Markets

Transmission rights should be feasible for the grid. The verification of the property both in the forward and day-ahead markets is an obvious necessary condition for firmness. The DC representation of the load flows, which implies the linearity of the PTDF relations describing the use of the lines simplifies this verification in the nodal system: transactions can be added and subtracted and the use of the lines by a portfolio of transactions can be derived from the use of the lines by individual transactions. We have seen that this superposition property only holds in the FB model for constant GDSKs. One can infer that the failure to satisfy the superposition property in real time in the zonal Market Coupling also implies its failure in the forward market. In other words, supposing that one knows the dependence of ZPTDFs on import/export in the real time market and assuming that this dependence carries through to the forward market, the nonlinearity of ZPTDFs complicates the construction of portfolio of hedging instruments of congestion charges in the forward market even if it does not make it strictly impossible (composing transmission rights using non-linear ZPTDFs is effectively possible). TSOs do not consider non-linear GDSKs. The capability to construct portfolios of hedging instruments by adding and subtracting individual instruments therefore depends on whether their assumption of constant GDSKs is valid in practice. This is an empirical question that only TSOS can respond. But the response to the question is crucial: the constitution of portfolios of hedging instruments may be impossible if the assumption of constant GDSKs is seriously violated.

### Simultaneous Feasibility in the Real Time and Forward Market

Assume that GDSKs are constant and hence that ZPTDFs are also constant. Simultaneous feasibility requires that the portfolio of transmission rights in the forward market be physically feasible for the grid in real time. This is an obvious necessary condition for firmness. Hogan (1992) also shows that it is a sufficient condition in the sense that the congestion costs collected by the ISO in real-time (day-ahead for Market Coupling) suffice to refund the congestion costs paid by the holders of transmission contracts. Holders of transmission rights are thus protected from congestion costs by an adequate portfolio of contracts. Because the real time conditions of the grid are not known at the contract time, simultaneous feasibility at the time of auctioning of transmission rights is imposed in nodal pricing by a set of scenarios of the grid topology that likely embed the real-time grid scenario. Firmness is thus not guaranteed if real-time grid conditions differ from what has been assumed when granting transmission rights. This is reported to be infrequent and not to cause solvency questions for ISOs. The question is whether this reasoning applies to FB transmission rights.

Simultaneous feasibility condition in the FB model would require that the set of transmission rights is feasible for day-ahead ZPTDFs. This is a drastic strengthening of the simultaneous feasibility condition of the nodal system that only involves grid characteristics. Simultaneous feasibility in the FB model would involve both several topologies of the grid (the PTDFs) and several GDSKs in day-ahead. This latter information is first inexistent in the forward market but were it known it is in principle entirely under the responsibility of the PXs or becomes a joint responsibility of PXs and TSOs if the latter manipulate GSKs as described in EnBW (undated) Sect. 5.7. The TSOs would thus take on risk created by PX operations. This is unlikely to be acceptable in practice and would raise problems of asymmetry of information between TSOs and PXs in theory. By construction the FB model makes it at best very difficult and possibly impossible for the TSOs to extend the simultaneous feasibility condition of the nodal system to the zonal Market Coupling.

The Central Auctioning of Transmission Rights

Notwithstanding the already negative above remarks, central auctioning of all transmission rights remains a central part of the simultaneous feasibility constraint and hence another necessary condition to guarantee firmness. Long-term transmission rights are also auctioned in the European Market Coupling (see EnBW undated for a description of the current platform). The existing European platform offers coordinated auction services on different interconnection but this coordination is purely administrative and deals with procedure. It does not offer anything like a central auctioning under simultaneous feasibility.

Conclusion

The above discussion shows that none of the condition that are considered necessary or sufficient to guarantee firmness in the nodal model can be transposed in substance to the FB model of Market Coupling.

#### 14.5.2.3 Firmness in the TC System

As has often been mentioned the TC model is mainly conventional and does not reproduce fundamental properties of the DC load flow model used in nodal pricing. The superposition of transactions is violated to the point that even netting is not possible. It is thus impossible to construct portfolios of transmission rights. TSOs had cautioned very early (ETSO 2001) against the very strange properties of TCs. Because TCs also depend on discretionary decision of TSOs, "simultaneous feasibility" becomes impossible to implement in practice except by drastically reducing the scope of offered transmission rights. The sole idea of a auctioning a large portfolio of rights that satisfy simultaneous feasibility is thus incompatible with the TC model.

### *14.5.3 Transmission Rights in the Framework Guidelines*

#### 14.5.3.1 Problem Statement

The Framework Guidelines allow for physical (PTR) and financial (FTR) transmission rights. Their description is brief: PTRs must be options subject to the "Use-It Or Sell it" (UIOSI) clause[6]; FTRs are options or obligations. Mixes of PTRs and

[6] Rights that have not been nominated before the opening of the energy market in day-ahead are released to the implicit auction of Market Coupling and the former owner of these rights receives the price determined in the auction.

FTRs are permitted in the internal market but not on one zonal border where transmission rights must be of only one type. PTRs should be harmonized in the internal market; similarly FTRs should also be harmonized. This harmonization is not described any further.

The Framework Guidelines provide scant additional comments. PTRs or FTRs are not necessary if a cross-border financial market already exists on both sides of the interconnector. This probably takes care of Nordpool that developed a market of Contract for Differences (NordREG 2010), even if Nordic regulators acknowledge that this market is no really liquid. The Framework Guidelines also provide for a common platform for the allocation of long-term transmission rights. Such a platform already exists (EnBW undated) and it is not clear whether the Framework Guidelines want to go beyond what is in place. The reference to the FTRs is thus the real novelty of the Framework Guidelines even if guidance is missing on how these could be organized.

Physical transmission rights are currently allocated on a yearly and monthly basis through explicit auctions. They are allocated year-ahead for a fraction of what is expected to be the capacity with additional tranches being released as one moves on and information on plausible transactions accumulates. They must be nominated before play-ahead to be used or will be sold back to the day-ahead market because of the UIOSI clause. Traders require firmness of transmission rights (e.g. EFET 2008). The Framework Guidelines also impose firmness but do not say on which horizon it should be guaranteed. TSOs have interpreted the requirement for firmness in a very restrictive way: they guarantee it (that is they compensate an interruption at the difference of zonal prices) after nomination in the day-ahead market but may restrict it for reason of grid security before that. This is certainly much less than what traders desire but is possibly the best one can do in the TC system. There is a common platform with uniform rules for acquiring these long-term rights. The description of that platform does not suggest anything like a simultaneous feasibility constraint. There is also no indication of a combinatorial auction that would allow traders to express a preference for bundles of rights to construct a portfolio. Rights are thus for individual links offered in a platform that essentially harmonizes procedures.

The contrast with the US FTR experience could not be more striking. The EU has so far reasoned in terms of congestion on interconnections and the Framework Guidelines retain that tradition. This could suggest a view more akin to the flowgate approach discussed in Oren (2012) and now abandoned in the USA. But even this resemblance is superficial. There is no notion of portfolio of interconnection rights to hedge zone-to-zone congestion and as we have seen, even the definition of congestion on a line as the difference between zonal prices in the Framework Guidelines is incompatible with the zone-to-zone definition in a FB model. The absence of theory of these rights, the disqualification of the TCs and contract path approach in meshed systems, the fact that the FB model has never been implemented anywhere so far and that the related flowgate model has been abandoned where it has been implemented, all this makes it difficult to comment on the effectiveness of current proposals. But the demand for firmness and the way firmness is handled in the US system can provide a beginning of discussion.

#### 14.5.3.2 Firmness of Financial Rights

The notion of simultaneous feasibility is central to the firmness of financial rights and the possibility to implement it might be the yardstick to assess the TC and FB models. Simultaneous feasibility requires that transmission rights allocated in the forward market be physically feasible in day-ahead (recall that we refer to day ahead and not real time in Market Coupling). This can never be totally guaranteed (just think of the extreme case of a collapse of the grid), but there are degrees in the extent to which one can get close to the objective. The nodal system guarantees simultaneous feasibility on the sole basis of the physical characteristics of the grid. The N-1 criterion allows one to define a grid topology in the forward market that should remain valid in the day-ahead in most circumstances. We have seen that the FB model would require simultaneously feasibility with respect to topologies constructed from a combination of grid characteristics and market information (the GDSKs). This combination makes it much more difficult to find a topology in the forward market that remains valid in the day-ahead market. Simultaneous feasibility thus requires enlarging the set of possible FB topologies, which implies restricting the set of guaranteed transmission rights. The situation is worsened if TSOs manipulate GDSKs to introduce congestion management considerations. The TC model goes one step further and requires simultaneous feasibility with respect to a set of topologies constructed from grid characteristics, market information (the worst case analysis) and congestion management decisions such as zone splitting in the Nordic system (continental TSOs mention remedies in their documents but do not provide information). This again increases the set of topologies in the feasibility constraints of the forward market and as a direct consequence reduces the set of guaranteed transmission rights offered to the market.

#### 14.5.3.3 Firmness of Physical Rights

Transmission rights are currently physical; they are equipped with the Use It Or Sell It (UIOSI) clause, which makes them equivalent to options. These rights seem to have the preference of all stakeholders except for large industrial consumers. Physical rights will thus probably prevail. Can they be made firm? It is here useful to recall that the nodal system created financial rights because it turned out to be impossible to make physical rights firm. The reason was that combination of rights allocated in the forward market could induce difficulties of dispatch in the spot market that could only be relieved by cancelling some of the allocated rights. In contrast financial rights do not impose dispatch constraints in real-time because it suffices to satisfy the simultaneous feasibility conditions in the forward market to guarantee the hedging of congestion costs in real-time. TSOs argue that real-time remedies will solve the problem of coordination in real-time (Amprion et al. 2011b). This maybe possible but reserving remedies for real-time amounts to restricting the set of transmission contracts allocated in the forward market. As the discussion of simultaneously feasibility with financial transmission rights shows optional rights further complicate coordination problem with the effect that they also reduce transmission possibilities. In short, besides the undocumented recourse to remedies, existing proposals do not suggest any reason to believe in the firmness of the proposed transmission rights.

## 14.6 Conclusion

Market Coupling is currently implemented in the so-called "Central West Europe" region that includes Belgium, France, Germany and The Netherlands. It is intended to expand to the whole of Europe. CWE is today larger than the combined US PJM and MISO system. Market coupling is conceptually a significant simplification of the nodal system (which does not mean that its operational implementation is simpler than nodal pricing) that underlies PJM and MISO. A natural question is how a conceptually less sophisticated organization can cope with a larger power system. A possible response is that it can't and that flaws will be revealed under stress (as was the case in California) if and when demand picks up again and wind power penetrates. This paper concentrates on one question of Market Coupling namely its potential to provide a working market of firm and tradable transmission rights.

Transmission rights seem to be a hard part of the design of the short term electricity markets. The physical rights of the US OASIS system were abandoned after they revealed difficult coordination problems. Financial rights on flowgates were tried in Texas and abandoned because of the considerable costs that they implied. Financial rights for node-to-node models are today the best solution. Contract for Differences that substitute FTRs in the Nordic system remain relatively illiquid in contrast with the financial energy market, but there is effectively less demand for congestion hedges than financial contracts and demand is mainly in Finland, Sweden and Denmark. In short, because transmission is a difficult problem, financial markets on congestion charges due to transmission bottleneck remain particularly challenging. ACER (2011) proposals allow for optional physical rights and financial rights on interconnections. The first have generally failed and the second proved costly.

Traders require firmness but provide no hint on how to get it. TSOs offer firmness under restricted conditions that make it useless for long term contracts. The Nordic system offers a bilateral market for hedging transmission. The US restructuring offers a concept of "simultaneous feasibility" that has proved operational for providing firmness of Financial Transmission Rights. The Framework Guidelines do not offer much hope of technically implementing simultaneous feasibility except by drastically reducing these rights and hence making their market illiquid. The joint implication of TSOs and PXs in these rights add problems of asymmetry of information that will further restrict liquidity. Illiquid markets of transmission rights will segment the geographic market, and hence hamper the long waited completion of the internal electricity market. The European power system is currently subject to a flurry of changing regulations that degrade the investment climate. The possibility of contracting long term on transmission would help. It is not clear how the Framework Guidelines will enhance the scope for long term contracts and hence how they will help.

## Appendix 14.1: Flows on the Interconnection in the Two-Node Model Before Cross-Border Trade

**Table 14.24** PTDF

| Node | Line 1–6 | Line 2–5 |
|---|---|---|
| 1 | 0.625 | 0.375 |
| 2 | 0.5 | 0.5 |
| 3 | 0.5625 | 0.4375 |
| 4 | 0.0625 | −0.0625 |
| 5 | 0.125 | −0.125 |
| 6 (hub) | 0 | 0 |

**Table 14.25** Flows

| Node | Injection | Line 1–6 | Line 2–5 |
|---|---|---|---|
| 1 | 216.6 | 135.4 | 81.2 |
| 2 | 116.6 | 58.3 | 58.3 |
| 3 | −333.3 | −187.5 | −145.8 |
| 4 | 466.6 | 29.1 | −29.2 |
| 5 | −258.3 | −32.3 | 32.3 |
| 6 | −208.3 | 0 | 0 |
| Total | 0 | 3.1 | −3.1 |

## Appendix 14.2: Incremental Flows Due to Exports in the North–South Model

**Table 14.26** PTDF

| Node | Line 1–6 | Line 2–5 |
|---|---|---|
| 1 | 0.625 | 0.375 |
| 2 | 0.5 | 0.5 |
| 3 | 0.5625 | 0.4375 |
| 4 | 0.0625 | −0.0625 |
| 5 | 0.125 | −0.125 |
| 6 (hub) | 0 | 0 |

**Table 14.27** Flows

| Node | Injection | Line 1–6 | Line 2–5 |
|---|---|---|---|
| 1 | 0.3 | 0.2 | 0.1 |
| 2 | 0.3 | −0.2 | −0.1 |
| 3 | 0.3 | 0.2 | −0.1 |
| 4 | −0.6 | −0.0 | −0.0 |
| 5 | −0.2 | −0.2 | 0.2 |
| 6 | −0.2 | 0 | 0 |
| Total | 0.5 | 0.5 | 1 |

## Appendix 14.3: Change of ZPTDF Due to a Limitation of Generation Capacity

**Table 14.28** PTDF

| Node | Line 1–6 | Line 2–5 |
|---|---|---|
| 1 | 0.625 | 0.375 |
| 2 | 0.5 | 0.5 |
| 3 | 0.5625 | 0.4375 |
| 4 | 0.0625 | −0.0625 |
| 5 | 0.125 | −0.125 |
| 6 (hub) | 0 | 0 |

**Table 14.29** Flows

| Node | Injection | Line 1–6 | Line 2–5 |
|---|---|---|---|
| 1 | 0.5 | 0.31 | 0.19 |
| 2 | 0 | 0 | 0 |
| 3 | 0.5 | 0.28 | 0.22 |
| 4 | 0 | 0 | 0 |
| 5 | 1.6 | 0.2 | −0.21 |
| 6 | 1.6 | 0 | 0 |
| Total | 4.3 | 0.80 | 0.20 |

## Appendix 14.4: Computation of ZPTDF in the Three Zone Model

**Table 14.30** PTDF

| Node | Line 1–6 | Line 2–5 |
|---|---|---|
| 1 | 0.625 | 0.375 |
| 2 | 0.5 | 0.5 |
| 3 | 0.5625 | 0.4375 |
| 4 | 0.0625 | −0.0625 |
| 5 | 0.125 | −0.125 |
| 6 (hub) | 0 | 0 |

**Table 14.31** Flows

| Node | Inflow | Line 1–6 | Line 2–5 |
|---|---|---|---|
| **NW** 1 | 0.5 | 0.3 | 0.1875 |
| 2 | 0 | 0 | 0 |
| 3 | 0.5 | 0.28125 | 0.21875 |
| 4 | 0 | 0 | 0 |
| 5 | 0 | 0 | 0 |
| 6 | 0 | 0 | 0 |
| Total | 1 | 0.59375 | 0.40625 |
| **SW** 1 | 0 | 0 | 0 |
| 2 | 0 | 0 | 0 |
| 3 | 0 | 0 | 0 |
| 4 | 0.8 | 0.05 | −0.05 |
| 5 | 0 | 0 | 0 |
| 6 | 0.2 | 0 | 0 |
| Total | 1 | 0.05 | −0.05 |
| **E** 1 | 0 | 0 | 0 |
| 2 | 0.66 | 0.33 | 0.33 |
| 3 | 0 | 0 | 0 |
| 4 | 0 | 0 | 0 |
| 5 | 0.3 | 0.04 | −0.04 |
| 6 | 0 | 0 | 0 |
| Total | 1 | 0.375 | 0.29 |

## Appendix 14.5: Flows on Interconnection Due to Intra-zone Market Clearing in the Three-Zone Model

**Table 14.32** PTDF

| Node | Line 1–6 | Line 2–5 |
|---|---|---|
| 1 | 0.625 | 0.375 |
| 2 | 0.5 | 0.5 |
| 3 | 0.5625 | 0.4375 |
| 4 | 0.0625 | −0.0625 |
| 5 | 0.125 | −0.125 |
| 6 (hub) | 0 | 0 |

**Table 14.33** Flows

| Node | Injection | Line 1–6 | Line 2–5 |
|---|---|---|---|
| 1 | 275 | 171.9 | 103.1 |
| 2 | 400 | 200 | 200 |
| 3 | −275 | −154.7 | −120.3 |
| 4 | 300 | 19 | −18.7 |
| 5 | −400 | −50 | 50 |
| 6 | −300 | 0 | 0 |
| Total | 0 | 186.9 | 214.0 |

## References


ACER (2011) Framework guidelines on capacity allocation and congestion management for electricity. Document FG-2011-E-002, Agency for the Cooperation of Energy Regulators, Ljubliana, 29 July 2011

Amprion, Apxendex, Belpex, Creos, Elia, Enbw, Epexspot, Rte, Tennet (2011a) CWE enhanced flow based MC feasibility report, 15 Mar 2011. Available at www.apxendex.com

Amprion, Apxendex, Belpex, Creos, Elia, Enbw, Epexspot, Rte, Tennet (2011b) Rules for capacity allocation by explicit auctions, 19 Oct 2011. Available at www.apxendex.com

Bjørndal M, Jørnsten K (2001) Zonal pricing in a deregulated power market. Energy J 22(1):51–73

Bjørndal M, Jørnsten K, Pignon V (2003) Congestion management in the Nordic power market-counter purchase and zonal pricing. Compet Regul Netw Ind 4(3):271–293

Booz & Co, Newbery D, Strbac G (2012) Physical and financial capacity rights for cross-border trade. Booz and Company, London

Buglione G, Cervigni E, Fumagalli E, Poletti C (2009) In tegrating European electricity market. Research Report 2, IEFE, Centre for Research on Energy and Environmental Economics and Policy, Bocconi University

Chao H-P, Peck S (1998) Reliability management in competitive electricity markets. J Regul Econ 14:198–200

Chauve P, Glowicka E, Godfried M, Leduc E, Siebert S (2010) Swedish interconnector case/ improving electricity cross border trade. Competition Policy Newsletter Number 2

De Jong HM, Hakvoort RA, Sharna M (2007) Effects of flow based market coupling for the CWE region. In: CENERTEC (Ed.) Proceedings of the 4th European congress economics and management of energy in industry (ECEMEI 2007), pp 1–9

DG Competition (2009) Antitrust: commission opens proceedings against Swedish electricity transmission system operator concerning limiting interconnector capacity for electricity exports. Memo 09/191. Available at http://europa.eu/rapid/press-release_MEMO-09-191_en.htm@PR-metaPressRelease_bottom

Dijkgraaf E, Janssen MCW (2009) Defining European wholesale electricity markets: an "and/or" approach. Tinbergen Institute Discussion Paper TI2009-079/3

Djabali R, Hoeksena J, Langer Y (2010) COSMOS description CWE Market Coupling algorithm, 20 June 2010, apx-endex

Duthaler C, Finger M (2008) Financial transmission rights in Europe's electricity market. Ecole Polytechnique Fédérale de Lausanne, November 2008

Duthaler C, Finger M (2009). Simulation of the European Electricity market on the full network model (nodal pricing). In: International conference on competition and regulation in network industries, Brussels. (www.crninet.com)

Duthaler C, Emery M, Anderson G, Kurziden M (2008) Analysis of the use of Power Transfer Distribution factors (PTDF) in the UCTE transmission grid. In: Power system computation conference, Glasgow. (www.pscc08.org)

EnBW, Creos, Elia, Transpower, Amprion, Rte, Tennet (undated) Rules for capacity allocation by explicit aucitons within Central West Europe Region (CWE Auction Rules)

EFET (2008) Dual purpose transmission rights. Discussion paper, European Federation of Energy Traders. November 2008

ETSO (2001) Definitions of transfer capacities in liberalized electricity markets. Final report, April 2001

ETSO (2006) Transmission risk hedging products. ETSO background paper, April 2006. Now available from ENTSO-E website

ETSO-EuroPEX (2004) Flow-based market coupling, ETSO-EuroPEX proposal: a joint ETSO-EuroPEX proposal for cross-border congestion management and integration of electricity markets in Europe. Internal report, September 2004. Now available from ENTSO-E website

ETSO-EuroPEX (2009) Development and implementation of a coordinated model for regional and inter-regional congestion management. Final report, January 2009. Now available from ENTSO-E website

European Commission (2009) Regulation no. 714/2009 of the European Parliament and of the Council of 13 July 2009 on conditions for access to the network for cross-border exchanges in electricity and repealing regulation, EC No. 1228/2003 Text with EEA relevance

European Commission (2010) Commission decision of 14.4.2010 relating to a proceeding under Article 102 of the Treaty of the Functioning of the European Union and Article 54 of the EEA Agreement (Case COM/39351-Swedish Interconnector)

European Commission (2011) 2009–2010 report on progress in creating the internal gas and electricity market. Commission Staff, working document, Brussels, June 2011

Glachant J-M (2010) The achievement of the EU electricity internal market through market coupling. RSCAS working paper 2011/11, European University Institute

Hagman B, Bjorndalen J (2011) FTRs in the Nordic electricity markets. Pros and cons compared to the present system with CfDs. Elforsk rapport 11:16, April 2011

Hogan WW (1992) Contract networks for electric power transmission. J Regul Econ 4:211–242

Huisman R, Kilic M (2011) A history of European electricity day-ahead prices. Erasmus School of Economics, Erasmus Universiteit, Rotterdam

Janssen M, Niedrig T, Wobben M (2011) Towards a better design of electricity transmission rights. Foundation for Research on Market Design and Energy Trading, Cologne

Kurziden MJ (2010) Analysis of flow-based market coupling in oligopolistic power markets. Ph.D. dissertation, Eidgenössische Technische Hochschule Zürich

NordREG (2010) The Nordic financial electricity market. Nordic Energy Regulators, report 8/2010

Oren S (2012) Point to point and flow-based financial transmission rights: revenue adequacy and performance incentives (this volume)

Parisio L, Bosco B (2008) Electricity prices and cross-border trade: volume and strategy effects. Energy Econ 30:1760–1775

Pellini E (2011) Measuring the impact of market coupling on the Italian electricity market using ELFO++, Surrey Energy Economics Centre, School of Economics discussion paper 133

Rious V, Usaola J, Saguan M, Glachant J-M, Dessante P (2008) Assessing available transfer capacity on a realistic European network: impact of assumptions on wind power generation. Author manuscript, published the "1st international scientific conference building networks for a brighter future", Rotterdam, Netherlands. Available at http://hal-supelec.archives-ouvertes.fr/docs/00/33/87/49/PDF/WindPowerATC_VR-MS-JU-JMG-PD4.pdf

Van Vyve M (2011) Linear prices for non-convex electricity markets: models and algorithms. CORE discussion paper 2011/50, Université catholique de Louvain, Louvain-la-Neuve, Belgium

Wobben M (2009) Valuation of physical transmission rights. An analysis of electricity cross border capacities between Germany and the Netherlands

Zachmann G (2008) Electricity wholesale market prices in Europe: convergence? Energy Econ 30:1659–1671

# Chapter 15
# Incentives for Transmission Investment in the PJM Electricity Market: FTRs or Regulation (or Both?)

**Juan Rosellón, Zdeňka Myslíková, and Eric Zenón**

## 15.1 Introduction

Government led reforms of the electric industry have taken place in the United States of America (USA) since the 1990s. The restructuring of the industry was concerned with changing the system historically treated as a natural monopoly to a free market industry. The generation and the distribution segments of the system were opened to competition. Transmission services, because of its characteristics, stayed as a monopoly under regulation. While the generation and distribution sectors were thus flourishing under the reforms, the transmission sector experienced a shortfall in necessary investment because it lacked incentives for development. The system has become congested in various areas as growth in electricity demand

This paper was originally published as: Rosellón, J., Mysíková, Z., and E. Zenón (2011), Incentives for transmission investment in the PJM electricity market: FTRs or regulation (or both?). Utilities Policy January 2011, 3–13. We thank Jeff Pavlovic for valuable help with data processing, and Hannes Weigt for helpful comments. Juan Rosellón acknowledges support from Pieran_Colegio de México, the Alexander von Humboldt Foundation, and Conacyt (p. 60334). The usual disclaimer applies.

J. Rosellón (✉)
División de Economía, Centro de Investigación y Docencia Económicas (CIDE), and German Institute for Economic Research (DIW-Berlin), Carretera México-Toluca 3655, Mexico, DF 01210, Mexico
e-mail: juan.rosellon@cide.edu

Z. Myslíková
Centro de Investigación y Docencia Económicas (CIDE), Carretera México-Toluca 3655, Mexico, DF 01210, Mexico

E. Zenón
Facultad de Ingeniería, Universidad Nacional Autónoma de México (UNAM), Mexico, Mexico

J. Rosellón and T. Kristiansen (eds.), *Financial Transmission Rights*, Lecture Notes in Energy 7, DOI 10.1007/978-1-4471-4787-9_15,

and investment in new generation facilities have not been matched by investment in new transmission facilities.[1]

The transmission network is a critical part of the system, and in the last decade transmission expansion became a crucial issue for the Federal Energy Regulatory Commission (FERC) and the US Department of Energy (US Department of Energy 2002, 2006). It was then understood that without efficient transmission expansion, the electric grid in the near future would be stretched far beyond its capacity increasing dramatically the final cost of electric energy, and negatively affecting the entire economy. Present-day reforms are searching for optimal mechanisms that would provide adequate transmission investment incentives to guarantee expanding the capacity of the network and relieve congestion problems. One area with congestion problems in its electricity networks is the US region known as PJM.[2] Our paper proposes and applies a mechanism that provides adequate incentives to promote expansion of the network in this area.

This chapter is organized as follows. In Sect. 15.2, we review the literature on incentive mechanisms for the expansion of electric transmission networks. Section 15.3 reviews the features of the PJM electricity transmission market, its current transmission pricing and investment policies. Section 15.4 provides description of the mechanism used in this paper for transmission expansion in the PJM region. It is an application of a merchant-regulatory mechanism where the optimization problem is treated as a two-level (or bi-level) programming problem of a Transmission Company (Transco), and an Independent System Operator (ISO). The Transco maximizes its benefit subject to a regulatory constraint (upper level problem). The ISO solves an optimal dispatching problem maximizing the social welfare (lower level problem). The two levels are solved simultaneously. In Sect. 15.5, the details of the simulation of the model are explained. The mechanism is tested for 17-node geographical coverage area of PJM, divided into zones according to the historical utilities control areas. The analysis addresses market efficiency and changes in social welfare caused by changes in nodal prices (an extension of the analysis for a modified region is provided in the appendix). Section 15.6 concludes.

[1] For a detailed analysis see the National Transmission Grid Study (NTGS) from US Department of Energy (2002) or Joskow (2005a).

[2] PJM is an abbreviation for the region operated by PJM Interconnection. The letters P-J-M represent names of its three original principal member states: Pennsylvania, New Jersey and Maryland.

## 15.2 Review of Literature: Incentive Mechanisms in Electricity Transmission

This section presents a survey of current research paths on transmission expansion mechanisms. We survey three approaches according to basic assumptions about whether the transmission sector can sustain competition, and according to the tools – regulatory, merchant and combined merchant-regulatory tools – employed by the mechanism.[3]

There are two basic regulatory approaches suggested in the literature. First, there is a regulatory mechanism based on price regulation (Vogelsang 2001). This mechanism relies on the rebalancing of the two (fixed and variable) parts of a price-capped tariff. The fixed part of the tariff is an instrument through which long-term costs are recuperated (i.e., it is a complementary charge). The variable part can be understood as a nodal price difference in the sense of the financial transmission right (FTR) literature (Rosellón 2003). The Transco rebalances over time the two parts of the tariff while meeting the price cap established by the regulator, and efficiently expands the network. The expansion process takes place so that incentives to keep the network congested are broken and, under certain conditions, there will be convergence to a steady-state Ramsey-type of equilibrium.[4] However, a critical aspect of the regulatory mechanism is its definition of the transmission output as the capacity flow between two points, and also that its reliance on assumed smoothly behaved properties of production and cost functions of transmission services which –both in theory and practice – are difficult to establish. Hogan (2002) argues that the properties of these functions are not well known (the functions are considered not linear), and are suspected to be generally non-differentiable and even discontinuous. Also, the definition of transmission output in meshed networks is a difficult issue. Under the definition of transmission output that he uses, the Vogelsang (2001) mechanism can typically be applied to radial lines only.

The second regulatory approach is based on a measure of welfare loss with respect to the Transco's performance. The basic approach used in Léautier (2000) and similarly in Joskow and Tirole (2002) is that the regulator rewards the Transco

---

[3] Apart from the three main approaches, usually one more is mentioned in the literature. This approach defines optimal expansion of the transmission network according to the strategic behavior of generators, and considers conjectures made by each generator on other generators' marginal costs due to the expansion. It explicitly models the existing interdependence of generation investment and transmission investment. However, it also relies on a transportation model with no network loop flows.

[4] The model reconciles allocative, productive and even distributive efficiencies as well as promotes convergence to Ramsey prices. Likewise, the expansion process is incentivated since, with the use of the mechanism, the expected revenues from expanding the network become greater than or equal to the revenues from keeping the network congested. Convergence to a "congestion" equilibrium –where the marginal cost of expanding the network equals the congestion cost of not adding an additional unit of capacity – is also achieved (see Crew et al. 1995; Vogelsang 2001; Hogan et al. 2010).

when the capacity of the network is increased so that congestion rents are decreased. On the other hand, the regulator can punish the Transco for taking advantage of a congested network by charging increasing fees, and accumulating higher congestion rents. Another variation is an "out-turn" based regulation. The out-turn is defined as the difference between the price for electricity actually paid to generators and the price that would have been paid absent congestion (Léautier 2000). The Transco is made responsible for the full cost of out-turn, plus any transmission losses.

The merchant approach to transmission expansion aims to bring competition into the transmission expansion process through the assignment of property rights specified as FTRs. An FTR is a financial instrument that allows the value of increased transmission capacity to be security and auction competitive, facilitating the entry of the private sector into transmission expansion investment (Hogan 2002). FTRs are defined according to transmission capacity between nodes with different prices, and grant their owner the right to collect the difference between the nodal prices. This process motivates investment. The assignment of FTRs is managed by the ISO. Under loop flows within a meshed transmission network, negative externalities might arise on property-right holders since the expansion of one link in the network might affect the capacities of other links. Kristiansen and Rosellón (2006) suggest a solution to this issue where the ISO retains some "unallocated FTRs" to use in case that negative externalities arise during the expansion process. They argue that using unallocated FTRs prevents a gaming-behavior of investors.

The last approach to transmission expansion aims to bring together the main tools of both the merchant and regulatory mechanisms. Hogan et al. (2010) design a combined model where price-cap regulation is merged with a redefinition of transmission output in terms of FTRs. This allows that FTR auctions inherit the regulatory logic in Vogelsang (2001). Conversely, the combined approach upgrades the Vogelsang model into a bi-level programming model where an ISO maximizes dispatch through a power-flow model providing the optimal loads and nodal prices needed to achieve expansion in meshed networks according to the rebalancing of each part of the two-part tariff. Rosellón and Weigt (2008) further combine the merchant and regulatory price-cap mechanisms with an engineering approach to calculating locational marginal prices (LMPs). They prove that this approach is effective in incentivizing investment in a real transmission network in Northwestern Europe.

## 15.3 The PJM Electricity Market

The US transmission network is a part of the North American electricity transmission system which consists of three interconnected systems – the Western Interconnect, the Eastern Interconnect, and the Electric Reliability Council of Texas (ERCOT). Together they comprise the bulk power system in the USA, much of Canada and a small portion of Mexico. Each system is coordinated independently within its power grid and the three systems are not synchronized together (electricity cannot flow between them except through the use of asynchronous tie lines). The current day organization of the electric industry in the USA differs across the states. In general there is no agreement or policies (or mechanism employed) that would establish how appropriate transmission investments should be identified, who bears the responsibility for making the investments, and who pays for the associated costs (Joskow 2005b). While in some states (or regions)[5] the operation via wholesale competitive market was accepted, other regions keep the industry under a completely regulated system without any marks of competitive market. No pure merchant system exists in any state. Even if FERC maintains the function of the regulator of "last instance" (exercising principal regulatory authority over interstate wholesale trade, and the associated transmission interconnection) the electric power industry in the USA has historically been regulated primarily by the states.[6] The legal responsibilities for important aspects of transmission policy are split between the federal government and the states. Each state or region has unique circumstances and organization of the transmission sector, and applied transmission investment policies.

Investor-owned utilities (IOUs) own 73 % of the transmission lines, federally owned utilities own 13 %, and public utilities and cooperative utilities own 14 %.[7] On one hand, in regions with wholesale markets (such as PJM, New York and New England), LMPs are widely used and FTRs could be used as a risk hedging tool.[8] Considering the investment to the transmission network, it is not always clear who should pay for it. When a new generator is included to the interconnection, reliability of the grid could be threatened, and new investment could be necessary to upgrade the grid. The new transmission investment costs could be projected into the basic charges for the transmission service reflected in their tariffs, or generators bear the costs. The exact policies differ from one market to another. On the other hand, in regions with pure regulation, transmission pricing and retail electricity power prices are usually calculated based on cost of service or a utility's embedded costs plus a negotiated rate of return on their investments, and the transmission network expansion policy is planned by state. From the point of view of expansion of interconnection capacity between control area operators, there is no process in place that would systematically evaluate opportunities to expand transmission capacity on both sides of the borders between them (Joskow 2005b).

[5] For example in PJM area, New England, New York or California.

[6] Joskow (2005b) argues that states in the USA have a variety of different views on the desirability of transitioning to competitive wholesale and retail electricity markets, and that there are has no clear and coherent national laws that adopt a competitive wholesale and retail market model as national policy.

[7] The values correspond to the year 2000 (Department of Energy, Energy Information Administration 1).

[8] In the New York Independent System Operator (NYISO)'s region FTRs are also known as long-term transmission rights or firm transmission rights.

PJM Interconnection is a part of the eastern-interconnect grid nowadays managing high-voltage electric networks as well as the wholesale electricity market in which 13 states[9] and the District of Columbia were included in 2008. It provides service to a population of approximately 51 million.[10] PJM is a Regional Transmission Organization (RTO). It is federally regulated, with the service in the area provided by IOUs and Public Owned Utilities (POUs).

As an RTO, PJM coordinates the movement of power within its region and is responsible for the operational and planning functions of the PJM bulk power system on behalf of participant members.[11] It also administers an open access transmission tariff that establishes prices for various categories of transmission services available to the third party transmission users, and defines how the associated revenues are distributed to the transmission owners (Joskow 2005b). It is not engaged in wholesale or retail marketing, and does not own generation, transmission or distribution assets. PJM actually operates four major product markets: energy,[12] capacity, FTRs, and the ancillary services markets. The price of transmission service offered by PJM is based on traditional regulatory cost-of-service (rate-of-return) formulas applied to one or more transmission owners.

The main features characterizing PJM markets are the use of LMPs and the existence of FTRs as a tool for hedging against the congestion costs.[13] LMPs in PJM are defined as "the cost to serve the next MW of load at a specific location, using the lowest production costs of all available generation while observing all transmission limits" (PJM Member Training Department 2007). In this way, the LMP reflects an equilibrium price including not only the value of available generation but the marginal losses and marginal cost of transmission congestion at each location as well. The LMPs in PJM are collected from 10 main hubs.[14] The FTRs market provides the market participants an opportunity to hedge themselves against congestion in the energy market. FTRs are obtained through annual and monthly auctions and bilateral trading.[15] It has a form of a financial contract which enables

[9] All or parts of Delaware, Illinois, Indiana, Kentucky, Maryland, Michigan, New Jersey, North Carolina, Ohio, Pennsylvania, Tennessee, Virginia, West Virginia.

[10] After establishing competition in wholesale markets in the USA, PJM was the first largest wholesale competitive operating market in the world. Currently it is one of the biggest Operators in the USA together with NYISO, New England ISO, California ISO, and the Midwest ISO (MISO).

[11] It is also responsible for maintaining the integrity of the regional power grid and for managing a regional planning process for generation expansion needed to ensure the reliability of the electric system (PJM Interconnection).

[12] The administrated energy markets consist of real time and day-ahead markets.

[13] Also the financial trading hubs, bilateral markets, day-ahead markets, real-time markets, ancillary services and installed capacity.

[14] The ten hubs for which PJM posts prices are: AEP Gen (all generator buses in AEP), AEP-Dayton (all buses in AEP and Dayton), Chicago Gen, Chicago, Eastern, N Illinois, New Jersey, Ohio, West Int., and Western.

[15] Parallel to FTRs, another tool exists on FTR markets – it is called an Auction Revenue Right (ARR). ARRs are allocated annually and provide their holders with revenue based on locational

the holder to receive revenues based on the day-ahead hourly energy price differences across a specified transmission path, and so give their holders the right to a proportionate share of annual congestion charges.

The transmission expansion planning is prepared by the RTO. There are several categories of transmission investments in PJM. When a new generating unit seeks to connect to the PJM network, the reliability criteria could be violated and an investment to the new transmission capacity could be needed. Also "merchant investment projects" (motivated by appearance of FTRs when a project is implemented) or "economic transmission projects" (which are investments whose expected economic benefits are associated with reductions in congestion costs) exist (Joskow 2005b). In general, PJM develops an annual regional transmission expansion plan that identifies transmission system enhancement requirements. The transmission companies propose their plans about the construction of new transmission lines or capacity increase to the RTO, FERC and the Department of Energy (DOE). When a transmission expansion plan is approved, FERC can offer incentive-rate treatment to reduce regulatory risk. The costs for investment made in order to reestablish reliability after connecting a new generation unit are generally paid by the generation unit.

According to the US Department of Energy (2006), the congested zones were identified in both Eastern and Western interconnected systems. PJM is one of the regions where one of the two principal critical congestion areas within the Eastern Interconnect Grid has been identified.[16] The area includes the eastern coast of the PJM region – beginning at metropolitan New York continuing southwards through Washington D.C. to Northern Virginia. Historically, the concern has always been how to move the electric energy from the lower-cost western part of the market to the eastern part of the market where the major load far away from the low cost generation is situated. The congestion in the PJM region is caused mainly because of the growing load together with plant retirements. Limited new generation investment near loads is another cause of congestion there. Even if there is a low-cost coal and nuclear power generation in Midwest, the east parts of PJM cannot use it because the capacity of the transmission network does not allow it.[17]

The installed capacity of PJM at the end of 2006 was 162,143 MW. Table 15.1 provides an overview of the generation plants in PJM, installed capacities

price difference between ARR sources and sink determined in the annual FTR auction (see Frayer et al. 2007).

[16] The critical congestion area is defined as a place where it is critically important to remedy existing or growing congestion problems because the current and/or projected effects of the congestion are severe. In these locations of the network it has frequently been necessary to interrupt electric transactions or redirect electricity flows because the existing transmission capacity was insufficient to deliver the desired energy without compromising grid reliability (US Department of Energy 2006, p. 21).

[17] The nodal prices reflect the described congestion problem for the west-east deliveries. For instance, at the western AEP-Dayton hub the nodal price in given moment in 2005 was $46/MWh while at PJM Eastern Hub it was $66/MWh at the same time (PJM Interconnection 2006, and PJM Summer 2007, Reliability Assessment).

**Table 15.1** Plant characteristics and price structure in PJM

| Plant type | % of total installed capacity in PJM | Part of total average weighted LMP PJM price in the 2006 | % of total generation |
|---|---|---|---|
| Coal | 41 % | 38.7 % | 56.8 % |
| Nuclear | 18.5 % | 0 % | 34.6 % |
| Natural gas | 29 % | 32.3 % | 5.5 % |
| Oil | 6.6 % | 5 % | 0.3 % |
| Hydroelectric | 4.4 % | 0 % | 2 % |
| Solid waste | 0.4 % | NA | 0.7 % |
| Wind | 0.19 % | 0 % | 0.1 % |

Source: Own calculations with from the PJM Interconnection (2006)

(in percentage terms), structure of average weighted LMP[18] (how the fuel prices influence the final LMP, in percentage terms), and percentage of total real generation. The PJM region could also be a power source for the neighboring regions (especially the New York metropolitan area) as long as transmission cross-border constraints are relieved.[19]

## 15.4 The Model

The model of transmission expansion that we apply to the PJM transmission network integrates the key concepts of incentive mechanisms presented in Sect. 15.2 of this paper, and relies on the modeling logic in Vogelsang (2001), Hogan et al. (2010), and Rosellón and Weigt (2008). The approach is then a combination of the merchant and regulatory mechanisms with an engineering approach – it merges the tools of the two main models for the adequate transmission expansion problem: a welfare optimization dispatch power-flow problem (lower-level problem) with a two-part tariff cap regulatory model (upper level). The way it is constructed simulates the real transmission operation and planning issues faced by an ISO, and a Transco. It has power to model many crucial aspects of practical cases where (1) a central authority applies certain kind of regulation, imposing a regulation constraint, (2) the Transco, subject to the regulation constraint, charges a fee for the transmission service and plans the transmission expansion, and (3) the ISO, operating the wholesale market, manages the electric dispatch, subject to the characteristics and capacity limitations of the transmission network. Its goal is to dispatch electric power in an efficient way.

[18] The other components of the average weighted nodal price are the price corresponding to generating $NOx$, $SO_2$, $VOM$ and markup.

[19] Figure 15.3 in the Sect. 15.5 shows some of the transmission links within the PJM region subject to congestion.

The combination of the last three concepts is modeled in the following way:

1. The merchant mechanism is introduced via system of nodal pricing and FTRs. Transmission expansion is carried out through the sale of FTRs. FTRs are defined according to node pairs that suffer congestion, and are commercialized via auctions where the participants enter voluntarily.
2. The regulatory part of the mechanism is based on Vogelsang (2001) regulatory mechanism – a cap constraint is intertemporally applied over a two-part tariff.
3. Dispatching is modeled through a welfare optimization program, subject to the engineering restrictions reflecting the transmission network's technical limitations. It defines the wholesale market prices in each short-run period.

The crucial step which enables the combination of the merchant and the regulatory approach is the definition of the transmission output in terms of FTRs. It is an approach originally introduced by Hogan et al. (2010), and solves the shortcoming of Vogelsang (2001) with an exact and convenient measure of transmission output as point-to-point transactions or FTR obligations. Hogan et al. (2010) show that, under certain conditions, convergence to Ramsey prices might be reached. In the case of PJM, the transmission sector bears parts of regulation as well as merchant elements. The structure in PJM region is similar to a theoretic "centralized ISO" structure.[20] The features of our model are in general compatible with the institutional setup in the PJM region. In particular, the existence of a competitive wholesale market with FTRs in PJM facilitates the application of our model.

Mathematically, the model is divided into two levels of optimization. The upper level represents a dynamic profit maximization problem solved by a Transco when considering transmission expansion. It reflects the opposite incentives that the Transco faces – to expand the transmission network which releases congestion and produces long term benefits for the society (given the growing demand for electricity and need for higher capacity), or to keep congestion in the network and get high congestion rents. The lower level problem reflects the optimization problem faced by an ISO operating the wholesale market, and dispatching the generation and transmission optimally. The lower problem, hence, defines the wholesale market outcome. The two-part tariff maximization forms a dynamic optimization problem running thru $T$ periods, subject to complementarity constraints. The two levels of the optimization are solved simultaneously.

### 15.4.1 Upper Level Problem

The Transco maximizes its objective function (the intertemporal flow of profits) subject to a price cap constraint:

---

[20] Wilson (2002) defines two possible structures for an ISO: a centralized structure and a decentralized structure. Generally speaking, in the former structure the ISO coordinates the equilibrium of the various electricity markets as a central planner, while the latter approach would reach such equilibrium in a sequential way through the free participation of economic agents. No electricity market has been proven to work in practice under a decentralized ISO.

$$\max_{k,F} \quad \pi = \sum_{t}^{T} \left[ \sum_{ij} \tau_{ij}^{t}(k^{t}) q_{ij}^{t}(k^{t}) + F^{t}N^{t} - \sum_{i,j} c\left(k_{ij}^{t}\right) \right] i \neq j \tag{15.1}$$

*s.t.:*

$$\frac{\sum_{ij} \tau_{ij}^{t}(k^{t}) q_{ij}^{w} + F^{t}N^{t}}{\sum_{ij} \tau_{ij}^{t-1} q_{ij}^{w} + F^{t-1}N^{t}} \leq 1 + RPI + X \tag{15.2}$$

The profit function allows for two basic sources of revenue – the first term of the profit function represents the congestion rent. In the FTR literature the congestion rent is generally defined as point-to-point FTRs, $q\mathrm{B}_{ij\mathrm{B}}$, between two nodes $i$ and $j$, multiplied by the FTR price, $\tau\mathrm{B}_{ij\mathrm{B}}$, which is set on the FTR auction. The congestion rent is only charged in the lines that generate "space" for new FTRs. If the limit of the overall capacity of a line is not reached during the transmission process in the period $t$, there are no FTRs generated on the line in $t$, and no congestion rent charged by the Transco.[21] The second term is a fixed fee $F$ charged to each of $N$ users of the transmission grid. It represents a fixed payment for the access to the transmission network. The last term in the maximization problem is the cost function, $c(k)$, which represents the costs of transmission-line capacity expansion between the nodes $i$ and $j$ incurred by Transco.

The restriction on revenue is the regulatory constraint set by the regulatory authority. The constraint is built as a two part tariff cap. The opportunity to rebalance the parts of the tariff guarantees that the Transco will not lose income through the diminishing of the congestion rent when the transmission network is expanded. A lower congestion rent will in turn decrease profits. This is offset as the Transco counters the diminishing congestion rent by increasing the fixed fee.

The weights $w$ used in the price tariff are the Laspeyres weights. According to Rosellón (2007), the Laspeyres weights applied to the Vogelsang (2001) two-part tariff mechanism grant a solution that will converge to an optimum under stable cost and demand functions. The price cap also adjusts for an efficiency factor, $X$, and an inflation factor, $RPI$. The Transco maximizes its profit subject to the regulatory restriction, through $T$ periods, considering the transmission lines between all the nodes $i$ and $j$ within the grid. Perfect information is assumed and there is no uncertainty about demand and generation capacity.[22]

---

[21] The idea that the throughput has to reach the capacity upper limit of the line to be congested is simplified. In reality, an important factor in congestion is also the susceptance of the transmission lines. Certain susceptance of a line can cause the line to be a source of congestion even though the throughput in the line has not reached the upper limit capacity of the line. This is considered in the constraints of the lower level problem.

[22] The model relaxes from an auction FTR price setting and the distribution of FTRs to the specific market participants.

In order to find the first-order optimality conditions, ignoring inflation and the efficiency factors, the derivative of the objective function (15.1) subject to the constraint (15.2) is:

$$\nabla q_{ij}^t \tau_{ij}^t (k^t) - \nabla c^* = (q_{ij}^w - q_{ij}^t (k^t)) \nabla \tau_{ij}^t \tag{15.3}$$

In order to simplify the application of this model to actual electricity networks Rosellón and Weigt (2008) avoid the FTR. They redefine the system of (15.1) and (15.2), so that the profit maximization problem can be rewritten as:

$$\max_{k,F} \quad \pi = \sum_t^T \left[ \sum_i (p_i^t d_i^t - p_i^t g_i^t) + F^t N^t - \sum_{i,j} c\left(k_{ij}^t\right) \right] \quad i \neq j \tag{15.4}$$

*s.t.*

$$\frac{\sum_i (p_i^t d_i^w - p_i^t g_i^w) + F^t N^t}{\sum_i (p_i^{t-1} d_i^w - p_i^{t-1} g_i^w) + F^{t-1} N^t} \leq 1 + RPI + X \tag{15.5}$$

The first term of (15.4) represents an alternative way to define the congestion rent. Instead of a congestion rent expressed in terms of FTRs multiplied by their price corresponding to each part of the grid, this is now defined in terms of the market clearing prices, demand and generation at every node. More exactly, it is defined as the difference between the payments from the loads, $p_i d_i$ , and the payments to the generators, $p_i g_i$. When the loads pay the generators precisely the price that energy costs at the place it was generated, no congestion and congestion rent exists. The relationship between the market clearing prices, $p_i$, and the FTR prices used in the original maximization problem is $\tau \mathrm{B}_{ij\mathrm{B}} = p\mathrm{B}_{j\mathrm{B}} - p\mathrm{B}_{i\mathrm{B}}$ . The regulation constraint is written in the same manner. It substitutes the FTR revenue with congestion rents arising from the differences in nodal market clearing prices.

### *15.4.2 Lower Level Problem*

This is a welfare maximization problem, and determines the wholesale market outcome. The optimization of electric dispatch undertaken by the ISO is subject to the technical restrictions of the network and power flows. There is a perfectly competitive environment assumed where the ISO maximizes social welfare *W*. Following Rosellón and Weigt (2008), the social welfare is defined as a difference between the gross consumer surplus and the total generation costs[23]:

---

[23] Rosellón and Weigt (2008) use this approach in order to obtain a more straightforward expression of the consumer rent and generators' rent.

$$\max_{d,g} \quad W = \sum_{i,t} \left( \int_{0}^{d_i^t} p_i(d_i^t) \mathrm{d}d_i^t \right) - \sum_{i,t} mc_i g_i^t \tag{15.6}$$

*s.t.:*

$$g_i^t \leq g_i^{t,\max} \quad \forall i, t \tag{15.7}$$

$$\left| pf_{ij}^t \right| \leq k_{ij}^t \quad \forall ij \tag{15.8}$$

$$g_i^t + q_i^t = d_i^t \quad \forall i, t \tag{15.9}$$

The first restriction to the welfare optimization, (15.7), is a capacity constraint that does not let any generation in any node $i$ exceed its generation capacity. Equation (15.8) reflects the restriction that the power flow $pfB_{ijB}$ between the nodes $i$ and $j$ cannot exceed the transmission capacity $k\mathrm{B}_{ij_\mathrm{B}}$ of the line. The constraint described by (15.9) imposes that demand at each node is satisfied by local generation or by a net injection $k\mathrm{B}_i$.

Then, in the same manner as in Hogan et al. (2010) and Rosellón and Weigt (2008), a DC-Load-Flow approach is applied in order to get the power flow within the meshed network. Simulation of the optimization of both levels simultaneously leads to iteration of efficient solution values. From the lower level optimization process, the vectors of optimal values of $d$ and $g$, as well as nodal prices $p$, are obtained and substituted into the upper level problem. Then the optimal values of capacity $k$ and fixed fee $F$ are in turn obtained.

## 15.5 Transmission-Expansion Simulation for the PJM Network

The data used for the simulation are obtained from a "snap shot" of a power flow during a non-peak demand period in the USA in 2006. The database information is organized according to the transmission operators of six main regions within the Eastern Interconnection in the USA, and a part of Canada. A more detailed subdivision of the data is presented according to the historic control areas in each region. In the system modeling for PJM, each of the historic control areas is called a zone. Every zone is characterized by number of generators, total generation potential, transmission lines and instantaneous demand of load centers within the zone. The total area operated by PJM (and included in the database) is divided into

17 zones.[24] For the purpose of modeling the PJM network topology, one node is assigned to each zone.[25]

Since the region that PJM operates has expanded significantly during various years, there are two data sets considered for the simulation. The first data set covers a region operated by PJM until 2006. The topology corresponding to this area is tested for original non-peak demand obtained from the database. The second data set is reduced to a region known as PJM-Classic which is an area operated by PJM until 2001. This data set is tested for peak demand. The basic difference in peak and non-peak demands will be reflected in the level of congestion within the network, and in the level of the nodal prices. When peak demand has to be satisfied, higher levels of energy are being transported among the nodes, and there is a higher load for some lines in the grid. Hence, the lines are more prone to congestion. Moreover, to satisfy higher demand it is more probable that higher cost generators would have to be turned on. Together with higher congestion levels in the network, this is a cause for higher peak-demand LMPs in comparison with the LMPs during the non-peak demand periods. Details of the PJM Classic topology – and the corresponding results for peak-demand data simulation – are included in the first part of the appendix.

### 15.5.1 Topology of the Network

The first data set includes the area of PJM until 2006.[26] Figure 15.1 represents the simplified topology of its Transmission Network. There are 17 nodes in total, where thirteen nodes are connected with more than two other nodes and the rest is connected to one or two other nodes. In two cases, where a single historic control area is divided in two parts without a common border, the topology follows this division and two sub-zones per one control zone are considered. Each sub-zone has its own node assigned in the model (nodes N11, N12 and N4, N5).

---

[24] The analysis assumes a closed area with a closed system of transmission lines. While in reality PJM trades energy to NYISO to the north, MISO to the west, and also to states in the south, congestion linked to these exchanges is not considered in the topology.

[25] The decision to assign one node to each zone comes from the fact that each utility owner within the region of PJM is given monopoly over the zone where it operates.

[26] The original PJM-West region was modified for the purpose of the simulation. First, it excludes the territory nowadays corresponding to Virginia Electric and Power Company which was added to PJM Interconnection in 2004 under the name of "Dominion Power". This territory is considered neither in the topology (and consequently nor in the simulation) because the data base does not include it. Second, given that the analysis is for a closed area only (so as to preserve integrity of the topology and avoid bias of results), the zone corresponding to Commonwealth Edison Company – which is a part of PJM-West situated in the state Illinois – is excluded from the data set. The exclusion was made because the zone has stronger transmission connections and commerce with zones which are parts of different ISOs' regions, and does not have common frontiers with any part of the remainder area of PJM.

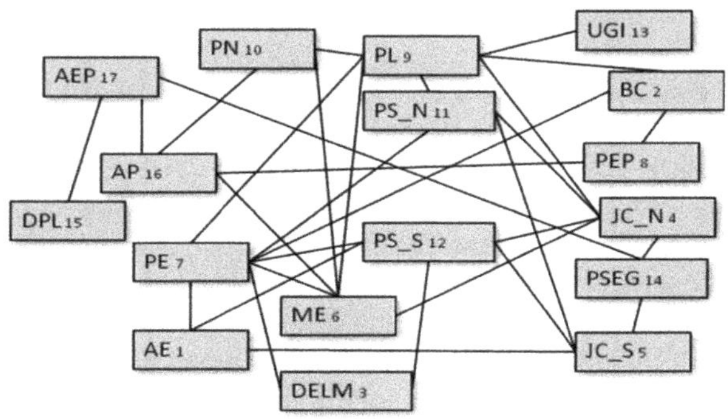


**Fig. 15.1** Topology of PJM (An explication of the abbreviations and precise location corresponding to the nodes is shown in Fig. 15.7 in the Appendix). (Source: Own elaboration with information from PJM Interconnection)

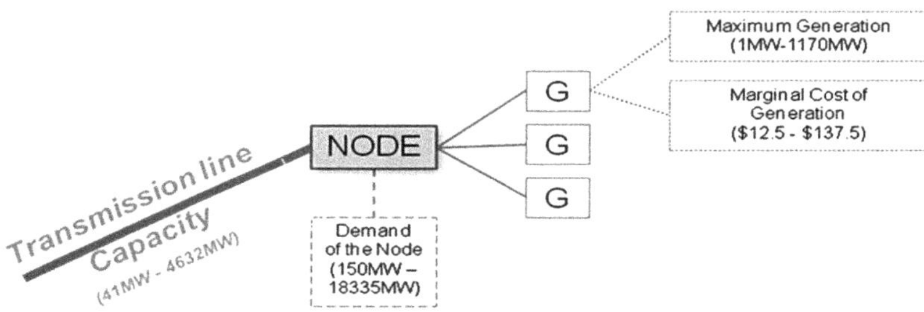


**Fig. 15.2** Detailed scheme of transmission network (Source: Own elaboration)

The transmission lines between the connected zones were aggregated in a way to obtain the total maximum capacity that can be transmitted between each two connected zones. These total connected capacities are represented in the model as single lines between the two zones. Because of the scale of aggregation, each aggregated area is considerably large, and consists both of load and generator centers. All nodes but node N14 (which has zero demand in the moment the snapshot was taken) are considered to be load nodes.

A detail of the transmission network topology is shown in Fig. 15.2. It is a scheme of variables, and their concrete values that are needed for the simulation. Each node in the topology has associated its maximum generation capacities, a reference (starting) demand, the cost of generation per MW, and the capacity of the transmission lines that connect it with other nodes.

The distinction and the assignation of the fuel type used by the generation units were made according to the maximum generation limit of the plant. This way the distribution and classification of the generation units in PJM – the types of generating plants and marginal cost of generating MWh corresponding to each kind – were obtained, and are shown in Table 15.2. An equal marginal cost level is assumed for each type of generation unit.

**Table 15.2** Generation plant characteristics[a]

| Assumed technology | MW cap for the generation plant | Fuel | Price for MWh |
|---|---|---|---|
| Internal combustion | 1–20 MW | Diesel | $137.5 |
| Turbine simple cycle | 21–199 MW | Natural gas | $72.5 |
| Turbine combined cycle | 200–499 MW | Natural gas | $45 |
| Coal | 500–800 MW | Coal | $20 |
| Nuclear | 801–9,999 MW | Uranium | $12.5 |

Source: Own elaboration with information from PJM Interconnection
[a]The fuel prices were obtained as an average cost reported in PJM Interconnection (2007) and Edison Electric Institute data reviews (www.eei.org).

### *15.5.2 Initial Conditions*

Our simulator works in such a way that, given the technical restrictions of the network, the demand is satisfied employing the low cost generators first. On the other hand, the total demand has to be satisfied completely (see (15.9) in Sect. 15.4.2) even if the last activated generator produces energy for double, triple or even higher costs compared to the first generator employed.[27] The functional forms – and if necessary also starting values of the parameters used in the simulation – are assumed according to the values in Table 15.3.

The demand function for each node is derived from the load level for each node, a reference price derived from the weighted average marginal cost[28] corresponding to every zone, and an assumed price elasticity of 0.25 at the reference point. The demands are assumed to be linear. Uniform reactance values $x_{ij}^{0} = 42.5$ for all the lines are assumed in $t = 0$ and individually change according to the expansion of each line. A depreciation factor of 8 % is assumed.[29]

The tariff cap is formed using a Laspeyres index in the regulatory tariff where the weights are the $(t - 1)$ period amounts. In the simulation there are 20 periods of time considered. The derived market results for one time period represent 1 h.[30] Even if the analysis of the transmission-network power flow is based on various simplifying assumptions, in a simulation with three-node network simplifying assumptions will not influence the general properties of the mechanism outcome. When relaxing simplifying constraints, the robustness of the mechanism is not affected – there is no effect on the desired properties of the mechanism. This result

[27] We only consider in this paper the case where new capacity can only be added to already existing transmission lines.

[28] Weights for each level of marginal cost are settled according to the proportion of the maximum generating potential of each plant type within the node.

[29] The value of the depreciation factor is taken from Rosellón and Weigt (2008). Twenty years are supposed to represent the depreciation time of assets in electricity markets and 8% represent an investment with rather low risk. For simplification, we do not account for inflation or efficiency factors within the Transco's price cap.

[30] As the values are obtained in hours, the Transco's revenue is multiplied by 8,760 for each period so as to represent yearly income.

**Table 15.3** Simulation values

| Simulation values | |
|---|---|
| **Number of periods** | 20 |
| **Costs** | Linear |
| Cost function | $c_{ij}^t = c_0 \cdot (k_{ij}^t - k_{ij}^{t-1})$ |
| $C_0$ (line expansion cost) | 130 $/MW |
| **Demand** | Linear |
| Assumed elasticity | 0.25 |
| **Reactance in t = 0** | $x_{ij}^0 = 42.5$ |

Source: Own elaboration with information from PJM Interconnection

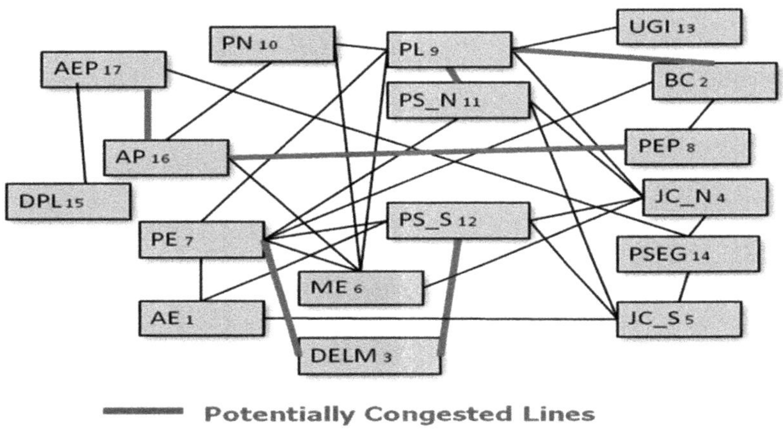


**Fig. 15.3** Potentially congested lines (Source: Own elaboration with information from PJM Interconnection)

can be extended to a more complicated transmission network topology (see Rosellón and Weigt 2008).

As mentioned in Sect. 15.3, there is an extended part of PJM that suffers high grade of congestion. 12 "zones" suffering from congestion were identified (US Department of Energy 2006). These congested paths within the PJM topology are shown in Fig. 15.3 as the thicker lines connecting the nodes. Because of the scale of aggregation, some of the congested parts inside the zones do not appear separately but will be identified during the simulation in aggregation in a particular line.

The highest nodal prices correspond to the nodes on the eastern part of the topology. These nodes correspond to an area that historically has high demand given by high population density and – compared to the generation situated in the west part of the region – with high cost electricity generation. Due to transmission bottlenecks, it is not possible to transport cheap energy from the west to the eastern part. The simulation will show if an application of the incentive mechanism would lead to price arbitrage, and decrease of nodal prices.

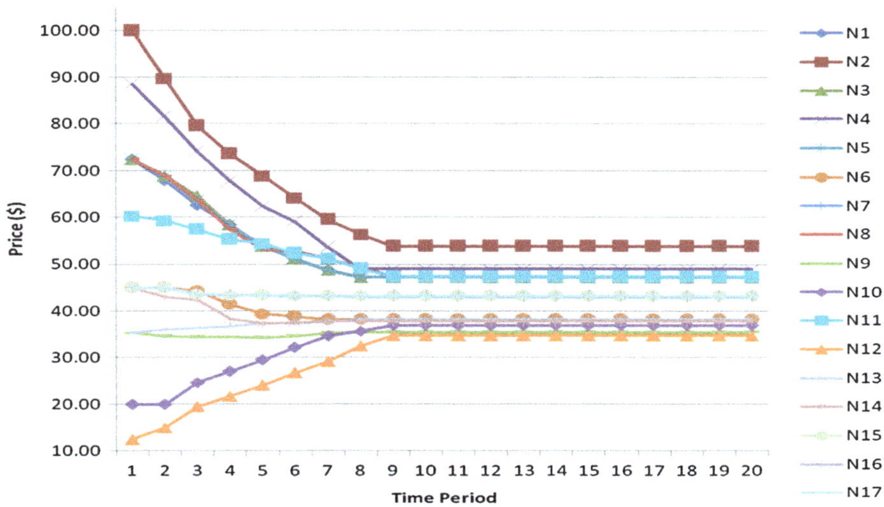


**Fig. 15.4** Price development for the PJM region (Source: Own elaboration)

### *15.5.3 Results: Price Development and Welfare Properties*

The mechanism seeks to promote for capacity increase of the transmission lines, which should then permit transmission of lower cost energy from the western part of the region to the eastern-coast area. To test scope of the mechanism, the development of nodal prices and welfare properties are considered as well.

Figure 15.4 shows price development in the PJM nodes over 20 periods. In the first period the nodal prices differ substantially as they are subject to a high level of congestion. Eastern node N2 has the highest nodal price ($100). The average price of the nodal prices in the first period is $53.64. However, convergence towards a common price level occurs fast within the first nine periods. The average price after the first nine periods is 17 % lower compared to the average nodal price at the beginning of the simulation. If the average level of the five highest nodal prices at the beginning of the simulation is compared to the average price of the same nodes after the first six periods of simulation, a decrease of 32 % can be observed. During the rest of the periods, most of the nodal nodal prices change only marginally.

The extension of the grid follows similar dynamics – the grid is expanded extensively during the first nine periods, and after the ninth period the grid expansion is relatively small. The striking fall of the prices is visible mainly for the nodes N2, N4, and N8. All of them are situated in the eastern area of PJM. This reflects the current problem mentioned in the Sect. 15.3. Transmission congestion separates the eastern part of the market from the remainder of the grid, and electricity prices on the east coast are higher compared to the rest of the region. Transmission congestion does not allow bringing cheaper energy produced in the western part of the region to the east.

If the grid is expanded, cheap nuclear and carbon energy that can be produced and transported mainly from nodes N10 and N12 is utilized to satisfy demand at other nodes, and nodal prices in nodes N10 and N12 increase. The average nodal price at the end of the simulation decreases to \$43.11, which is 20 % lower than at the beginning of the simulation. An arbitrage of nodal prices occurs and the former difference of \$87.5 between the highest and the lowest price at the beginning of the simulation is reduced to \$19.22 after the 20 periods.

### 15.5.4 Welfare Properties

The nodal price development brings about welfare changes. The purpose of the mechanism is to permit arbitrage of prices, and an increase in social welfare, through transmission expansion. When comparing social welfare, only changes that are caused by nodal prices changes are considered. As argued in Vogelsang (2001), the fixed fee acts as a lump-sum tax. The major concern is centered on the development of the nodal prices which converge to marginal costs. Figure 15.8 in the appendix shows the general development of the fixed fee when nodal prices increase.

In order to assess the performance of the mechanism ("Regulatory Approach"), the results from the simulation are compared to the benchmark case without network extension, and to a benevolent ISO case[31] ("Welfare Maximization"). Table 15.4 shows the welfare characteristics of the mechanism. The basis for the estimation of the corresponding rents are the demand function of each node, the congestion rent (first part in (15.4)), and consumer and producer surpluses (15.6).

An increase in consumer rent is observed after the mechanism is applied. Consumers pay lower congestion costs. Even if the nodal prices increase in two cases, the consumer surplus reduction is offset by a price decrease in the other 15 nodes. Note that the sum of the demands in the two nodes that experienced price increase is not higher than the sum of the demands in the remainder part of the system. Since, after the adjustment, prices lie above its marginal cost the producer surplus increases as well as a significant part of total generation that corresponds to nuclear and carbon generation.

The new installed capacity is 42 % higher than the capacity at the beginning of the simulation. As expected, the congestion rent is not equal to zero but its level decreases substantially. The original level of the congestion rent is reduced to 15 % within the 20 periods. The regulatory approach then produces results that are relatively close to a pure welfare-maximizing outcome, and suggest convergence to the welfare optimum levels. Comparing the results for the European model tested by Rosellón and Weigt (2008), the results for PJM show a similar tendency.

[31] The benevolent ISO case is obtained from the maximization problem: $\max_{d,g} \quad W = \sum_{i,t} \left( \int_0^{d*} p_i(d_i^t) \mathrm{d}d_i^t \right) - \sum_{i,t} mc_i g_i^t - \sum_{i,j} c\left(k_{ij}^t\right)$, subject to the restrictions in the lower level problem.

**Table 15.4** Comparison of the regulatory and benevolent ISO approach for PJM region

| | No grid extension | Regulatory approach | Welfare maximization |
|---|---|---|---|
| Consumer rent (MioUSD/h) | 6.53 | 6.63 | 6.67 |
| Producer rent (MioUSD/h) | 0.36 | 0.59 | 0.64 |
| Congestion rent (MioUSD/h) | 0.067 | 0.01 | 0.006 |
| Total welfare (MioUSD/h) | 6.95 | 7.23 | 7.32 |
| Total grid capacity (GW) | 35.8 | 50.83 | 52.83 |
| Average price (USD/MWh) | 53.64 | 43.11 | 42.97 |

Source: Own elaboration

**Table 15.5** Comparison of the non-peak and peak demand nodal prices for the 17-and 14-node topology

| | Non-peak demand (17 node topology) | | Peak demand (14 node topology) | |
|---|---|---|---|---|
| Number of the node | 1. Period nodal price | Final nodal price | 1. Period nodal price | Final nodal price |
| 1 | $72.5 | $47.23 | $137 | $49.70 |
| 2 | $100 | $53.84 | $137 | $59.30 |
| 3 | $72.5 | $47.23 | $72.50 | $46.20 |
| 4 | $88.53 | $49.01 | $137 | $51.97 |
| 5 | $72.50 | $47.23 | $72.50 | $46.76 |
| 6 | $45.00 | $38.16 | $45.00 | $46.70 |
| 7 | $45.00 | $43.04 | $72.50 | $46.15 |
| 8 | $72.50 | $47.43 | $137 | $59.30 |
| 9 | $35.27 | $35.33 | $20.00 | $39.60 |
| 10 | $20.00 | $36.86 | $20.00 | $39.30 |
| 11 | $60.33 | $47.36 | $72.50 | $46.80 |
| 12 | $12.50 | $34.62 | $45.00 | $42.70 |
| 13 | $35.27 | $38.12 | $20.00 | $39.40 |
| 14 | $45.00 | $37.72 | $20.00 | $39.70 |
| 15 | $45.00 | $43.21 | – | – |
| 16 | $45.00 | $43.21 | – | – |
| 17 | $45.00 | $43.21 | – | – |

Source: Own elaboration

As mentioned at the beginning of this section, even more pronounced fall of the nodal prices and bigger increase of the rents could be experienced if the demand tested in the simulation were a peak one. In Table 15.5, results from the non-peak demand and peak demand testing are compared. Details of peak-demand testing within the smaller region of PJM called PJM-Classic are presented in the appendix. The first period nodal prices for the peak demand testing are in several nodes higher than in the case of non-peak demand. In general, this is given so as to satisfy the peak demand. Apart from the cheapest generators that provide energy when satisfying non-peak demand, more expensive generators have to be turned on. Another factor that increases the total cost of providing energy for peak demand is higher congestion. For the majority of the nodes, the final level of the nodal prices is higher when peak demand is satisfied. For example, in the case of nodesN13 and

N14, even if the first period nodal prices were higher for the non-peak demand, at the end of the simulation their nodal prices are higher when the peak demand is satisfied. However, when comparing the nodal price levels for the peak and non-peak demand situations, it has to be taken into account that differences in topologies influence the differences in the nodal prices as well.

## 15.6 Conclusions

This paper presents an application of a merchant-regulatory mechanism for transmission grid expansion to the transmission network in the PJM region. The theoretical model is based on a structure with regulated profit-maximizing Transco, and a competitive wholesale market with nodal price setting and FTRs. Regulation is applied through a price cap imposed on a two-part price tariff that the Transco can charge to users of the transmission network. The regulatory constraint allows for the rebalancing of the variable and fixed parts of the fee in order to let the Transco preserve its benefits when congestion rents decrease due to the increased transmission-grid capacity. The Laspeyres weights are used in the two-part tariff mechanism. The wholesale market is operated by an ISO that coordinates generation and transmission, maximizing the social welfare. FTRs signal the need for new transmission capacity.

The purpose of the mechanism used for the simulation is to arbitrage nodal prices and to foster their convergence to an steady-state equilibrium state with lower congestion rents and higher total welfare. The capacity increases of the transmission lines permit transmission of lower-cost energy to the zones with higher demand and more expensive energy generation. The mechanism is applied to the region that suffers critical levels of congestion combined with growing demand. To date, no coherent mechanism that promotes adequate expansion of the PJM transmission network exists. Moreover, the PJM network is a complicated system of loads and generators covering a considerably large part of the US area. Transmission services are getting unreliable in PJM, and the congestion costs are a significant part of the energy price charged in the region.

A 17-node and 14-node network topology was designed for PJM, and the mechanism is tested for both non-peak and peak demands. Starting with a grid that suffers critical levels of congestion in various zones, the simulation of the mechanism proves that after first nine periods the congestion is relieved, nodal prices converge to a common lower average level resembling the marginal cost of energy generation, the consumers pay lower congestion costs and both consumer and producers surplus increase. In general, the nodal prices for peak-demand periods are higher than for the non-peak period, given that more of the high-cost generators are turned on and also because the higher demand could cause higher congestion in the transmission lines. The simulation proves that the mechanism works for a quite complicated meshed topology such as the PJM one. The installed capacity of the 17-node transmission network after the simulation is 42 % higher than the capacity of the original grid, and the congestion rent decreases to 15 % of its original level. Total welfare increases. Given that the various composing elements of our mechanism and its features are compatible with the FTR-based competitive wholesale market in PJM region, we believe that our mechanism holds promise for being applied in practice.

Next steps in modeling the PJM electric transmission system would implement some new elements to the model. The purpose of future research would improve on the engineering – lower level problem – part of the optimization, and focus in a more detailed geographical division of the PJM region. The intention would be to create different zonal divisions which could reflect the set of zonal areas that is actually used in the internal PJM modeling. Additionally, we would also like to improve on the data set on marginal costs. In the actual operation of markets, marginal costs can be much higher due to imperfect competition.

# Appendix

## *PJM Classic: Peak Demand Testing*

The second data set includes the zones which comprised PJM prior to 2006, referred to by the term "PJM Classic". It takes account of the PJM region after the establishment of a competitive wholesale power market and before it expanded, when its operating territory consisted of eastern Pennsylvania, New Jersey, and part of Maryland, Delaware and District of Columbia. Figure 15.5 represents the simplified topology of the transmission network of PJM-Classic which has 14 nodes, and 26 transmission lines connecting the nodes.

Compared to the 17-node PJM region, this data sample excludes the zones corresponding to nodes 15, 16 and 17. The PJM Classic topology is used in order to test the mechanism facing a peak demand conditions.[32] If not specified differently, the starting conditions and all the details of the simulation are the same as in the case of simulation of the mechanism for 17-node PJM topology.

The results of nodal price development are shown in Fig. 15.6. In Table 15.6, the welfare properties results are specified.

The general results are the same for both topologies – the nodal prices converge to an equilibrium level after the first six periods of the transmission network expansion. However, when comparing the welfare properties of the mechanism for the simulation of the peak demand, the results are more pronounced, highlighting the power of the mechanism. The average nodal price is almost 36 % lower after the mechanism is applied, the transmission network capacity is doubled compared to the first period, and both consumer and producer surplus increase. The price fall is steeper and, given that the demand is higher, the consumers' surplus increase is

[32] The peak demand values were obtained adjusting the original demand data according to the February 2006 peak values reported in "PJM Summer 2007 Reliability Assessment (2007)" for the zones at PJM Classic.

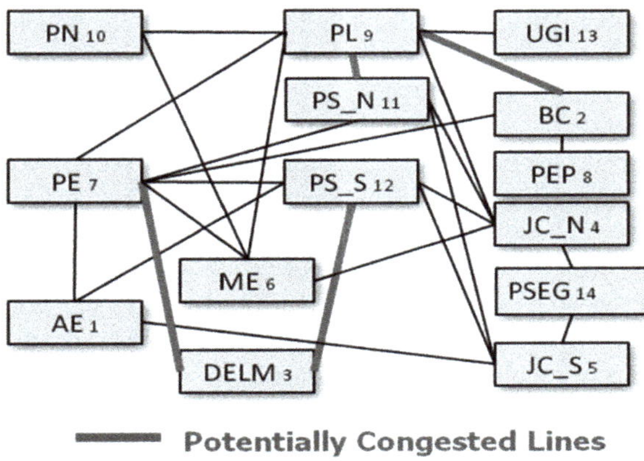


**Fig. 15.5** Topology of PJM Classic region (Source: Own elaboration with information from PJM Interconnection)

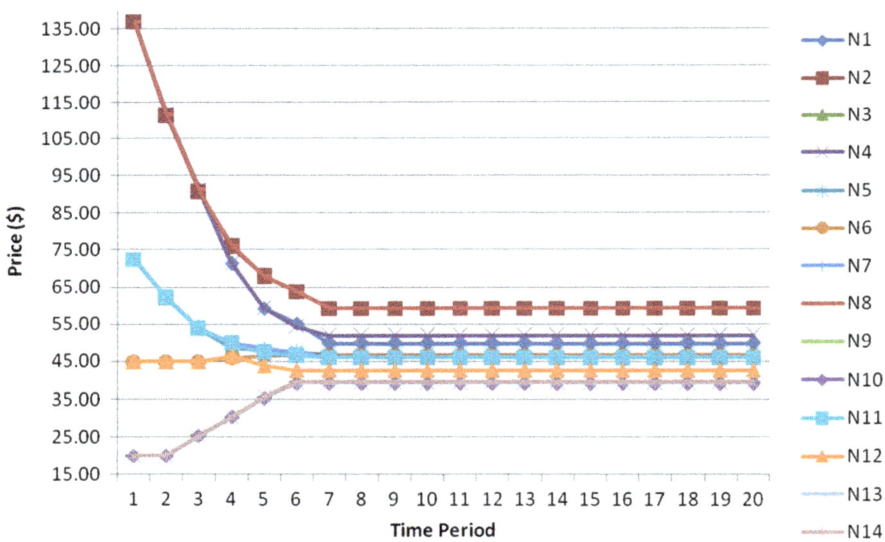


**Fig. 15.6** Price development for PJM Classic region (Source: Own elaboration)

higher than in the case of 17-node PJM case.[33] The congestion rent after the 20 periods of simulation decreases to 16 % of its original level.

In general, the welfare properties in case of higher demand are expected to be more pronounced as the need for transmission network expansion in the network that suffers high levels of congestion could be higher.

[33] However, a comparison with the results for 17-node PJM topology should be made with precaution as there are some significant differences between the cases. The PJM Classic topology does not include three nodes with quite high demands and generation potential. Another important detail is that it is tested for demand in different periods of the year and day.

**Table 15.6** Comparison of the regulatory and benevolent ISO approach for PJM Classic region

| | No grid extension | Regulatory approach | Welfare maximization |
|---|---|---|---|
| Consumer rent (MioUSD/h) | 8.01 | 8.13 | 8.17 |
| Producer rent (MioUSD/h) | 0.44 | 0.68 | 0.73 |
| Congestion rent (MioUSD/h) | 0.076 | 0.012 | 0.0076 |
| Total welfare (MioUSD/h) | 8.53 | 8.82 | 8.91 |
| Total grid capacity (GW) | 26.91 | 49.88 | 52.63 |
| Average price (USD/MWh) | 72.0 | 46.63 | 46.21 |

Source: Own elaboration

## *PJM Zones*

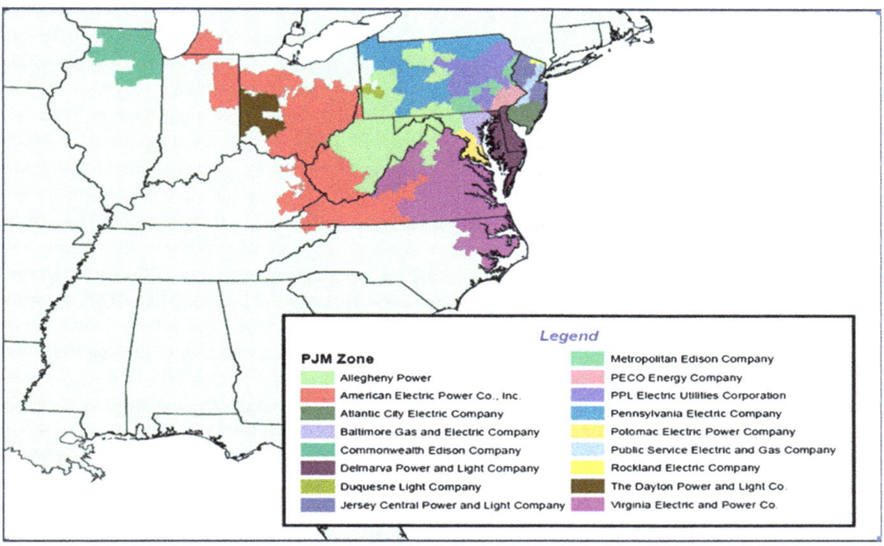


**Fig. 15.7** Map of PJM region and the utilities operating in each zone in the year 2008 (The map was obtained from PJM Interconnection (http://www.pjm.com). The correspondence with the abbreviations used in the topology are the following: *AE* Atlantic City Electric, *BC* Baltimore Gas and Electric Company, *DELM* Delmarva Light and Power Company, *JC_N* Jersey Central Power and Light Company (North), *JC_S* Jersey Central Power and Light Company (South), *ME* Metropolitan Edison Company, *PE* PECO Energy Company, *PEP* Potomac Electric Power Company, *PL* and *PN* Pennsylvania Electric Company, *PS_N* Pennsylvania Electric Company (North), *PS_S* Pennsylvania Electric Company (South), *UGI* Public Service Electric and Gas Company.) (Source: PJM Interconnection)

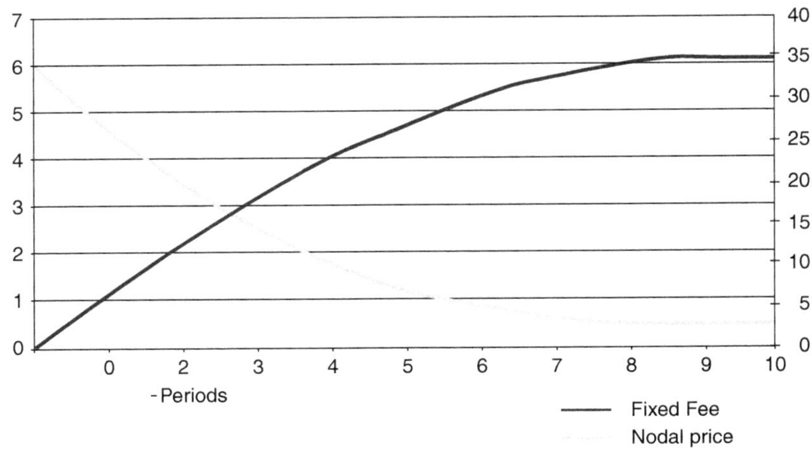


**Fig. 15.8** Fixed fee development (Source: Own elaboration)

# References


Department of Energy, Energy Information Administration 1. (http://www.eia.doe.gov), consulted in January 2009

Frayer J, Ibrahim A, Bahceci S, Pecenkovic S (2007) A comparative analysis of actual locational marginal prices in the PJM market and estimated short-run marginal sosts: 2003–2006. London Economics International LLC, Boston

Hogan W (2002) Financial transmission right formulations. JFK School of Government: Harvard Electricity Policy Group Harvard University, Mimeo

Hogan W, Rosellón J, Vogelsang I (2010) Toward a combined merchant-regulatory mechanism for electricity transmission expansion. J Regul Econ 38(2):113–143

Joskow P (2005a) The difficult transition to competitive electricity markets in the U.S. In: Griffin J, Puller S (eds) Electricity regulation: choices and challenges. University of Chicago Press, Chicago

Joskow PL (2005b) Transmission policy in the United States. Utilit Policy 13:95–115

Joskow P, Tirole J (2002) Transmission investment: alternative institutional frameworks. J Ind Econ 53:233–264

Kristiansen T, Rosellón J (2006) A merchant mechanism for electricity transmission expansion. J Regul Econ 29:167–193

Léautier T (2000) Regulation of an electric power transmission company. Energy J 21:61–92

PJM Interconnection (http://www.pjm.com), consulted in January 2009.

PJM Interconnection, 2006 State of the Market Report, Market Monitoring Unit; March 2007

PJM Member Training Department (2007) Locational marginal pricing 101, LMP-101 training Available online on https://pjm.adobeconnect.com/_a16103949/p20016248/

Rosellón J (2003) Different approaches towards electricity transmission expansion. Rev Netw Econ 2:238–269

Rosellón J (2007) An incentive mechanism for electricity transmission expansion in Mexico. Energy Policy 35:3003–3014

Rosellón J, Weigt H (2008) A dynamic incentive mechanism for transmission expansion in electricity networks – theory, modeling and application. Documento de trabajo CIDE 424, Mexico

PJM Summer 2007 Reliability Assessment, Indiana Utility Regulatory Commission, May 30, 2007
U.S. Department of Energy, National Electric Transmission Congestion Study, August 2006
U.S. Department of Energy, National Transmission Grid Study, May 2002
Vogelsang I (2001) Price regulation for independent transmission companies. J Regul Econ 20:141–165
Wilson R (2002) Architecture of power markets. Econometrica 70(4):1299–1340

# Concluding Remarks

This book analyzed a variety of detailed concepts, theories and practical experiences on FTRs. It should have provided its readers with an overview of financial transmission rights, and their advantages and disadvantages. The theoretical part of the book addressed the formal definition of FTRs and their different modalities, going through the various mathematical formulations discussed by Hogan (Chap. 1). Transmission pricing was studied as a basis for the derivation of FTRs from nodal-price differences. Likewise, issues such as FGRs, forward and spot auction markets, market power, FTR-based merchant and combined merchant-regulatory mechanisms for transmission expansion, and FTRs in an experimental economics framework were also covered in detail. The practical part of the book dealt with issues such as real-world revenue adequacy in markets using LMPs, practical aspects regarding bidding of FTRs, financial hedging and risk management strategies. Likewise, it presented a comprehensive update on the most recent status in countries implementing FTRs

The experience with FTRs the last decade has evolved substantially with more markets to trade and more sophisticated FTR operations. FTRs have served their purpose as transmission congestion hedging instruments. Additionally FTRs have provided revenue sufficiency for contracts for differences, distributed the merchandizing surplus an ISO or RTO accrues in market operations, and provided a price signal for transmission developers. Moreover, as discussed by Arce (Chap. 11), FTRs have become a very appealing financial instrument both to financial institutions as well as to investors. Likewise, as described by Adamson and Parker (Chap. 12), the experience with implementation of FTRs in the NYISO over the period 2000–2010 shows that the market quickly reduced the transmission congestion spreads between forward and spot prices after a relative inefficiency at inception. However, FTRs may not always serve as a perfect hedge against congestion charges, as alerted by Benjamin (Chap. 9).

Outside the US FTRs are still in their infancy. As claimed by Read and Jackson (Chap. 13), in New Zealand and Australia FTR proposals have been mainly used to hedge LMP risk. New Zealand has designed FTRs so as to deal with locational price differentials resulting from losses and ancillary service requirements. Australia has

J. Rosellón and T. Kristiansen (eds.), *Financial Transmission Rights*, Lecture Notes in Energy 7, DOI 10.1007/978-1-4471-4787-9,

issued FTRs between zones, but they are not as firm as the underlying capacity. Likewise Europe discusses proposals similar to those of Australia including linking FTRs to market coupling. However, Aertrycke and Smeers argue (Chap. 14) that the organization of both transfer capacity and flow-based European models makes it unlikely that firmness of FTRs can be guaranteed without restricting the possibilities of the transmission grid.

Revenue adequacy is important for guaranteeing the firmness of FTR payments. Yet, Bautista Alderete points out (Chap. 10) that attaining revenue adequacy is difficult in practice due to the changing nature of the variables (such as derates and outages) that impact both the issue of FTRs and funds gathered in the energy market. The research by Oren (Chap. 3) demonstrated that in FTR/FGR markets potential short positions by FGR owners might capture some of the FTR auction revenues in exchange for assuming liability in the FTR market revenue shortfalls. Moreover improvements in line ratings would be a way to reduce revenue shortfalls.

One common allocation method for FTRs is auctions. O'Neill et al. propose (Chap. 4) that the auction design might be improved by implementing a non-linear model including forward auctions for FTRs in an AC load flow model, with reactive power as well as auctions for FTRs on a DC load flow model with hedging for losses. FTRs are also in principle enhancing social welfare. Henze et al. show (Chap. 8) that introducing FTRs in an appropriate manner may reduce the physical capacity needed for the full benefits of competition. The experimental-economics analysis by Henze et al. on FTRs measures spot and LTFTR prices, capacity, and welfare, and compares it to a simulated benchmark. These results demonstrate that, overall, LTFTRs perform well, though showing some heterogeneity.

The main area where FTRs demonstrate some shortfall is transmission investments. In PJM, the welfare efficient expansion of the network might be achieved through a combined merchant-regulatory mechanism that includes the FTR biddings. To mitigate these shortfalls Perez-Arriaga et al. propose (Chap. 2) that transmission charges are to be calculated according to transmission investor responsibilities. Furthermore independency from short-run commercial transactions is required. Complimentary charges are to be calculated once and for all in advance of construction of new transmission capacity. This further provides solid ground for the further calculation and trading of FTRs. Kristiansen and Rosellón propose (Chap. 6) a merchant mechanism for transmission investment depending on investor preferences and simultaneous feasibility. Such a model considers existing FTRs and FTR reserves for possible negative externalities. Rosellón (Chap. 7) and Rosellón et al. (Chap. 15) suggest a combined FTR based merchant-regulatory mechanisms to incentivize transmission expansion which can be implemented as a price cap on the two-part tariffs of the transmission owner. Such a mechanism converges over time to an efficient steady state Ramsey equilibrium. However, a potential weakness might be the way FTRs are allocated for the existing network and their impact on retail rates, as discussed by Benjamin. Furthermore, generators' ownership of FTRs may influence the effects of transmission lines on competition. Joung et al. (Chap. 5) show that introducing FTR options or FTR obligations in an appropriate manner may reduce the physical capacity needed for the full benefits of competition.

To sum up, we feel that the overall overview, both of the theory and practice of FTRs, is however an optimistic one. FTRs contribute to reduce the transmission congestion spreads between forward and spot prices, generally implying welfare improvements. However, as discussed in Chaps. 2, 6, 7, and 15, FTRs alone are not enough to provide sufficient incentives for efficient expansion of transmission capacity. They need to be combined with adequate regulation. But again, the introduction of FTRs together with LMPs in the framework of a competitive electricity market provide adequate hedging of transmission price risk, grant revenue adequacy for contracts for differences, efficiently redistributes congestion rents that the system operator collects, and provides efficient price signals for transmission and generation investors. Of course, the further practical implementation of FTRs faces the variety of challenges discussed throughout the book. We hope that our text contributes in overcoming such challenges, as well as in inspiring future research on FTRs.

# Index




J. Rosellón and T. Kristiansen (eds.), *Financial Transmission Rights*, Lecture Notes in Energy 7, DOI 10.1007/978-1-4471-4787-9,

**F**

**R**

www.ingramcontent.com/pod-product-compliance
Ingram Content Group UK Ltd.
Pitfield, Milton Keynes, MK11 3LW, UK
UKHW022225051026
15624UKWH00002B/363

* 9 7 8 1 4 4 7 1 4 7 8 6 2 *